The Twentieth-Century Sciences

Essays by

Erik H. Erikson
Edward Shils
Talcott Parsons
Paul A. Samuelson
Curt Stern
Gunther S. Stent
Robert Olby
Linus Pauling
Saul Benison
Gerald Holton
Alvin M. Weinberg
A. Hunter Dupree
R. R. Wilson
W. O. Baker

The Twentieth-Century Sciences

Studies in the Biography of Ideas

Edited by GERALD HOLTON

W · W · NORTON & COMPANY · INC · *New York*

 Published simultaneously in Canada by George J. McLeod Limited, Toronto. Printed in the United States of America.

Library of Congress Catalog Card No. 78-163369

SBN 393 06384 4

1 2 3 4 5 6 7 8 9 0

CONTENTS

INTRODUCTION

For a collection of essays such as this, the introduction should not be a roadblock. On the contrary, all I really need say is that the essays speak for themselves, and that there is little reason to delay reading them. Some readers, however, may be interested in the aims of this collection, in the reasons that led us to assemble it, and in some of the preconceptions that helped give it form.

In planning, we had three chief objectives in mind. First, we wanted to give an account of selected contemporary disciplines—in both the natural and the social sciences—including the biography of their transforming concepts or theories. Second, we hoped to explore the intellectual biography of certain leading scholars of our century who have profoundly shaped current thought in science, and with it today's conception of nature and man. This is an area which is still too little developed. While political and literary autobiography and biography flourish, biography of scholars remains fairly uncommon. As a result, one of the major forms of evidence relating to the intellectual activity of our times risks being permanently incomplete.

Finally, our object was to indicate something of the nature of the institutional settings that have been hospitable to learning and scholarship in this century. At a time when the appropriateness of universities as centers of research is being actively reexamined and when there is a new kind of debate about the virtues of collaborative research, it is important to consider what formal and informal institutional provisions have contributed most to contemporary scientific advance.

The chief occasion that led us to plan this collection was the realization of a paradoxical situation that should be attacked head-on. It is often acknowledged that our age has been transformed by major ideas of modern science—for example, those in quantum mechanics and genetics, in psychology and sociology—as well as by the organization of scientific research. What scientists have been doing at the frontiers of research has helped define what is modern about our time, and figures among its chief excitements. Indeed their work increasingly interests a number of fairly new professions—the sociologists of science, the philosophers of science, the historians of science, the new group interested in science and public policy. And yet, large numbers of intellectuals find themselves without access to the intellectual biography of scientific ideas. What is worse, they are buffeted by the new flood of largely uninformed, antirational, and antiscientific tracts that all but drown out the appropriate, corrective criticisms that we shall always need.

In such a situation, it is doubly urgent to celebrate what is worth celebrating, to tell what the state of science is today, to clarify problems in the life of science where these exist, and above all to let a variety of thoughtful scientists speak about these things in their own voice. We therefore urged our contributors to adopt any style congenial to them, but to express in each case their sense of science as a developing, sometimes inchoate private achievement that may have deep consequences, rather than to stress only the more usual sense of science as a finished institutional product, as embalmed in textbooks, in dry success-story terms without a hint of its effects on other fields of thought or activity.

Taken as a whole, the collection does, I think, meet this primary hope. While none of the essays is intended to replace an introductory textbook on the subject, they should help to provide the reader with an explanation of why a particular contribution became a major theme of twentieth-century thought. At the same time they also speak of origins (including those which may have been outside the sciences as narrowly defined); of influences of teachers, colleagues, and institutions; of failures and unrealized ambitions; and of unforeseen effects that shaped the final outcome.

In some cases, undoubtedly, a nonspecialist will have to take more of the technical material on faith than he would perhaps want. Thus at a conference of most of the authors, meeting over their preliminary drafts, the objection was raised about one of the papers that without understanding the technical details one could not follow it fully. But at that point, one of the participants summed up the consensus when he exclaimed that he did not agree. Enough could be appreciated without going through the hopeless task of providing a technical primer: "I am not a physicist either, and I have great difficulties in understanding some of this. But I felt somewhat as though I was in a Kabuki theater, entranced by the motion and action and highlights and excitement, and the lack of a technical background was not so serious that I could not understand what was going on." In spite of some technical difficulties, in most cases there is gained a basic authoritative document which only that particular author could have written, and to which scholars will probably continue to refer in the distant future.

Some of the essays are first-hand accounts of the "nascent moment." Others give a wider, analytic view of the transforming influence of a conception, usually showing a new scientific field in its heroic phase. This double structure was intended in the design of the issue. Among the many possibilities clamoring for attention, we selected a few chief areas in which there is some degree of overlap. Where it was possible within the limits of space, we have tried to obtain both an autobiography or biography of an outstanding individual whose contribution will be clearly identified with a major creative achievement of twentieth-century science, and also a second, related essay from a scholar who, in an analytic article on the same field, could provide background on the larger context surrounding the contribution.

Thus (arranged in an order from the least to the most mathematical sciences), Erik Erikson's autobiographical study of his work in psychoanalysis is followed by the sociological essays of Edward Shils and Talcott Parsons, and Paul Samuelson's on economics; a pair of essays by Curt Stern and Gunther Stent on genetics leads into two essays by Robert Olby and Linus Pauling, covering related developments in molecular biology, with strong links to chemistry and physics in Pauling's paper. Saul Benison describes a case study of medical research and its support. My own two essays, on the philosophical pilgrimage of Einstein, and on the origins of Niels Bohr's complementarity point of view in quantum mechanics, provide a bridge to the subject matter of Alvin Weinberg's, A. Hunter Dupree's, and Robert Wilson's essays on the organization of scientific work in its typically twentieth-century modes—namely, large-scale and often interdisciplinary—and on all the advantages and pitfalls of teamwork in science. Finally, W. O. Baker writes of the influence on science of both the computer and the cybernetic notions that it has helped to inject into our work and thought.

One regrets that even a book of substantial size cannot also do justice to the number of other fields that fully deserve to be represented. For example, advances in this century in fields such as mathematics, areas of psychology other than those discussed here, linguistics, cosmology, ecology, and anthropology certainly merit our attention. Then, too, there are important topics such as the effects scientists have had upon technology, including some of the pathologies and abuses of scientific knowledge. In our planning meetings we also thought of other ways of shaping the material: the biography of a laboratory during a seminal period, for example.

So much, then, for the primary purpose of the issue; but there are a number of secondary purposes. For example, a series of essays such as these provides a base for responding to the current cry, now almost a cliché, for the demonstration of the relevance of the sciences. There are at least four aspects to this search for relevance, and the present collection seems to me to address itself to all of them. In its most basic sense the relevance of the sciences is intellectual excitement that comes from understanding an impressive feat of human achievement; the sort of satisfaction that can be derived from the discovery of the structure of a molecule is as self-validating as the experience of discovering the internal structure of a major literary work. If we do not welcome them in one field they will soon not be welcomed in any other, and such a society is doomed to grow dull and decay. The joy of intellectual engagement in the deep understanding of phenomena of the material world, and the joy of discovering thereby the success of one's rational and intuitive faculties—these are surely among the most relevant activities that can be pursued.

Secondly, the sciences are relevant because of their influence on the intellectual and material climate of the time. Everyone is familiar with examples of such long-range effects as Newton's influence on the imagination of eighteenth-century poets and theologians, or the genetic role of Faraday's "toys" in the development of electric motors and generators and then in the growth of cities, once subways, tramways, and elevators could be built that run on electricity. But there is no doubt that the same long-range effects on social organizations and values will arise from the work of many of the scientists here discussed or, indeed, who are contributing essays to this book.

The third response to the cry for relevance, and the most usual one, is to point to immediate benefits rather than the long-term ones. The intentional use of basic science "for the relief of man's estate," in the phrase of Francis Bacon (or to provide a quick "spin-off," to use the modern term) are indeed goals that sometimes involve science in its basic sense. But here a warning is in order. It would be quite wrong to settle merely for the assistance that, say, chemistry can give in solving problems such as atmospheric pollution. It would be wrong for two reasons. First of all there really is and need be relatively little connection between today's basic research and current technological advance. Only rarely is a fundamental discovery made consciously as a prelude to a major technical improvement. And, indeed, the relevance of science to a society demanding immediate benefits is best seen in a completely different way. Technological advance all too often brings with it major social problems that arise as unforeseen by-products, and these problems cannot be cured or even properly understood through existing scientific or technological or political means alone. Rather, such cures depend to a large extent on basic scientific advances that are yet to be made. To put it differently, at the heart of social problems created by technological advances is the *absence* of specific, basic scientific knowledge. This fact gives the sciences a whole new mandate and opens a new range of expectation for basic research.

Examples come readily to mind. It is quite customary to say that the population explosion is in part caused by the advance of medical science. But one can equally well claim that the population explosion is bound to overwhelm us precisely because we do not yet have sufficient knowledge in an area of pure science. That is to say, the complex problem of overpopulation is due in some degree to current ignorance of the basic process of conception—its biophysics, biochemistry, physiology; no wonder therefore that attempts at controlling population are so halting.

Similarly, the problem of bringing food to hungry people in arid lands that are near the sea is to a large extent political, as are most of the problems which science, in its recent progress, is said to have acerbated. But it is also a problem for basic science: before it is possible to design far more economical desalination plants, a more fundamental understanding of the structure of liquids—one of the much neglected problems in current physics and chemistry—and of the way materials move through membranes will be needed. This is one of those wonderfully complex areas of work in which research, to be effective, appears to be of necessity interdisciplinary and large-scale. Simply having learned to do team research on the scale of twentieth-century installations described in two of the articles may turn out to be half the battle.

Yet another response to the question of relevance may be just this: science is a style of life that, as several of the essays here show, has values that are to be jealously preserved if we intend to build bulwarks against a rising anti-intellectualism. An example that comes out of several presentations is the way, by demonstration of scientific talent, a certain "minority group" has broken through and found a welcome—not without its preliminary trials—in the more established community. The minority group I speak of is, of course, that made up of young people. A whole set of social inventions and devices operates in the life of science to recognize and reward talent early. There are in these essays some touching evidences of this sort of beneficent patronage.

It may well be that one of the bonds among scientists is forged by just sharing a style of life that starts with their early experiences as students. The personal development of such diverse scientists as Erikson, Parsons, Samuelson, Crick, Pauling, and Wilson, among others here presented, shows that a set of ingenious social devices exists to seek out special scientific talent and to bring the acolyte quickly to the most fruitful frontiers of research. In the process the young scientist usually has both the opportunity of training and companionship with a team and also, in the best cases, the opportunity of developing even his most idiosyncratic and iconoclastic ideas.

To the three reasons for bringing together these essays—for their own sake in order that they be read by those who care to be literate about what is happening in the sciences; as a survey of certain chief areas in terms both of individual contributions and the progress of the field as a whole, with some attention to the points where such areas overlap; and as contributions toward the current discussion on the relevance of science—there is another yet to be added. These essays should, for some scholars, contain raw material for their work. During the past two decades historical scholarship in modern science has been dramatically increasing in quantity and quality. Historical, sociological, philosophical studies which use a scientific contribution as a jumping-off point are beginning to flourish. Yet clearly there is room for additional conceptions and new approaches to historiography—for example, above all in the detailed study of actual cases in the genesis of modern theory.

The modern scholar has moved away from the somewhat simplified view of the formulation of scientific theory that was lately dominant. He is likely to look at cases such as those presented here in terms of at least several components: the differences between the pursuit of science as a private activity (what Einstein called the "personal struggle"), and science as a public institution of sharable discourse; the intersect between the chronological development of a field of study and the chronological development of the contributor himself; such sociological questions as the role of colleagueship, of the formal or informal institutional setting within which an individual contribution was made, or of the resistance to new ideas; the epistemological and other elements of the intellectual inheritance of the creator of a new field; the basis for those quasiaesthetic choices that are often made in actual work (for example, when some ad hoc hypothesis is accepted or rejected with a remark such as that of P. A. M. Dirac: "a theory that has some mathematical beauty is more likely to be correct than an ugly one").

All these elements are in fact illuminated by passages in the essays that follow. But I would also draw attention to another component in the analysis of specific case histories. It is what can be called the thematic aspects: the unknown or unconfessed preconceptions and suppositions which are at the base of every major contribution. As several authors indicate, a number of the most active themata in current science are in fact inherited from earlier centuries. (This is not surprising since, regardless of their successes or failures, it seems that we rely on relatively few themata for most of the sciences.) Thus the efficacy of quantification in the treatment of natural phenomena, and particularly the explanatory role associated with small integers (as in quantum physics or genetics) would not surprise a Pythagorean. It may remain eternally fascinating that the contrasting traits the early geneticist dealt with could be put in terms of ratios of small whole numbers. But one has the impression that it was almost unavoidable, given a prejudice in favor of quantification and particularly in favor of small whole numbers, that they would sooner or later pave the way for some forms of genetics—and for atomic physics. It was Niels Bohr himself who said, "This interpretation of the atomic number [as the number of orbital electrons] may be said to signify an important step toward the solution of one of the boldest dreams of natural science, namely, to build up an understanding of the regularities of nature upon the consideration of pure number."

Such nineteenth-century themes as evolution are of course still very much with us (in psychological and sociological research as well as in genetics and astrophysics), and they have reached their most potent stage in our century. One thinks here, for example, of the evolutionary aspect of Erikson's life-cycle concept, and indeed of the possibility that the thema of evolution in the first instance derives from the life-cycle perception. The taming of randomness appears in these essays in several guises, as does the conscious or unconscious search for symmetry. The struggle between holism and reductionism is not new either. An alternative to reductionism, namely the acceptance of dualities, is a chief theme that runs through at least two of the essays. There is surely something like a scientific imagination shared by all scientists, which forms one of the ties among them and which makes possible the interdisciplinary approach that characterizes almost all the developments here described.

Indeed, what is perhaps most impressive is that the transforming conceptions of twentieth-century science are based primarily on more sophisticated versions of previously established themata. That seems to be our strength as well as our limitation. It appears that the advances that have allowed modern scientists to deal with greater and greater complexity rely not so much on utterly novel conceptual tools at the level of preconception and fundamental thematic attitude, but rather on a higher order of intellectual daring.

As the materials here show, this spirit is in no danger of decline. On the contrary, given a sufficiently high level of opportunity and morale, one may confidently expect the next decades to bring a flowering of scientific progress for which the successes of the past may turn out to have been simply a preparation.

G. H.

ACKNOWLEDGMENTS

MANY DEBTS were incurred in organizing this volume, and it is a pleasure to acknowledge them. First of all I thank Professor Stephen R. Graubard, the Editor of the American Academy of Arts and Sciences and of its journal, *Dædalus.* All of these essays appeared in *Dædalus* (except those by Benison, Dupree, and Samuelson); Mr. Graubard and his staff were of incalculable help every step of the way, from expediting the issue to bringing the authors together for a conference at Villa Serbelloni at Como Lake.* The Rockefeller Foundation kindly made this beautiful spot available to us, and Mr. and Mrs. John Marshall received us with great hospitality at the Villa. The Ford Foundation has our deep gratitude for giving us the financial support needed to commission and assemble these essays. Last but not least, our thanks are due to members of the planning committee for this work, and to both Mrs. Diana Menkes and Miss Helen Stewart, who ably assisted the Editor in the preparation of the volume.

Gerald Holton

* Those who attended the conference or were members of the Advisory Committee included W. O. Baker, Saul Benison, René Dubos, A. Hunter Dupree, Erik H. Erikson, Clifford J. Geertz, Charles C. Gillispie, Gerald Holton, Salvador E. Luria, Guido Majno, Ernst Mayr, Henry A. Murray, Robert Olby, Talcott Parsons, Linus Pauling, Don Price, Gardner C. Quarton, Willard Van Orman Quine, Isador I. Rabi, Léon Rosenfeld, Edward Shils, B. F. Skinner, Gunther S. Stent, Curt Stern, Alvin M. Weinberg, Charles Weiner, R. R. Wilson.

The Twentieth-Century Sciences

ERIK H. ERIKSON

Autobiographic Notes on the Identity Crisis

I HAVE been encouraged to put the concept of "identity crisis" in the center of an autobiographic essay. This, quite fittingly, raises the question as to what kind of identity this concept can claim in a discussion of transforming conceptions of modern science. Whether or not identity crisis is itself such a conception in my own field, psychoanalysis, is difficult to decide in the absence of a representative of my profession who might well assert that the concept is not, strictly speaking, psychoanalytic, because it deals with matters too close to the "social surface" to preserve the essence of depth psychology. I would, at any rate, prefer the more modest assumption that I have been invited to write this essay because *Dædalus,* on the basis of some related past symposia,[1] takes it for granted that the problem of psychosocial identity is relevant to any discussion of innovation and builds a bridge between psychoanalytic and other approaches to it.

That I have attempted to demonstrate this bridge in two studies dealing with the interrelation of the life histories and the historical period of great innovators[2] also seems to suggest to some that I should be willing to confess some of the possible reasons for my having been the person who, at a given time in his life and in the history of psychoanalysis, came to observe and to name something by now so self-evident as the identity crisis and to explain, in fact, why it now seems so self-evident. For identity concepts have immediately secured for themselves, if not the clear status of scientific innovation, yet the onus of a certain novelty in the thinking or, at any rate, the vocabulary, of a wide range of persons in many countries. In my recent collection of essays on the subject[3] I reported fondly that not long ago a Catholic student organization at Harvard announced in the paper that it would "hold an identity crisis" in a given place on a certain evening at 8:00 P.M. sharp. More recently, it was reported that President Nixon's son-in-law, being also President Eisenhower's grandson, on that very score felt an identity crisis coming on, whereupon he consulted his father-in-law who in turn asked a minister to discuss the matter in a Sunday sermon at the White House. Finally, to continue on this high level, the Pope in a recent speech recommended the image of a newly sainted sixteenth-century Spaniard of Jewish descent to the young priests of our time—"a time when they say the priesthood itself suffers . . . a crisis of identity." The varied uses of the term naturally suggest, to many serious workers, a popularization beyond redemption, while others continue to insist that the term and what it really stands for is firmly anchored in conceptual necessities. Looking around for some confirmation, I can point to the fact that the items "Identity, psychosocial," and even "Lifecycle, human" have appeared in the newest edition of the *Encyclopedia of the Social Sciences*.[4]

But as to the suggestion that I should go further and demonstrate how I would use my own tools in an attempt to look at my own life, let me say from the onset that I do not feel obliged to be as effusive in my self-revelations as were the two religious activists who are the subjects of my psychoanalytic studies. They, of their own volition, left passionate confessional data behind them, and this apparently as part of what they considered their over-all mission. Their confessions of overwhelming experiences and of singular usurpations literally "call" for the enlightened interpretation of subsequent generations. We psychoanalysts, on the other hand, have prepared ourselves for just such a job by undergoing didactic psychoanalyses which cooled our self-revelatory fervor and made it clear to us that all voluntary confession can serve both as ingenious cover up and as disarming propaganda: any really "true" confession quickly leads to the unspeakable. My position, then, seems to be midway between the effusive subjects of my biographic efforts, and the representatives of the natural and social sciences in this volume who, it seems, are reluctant to use sentences beginning with "I." My job, then, is to use my own life in order to demonstrate some motivational dimensions in the total life situation of an individual who comes to formulate something "new."

Let me present, at this point, a kind of glossary which will, if not define, at least circumscribe what an identity crisis *is*. Here I take heart from the reassurance of Stuart Hampshire, who states approvingly that I "leave my much misused concept of identity undefined" because it primarily "serves to group together a range of phenomena which could profitably be investigated together."[5] He understood, it seems, the difficulty of establishing the nature and the position of something that is *psycho* and *social*. For we have as yet no one social science comparable to the natural variety. In each of the social sciences, in fact, the workings of identity appear in different contexts of verifiability. To say then that the identity crisis is psycho *and* social means, when approached psychoanalytically, that the "psycho" side of it:

1. Is partially *conscious* and partially *unconscious*. It is a *sense* of personal continuity and sameness, but it is also a *quality* of unself-con-

scious living, as can be so gloriously obvious in a young person who is finding himself as he is finding his communality. In him we see emerge a unification of what is irreversibly given (that is, body type and temperament, giftedness and vulnerability, infantile models and ingrained prejudices) with the open choices provided (available roles, occupational possibilities, values offered, friendships made, sexual encounters), and all this within traditional or emerging cultural and historical patterns.

2. It is beset with the dynamics of *conflict,* and especially at its climax can lead to contradictory mental states such as a sense of aggravated vulnerability and, alternatively, one of grand individual promise.

3. It has its own *developmental period,* before which it could not come to a crisis, because the somatic, cognitive, and social preconditions are not yet given; and beyond which it must not be delayed, because the next and all future developments depend on it. This developmental period is, of course, *adolescence and youth,* which also means that the identity crisis partially depends on *psychobiological* factors, which secure the somatic basis for an organism's coherent sense of vital selfhood.

4. It reaches both into the *past* and toward the *future:* it is grounded in the stages of childhood and will depend for its preservation and renewal on each subsequent stage of life.

The "socio" part of identity, on the other hand, must be accounted for in that communality within which an individual must find himself. No ego is an island to itself. Throughout life the establishment and maintenance of that strength which can reconcile discontinuities and ambiguities depends on the support first of parental and then of communal models. Youth, in particular, depends on the ideological coherence of the world it is meant to take over, and therefore is highly aware of whether the system is strong enough in its traditional form to be confirmed by the identity process even as it confirms it, or sufficiently weakened to suggest renovation, reformation, or revolution. Psychosocial identity, then, also has a *psychohistorical* side, and life histories are inextricably interwoven with history.

All this sounds probable enough and, especially when shorn of its unconscious dimension, appears to be widely and faddishly acceptable in our day. The unconscious complexities usually ignored are outlined in my encyclopedia article, from which I paraphrase:

1. The crisis is sometimes hardly noticeable and sometimes very much so: in some young people, in some classes, at some periods in history, the identity crisis will be noiseless; in other people, classes, and periods, the crisis will be clearly marked off as a critical period, a kind of "second birth," institutionalized by ceremonial procedure, or intensified by collective strife or individual conflict.

2. Identity formation normatively has its negative side which through-

out life can remain an unruly part of the total identity. The *negative identity* is the sum of all those identifications and identity fragments which the individual had to submerge in himself as undesirable or irreconcilable. It is, of course, especially complex in atypical individuals and in marked minorities which are made to feel "different." In the event of aggravated crises, an individual (or, indeed, a group) may despair of the ability to contain these negative elements in a positive identity. A *specific rage* can be aroused wherever identity development loses the promise of a traditionally assured wholeness: thus an as yet uncommitted delinquent may become a criminal, or a rebel a mortal enemy. Such potential rage is easily exploited by psychopathic leaders, who become the models of a sudden surrender to total doctrines and dogmas in which the negative identity becomes the dominant one: the Nazis fanatically cultivated what the victorious West as well as the refined Germans had come to decry as "typically German." The rage aroused by threatened identity loss can explode in the arbitrary destructiveness of mobs, or it can serve the systematic violence of organized machines of destruction.

3. The depth of the identity conflict often depends on the latent panic pervading a historical period. Some periods in history become identity vacua caused by three basic forms of human apprehension: *fears* aroused by new facts, such as discoveries and inventions (including weapons) which radically expand and change the whole world-image; *anxieties* aroused by symbolic dangers vaguely perceived as a consequence of the decay of existing ideologies; and the *dread* of an existential abyss devoid of spiritual meaning.

But why would some decisive insights concerning these universal matters first come from psychoanalysis? The fact is that psychoanalysis as a clinical science discovers new aspects of man's nature by trying to cure previously obscure disorders which, at a given time, suddenly seem to assume epidemiological significance—as did hysteria in Freud's early years. In our time a state of *identity confusion,* not abnormal in itself, often seems to be accompanied by all the neurotic or near-psychotic symptoms to which the person is prone on the basis of constitution, early fate, and malignant circumstance. In fact, young individuals undergoing such a confusion are subject to a more malignant disturbance than might have manifested itself during the rest of their lives, because it is a characteristic of the adolescent process that the individual should semideliberately give in to some of his most regressed or repressed tendencies in order, as it were, to test rock bottom and to recover some of his as yet undeveloped childhood strengths. Clinically, it has proven important to recognize that many young people have in the past been judged to suffer from a chronic malignant disturbance, where a severe developmental crisis was in fact, indicated. At the same time, the epidemiological variations of such crises over the decades strongly suggest some relation to changing history. This, then, is the clinical anchorage for the conception of an identity crisis. To make the emergence of this conception plausible in autobiographic terms I will first describe my encounter with psychoanalysis in my young adulthood and then go back into my childhood and youth to see what may have prepared me for this crucial experience.

I am about two years younger than the twentieth century, and I therefore can speak roughly of the decades of my life as coinciding with those of the calendar. I was graduated from the Vienna Psychoanalytic Institute in my (and the century's) early thirties.

The very beginning of my career marks me as one of those early workers in my field who had quite heterogeneous professional origins. I was an artist before I studied psychoanalysis, and can otherwise boast only of a Montessori diploma. In fact, if William James could say that the first lecture in psychology he ever took was the first one he gave, I must concede that the first course in psychology I ever took was also the first and the last I flunked. But, for reasons to be given presently, I seemed acceptable to the Vienna Psychoanalytic Institute, which was the training arm of a private society not connected with and often opposed and belittled by both academic departments and professional organizations. Although Sigmund Freud was a doctor and medical school lecturer, he had taken it upon himself to select the original circle of men and women (medically trained, as most were, or not) who were willing to learn his methods according to his prescription. For this they had to be gifted with a certain rare perceptiveness for the nonrational, and both sane and just a bit insane enough to *want* to study it. For them, Freud created simultaneously a training institute, a clientele, a publishing house, and, of course, a new professional identity.

The truly transforming scientific and therapeutic orientation created by Freud was based on a radical change in the role of the doctor as well as that of the patient, and thus in the nature of the clinical laboratory. If the "classical" hysterical patients, as Freud had concluded, were far from being the degenerates they were judged to be by his medical contemporaries, he could only conclude that the authoritative methods used to cure them violated, in fact, what alone can free a man from inner bondage, namely, the conscious acceptance of certain truths about himself and others. He advocated that the psychoanalytic practitioner himself undergo treatment so as to come to terms with his own unconscious and to acquire the capacity to explain, and not to evade or to condemn. The patient in turn was asked to verbalize all his available thought processes, and thus to become co-observer as well as client; while the analyst, as observer, continued to observe himself as he observed the patient's trend of thought. They were thus to become collaborators in the job of becoming conscious of (and of classifying) that reservoir of unconscious imagery and affect which soon proved not only to be festering in individual patients, but also to have been repressed in all past history by the whole race—except for occasional seers and prophets, creative writers and philosphers. The atmosphere thus created in the Vienna circle was one of intense mutual loyalty and of a deep devotion to a truly liberating idea, if often also of morbid ambivalencies and of deep and unforeseeable mental upset. Here much fascinating material is awaiting the historian who manages to become a historian of ideas and to resist the temptation merely to turn early psychoanalysis upon itself.

My own training psychoanalysis was conducted by Anna Freud, who accepted me as a fellowship candidate, and this on the basis of the fact that she and her friends had approvingly witnessed my work with children both as a private tutor and as a teacher in a small private school. Anna Freud had founded the Vienna version of the subspecialty of child analysis, and I, too, was to be primarily trained as a psychoanalyst of children, although the general training program included the treatment of adolescent patients and of adults. The reader will appreciate the complex feelings aroused by the fact that my psychoanalyst was the daughter of the then already mythical founder who often appeared in the door of their common waiting room, in order to invite *his* analysand into *his* study. But this is only a special circumstance within the peculiar bonds and burdens of a training psychoanalysis. While such a preparation is of the essence in this kind of work, and while this essence can up to a point be comprehended in a systematic and disciplined way, it stands to reason that the generational succession of teachers, so important in all fields, will retain a special quality in the life of any maturing and even of the trained psychoanalyst—a burden reaching well into the older conceptualizations he may later feel called upon to confirm or to disavow. Even the most venerable training psychoanalyst is (and must be) both welcomed as a liberating agent and resisted as a potential indoctrinator, and thus be both accepted and rejected as an identity model: for in all pursuits which attempt to gain a rational foothold in man's pervasive irrationality, insights retain an unconscious involvement which can only be clarified by lifelong maturation. At the same time, however, only the systematic clinical acquaintance with the unconscious can convey a certain ceaseless surprise over the creative order that governs affects never faced before and a certain sense of liberated sanity just because of the chaos faced. This is probably hard to comprehend except through the experience itself; and yet some of it must be visible in the impact of psychoanalysis on other fields and the augmented access it has provided to a wealth of data previously not visualized or not seen in their relation to each other. Considering the total newness of such data, it should not be surprising that it will take many generations to find the proper forms of verification and suitable methods of application. In the meantime, however, it should be obvious that this novel training situation, extreme in what we may call a *disciplined subjectivity*, only brings to the fore latent factors also present in more objective courses of study.

I should add here that during the years of my training Freud himself no longer taught and never appeared at public functions. I sometimes met him in his or some friends' house or garden, but never addressed him, not only out of shyness but because of the pain that all attempts at speaking could cause him. He was in his early seventies; a radical upper-jaw operation had done away with a cancerous affliction years before, but the "infernal prothesis" which covered the roof of his mouth led to repeated fresh outbreaks and fresh operations. His daughter was also his nurse and his private secretary, his companion and his ambassador both to the old guard of the psychoanalytic movement and on rare ceremonial occasions. Some of the writings which Freud published during these years were of a markedly philosophical bent, or what the Germans call *weltanschaulich*. In his *Autobiographic Study*,[6] which he wrote in his sixty-eighth year, he himself ascribed this trend to "a phase of regressive development" back to his premedical and, in fact, adolescent interest in problems of culture, or rather *Kultur*. Theoretically, it all culminated in the concept of the Death Instinct—a grandiose contradiction in terms.

In the meantime, our training continued to impress on us in all clinical immediacy the five conceptions which Freud in the same autobiographic study had called "the principal constituents of psychoanalysis." These five have remained fundamental to the modifications of psychoanalytic technique and to its application to other fields throughout this century. The most fundamental is "inner resistance," a term which was never meant to be judgmental, but physicalistic. Memories and thoughts are resisted even by him who wants to recover them, whether out of despair or out of curiosity. For such inner resistance, Freud blamed "repression," a defensive quality of the mind which marks the "unconscious" as much more than *not* conscious. Its drives and wishes, memories and phantasies, in fact, reassert their right to awareness and resolution, if only in indirect ways: in the symbolic disguise of dreams and day dreams, in symptoms of commission (acts alien to the actor or unintended in their consequences), and in symptoms of omission (inhibitions, avoidances, and so forth). If Freud, on the basis of his Victorian data, found in his patients' special repressions and resistance primarily what he called the "aetiological significance of sexual life"—that is, the pathogenic power of repressed sexual impulses—he, of course, called sexual a wide assortment of impulses and affects never previously included in that definition. It was the burden of the libido theory to show how much in life is codetermined by derivatives of the previously unrecognized infantile sexuality. Freud, therefore, considered systematic attention to the "importance of infantile experiences" an intrinsic part of his method and his theory: and we know now how his and Karl Abraham's first crude findings opened up a whole new view of the stages of life.

I would add to these five points the prime importance of what Freud calls "transference"—that is, a universal tendency to experience another person (unconsciously, of course) as equivalent to an important figure of the preadult past. Such transference serves the inadvertent reenactment of infantile and juvenile wishes and fears, hopes and apprehension; and this always with a bewildering "ambivalence"—that is, a ratio of loving and hateful tendencies which under certain conditions coincide dangerously or alternate radically.

Tranference, of course, plays a singularly important role in the clinical encounter; but it must be clear that all these interpersonal tendencies, which in the training analysis and in clinical work move into the center of attention, also pervade daily life, and especially where work arrangements are the basis for intense experiences of leader- and followerships and of fraternal or sororal rivalry. The question as to where and when such tensions support or hinder the inventiveness, solidarity, and altruism demanded of scientific workers has been asked repeatedly in these discussions. The answer is that mutual emotional involvement, even of the kind called transference, can, under favorable conditions, evoke powers of filial loyalty and sharpen issues of ambivalence, and can thus support personal growth and creative innovation. If psychoanalysis uses this power (which David McClelland unhesitatingly calls religious) for therapeutic purposes, it only puts to systematic use what Mircea Eliade[7] has described in primitive rites which seek rebirth in a return to origins: and Eliade rightly recognized a parallel in the (often quite ritualized) faith on which modern psychotherapy was founded.

This, however, does not justify the frequent assumption that psychoanalysis is primarily faith healing. To Freud and his followers the psychoanalyst's office has always been not only a healer's den, but also a psychologist's laboratory. What was thus observed had to find a classification, a terminology, and a methodology which would make therapeutic techniques ever more adequate in dealing with a widening range of pathological conditions and would help formulate a body of insights amounting to a communicable "field." That in some of Freud's most gifted followers idiosyncratic gifts as well as ideological predilections often seemed to obscure the ground plan for which he thought he alone had a firm sense of schedule—that is certainly not surprising. A truly historical study of these developments is probably far off; but it could throw much light on the influence of such personal passions and such residual pathology as are aroused when man's central motivations are submitted to observation and conceptualization. Yet it sometimes seems that such passions are never totally absent even from the most controlled laboratory work, especially when assumptions that have contributed to the stability of a classical world-image are being questioned, and when the necessity of transforming ideas becomes a matter of competition between individuals and schools.

At that time, in Vienna, all theoretical training took place in the evenings in what today might be called something like a free psychiatric university. Again, only those who have attended similarly independent study groups of men and women serving what is felt to be a truly transforming idea—and serving at the sacrifice of income, professional status, and mental peace—will know of the devotional atmosphere in which (as every reader of Freud's work knows) no clinical detail was too small and no theoretical insight too big to merit intensive presentation and debate.

Psychoanalytic writings do not always reflect the high degree of medical and scientific common sense and humane humor pervading the actual study of ongoing treatments. The clinical laboratory includes the regular and exhaustive comparison of treatment histories. Clinical conferences, close to the data, are the heart of the matter. Freud's method, in fact, forced clinicians to attempt to "locate" any given clinical observation on a number of coordinates which he called "points of view." I have discussed these in detail in my paper on clinical evidence.[8] Here I can only indicate that these points of view include a "structural" one which would locate a given item in a model of the mind, the main compartments of which are the "id" (a cauldron of primeval drives and primal wishes), the "ego" (the mental organization mediating between the id and the outerworld), and the "super-ego" (the internalized moral standards which guard the ego's mediation). There is also a "dynamic" point of view, which takes account of the tension and conflict between these inner domains. An "economic" point of view, in turn, attempts to conceptualize the householding of mental energy in man's precarious inner balance. Finally, the "genetic" point of view permits the reconstruction of the origin and the development of all these structures, functions, and energies.

Although Freud called these points of view his "metapsychology," and assigned to them a level of abstraction not accessible to direct observation, it is hard to overcome the impression that they served as the bridge by which Freud, a fervent and painstaking medical researcher before he became the first psychoanalyst, could apply his anatomical, physiological, pathological, and developmental modes of thinking to the workings of the mind. I do not mean here only Freud's early attempts, in tune with the dominant neurology of his medical school years, to link mental associations to concrete processes in the nervous system. These observations only served to open for him vast promises of finding by a new clinical method an approach to the very secrets of man's conflicted nature—and of making these secrets accessible to science. But as though the elemental sweep of the resulting discoveries had aroused a sense of hybris in the medical researcher, he seems to have determined, in his metapsychology, to reconnect his vast findings with the disciplined thought patterns which in his young manhood had commanded his fidelity and helped establish his occupational identity. It is often overlooked that even his preoccupation with an all-pervasive libido was related to a scientific commitment to think in terms of energies *"equal in dignity"* to the forces found in physics and in chemistry. On their neurological homeground, however, such modes of thinking had been based on visible, observable, and verifiable facts, while in the study of the mind they sooner or later served, especially in the hands of dogmatic followers, as unchecked reifications—as though the "libido" or the "ego" had, after all, become observable entities.

The intellectual milieu governing the many evenings spent in small, intensive seminars and "continuous case" discussions (and some were so small that we could comfortably meet in our teachers' homes) is best characterized by a listing of these teachers. All but the first two later came to this country to preside over the strange fate of a psychoanalysis influential in medical training, lucrative in practice, and popularized in the media.

My training in child analysis took place in the famous *Kinderseminar* led by Anna Freud; that in the treatment of juvenile disturbances (including delinquency) was directed by August Aichhorn. I reported my first adult case in Helene Deutsch's clinical seminar. Heinz Hartmann was the leading theoretician, and his thinking which later culminated in his monograph on the adaptive function of the ego influenced me deeply. The basic theoretical struggle at the time was between Anna Freud's clarification of the defensive mechanisms employed by the ego against the drives[9] and Hartmann's explorations of the ego's adaptive response to the environment.[10] One of the most obscure and yet fascinating teachers was Paul Federn, and it is quite possible that in his seminar I first heard the term identity mentioned in one of its earlier usages. The preoccupation with the ego was then replacing that earlier pansexual attention to the id which was based on Freud's original determination to find the whole extent of man's enslavement to sexuality. What survived from that first period was a strangely ascetic veneration of Eros, a kind of intellectual bacchanalia. Nothing seemed to be further from the mind of those early workers (not even the followers of Wilhelm Reich) than that psychoanalysis might someday be used as an argument for sexual freedom outside the rules of a bourgeois or, for that matter, a proletarian convention. Both Reich and, as visitor, Siegfried Bernfeld, who was deeply involved in problems of youth, I remember as very inspiring teachers already then driven to a certain tragic isolation by their belief that Freud's "libido," which sounded so tangibly quantitative, would *have* to be found and isolated physically. In general, the student could not help sensing in the didactic milieu a growing conservatism and especially a pervasive interdiction of certain trends of thought. This concerned primarily any idea which might be reminiscent of the deviations perpetrated by those earliest and most brilliant of Freud's co-workers (such as Rank, Adler, and Jung) who had been separated from the movement already before World War I. In other words, the psychoanalytic movement was now already working under the impact of its own historical trauma, its mythical rebellion against the founder. The possible merits and decisive demerits of those deviations the student could not judge. I must admit that, after such intense training under such complex conditions, the idea of moving on and working independently seemed an invigorating idea.

If I should now briefly indicate what uncertainty and curiosity I took with me when my graduation permitted my emigration from Europe, I would oversimplify it in the following way. Psychoanalysis had broken through to much that had been totally neglected or denied in all previous models of man: it had turned *inward* to open up man's inner world to systematic study; it searched *backwards* to the ontogenetic origins of the mind and of its disturbances; and it pressed *downwards* into those instinctual tendencies which man thought he had overcome when he had repressed or denied the infancy of individuals and the evolution of the race. That unconscious substratum (as Darwin, too, had discovered) was the territory to be conquered, the origins to be acknowledged. But conquerors so easily lose themselves in the discoveries of the new territory; how to reassimilate them to what is already known—that is the job of the second stage. The question remained, I felt dimly, whether an image of man reconstructed primarily on the basis of observation in the clinical laboratory might not lack what, in man's total existence, leads *outward* from self-centeredness to the mutuality of love and communality, *forward* from the enslaving past to the utopian anticipation of new potentialities, and *upwards* from the unconscious to the enigma of consciousness. While each of these conceptual necessities would have to await opportune conditions of study, they seemed to me always implicit in Freud's own writings: if not in the *content* he grimly pursued, then in the grand *style* of this pursuit—the style for which, in those very days, he received the Goethe Prize as the best scientific writer in the German language. To Freud, the *via regna* to mental life had been the dream. For me, children's play became the first avenue to an understanding of growing man's conflicts and triumphs, his repetitive working through of the past and his creative self-renewal in truly playful moments. I identified with Freud, then, not as the former laboratory worker who insisted on a terminology made for the observation of transformable quantities of drive enlivening inner structures, but as the sharp discerner of verbal and visual configurations which revealed what was clearly suggested or flagrantly omitted, that meant what it said or meant the opposite. To put it bluntly, I have always suspected (maybe because I do not really understand these things) that what sounded most scientific in psychoanalysis in terms of nineteenth-century physicalism was more scientism than science, even though I understood that psychology and social science in attempting to free themselves from philosophy and theology had no choice but to try, for a while, to think in the scientific imagery of the century. But Freud's phenomenological and literary approach, which seemed to reflect the very creativity of the unconscious, held in itself a promise without which psychoanalytic theory would have meant little to me. This may be one reason why, in later years, I proved inept in the customary kind of theoretical discussion and was apt to neglect ruefully the work of my colleagues—and not only where they seemed to take Freud at his most atomistic and mechanical word, or where they turned neo-Freudian. All this may well have an admixture of a particular transference on Freud. But then, he was the father of it all—a fact which I probably tried to objectify in my later studies of great men, as well as in a few essays on Freud himself.[11]

But before I sketch the direction of my own work which, in a new era and on a new continent, brought me to such concepts as the "identity crisis," I will briefly account for such aspects of my childhood and youth as made the encounter with Freud's circle so personally significant.

In the Europe of my youth the choice of the occupational identity of "artist" meant, for many, a way of life rather than a way of making a living; and, as today, it could mean primarily an anti-Establishment way of life—except that the European Establishment had created a well-institutionalized social niche for such idiosyncratic needs. A certain adolescent and neurotic shiftlessness could be contained in the custom of *Wanderschaft;* and if the individual had some gifts into the bargain, he could convince himself and others that he should have a chance to show that he might have a touch of genius. There were, of course, youth movements for those who wanted to abandon themselves to some collective utopia; but much of what young people today display in alienated and concerned groups was then more commonly experienced in an isolation shared only with a few equally special friends. To be an artist, then, meant to have at least a passing identity, and I had enough talent to consider it for a while an occupational one. The trouble was, I often had a kind of work disturbance and needed time. *Wanderschaft* under those conditions meant neurotic drivenness as well as deliberate search, even as today dropping out can be a time of tuning in, or of aimless negativism. But somehow, when we did not work, we had a deep and trusting relationship (often called "romantic" today) to what was still a peasants' Nature; we kept physically fit by interminable hiking; we trained our senses to changing perspectives, and our thoughts to distilled passages of, say, Angelus Silesius and Laotse, Nietzche and Schopenhauer, which we carried in our knapsacks. I will not describe the pathological side of my identity confusion for which psychoanalysis later seemed, indeed, the treatment of choice, while it may have assumed at times what some of us today would call a "borderline" character—that is, the borderline between neurosis and psychosis. But then, it is exactly this kind of diagnosis to which I later undertook to give a developmental perspective. And indeed, some of my friends will insist that I needed to name this crisis and to see it in everybody else in order to really come to terms with it in myself. And they can quote a whole roster of problems related to my *personal* identity.

There is first of all the question of origin which often looms large in persons who are driven to be original. I grew up in Karlsruhe in Baden as the son of a pediatrician, Dr. Theodor Homburger, and his wife Karla, née Abrahamsen, a native of Copenhagen, Denmark. All through my earlier childhood they kept secret from me the fact that my mother had been married previously and that I was the son of a Dane who had abandoned her before my birth. They apparently thought that such secretiveness was not only possible (because children then were not meant to know what they had not been told) but also advisable, so that I would feel thoroughly at home in their home. As children will do, I played in with this and more or less forgot the period before the age of three when my mother and I had lived alone. Then, her friends had been artists working in the folk style of Hans Thoma of the Black Forest. They, I believe, provided my first male imprinting before I had to come to terms with that intruder, the bearded doctor, with his healing love and mysterious instruments. Later, I enjoyed going back and forth between the painters' studios and our house, the first floor of which, in the afternoons, was filled with tense and trusting mothers and children. My sense of being "different" took refuge (as it is apt to do even in children without such acute life problems) in phantasies of how I, the son of much better parents, had been altogether a foundling. In the meantime, however, my adoptive father was anything but the proverbial stepfather. He had given me his last name (which I have retained as a middle name) and expected me to become a doctor like himself.

Identity problems sharpen with that turn in puberty when the contradictory identifications of the past must be reconciled and when the images of future roles become inescapable. My stepfather was the only professional man (and a highly respected one) in an intensely Jewish small bourgeoisie family, while I (coming from a racially mixed Scandinavian background) was blond and blue-eyed, and grew flagrantly tall. Before long, then, I acquired the nickname "goy" in my stepfather's temple; while to my schoolmates, I was a "Jew." Although I had tried desperately to be a good German chauvinist, I became a "Dane" when Denmark remained neutral during the First World War. Such designations, even if occasional, can underscore a sense of alienation and, in fact, become for a while more representative of one's "true" self than the simple and stable social facts making up one's *personal identity.* As to the admixture of German, Jewish, and Danish elements in my background, I should specify that I am speaking of a period before all Jewishness received transvaluations both by National Socialism and by Zionism. My mother could still be nostalgically Danish and cultivate in me a somewhat mythological Danishness without feeling constrained to emphasize that her family was also Jewish. On the other hand, the reformed Judaism in our social life was not conducive to strong identification; on the contrary, it was part of the transparent ceremonialism of a *Buergertum* which young people yearning for relevance, then as now, vowed early to leave behind with a vengeance. At the end, the Jewish as well as the Danish elements in my background combined with German schooling most positively where they together transcended burgherly confines in pursuit of a universal *Bildung* for which individual teachers, writers, and artists were the models.

Karlsruhe was the old state capital of a Lutheran grand-dukedom with a sizeable Catholic population. Some Jewish families such as the Homburgers had lived there almost since its founding, early in the eighteenth century. Strangely enough, I do not remember having been interested in the old Luther of the state church or, for that matter, in the young one; and yet years later I was to choose young Luther as a matter of course for the presentation of a dramatic identity crisis of historical import, and related both to the emergent German language and to the Christianity of the Gospels to which I early felt inescapably drawn. After graduation from the type of high school called a humanistic gymnasium (where one then acquired a solid classical *Bildung* and a sense for the common roots of the Western languages), I went to art school, but always set out again to be a wandering artist—as described. I now consider those years an important part of my training. Sketching (as even a man like William James experienced) can be a good exercise in tracing impressions. And I enjoyed making very large woodcuts: to carve stark images of nature on this primary material conveyed an elemental sense of both art and craft. And in those days every self-respecting stranger in his own (northern) culture, drifted sooner or later to Italy, where endless time was spent soaking up the southern sun and the ubiquitous sights with their grand blend of artifact and nature. But if this was a "moratorium," it certainly also was a period of total neglect of the military, political, and economic disasters then racking mankind: as long as one could expect some financial support from home, and was not suddenly lost in some cataclysm, one lived (or so one thought) by the measure of centuries, not of decades. Come to think of it, we lived in a rampantly patriarchal universe tempered by maternal forces. There was God and his Judeo-Christians, territorial and always ready for punitive wars; and there was Nature and the Greeks, eternal, healing, and permissive as long as a certain measure was preserved. Such narcissism obviously could be a young person's downfall unless he found an overweening idea and the stamina to work for it.

It was my friend Peter Blos (today a New York psychoanalyst best known for his classical writings on adolescence)[12] who came to my rescue. During our later childhood in Karlsruhe, he had shared his father with me, a doctor both prophetic and eccentric (he first told us about Gandhi), and we had been friends in Florence. Now he invited me, with the encouragement of the founder, Dorothy Burlingham, to join him in that small school in Vienna. With their help I learned to work regular hours, and I met the circle around Freud.

It must be more obvious now what Freud meant to me, although I would not have had words for it at the time. Here was a mythical figure and a great doctor who had rebelled against the medical profession. Here also was a circle which admitted me to the kind of training that came as close to the role of a children's doctor as one could possibly come without going to medical school. What, in me, responded to this situation was, I think, an ambivalent identification with my stepfather, the pediatrician, mixed with a search for my own mythical father. And if I ask myself in what spirit I accepted my truly astounding adoption by the Freudian circle, I can only surmise (not without embarrassment) that it was a kind of favored stepson identity that made me take for granted that I should be accepted where I did not quite belong. By the same token, however, I had to cultivate not-belonging and keep contact with the artist in me: my psychoanalytic identity therefore was not quite settled until much later, when with the help of my American wife, I became a writing psychoanalyst, if again in a language which had not been my own.

One could well suspect that later on I succeeded in making a professional life style out of being a stepson when I, throughout, worked in institutional contexts for which I did not have the usual credentials.

But, as pointed out, psychoanalysis then almost methodically attracted and collected men and women who did not quite belong elsewhere, and some of my most outstanding colleagues have followed similarly irregular life plans. What is to be demonstrated here, then, is not singularity of achievement, but the configurational affinity of life plan and choice of concepts.

While I am on the stepson theme, I must also, with due surgical brevity, expose the dangers of such a development both for a person's character and his concepts. That a stepson's negative identity is that of a bastard need only be acknowledged here in passing. But a habitual stepson might also use his talents to avoid belonging anywhere quite irreversibly; working between the established fields can mean avoiding the disciplines necessary for any one field; and being enamored with the aesthetic order of things, one may well come to avoid their ethical and political as well as their conceptual implications. If one can find all these weaknesses in my work, there are also some energetic attempts to balance them, and this exactly in a serious and methodical turn to social and political conditions and—possibly inspired by my great compatriot Kierkegaard—to religious actualists such as Gandhi, certainly a man almost totally devoid of the aesthetic dimension.

That much about the circumstances of my life. If it seems obvious that such a life would predispose a person to a malignant identity crisis, this must be said to be only partially true; for in my instance the more obvious identity conflicts concerned my *personal* identity and *psychosocial* choices which were relatively clearly delineated. If the malignancy of the identity crisis is determined both by defects in a person's early relationship to his mother and on the incompatibility or irrelevance of the values available in adolescence, I must say that I was fortunate in both respects. Even as I remember the mother of my early years as pervasively sad, I also visualize her as deeply involved in reading (what I later found to have been such authors as Brandes, Kierkegaard, Emerson) and I could never doubt that her ambitions for me transcended the conventions which she, nevertheless, faithfully served. On the other hand, she and my stepfather had the fortitude to let me find my way unhurriedly in a world which, for all the years of war and revolution, still seemed oriented toward traditional alternatives, so that the threatening cataclysms could still be ascribed to criminal men and evil nations or classes. What I eventually came to describe as more malignant forms of identity crises both in groups and in individuals was probably of quite a different order than what we then experienced. All the warring ideologies of our young years harbored *some saving scheme* which was to dominate forever after just one more war, just one more revolution, just one more new deal. It is only in our lifetime that the faith in change has gradually given way to a widespread fear of change itself—and a suspicion of faith itself. Identity problems and even the symptoms of identity confusion probably have changed accordingly. Incidentally, the comparative study of the nature of identity crises at different periods of history (and in different groups during the same period) may well turn out to be a historical as well as a clinical tool, provided that the uses of the concept itself are submitted to historical scrutiny.

The relevance of a man's work, then, always derives from history as well as from his life history. The German holocaust opened other countries to the migration of psychoanalysts who not only survived but, in fact, succeeded in establishing power spheres in the cities of their choice. In adapting to new classes of patients in a variety of national and cultural settings (and also by reflecting on their own fate and on the fate of those who had perished), some psychoanalysts found it mandatory to revise their model of human functioning.

My training psychoanalysis had ended when I met and married Joan Serson, then a dancer and teacher, later also a craftsman and writer; she, too, was a member of our small school. At about the time when Hitler came to power in Germany, I graduated and we left Vienna with our two small sons. Vienna, at that time, chose not to foresee the total disruption that would soon separate the regions of Europe, not to speak of the old country and the New World. I first attempted to regain my Danish citizenship and to help establish a psychoanalytic training center in Copenhagen. When this proved impracticable, we emigrated to the United States and settled in Boston where a psychoanalytic society had been founded the year before. Since my graduation in Vienna had made me a member of the International Psychoanalytic Association I was welcomed in the American Association as well. Although the medical professionalization of psychoanalysis in the United States would soon thereafter lead to the exclusion of further nonmedical candidates from clinical training, I remained as one of its very few nonmedical training analysts, always in the hope that the quiet contributions of nonphysicians would sooner or later impress American psychoanalysts with the wisdom of Freud's conviction that this field should not be entirely subordinated to medical professionalism. Personally I was, of course, ready to abide by the medical and legal cautions necessary in therapeutic work; and I cannot say that my being a nonphysician has ever interfered with my work.

In fact, to an immigrant with a specialty (and the term immigrant had not yet given way to that of refugee) this country proved, indeed, a land of unlimited possibilities. Harvard, and later Yale, did not hesitate to provide medical school appointments, and thus an expanded clinical experience. At Harvard there was also Harry Murray's Psychological

Clinic, where an intensive study of students proved a valuable guide to the characteristics and values of American academic youth, while Murray's style of thinking conveyed something of the grand tradition of William James. And there was a flowering of interdisciplinary groups, led and financed by imaginative men like Lawrence K. Frank of the General Education Board and Frank Freemont-Smith of the Josiah Macy Jr. Foundation, and vigorously inspired by such wide-ranging observers as Margaret Mead and Kurt Lewin. Each participant was expected to make himself understood at these small and intense meetings, and I think that this also taught me (as I slowly learned to speak and write in English) to write for an interdisciplinary audience, an effort which, in turn, may have had some influence on my choice of concepts. After the Second World War, we made contact again with receptive colleagues from all over Europe—J. Huxley, Lorenz, Piaget—in the Child Study Group of the World Health Organization. The conferences of the American Academy of Arts and Sciences have since assumed the role of interdisciplinary meetings essential to my professional life.

To summarize in decades: the *thirties,* I was first of all a practicing psychoanalyst, primarily with children, making frequent excursions to clinical conferences in the medical area of Harvard. I did some graduate work in psychology on the side, but when the Yale Medical School gave me a full-time research appointment, I decided to weather the future without belated degrees. The Yale Institute of Human Relations offered then a remarkable interdisciplinary stimulation under the leadership of John Dollard; and my job permitted me my first field trip (with Scudder Mekeel) to the Sioux Indians in South Dakota. I spent the *forties* in California, having been invited to analyze the ongoing records of a longitudinal study (led by Jean Macfarlane) of a cross section of Berkeley children. From there I made my second field trip (with Alfred Kroeber) to the Yurok Indians in California. Later, having been appointed a training psychoanalyst, I resumed private practice in San Francisco, but continued to act as a consultant in various public clinics, including a veteran rehabilitation clinic at the conclusion of the Second World War. My first professorship, in Berkeley, was short-lived because of the loyalty oath controversy during the McCarthy era. I was fired before the first year was up, and, after being reinstated as politically dependable, I resigned because of the firing of others who were not so judged. As I think back on that controversy now, it was a test of my American identity; for when we foreign born among the nonsigners were told to "go back where we came from," we suddenly felt quite certain that the ideals behind our apparent disloyalty to the soldiers in Korea were, in fact, quite in line with what they were said to be fighting for. In the *fifties,* having returned east and to therapeutic work in a hospital, I was able to verify on my clinical homeground what by then I had learned about the identity crisis.

I think I have now said enough about myself to come to the question of how the concepts of "identity" and "identity crisis" emerged from my personal, clinical, and anthropological observations in the thirties and forties. I do not remember when I started to use these terms; they seemed naturally grounded in the experience of immigration and Americanization. As I summed the matter up in my first book which appeared in 1950:

> We began to conceptualize matters of identity at the very time in history when they become a problem. For we do so in a country which attempts to make a super-identity out of all the identities imported by its constituent immigrants; and we do so at a time when rapidly increasing industrialization threatens these essentially agrarian and patrician identities in their lands of origin as well.
>
> The study of identity, then, becomes as strategic in our time as the study of sexuality was in Freud's time. Such historical relativity in the development of a field, however, does not seem to preclude consistency of ground plan and continued closeness to observable fact. Freud's findings regarding sexual etiology of a mental disturbance are as true for our patients as they were for his; while the burden of identity loss which stands out in our considerations probably burdened Freud's patients as well as ours, as re-interpretations would show. Different periods thus permit us to see in temporary exaggeration different aspects of an essentially inseparable whole.[13]

Identity problems were in the mental baggage of generations of new Americans, who left their motherlands and fatherlands behind to merge their ancestral identities in the common one of self-made men. Emigration can be a hard and heartless matter, in terms of what is abandoned in the old country and what is usurped in the new one. Migration means cruel survival in identity terms, too, for the very cataclysms in which millions perish open up new forms of identity to the survivors. In the Roosevelt era, we immigrants could tell ourselves that America was once more helping to save the Atlantic world from tyranny; and were we not hard at work as members of a helping profession which—beyond the living standards it made us accustomed to—contributed to a transforming enlightenment apt to diminish both the inner and the outer oppression of mankind? What now demanded to be conceptualized, however, called for a whole new orientation which fused a new world-image (and, in fact, a New World image) with traditional theoretical assumptions. I could not look at my patients' troubles any more in (what I later came to call) "originological" terms—that is, on the basis of where, when, and how "it all started." The question was also, where were they going from where they were and who was going with them? And if something like an identity crisis gradually appeared to be a normative problem in adolescence and youth, there also seemed to be enough of an adolescent in every American to suggest that in this country's history, fate had chosen to highlight identity questions together with a strangely adolescent style of adulthood—that is, one remaining expansively open for new roles and stances—in what at the time was called a "national character." This, incidentally, is not contradicted by the fact that today some young adults are forcefully questioning the nation as to what generations of Americans *have*, indeed, *made* of themselves by claiming so irreverently to be self-made, and what they have made of their continent, of their technology, and of the world under their influence. But this also means that problems of identity become urgent wherever Americanization spreads, and that some of the young, especially in Americanized countries, begin to take seriously not only the stance of self-made men, but also the question of adulthood, namely, how to *take care* of what is being appropriated in the establishment of an industrial identity.

At any rate, the variety of my clinical and "applied" observations now helped me to see a nexus of individual and history as well as of past and future: the Berkeley children, in the particular setting of their parents' Californization, could be seen approaching a special and yet also normative identity crisis, such as seemed to be built into the human life plan. A different version of such a crisis could be seen in the American Indians, whose expensive "reeducation" only made them fatalistically aware of the fact that they were denied both the right to remain themselves or to join America. I learned to see traumatically renewed identity crises in those returning veterans of World War II who had broken down with what were alternately called symptoms of shock or fatigue or of constitutional inferiority and malingering. In retrospect, I had by then recognized a national identity problem in that most surrounded great nation in Europe, once defeated and humiliated Germany, now hypnotized by a markedly adolescent leader promising a thousand years of unassailable super-identity. And I could later verify the symptoms of acute and aggravated identity confusion in the young patients of the Austen Riggs Center in the Berkshires where I had turned after the University of California debacle. There also I found my critic and friend David Rapaport, who professed to see a place for my concepts (next to Heinz Hartmann's) in the edifice of psychoanalytic ego psychology,[14] but not without having added to the dynamic, structural, economic, and genetic points of view an "adaptive" one, relating the ego to the environment. In the sixties I temporarily suspended my clinical work in order to learn how to teach my whole conception of the life cycle—including the identity crisis—to people normatively very much in it: Harvard undergraduates. But this is another chapter, as is the more systematic pursuit of psychohistorical problems by myself and by my colleagues during this decade. In this essay, I restrict myself to the emergence of the identity concepts.

Clinical verification, as I have indicated, is always of the essence in any conceptual shift in psychoanalysis because it confirms that (and why) a syndrome such as "identity confusion" is not just a matter of contradictory self-images or aspirations, roles or opportunities, but a central disturbance dangerous for the whole ecological interaction of mind organism with its "environment"; man's environment, after all, is the shared social universe. The symptoms of identity confusion, then, could be found in the psychosomatic sphere as well as in the psychosocial one, in the times as well as in the individual. Psychosocial identity proved to be "situated" in three orders in which man lives at all times.

1. The somatic order, by which an organism (as René Dubos has emphasized in our discussions) seeks to maintain its identity in a constant renewal of the mutual adaptation of the *milieu intérieur* and the environment.
2. The ego order, that is, the integration of personal experience and behavior.
3. The social order maintained together by ego-organisms sharing a geographic-historical setting.

These three orders, while relative to each other, yet maintain a certain autonomy justifying a different approach to each. Both man's triumphs and his debilitating conflicts, however, originate in the tension between these orders. For they seem to wholly support each other only in utopian schemes, which give man the visionary impetus to correct, at intervals, the accrued dangers to health, sanity, or social order. In this context, too, the study of the identity crisis in adolescence becomes strategic because at that stage of life, the organism is at the height of its vitality and potency; the ego must integrate new forms of intensive experience; and the social order must provide a renewed identity for its new members: no wonder that the ideological thinking of youth is given to total solutions of a utopian kind.

At this point, it must be rather obvious why the concept of identity crisis also helped me to recognize the transforming function of the "great man" at a certain junction of history. As I put it in my book on young Luther: deeply and pathologically upset, but possessed both by the vision of a new (or renewed) world-order and the need (and the gift) to transform masses of men, such a man makes his individual "patienthood" representative of a universal one, and promises "to solve for all what he could not solve for himself alone."

Finally, after all the new insights that totalitarianism, nuclear warfare, and mass communication have forced us to face, it can no longer escape us that in all his past man has based much of his identity on mutually exclusive group identities in the form of tribes, nations, castes, religions, and so on. We really suffer from an evolutionary identity problem: is man one species, or is he destined to remain divided into what I have referred to[16] as "pseudo-species" forever playing out one (necessarily incomplete) version of mankind against all the others until, in the glory of the nuclear age, one version will have the power and the luck to destroy all others just moments before it perishes itself?

If in a given field the "classical" approach is identified totally with one man, it stands to reason that in his lifetime or right after his death even the most necessary conceptual transformations become associated with emotions of sacrificial loyalty or, indeed, father murder. Thus, often quite in contrast to the founder's own irreverent beginnings, the followers' thinking comes under the shadow of the question of what "he" would have approved or would have disavowed, as if even the greatest among men could (if, indeed, they would) guarantee the purity of their creation once it enters the domain of historicity. Incidentally, to decide what the word "classical" really implies in all its logical and emotional ramifications would be well worth some psychohistorical study. In any given historical period there must be a mutual assimilation of all the concepts deemed classical in various fields; for even as the individual ego attempts to maintain a livable orientation in the multiplicity of experience, so must a civilization strive to receive and to integrate all transforming ideas into a coherent universe. Such a process is, of course, more obvious in religio-ideological and medico-humanitarian pursuits because they both simplify and unify in order to reassure, cure, and teach. But even in the hallowed objectivity of science, truly transforming ideas seem to have a fate which not only leads from scientific incredulity to verification, but also from philosophical repugnance to a new sense of classical coherence proving once more that "God does not play dice with mankind." And who could be greater, even if he only very partially understood, than the modest scientist who takes a daring new look behind appearances and returns to affirm that God knows what He is doing?

And being somewhat of an expert in play I must add that while we remember great men usually as severe creators, they are certainly also the ones in whom a divine playfulness is undiminished in its capacity to transcend in new formulae some of the traumatic discrepancies of the times. To me, Freud, too, had throughout his decades played so sovereignly with so many types of conceptualization (from nature philosophy to economics) that I could not see how he could have thought of himself as anybody but a man who had to *bring the thought patterns of the millennia up to the moment of scientific introspection* while utilizing the modes of hippocratic observation which had become part of his medical identity. Perhaps great transforming ideas always contain only a limited number of truly provable assumptions—enough to establish some lasting roots in observation while branching out into new world-images.

Because of the marked ideological-humanitarian aspects of pshychoanalysis, the development of a revolutionary approach into a classical and even an orthodox one—and into the resulting heresies—took little time. I, at least, could witness it in one lifetime; although I must admit that my primary interest in the flux of phenomena precluded any attempt to find safety in orthodoxy or escape in heresy.

As indicated, Freud had, to my mind, transferred the modes of medical research to psychology as he learned to "locate" early psychic traumata in a way analogous to the investigation of brain lesions. Such transfer of conceptual imagery could probably be demonstrated in the work of other originators of transforming ideas and found to be highly productive if balanced and counteracted by other modes of thought pervading the thinker's over-all identity. In Freud's case there was, no doubt, a continuous struggle between the doctor role and such powerful identity fragments as liberator, scholar, writer. These he permitted himself to cultivate only sparingly, for he was, above all, a physician who would cure man of the curse of the past whether it originated in evolution, in primitive history, or in early life history—all three aspects of man's life discovered in Freud's very century by the Darwinists, the archeologists and anthropologists, and by himself: his many analogies between the mind's layers and archeological research come to mind here. Freud's antiteleological concern with the past as incapsulated and entombed in the mind has led, both in practice and theory, to astonishing discoveries; and if it is true that the mood of this kind of mastery of the past resembled somewhat the ritual reenactment of beginnings (creation, spring, birth), then it would seem plausible that psychoanalysis appealed above all to people who had lost their origins in soil, ritual, and tradition. But as we have seen, it will not do to explain human phenomena by their origins in childhood without asking why and how the social environment initiates, reinforces, and aggravates selected childhood conflicts and makes their outcome part of the positive and negative identity fragments which will vie with each other in adolescence. By an exclusive emphasis on origins, psychoanalysis has in fact contributed to a world-image pervaded by a new sense of predestination, which, it then seems, can be alleviated only by a religious faith in psychoanalysis, or at least in its habitual vocabulary.

Freud's combination of a retrospective and introspective approach in psychoanalysis is often referred to as his "fatalism"; while I am only too aware of the fact that later concepts such as mine are welcomed by many as a more "optimistic" promise of life chances not doomed by childhood experience. But identity concepts only emphasize for one stage of life what is true for all, namely that periods of rapid growth and of a widening range of cognition permit, in interaction with living institutions, a renewal of old strengths as well as an initiation of new ones. That, however, does not in itself provide a more benign outlook. Rather it demands new and ruthless insights into the functioning of society and this especially in a world of rapid and unpredictable change.

Are such conceptions as the identity crisis, then, mere additions to the classical scheme, or do they call for a transformation in clinical and theoretical outlook? I can only introduce this question in conclusion with a few notes on what, over the years, I have found to be significant theoretical differences between the classical psychoanalytic outlook and newer perspectives such as my own. These differences will highlight what I meant when, earlier in this paper, I suggested that backward *and* forward, inward *and* outward, downward *and* upward may all be dimensions to be considered in the development of a psychoanalytic model of human existence.

1. *Super-ego and Identity.* In his contribution to this volume, Talcott Parsons relates his momentous integration of Freudian thought and modern sociology. He emphasizes the usefulness which the Freudian concept of the super-ego (as "the internalization of the social structures") has had for his attempts to link man's inner life and his social world. Let me, therefore, compare the super-ego as an earlier concept—both in the sense that the concept was created earlier in the history of psychoanalysis and in the sense that the super-ego originates earlier in life—with the identity concept.

The child internalizes into the super-ego most of all the prohibitions emanating from the social structure—prohibitions, furthermore, which are perceived and accepted with the limited cognitive means of early childhood and are preserved throughout life with a sadomasochism inherent in man's moralistic proclivities. (These are aggravated, of course, in cultures counting heavily on guilt as an inner governor.) Thus internalized infantile moralism becomes isolated from further experience, wherefore man is always ready to regress to and to fall back on a punitive attitude which not only helps him to re-repress his own impulses but which also encourages him to treat others with a righteous and often ferocious contempt, quite out of tune with his more advanced insights. Man could not become or remain moral without some such moralistic tendency; yet without a further development of truly *ethical* strivings, that is, a subordination of his moralism to the shared affirmation of values, man could never build the social structures which define his adult privileges and obligations.

Such further development, however, is not taken care of by Freud's "structural" point of view, which is useful primarily in analyzing the extent to which a person in his childhood has become a system to himself, unable in later stages to adapt to and to grow with the actual present. I have briefly accounted for the way in which a psychosocial theory can explain why young people must and can (as children cannot) join each other in cliques and "subcultures" and eventually join up with large-scale ideological trends of past or present. Here the strength of the ego seems to be dependent not only on the individual's preadolescent experience (including the contents of the super-ego), but also on the support it receives from adolescent subcultures and from the living historical process.

But in psychoanalysis, what has thus been learned about later stages of life must always be reapplied to previous observations on earlier stages and disturbances. It becomes obvious, then, that an intricate relation between inner (cognitive and emotional) development and a stimulating and encouraging environment must exist from the beginning of life, so that no stage and no crisis could be formulated without a characterization of the mutual fittedness of the individual's expanding capacity to relate to people and institutions, on the one hand, and, on the other, of the readiness of these people and institutions to include him in ongoing cultural concerns. All this, in fact, determines the nature of the identity crisis which, in turn, determines what happens to the remnants of the infantile super-ego, and the aggression stored in it.

2. *Clinical and Actual Reality.* Psychoanalytic treatment will reveal the "scarcity economics" of an impoverished condition, in which much of the available energy has been directed inward as anxiety spread and action became inhibited. But just because the "economic" point of view was invented to account for such simplified circumstances—simplified by the stereotypy of symptomatic behavior—they serve less well to understand the exchange of energies and, in fact, that shared multiplication of energies with a widening radius of individuals and institutions throughout life's stages. What is at stake here is not more and not less than psychological "reality." The clinical world view rightly takes the inability to accept reality as its baseline and endeavors to fill in blind spots, to correct distortions, and to do away with illusions caused by infantile fixations. And, indeed, in removing all these, therapy creates a minimum condition for dealing with the world "as it is." To that extent, the clinician must take it for granted that his own function within the established order is "reality oriented."

The experience of the identity crisis, so I have indicated, takes place when the world of childhood gives way to that of an ideological universe which for a while coexists with the accumulative knowledge of "reality." True, youth gradually becomes equipped with all the cognitive functions which adult man will ever call his own. And yet, it is often radically involved in world-outlooks which make the "facing of reality" a hazardous criterion of creative imagination. To give this whole matter another dimension, I have insisted that the German *Wirklichkeit* really combines "reality" with "actuality," that is, consensually validated world of facts with a *mutual activation* of likeminded people.[15] Even among the most intelligent and informed men, there is always a search for communality with those who not only think alike but also make each other feel active and masterly. And it must be obvious how often, in adolescence as well as in adult ideology, the most gifted minds must surrender their sense of verifiable fact to that of a mutual actualization through the sharing of a unified world image.

3. *The Inner and the Outer World.* The classical psychoanalytic technique was also the original research laboratory; how human nature works outside it was, by necessity, a matter of the speculative application of clinical findings to the "world outside." But psychoanalysis, as any other field, can renew itself only by gaining insight into the nature of its observational setting. The natural sciences have had to take account of the fact that not only the personal equation of old, but also all the details of the laboratory arrangements are apt to become inseparable from the bit of nature isolated for observation. In psychoanalysis, however, it is still claimed that only the clinical situation provides the setting that can reveal the true workings even of the "normal" mind. The authority often quoted in this context is Freud's statement that only a broken crystal reveals its structure which is invisible in intactness. But a living ego-organism is not a crystal, and even as anatomy and pathology must yield to physiology and biochemistry in the attempt to reconstruct intact functioning, psychoanalysis must complement its clinical findings with the study of psychosocial functioning.

It is, of course, true that no situation affords a better access to the workings of the unconscious than does the psychoanalytic one. Yet the greatest difficulty in the path of psychoanalysis as a general psychology probably consists in its first conceptualization of the environment. To a patient under observation, the "world" he records (and more often than not, complains about) easily becomes a hostile environment—"outer" as far as his most idiosyncratic wishes are concerned, and "outside" his relation to his therapist. This says much about him and about man; but it seems difficult to account for the nature of the clinical laboratory if the nature of the human environment is not included in the theory which guides the therapeutic encounter. The fact is, such neglect sooner or later leads to the adoption of a clinical world view. When I once heard a revered teacher refer to the world at large as the "outer-psychoanalytic outerworld" I experienced a conceptual as well as a philosophical shudder. For even though it is true that, by definition, "the world" must always resist psychoanalytic insight, the continued systematic study of such resistance in ever new historical disguises is of the essence of the psychoanalytic enterprise. Such considerations give due warning that a science so close to questions of health and ethics must include methods of observing its own functioning in the cultural-historical process and its (intended and unintentional) influence upon it.

A related problem is the application of what has been observed in the encounter of therapist and patient to other forms of relationships. In the psychoanalytic situation, early and earliest relations with significant persons is ever again transferred and relived, relost and renewed, and this (it seems probable) at least partially because of the technical choice of the basic couch arrangement by which the observer and the observed do not look each other in the face, communicate only with words, and thus avoid the earliest and the most lasting mutual affirmation: that by vision. Here, again, what is produced in the laboratory is highly instructive all around; but it would seem wise to apply it only with caution to the functioning infant, as well as to any other person outside clinical captivity, and especially also to the young person whose needs during the identity crisis demand (beside some "parent substitutes") a vigorous peer group, an ideologically integrated universe, and the experience of a chosen mutuality with newly met persons and groups.

A close study of the original clinical setting as a laboratory would, incidentally, also reveal the fact that at least one vital aspect of life—namely, violence—systematically escapes the treatment situation, as if the individual would participate in any uses of direct violence (from aimless riot to armed force) only when absorbed in the unenlightened and uncivilized mass that makes up the "outerworld." But then, as pointed out, the clinical laboratory channels all impulse to act into introspection, so that the cured patient may be prepared for rational action. Thus, the clinician learns much more about the nature of inhibited and symptomatic action than about that of concerted action in actuality—with all its shared a-rationalities.

4. *The Ego and the I.* A final methodological problem may or may not point beyond psychoanalysis. Freud's puritanical self-denial as an observer in the long run prejudiced the very act of observing. Freud, it is true, went further than any man before him in revealing publicly the role which his own conflicts played not only in his own dreams, but also in his dealings with his patients. But in doing so, he failed to analyze the role of the observer who chooses to observe his own conflict. He con-

ceptualized the ego as a psychological part of man's unconscious inner structure, but he did not question the "I," the observing core of consciousness. But who wields this consciousness, so vastly refined as a weapon and tool in dealings with the unconscious? In using the word puritanical, I meant to imply that Freud's self-restraint in this regard, though it could be expressed in antireligious terms, was deeply religious in nature: man had talked enough of his soul, and had for too long congratulated himself on being "the measure," or the conscious "center," of the universe. If a man like Kierkegaard could write about the leap of faith, a doctor and scientist could only grimly and rationally describe with the methods given to him what many of the most sublime men as well as the most depraved had previously refused to acknowledge.

It is one thing, however, to cultivate the proud rationality of the Enlightenment of which Freud was probably the last great representative, and which he crowned by insisting that irrationality and the unconscious be included in the sphere to be understood rationally. It is another matter to derive from such inquiries into man's conflicts a model of man. At any rate, I have found myself studying in the lives of religious innovators that border area where neurotic and existential conflict meet and where the "I" struggles for unencumbered awareness and ethical involvement. And again, is it not in adolescent experience that the "I" can first really perceive itself as separate? It does so, as it finds itself both involved and estranged in peculiar states which transcend the identity crisis in psychosocial terms, because they represent not only the fear of otherness and the anxiety of selfhood, but also the dread of individual existence bounded by death. All of this, of course, is easily forgotten when the young adult assumes his responsibilities and when he is forced to participate in the hierarchies of his society, with their organized beliefs and convincingly concealed irrationalities. But in the long run, the "I" transcends its limited identity and, sooner or later, faces the dilemma of existence versus politics which I have attempted to approach in my work on the Gandhian version of truth.

Behind all this may well be another and wider identity crisis. Psychoanalysis, in line with the Enlightenment, has debunked the belief (and the need to believe) in a deity. It has suggested that the god image "really" reflects the infantile image of the father, as, indeed, it does in transparent cultural variations. But may it not be for good phylogenetic reasons that the ontogenetic father is overendowed with an awe which can later be shared in common beliefs in god-images and in semidivine leaders? For a community of I's may well be able to believe in a common fund of grace only to the extent that all jointly acknowledge a super-I. This is a problem of such magnitude that mere intellectual denial hardly touches it, least of all if this denial is accompanied by a total fascination with the creator of one's field. At any rate, what religion calls grace and sin transcends the comfort of adaptation and the management of guilt.

All this takes us beyond the identity crisis in its developmental and psychosocial determinants. But, then, this crisis is (and makes sense as) only one of a series of life crises. What does happen, we may well ask in conclusion, to adults who have "found their identity" in the cultural consolidation of their day? Most adults, it is true, turn their backs on identity questions and attend to the inner cave of their familial, occupational, and civic concerns. But this cannot be taken as an assurance that they have either transcended or truly forgotten what they have once envisaged in the roamings of their youth. The question is: what have they done with it, and how ready are they to respond to the identity needs of the coming generations in the universal crisis of faith and power? In the end, it seems, psychoanalysis cannot claim to have exhausted its inquiry into man's unconscious without asking what may be the inner arrest peculiar to adulthood—not merely as a result of leftovers of infantile immaturity, but as a consequence of the adult condition as such. For it is only too obvious that, so far in man's total development, adulthood and maturity have rarely been synonymous. The study of the identity crisis, therefore, inexorably points to conflicts and conditions due to those specializations of man which make him efficient at a given stage of economy and culture at the expense of the denial of major aspects of existence. Having begun as a clinical art-and-science, psychoanalysis cannot shirk the question of what, from the point of view of an undivided human race, is "wrong" with the "normality" reached by groups of men under the conditions of pseudo-speciation. Does it not include pervasive group retrogressions which cannot be subsumed under the categories of neurotic regressions, but rather represent a joint fixation on historical formulae perhaps dangerous to human survival?

But I have promised to pursue the matter of the "identity crisis" only as far as autobiographic considerations would carry it. At the same time I have recapitualted my conceptual ancestry insofar as it originates in psychoanalysis, in order to indicate that psychoanalysis represents a very special admixture of "laboratory" conditions, methodological climate, and personal and ideological involvement. Other fields may claim to be governed by radically different admixtures and certainly by much less subjective kinds of evidence. But I wonder whether they could insist, at any time, on a total absence of any one of the ingredients described here.

REFERENCES

1. Erik H. Erikson, "On the Nature of Clinical Evidence," *Dædalus* (Fall 1958), pp. 65-87, revised in *Insight and Responsibility: Lectures on the Ethical Im-*

plications of Psychoanalysis (New York: W. W. Norton, 1964); "The Concept of Identity in Race Relations," *Dædalus* (Winter 1966), pp. 145–170; "The Nature of Psycho-Historical Evidence: In Search of Gandhi," *Dædalus* (Summer 1968); "Reflections on the Dissent of Contemporary Youth," *Dædalus* (Winter 1970).

2. Erikson, *Young Man Luther: A Study in Psychoanalysis and History,* Austen Riggs Center Monograph, 4 (New York: W. W. Norton, 1958); *Gandhi's Truth* (New York: W. W. Norton, 1969).

3. Erikson, *Identity: Youth and Crisis* (New York: W. W. Norton, 1968).

4. Erikson, "The Human Life Cycle," *International Encyclopedia of the Social Sciences* (New York: Crowell-Collier, 1968); "Psychosocial Identity," *International Encyclopedia of the Social Sciences* (New York: Crowell-Collier, 1968).

5. Stuart Hampshire, *The London Observer* (December 1, 1968).

6. Sigmund Freud, "An Autobiographical Study," *The Complete Works of Sigmund Freud* (London: Hogarth, 1953), XX, 40. Originally appeared in 1925, in a volume on the state of medicine as revealed in the autobiographies of its leaders. Also in Norton Library (New York: W. W. Norton, 1963).

7. Mircea Eliade, *The Myth of the Eternal Return,* Bollingen Series XLVI (New York: Pantheon Books, 1954).

8. Erikson, "On the Nature of Clinical Evidence."

9. Anna Freud, *The Ego and the Mechanisms of Defense* (New York: International Universities Press, 1946).

10. Heinz Hartmann, *Ego Psychology and the Problem of Adaptation* (New York: International Universities Press, 1958).

11. Erikson, "Freud's 'The Origins of Psychoanalysis,'" *International Journal of Psychoanalysis,* 36 (1955), 1-15; "The First Psychoanalyst," *Yale Review,* 46 (1956), 40-62, revised in *Insight and Responsibility;* Book review, "Thomas Woodrow Wilson, by Sigmund Freud and William C. Bullitt," *The New York Review of Books,* 8 (1967).

12. Peter Blos, *On Adolescence* (New York: The Free Press of Glencoe, 1962).

13. Erikson, *Childhood and Society* (New York: W. W. Norton, 1950), pp. 242-243.

14. Merton M. Gill, ed., *The Collected Papers of David Rapaport* (New York: Basic Books, 1967).

15. Erikson, *Insight and Responsibility,* p. 159.

16. "The Ontogeny of Ritualisation in Man," *Philosophical Transactions of the Royal Society of London,* Series B., No. 772, Vol. 25, 1966, pp. 147–526.

EDWARD SHILS

Tradition, Ecology, and Institution in the History of Sociology

Sociology at present is a heterogeneous aggregate of topics, related to each other by more or less common techniques, by a community of key words and conceptions, by a widely held aggregate of major interpretative ideas and schemes. It is held together, too, by a more or less common tradition—a heterogeneous one in which certain currents stand out—linked to common monuments or classical figures and works. The tradition lives in a self-image which links those now calling themselves sociologists with a sequence of authors running back into the nineteenth century. Although in fact the main ideas which live on in contemporary sociology have a much older history, sociologists do not generally see themselves as having an ancestry originating any earlier than the nineteenth century.

Most of sociology is not scientific in the sense in which this term is used in English-speaking countries. It contains little of importance that is rigorously demonstrated by commonly accepted procedures dealing with relatively reproducible observations. Its theories are not ineluctably bound to its data. The standards of proof are not stringent.

There are differences, of course, among the various substantive fields of sociology, some being more scientific than others, but on the whole the standard of scientific accomplishment is low. This does not mean that in those parts which are not scientific there is not some very substantial learning or that there is not an accumulated wisdom which merits regard and consideration. Nor does it mean that some of it, even in its present intellectual state, cannot contribute to the improvement of policy and administration and to civility in a broad sense. It only means what it says, namely, that much of contemporary social science is not very scientific in the sense in which the term has come to be understood.

Nonetheless, sociology does exist. It has a history which present-day sociologists regard as their history—although of course the image of this history varies somewhat from country to country and among sociologists within particular countries. It has a vast and rapidly expanding stock of works and a large and also rapidly expanding personnel working in and forming academic institutions of research, teaching, training, and consultation; governmental institutions of research, execution, and consultation; and private, nongovernmental, nonacademic institutions in which sociologists perform some of the same activities they perform in governmental institutions. The stock of works, heterogeneous though it is in its particular subject-matters, in its techniques of observation and analysis, and in particular interpretations, is characterized by a few widely pervasive major ideas or beliefs about society, by a few major concepts or delineations of significant variables.

The concepts of social system, of society and its constituent institutions, of social stratification and social mobility, of elites and ruling classes, of social status, of role, of bureaucracy and corporate organization, of kinships and local community, of history and tradition, of intellectuals and ideology, of consensus, anomie, and alienation, of conformity and deviance, of charisma, of the sacred and rebellion against institutions—all these concepts have been grouped into a conception of modern society (and its variant of mass society) which is further defined by contrast with a conceptual construction variously designated as folk society, traditional society, or *Gemeinschaft*.

How has all this come about? Why has the intellectual stock of sociology come to be what it is and why has it taken that form in particular places? Why have certain ideas which are now thought to be constitutive of sociology come to dominate the subject?

One of the older answers was that social science or sociology could not emerge until men were capable of sufficient detachment from involvement in their own affairs and in their beliefs about these affairs; a corollary of this was that sociology could arise only when authority had lost some of its sanctity and when traditional beliefs became somewhat discredited. This view has been put forward by a number of writers, most notably, Durkheim and Sombart. There is something in this view—but not enough. The heavens belonged to the gods and religious beliefs and to those who took those beliefs in their charge on behalf of the gods, but that did not prevent the emergence of astronomy. It might have impeded the emergence of astronomy, but it did not prevent it. Man's body in the West was God's creation, but that, too, did not prevent the study of anatomy or the understanding of the circulation of the blood. In Greece, Rome, and medieval Islam, in the writings of Aristotle, Polybius, Thucydides, and Ibn Khaldun, ideas were put forth which have reappeared in modern sociology, but the subject never became precipitated in the way that physics or mathematics became established in the seventeenth and eighteenth centuries. Ancient Greece and Rome and medieval Islam were not secularized societies, they were not traditionless societies in which authority and custom had receded from earlier strength. So the hypothesis is not quite satisfactory. There is, however, something in the hypothesis that sociology requires a cultural matrix or setting which is not so exigent in its demands on substantive belief that it stifles detachment. There is something in the hypothesis that sociology requires a loosening of the belief that divine or magical powers intervene at will in human affairs. It is also true that sociology as a body of generalized knowledge about society requires freedom from a jealous and pervasive ecclesiastical or secular authority which is apprehensive about the potential dangers of the formation of beliefs not identical with those it holds about itself. These conditions, however, are the conditions of any intellectual activity which is not wholly committed in advance to agreement with the views held by ecclesiastical and earthly authority on matters which that authority believes are vital.

What I wish to say here is that the recession of earthly and ecclesiastical authority, and a loosening of the grip of traditional beliefs which consecrate that authority, is only a very general precondition. It tells very little about the intellectual direction and the territorial location of the growth of sociology.

Sociology, even if it is not very scientific, is an intellectual accomplishment. The practice of sociology—that is, sociological research and sociological reflection—is an intellectual activity. As an intellectual activity, it operates within an intellectual tradition and as such each sociological action takes place within the framework of the tradition which it in turn affects and, in some important instances, modifies markedly. The tradition of sociology even now is not rigorously coherent and authoritative in its presentation and it was less so in its earlier stages in the nineteenth and twentieth centuries. It has offered to its recipients a variety of possibilities. There has been a process of selection. The selection has been a function of intellectual self-evidentness—which is characteristic of any tradition—of exposure through individual contact, further fostered by institutional consolidation. Institutions have not created sociology; it has been created by individual sociologists exercising their powers of observation[1] and analysis on social situations apprehended within the focusing framework of sociological traditions. Institutions concentrate attention and reinforce selected ways of perceiving and interpreting experience. Institutions foster the production of works, and the works—with what they contain in the way of interpretation of social reality—become part of the focusing tradition. Institutions create a resonant and echoing intellectual environment. The sociological ideas which undergo institutionalization are thereby given a greater weight in the competition of interpretations of social reality.

Institutionalization

By institutionalization of an intellectual activity I mean the relatively dense interaction of persons who perform that activity. The interaction has a structure; the more intense the interaction, the more its structure makes place for authority which makes decisions regarding assessment, admission, promotion, allocation. The high degree of institutionalization of an intellectual activity entails its teaching and investigation within a regulated, scheduled, and systematically administered organization. The organization regulates access through a scrutiny of qualification, provides for organized assessment of performance, and allocates facilities, opportunities, and rewards for performance—for example, study, teaching, investigation, publication, appointment, and so forth. It also entails the organized support of the activity from outside the particular institution and the reception or use of the results of the activity beyond the boundaries of the institution. An intellectual activity need not be equally institutionalized in all these respects. It should be remembered that an intellectual activity can be carried on fruitfully with only a rudimentary degree of institutionalization.

Sociology is more institutionalized where it can be studied as a major subject than where it can be studied only as an adjunct subject, where it has a specialized teaching staff of its own rather than teachers who do it only as a *Nebenfach,* where there are opportunities for the publication of sociological works in sociological journals rather than in journals devoted primarily to other subjects, where there is financial, administrative, and logistic provision for sociological investigation through established institutions rather than from the private resources of the investigator, where there are established and remunerated opportunities for the practice of sociology (teaching and research), and where there is a "demand" for the results of sociological research.

The social sciences today have become relatively highly institutionalized branches of study in the countries of Western Europe and of North America—more recently in the former than in the latter. They became established as academic subjects later than most of the other major academic disciplines, such as mathematics, physics, chemistry, zoology, botany, classics, oriental studies, modern languages, and literature. They have, however, now made up the distance. Degrees, undergraduate and postgraduate, are awarded for the completion of organized courses of study in them; research training under qualified teachers is provided, often as part of research teams organized and supervised by the teachers, reported preliminarily in seminars where work is criticized. Journals, with academically qualified referees and often supported by learned societies, are available for the publication of the results of research in each field and in some quite specialized subfields. In addition to more routine and periodic meetings of learned societies, there is an elaborate network of communication through the circulation of offprints, memoranda, and preliminary versions of research reports directed to both practitioners and apprentices in various fields and subfields, and even in the informally circumscribed domain of a particular problem. Practically all of these activities are firmly incorporated into the structure of universities and professional associations which have a momentum of their own.

In the first half of the nineteenth century, the social sciences scarcely existed as academic subjects anywhere, although of course they had forerunners in the *Staatswissenschaften,* in the juridical sciences, on the edge of historical-institutional studies, and in moral philosophy. Their academic establishment, which began slowly and unevenly in the last quarter of the nineteenth century, was made possible by a loosely articulated set of intellectual traditions which extend back to western antiquity and which acquired focus and delineation in the course of the nineteenth century.

The nodal points of these traditions were not generally university teachers, often they were not even university graduates. Indeed, not even in economics, which became an intellectually *orderly* discipline, relatively speaking, much earlier than sociology and political science, did the academic element have the field to itself. Adam Smith stands out as the first important academic contributor to the subject, although political economy was only one of the four subjects for which he was responsible within his jurisdiction as professor of, first, logic and, then, moral philosophy. (He had been out of academic life for thirteen years when *The Wealth of Nations* was published in 1776.) But Ricardo was never an academic, nor was James Mill, and the Reverend William Paley's teaching at Haileybury could not be regarded as strictly academic since that body was closer to a secondary school than a university. The major synthesist of the middle of the century, John Stuart Mill, had only the most marginal academic connections; he was rector of St. Andrews, a wholly honorific position and a very transient one. Nonetheless, Cairnes, Senior, Fawcett, Sidgwick, and Jevons were university teachers, and beginning not later than Alfred Marshall's first academic steps, serious young men could obtain at a few places in Great Britain an orderly and disciplined training in economic analysis. Empirical economic research was another matter, being usually in the hands of government officials, private amateurs, and voluntary bodies; it was largely avocational and minimally institutionalized. Even in this field, however, the possibility of acquiring guidance in the techniques of research existed within the academic frame. Thorold Rogers as professor at Oxford and Archdeacon Cunningham at Cambridge and King's College, London, could help a young man to learn to do economic research with a quantitative bent. If we compare the situation with that prevailing in France, where economics was taught seriously only in the *Conservatoire des arts et métiers,* at the *École des ponts et chaussées,* and at other technical institutions where young men were being prepared for technological careers, and at the *Collège de France,* where there were no pupils, properly speaking, and came into the *Faculté de droit* in Paris only in the 1870's, where it was suffocated under legal lumber, we see one of the reasons why British economics led the world in the nineteenth and early twentieth centuries.

It must be emphasized that reference to institutionalization does not, by any means, wholly explain the great ascendancy of British economic analysis. The conception of the problems to be studied—which was partly a function of prevailing public opinion—had a good deal to do with it; the contacts of economists with the worldly affairs of Parliament, investigative commissions, governmental departments, and leading politicians and businessmen also gave them experiences in which they could exercise the analytical powers nourished by a rich tradition and a lively publicistic discussion. The British economists had a more fruitful point of departure because they analyzed the equilibrium of an economic system and not the economic necessities of governments. The latter task—partly arising from intellectual tradition, partly from the institutionalization of economics in administration—held back the development of economics in the German universities because it impeded the discovery of the market. This discovery and its subsequent elaboration were the great achievements of British economics between about 1775 and 1875 and in this achievement the institutionalization of economic analysis in the universities, governmental institutions, and the press played an important role. Although in this period academic institutionalization was rudimentary in contrast to what it later became, it was more advanced than in any other country. It was, moreover, reinforced—that is, attention focused and intellectual interchange intensified—by the high quality and density of public discussion in journals, books, investigative commissions, and parliamentary debates.

Sociology had several important figures who established the name of the subject—a significant step—and who studied its subject-matter. Auguste Comte and Herbert Spencer were the greatest of these forerunners. Neither of them was a member of the academic profession at any time in his career.[2] Tocqueville and Marx, who in substance are qualified to join the gallery of ancestors of present-day sociology, only recently have been nominated for retroactive membership in the tradition. They were neither more nor less academic than Comte or Spencer:

they had no students, assistants, monograph series, or seminars; no dissertations were written under them.[3] Comte and Spencer installed themselves at the beginning of the tradition by giving a name and a vague set of boundaries to the subject still to be born.

Other originators—German historians and British and French statisticians—who in effect generated and impelled forward the tradition from which sociology as we know it emerged, have not been equally acknowledged. They are allowed to dwell in oblivion. But they are effective nonetheless.

1. The European Founders: Academic Cultivation Without Institutional Provision

The separateness of sociology from the universities during the first three quarters of the nineteenth century was reduced when founders replaced forerunners. The five European founders who stand between the forerunners and the academically established generations of the present century were all academic. Only one of them, however, was academically responsible for sociology. Ferdinand Tönnies was *Privatdozent* in philosophy and later, for several years, professor of economics and statistics; he taught sociology only after his retirement. Vilfredo Pareto never taught sociology; he was a professor of economics for about five years in his late forties. Max Weber, during his short academic career, was professor of economics. Georg Simmel taught philosophy as *Privatdozent* and *ausserordentlicher Professor* for most of his career and became *ordentlicher Professor,* but not of sociology, only a few years before his death. Emile Durkheim was the only one of the generation of founders who made a full academic career with an official responsibility for sociology. He was professor of sociology and education at Bordeaux from 1887 to 1902 and from then until his death in 1917, professor of the same subjects under various titles at the Sorbonne.

Of all the founders of sociology, only Durkheim was successful in institutionalizing his work during his lifetime. Only he had pupils and collaborators, regularly, over a long period. Sociology acquired an institutional form only around Durkheim and that was less through provision by the university system than through Durkheim's own organizational initiative and skill in the formation of the *Année Sociologique.*

Durkheim followed the German pattern: the professor of the subject covered the whole field himself, trained younger collaborators through intensive discussions, and published the results of their individual and joint work in the organ of the institute, which was reserved primarily for contributors from his personal circle. One difference was that in Germany the university provided funds for the institute, which was attached to the chair; in France there was no such provision for university professors. Another important difference between Germany and France was that in German universities those who worked with the professor in his institute were specializing in his discipline, were writing dissertations on it, and would depend on his sponsorship for *Habilitation* and subsequent academic appointment. There was provision for training in research at an advanced level in Germany from the early part of the century; in France training in research was provided only when the *École Pratique des Hautes Études* was founded in 1868. But neither in Germany nor in France did this affect sociologists in the beginning of the twentieth century. Sociology as part of the "human sciences" came into the EPHE only when the *sixième section* was formed after the Second World War. There were, furthermore, few opportunities under the French educational or administrative system for persons who had experience in sociological research. Those who "did" sociology did it because they were genuinely interested rather than because they expected to make a career out of it, because very few could. This meant that Durkheim's entourage consisted of intellectually deeply interested persons whose professional possibilities and aspirations lay outside sociology. While they were active they produced important work, but once they became active in their own careers, they on the whole produced very little which belonged to the core of sociology. It also meant that their reproductive capacity as sociologists was slight, since very few became teachers of sociologists. They did not train sociologists because there were very few students of sociology: there were very few students of sociology because there were practically no posts in sociology to which they could aspire.

In France, therefore, the institutionalization and the resultant expansion and continuity of production was generated almost entirely by the personal force of the individual professor. He was helped, of course, by the existence of the EPHE and the section on *sciences religieuses,* which was not dominated by professors.[4] Hence when Durkheim passed from the scene, the body of interpretation which had developed from his teaching and writing ceased to grow, and French sociology as the study of modern society practically disappeared for more than a quarter of a century. It is often said that the death in battle of some of the most distinguished of the younger collaborators of the *Année Sociologique* was the cause of the cessation of the Durkheim outlook in French sociology, and there is some truth in this. I think, however, that more important was the fact that the institutional structure built by Durkheim rested only on him and was not integrated into the institutional structure of the French university system.

The fate of sociology in Germany supports this interpretation. The three great founders of sociology in Germany were not professors of the subject. They had no research facilities such as German professors ordinarily had; they had no ongoing seminar for training students in research and for bringing their dissertations to the point where they could be published as research monographs. There was no place for students to go professionally once they finished their training—had formally instituted training been available in Germany. Before the First World War there were no professorships of sociology in technical or teacher training colleges. There were no junior sociology posts in the universities, just as there were no senior posts. There was no employment for sociologists, as sociologists, outside universities. They had until the 1920's no journal of their own in which they could publish their works and thereby define their identity. In consequence, there was no concentration of minds on the ideas of great potentiality which came from the three German figures to whom the sociological world now looks back as founders. The ideas of Max Weber, Georg Simmel, and Ferdinand Tönnies did not undergo the process of reinterpretation, partial assimilation, and elaboration which is the characteristic course of development of intellectual products. The failure to do this reinforced the causes of the failure. At the University of Cologne there was, it is true, an active professor of sociology, Leopold von Wiese, who had students, a research institute, and a journal, those three important constituents of the institutionalization of an academic subject. Unfortunately for German sociology, von Wiese's ideas were incapable of development. He was concerned primarily with nomenclature and taxonomy and gave his students no other tasks. Institutionalization in this case could only have consolidated intellectual sterility.[5]

Brief attention might be devoted to another "founder," L. T. Hobhouse, professor of sociology at the London School of Economics from 1907 until 1929. Hobhouse had a widely ranging comparative interest. The breadth of his knowledge was comparable to Weber's and he organized one major piece of research which was unique, until recently, for its integration of quantitative methods into comparative study. Yet today Hobhouse is almost entirely disregarded by sociologists.[6]

Part of the explanation for the oblivion into which Hobhouse has fallen lies in the fact that his interpretation of modern society was felt to be an evolutionary optimism. In fact, it was in some respects not too different from Weber's or Durkheim's—coming as it did from traditions closely akin to or overlapping those which presided over the ideas of the continental founders of sociology. A large part of the explanation, however, seems to me to lie in the very low degree of institutionalization of sociology in British universities. Hobhouse was, with Edward Westermarck,[7] the sole incumbent of a professional chair of sociology in the United Kingdom; his department was also the only department. Despite the sympathetic support of R. H. Tawney, there was no informal consensus among the more esteemed academic personalities regarding the value of the sociological approach to social matters. (Such a consensus began to appear only after the Second World War.) There were no undergraduates who read sociology for a degree in Hobhouse's time (it was not provided for in the London School of Economics syllabus), and British postgraduate students specializing in sociology must have been practically nonexistent. There were no academic appointments in sociology in Great Britain outside the London School of Economics.[8] There was one sociological periodical in England at the time—*The Sociological Review*—and it had no connection with the London School of Economics or any other academic institution. Its inspiration came from Patrick Geddes, Victor Branford, and ultimately LePlay, and its outlook was capable of intellectual development; it was, however, in the hands of a voluntary association of amateurs, enthusiasts, and cranks. There was no provision for research in sociology by students or teachers at the London School of Economics. The great tradition of empirical inquiry in Great Britain which had developed independently of the universities remained until well after the First World War and the work of the public-spirited men of means was done without the help of the universities.

Officially, the situation was not too different from that in France, but the strength of Durkheim's personality, the persuasiveness of his convictions and the superior prestige of the Sorbonne in the French university system—which the London School of Economics did not, at that time, possess—made a great difference. More important than the relatively low prestige of the London School of Economics during Hobhouse's career was the fact that the type of young collaborators available to Durkheim at the Sorbonne were not available to Hobhouse. Durkheim was able to draw to himself young scholars who were specialists in folklore, oriental studies, and ethnography, subjects which did not exist at the London School of Economics. Such studies were pursued in traditional ways at University College, London, and later at the School of Oriental and African Studies, where they had no contact with sociology. There were few postgraduate students and young lecturers at these other institutions and they had to make their careers by meeting the requirements of their own departments, which had a Germanic structure. Young Oxford and Cambridge dons in classics and oriental studies in the first quarter of the present century were unlikely to be drawn to a professor at the London School of Economics.

The fact that Hobhouse did not have Durkheim's imperious personality and lacked his organizing enterprise, and the fact that evolutionism, even of the more cautious type cultivated by Hobhouse, enjoyed little intellectual esteem at the time in British academic circles, added to the obstacles to any influence Hobhouse might have exerted. Hobhouse's one distinguished protégé was Professor Morris Ginsberg, who succeeded him, but Ginsberg was a shy person and lacked Durkheim's organizing skills. The result was that the French sociology of Durkheim's circle left behind a massive deposit. British academic sociology—Hobhouse's variant of sociological evolutionism—left behind only the one work on the material culture of simpler peoples.[9]

The eugenics movement, which was not academically institutionalized, came in a roundabout way to have a more powerful impact on British sociology. It did so through its influence on the development of sophisticated statistical techniques by Galton, Pearson, and Fisher and through the short-lived Department of Social Biology at the London School of Economics in the 1930's. British studies in social selection and mobility belong to this tradition rather than to Hobhouse's.[10]

2. *American Founders: The Beginnings of Academic Institutionalization*

Sociology had a different fate in American universities. It became institutionalized earlier in the United States than anywhere else in the world; it became institutionalized earlier at the University of Chicago than elsewhere. Provision for teaching sociology at Columbia and Yale universities was made at about the same time as at Chicago, but by the turn of the century, provision for the training and research began at Chicago—hesitantly but encouragingly. There was not much to show at first. Research was occasional and discontinuous. In its first decade Chicago produced only minor and scattered pieces of research. The University of Pennsylvania (which sponsored W. E. B. Du Bois' *The Negro in Philadelphia*) and Columbia University did not lag markedly behind Chicago at this time. But, from the very start, there was a propitious atmosphere for research at Chicago, created, in part, by its youth, the first president's determination to make it into a research university (Gilman had already taken the lead in this at Johns Hopkins), and the large number of vigorous young professors who shared the ideal embodied in the research-centered seminars of the German university. Furthermore, because all the departments were new, there was no "old guard" to hamper the immediate establishment of a department of sociology.

Nonetheless, even at the University of Chicago, progress was slow. Albion Small was steeped in the German historical thought of his time, which he regarded as the matrix of sociology. He knew, however, that something more than historical scholarship was required and he was sympathetic to direct observational studies of contemporary events. He believed in seminars and postgraduate studies. Henderson, who was primarily concerned with urban welfare problems, also appreciated the desirability of descriptions and analyses of urban life, particularly the life of the poor, the immigrants, the working classes, and so forth. The Germanic sociological sensibility, intellectual curiosity, and a belief that what had been inherited was not enough all made for the conviction that a new type of research was needed. Nonetheless, even W. I. Thomas, who had an original mind and who received the third Ph.D. granted by the Department of Sociology, continued to work (at first) under the influence of German *Volkskunde,* ethnography, and physical anthropology on the basis of published sources.[11] Direct observational studies and the use of human documents emerged only gradually.

W. I. Thomas was, more than any other person, responsible for this new direction in research. He refined further a subject which had been developed in Germany[12]—the *Volkskunde* of living German rural society, based on impressionistic field observation and interviewing—but which had never found a place in German university studies in the social sciences. Thomas' readiness to observe directly, to collect the "human documents" of living persons, was supported by Small and Henderson. They did not think it undignified for a professor or the professor's pupils to wander about the streets and interest themselves in "low life." This was very different from the German academic situation; even those senior academic figures who were members of the *Verein für Sozialpolitik* and thus very concerned about the condition of the working classes thought that information about them had to be obtained through *Sachverständige,* that is, from middle-class persons who in a professional capacity—for example, magistrates, clergymen, municipal administrators, physicians, and so forth—were in contact with the lower classes.

This accomplishment of the Chicago Department of Sociology hardly seems to be a great intellectual accomplishment. Yet it marked the beginning of empirical sociology; it was the beginning of sociology as a potentially scientific, serious intellectual subject cultivated in a relatively institutionalized manner. This distinctive development went further when, at the beginning of the second decade of the century, Thomas succeeded in persuading Robert Park, who until then had led a somewhat errant existence as a newspaperman, a student in Germany, an assistant in philosophy at Harvard, and a secretary to Booker T. Washington at Tuskegee Institute, to join the Department of Sociology at Chicago. As a newspaperman interested in urban life, he was already acquainted with the surveys by which civically concerned bodies and individuals had attempted to arouse public opinion regarding the condition of the poor in the great cities. These surveys were the intellectual heirs of the American muckrakers of the turn of the century and of the British surveys of the preceding century. The surveys which attracted Park's attention had been conceived and carried out without benefit of academic sociology, and, when they were finished, the organizations which had been created to carry them out were disbanded.

The Chicago Department of Sociology changed this situation. Although the department itself never conducted surveys of the types conducted in Pittsburgh, Cleveland, and Springfield, it assimilated their techniques of direct observation, interviewing, and quantitative treatment—albeit elementary—of results into the training of postgraduate students.

Park brought to Chicago, in addition to this concern for direct observation and interviewing—one variant of which was the participant-observer technique—a fresh and vivid sense of the essential themes of German analytical sociology. Park had attended Simmel's lectures at Berlin, he read his works, and he absorbed certain of his views of modern society,[13] particularly those regarding the relations of urbanization to social differentiation and individuality.

This combination came to fruit just after the end of the First World War. Privately financed, nonacademic bodies concerned with race relations provided *debouchés* for sociologists (for example, the Carnegie studies on "Americanization" and the report on *The Negro in Chicago* requested by the Commission on Race Relations, supervised by Park and conducted by Charles Johnson). Civic and municipal organizations concerned with juvenile delinquency likewise provided support and employment opportunities as well as willingness to open their records to sociologists. Although Thomas was removed from the academic scene as punishment for a peccadillo, he and Park had already established a technique of research and a rough conceptual scheme which made sense of urban phenomena.

During the 1920's the pattern of institutionalization of sociology became well established. It centered on a standard textbook which promulgated the main principles of analysis, postgraduate courses, lectures, seminars, examinations, individual supervision of small pieces of field research to be submitted as course and seminar papers, and dissertations done under close supervision fitting into the scheme of analysis developed by Park, Thomas, and Burgess. It was sustained by the publication of the main dissertations in the Chicago Sociological Series and the transformation of the *American Journal of Sociology* into an organ of University of Chicago research. It was reinforced by public authorities and civic groups which offered sponsorship and cooperation for research, and by financial support from the university and private philanthropists. It was nurtured by a growing public interest in the subject, employment opportunities for sociologists as teachers in colleges and universities (mainly in the Midwest and Northwest), by a professional association (The American Sociological Society) which provided a larger forum and public for Chicago sociology, and by the summer institute of the Society for Social Research which maintained the solidarity of Chicago sociologists at home and away.

3. An International Illustration of the Difference Made by Institutionalization: The Diverse Fates of Horkheimer and Mannheim

The significance of institutionalization, even in the limited measure in which it was possible in sociology in Germany in the 1920's and the early 1930's, may be seen in the divergent destinies of the ideas of Karl Mannheim on the one side and of Max Horkheimer on the other. The difference in impact of the ideas of Mannheim and Horkheimer after they left Gemany also attests to the differences between Great Britain and the United States with respect to the institutionalization of sociology.

Mannheim was the more original and many-sided of the two: he had a more differentiated perception of contemporary society, a more vivid apprehension of particulars, and at least the same breadth of interest in macrosociological transformations. His knowledge of contemporary empirical research was greater than Horkheimer's and his ideas could have been translated into concrete research problems more easily than Horkheimer's. He wrote more and on more particular topics than Horkheimer. Mannheim raised important problems regarding the conditions of political detachment and partisanship among intellectuals, regarding the conditions of different forms of conflict and consensus among generations, regarding the influence of types of political partisanship on conceptions of historical time. All of these problems were related to matters of great contemporary interest, and they were capable of being empirically investigated. Yet Mannheim has had little influence, and Horkheimer is in a certain sense one of the most influential of modern social thinkers. Mannheim was, I have been told by his former students, a scintillating teacher, but in Germany he was a professor for only four years, and, although several very interesting dissertations were produced under his supervision,[14] the output of his pupils during his Frankfurt period was neither massive nor concentrated enough to provide a focus of attention and to create a far-reaching consensus as to what ought to be done and how to go about it. In 1933, Mannheim left Germany and went to the London School of Economics where, as in the time of Hobhouse and Westermarck, there were still very few postgraduate students of sociology, where there was no institutional provision for the organization, support, and supervision of research in sociology, where there was no organ of publication, and where there were no opportunities for the employment of those who had been trained in the subject. The Great Depression blocked numerous potential academic careers in Britain in well-established subjects, while the labor market for students of a fledgling subject like sociology remained at a standstill. The war years brought further attrition of what was already meager. Mannheim left the London School of Economics in the middle forties and became professor of the sociology of education at the Institute of Education in London. He died in the beginning of 1947.

The fortunes of Horkheimer's ideas were very different. The ideas themselves were relatively simple: modern society has become increasingly destructive of individuality as authority has become more concentrated and as organization has become more inclusive and more depersonalized; man has become a pawn for manipulation by others: the capacity for and the use of reason have declined. In many respects his ideas were like Mannheim's, although they were simpler and exhibited a slighter intimacy with the facts of contemporary societies. Horkheimer became in the course of several decades one of the most influential sociological writers of his time. He has certainly had a much greater impact on sociological work than Mannheim. Why was this so?

Horkheimer had the advantage of taking over the professorship of Carl Grünberg, a very productive teacher who had since before the First World War produced an admirable scholarly journal, the *Archiv für die Geschichte des Sozialismus und der Arbeiterbewegung.* In the middle 1920's Grünberg began to publish a series of *Beihefte,* which contained monographs on the subjects which came within the terms of reference of his chair—terms of reference which were becoming broader throughout the 1920's. As far as I know, the same wealthy patrons who supported Grünberg's work on the history of socialism and the labor movement took over the responsibilities for the activities associated with the chair when Horkheimer succeeded to it. The title of the chair was changed from "the history of socialism" to "social philosophy" and the *Archiv für die Geschichte des Sozialismus* was replaced by the *Zeitschrift für Sozialforschung.* The *Institut für Sozialforschung* had in fact existed before Horkheimer's accession to the professorship and to the direction of the new journal. Friedrich Pollock was the editor, but Horkheimer was obviously the leading spirit intellectually in the determination of editorial policy. The institute already had to its credit a number of large monographs, one of which was a major work of sinological scholarship by the then Marxist scholar, K. A. Wittfogel (*Wirtschaft und Gesellschaft Chinas*), and another, under commission, was a Marxist study of European thought by Franz Borkenau (*Der Übergang von feudalen zum bürgerlichen Weltbild*). The journal was changed in content from the history of socialism and the labor movement to macrosociology with a very strong Marxist and a less pronounced psychoanalytic bent. A large-scale collective investigation into authority and family structure—internationally and historically comparative, sociological, and psychoanalytic—was launched. It had many collaborators, each working alone on one problem. Another piece of research on "punishment and social structure" was historical and sociological in character.

This dense exfoliation would have been impossible without substantial financial support and the institutional apparatus made possible by that support. When Hitler's regime dismissed Jewish and socialist teachers, the Horkheimer institute went into exile. At first it went to France where it was given transient hospitality at the *Centre de documentation sociale* which Charles Bouglé had created and attached to the *École Normale Supérieure*. Since his patrons apparently had much of their wealth abroad, Horkheimer was able to emigrate to the United States with his most devoted collaborators, and to reassemble some of the others once he got there. Thanks, presumably, to Horkheimer's diplomatic and organizational skill, the institute established itself within the framework of Columbia University. Horkheimer was granted special status as a member of the faculty of Columbia University and he cultivated close ties with certain other members of the university. A collective institutional life was maintained and the journal continued, this time under the title *Studies in Philosophy and Social Science*. When the United States entered the war, three institute members found employment in the Office of Strategic Services. Meanwhile, other members were making their way as authors and as academic scholars. Near the end of the war, Horkheimer was engaged as research director by the American Jewish Committee, which granted him a sum of money—for that time very large—to conduct a widely ranging study of anti-Semitism. This afforded employment for a number of the stipendaries of the institute. The arrangement permitted the institute to adapt its viewpoint—an amalgam of Marxism, psychoanalysis, and patrician disdain for mass society—to American society and to fuse its point of view with the techniques of American social psychology and with the idiom then prevailing in American sociology. Franz Neumann and Otto Kircheimer became members of the government department of Columbia University, and exercised much influence over younger staff members and postgraduate students.

The institute then returned to Germany, carrying with it what it had learned of American sociological and social psychological techniques and its own theory of mass society. It began at once a monograph series—not at all like the historical-sociological studies of the pre-Nazi period, but more based on field work, interviews, sample surveys, group discussions, and so forth. Horkheimer and Theodor Adorno came to rank with the leading intellectual figures of the German Federal Republic. Herbert Marcuse, after having played a minor role as a "Sovietologist" at the Columbia University Institute of Russian Studies, became a professor at Brandeis and then in California, and, by the cunning of history, if not of reason, became an intellectual idol of the New Left. After a career of warning against the dangers of freedom, Erich Fromm became the apostle of a society constituted by love and the sage of "socialist humanism." Leo Löwenthal became a professor at the University of California and an influential sociologist of mass culture, drawing on and applying in particular studies certain themes of the earlier writings of Horkheimer and Adorno. In Germany, the leading protégé of Horkheimer and Adorno became the main exponent of "critical sociology" and for a time one of the main intellectual inspirations of the *Sozialistische deutsche Studentenbund.* Karl Wittfogel joined the staff of the University of Washington in Seattle, and became a leading anti-Communist student of Chinese history, after having been a crude and aggressive Communist polemicist as a young man in Weimar Germany.

But the history of the *Institut für Sozialforschung* in Weimar Germany, the United States, and the Federal German Republic is not just the story of the cat that landed on its feet. It is a testimonial to the skill of a shrewd academic administrator, who by good luck and foresight inherited a favorable institutional situation and developed its connections within the various universities in which it was located, maintained its internal structure, *and* extended its connections outside the university. As a result it became the mechanism by which some of the most influential ideas of present-day social science developed. The doctrine of mass society, of the dehumanizing effects of the exercise of and the subordination to authority, of the role of the mass media, of the "power elite," owe a great deal to the *Institut für Sozialforschung.* It has not only provided the current ideas of the New Left, but it has also influenced and called forth a large amount of research among sociologists who had no direct connection with the institute. In contrast to this, Karl Mannheim, having created no following, has found none since his death, despite the repeated calls for a sociology of knowledge.

This digression about the *Institut für Sozialforschung* has been intended only to show the significance of institutionalization for the establishment of a subject. Institutionalization is not a guarantee of truthfulness: it only renders more probable the consolidation, elaboration, and diffusion of a set of ideas. It is not the sole determinant of the acceptance or diffusion of ideas. Intellectual persuasiveness, appropriateness to "interesting" problems, correspondence with certain prior dispositions and patterns of thought of the potential recipient are also very significant. Institutionalization serves however to make ideas more available to potential recipients, it renders possible concentration of effort on them, it fosters interaction about them, and it aids their communication. Insofar as it offers the possibility of a professional career in the cultivation of the particular intellectual activity, it both makes possible the continuity of exertion on a full-time basis and it adds a further motivation for its performance. The existence of practical or executive professions, which require the study of an intellectual discipline as a qualification and as a constituent of professional practice, provides a student body and teaching opportunities —and therewith research opportunities which develop in the interstices of teaching. In these ways, institutionalization makes a difference to the fate of ideas.

4. The Institution of Sociology and Academic Systems

The practice of sociology might be seen as the center of a series of concentric institutional circles. In Durkheim's case we saw the creation of an institution for sociological work, which did not attain, however, a full and formalized corporate structure in the production of the works themselves. Durkheim did not conduct seminars; he held informal discussions with his protégés. A corporate form was achieved in the organization and production of the *Année Sociologique* and the *Travaux de l'Année Sociologique.* The circle which Durkheim formed was a rudimentary institutionalization. It had various ad hoc connections with other institutions, for example, with the *École Pratique des Hautes Études* and the *École des Langues Orientales,* where the informally adherent members of the circle had their employment, and with Felix Alcan, the publisher of their works. Durkheim himself held a professorship at the Sorbonne and he therefore was an incorporated member of an academic institution: it was not, however, in that capacity that he organized the work of his circle. He created a proto-institution with only peripheral and fragmentary institutional connections.

Max Weber's activity as a sociologist was much less institutionalized. He did not teach sociology; he supervised for a limited period a set of research projects for the *Verein für Sozialpolitik;* he tried to institutionalize two research projects on the press and on voluntary associations through the *Deutsche Gesellschaft für Soziologie* (he failed); he wrote *Wirtschaft und Gesellschaft* as one section of a comprehensive series of handbooks on economics, organized by the publisher Siebeck; he edited a great journal of social science and social policy, very little of which was devoted to sociology. The connection of his sociological activities with institutions was peripheral, fragmentary, and transient.

Thomas and Park were more institutionalized in their practice of sociology: they taught the subject regularly within the framework of a systematic course of study, organized toward the granting of university degrees; they supervised the research of students who were working toward postgraduate degrees. They themselves conducted organized research projects, employing assistants or collaborators, supported by university or externally granted funds. Many of the students whom they trained went on to sociological careers as teachers and research workers. Thomas and Park occupied a constitutionally provided position in the structure of the university. They were linked relatively densely with civic, governmental, and private bodies which were interested in the results of their research and which encouraged them by their interest and occasionally by their financial support as well as by accommodating research workers in their midst or under their auspices. In the 1920's private philanthropic foundations established grants for sociology and the Social Science Research Council established predoctoral and, later, postdoctoral fellowships for the promotion of sociology, some of which were held at the University of Chicago. During the Great Depression, Chicago sociologists found employment as social statisticians in a number of government departments and agencies. Thus Chicago sociology, that is, the sociology of Thomas and Park, was institutional at its center and in a fairly dense network of connected institutions removed at various degrees from the primary educational and research processes.

The primary institutional system of sociology is thus affected by its linkages with the environing institutional context: the university itself, foundations, civic bodies, government, business firms, publishing enterprises. The availability of the latter is of some consequence for the internal, primary institutionalization of sociology, in the provision of legitimacy through the sponsorship of established institutions, resources, employment opportunities, and so forth.

The study of the establishment and diffusion of sociology and the influence of this process of institutionalization on the substantive composition of the sociological traditions cannot be confined to the study of primary institutionalization. It is necessary to ask why sociology was able to become institutionalized at a particular time in the United

States when it had not become equally institutionalized in Europe, although at the same time the intellectual accomplishment of European sociology was greater than that of American sociology.

To account for this we must go beyond the tradition of sociology and beyond the primary institutionalization of sociology. We must consider the social structures which permitted it or inhibited it or fostered it. The academic systems first of all. In Germany the creation of a new chair depended on the consensual decision of other professors in the same faculty—rivals—for resources and prestige and the approval of the university senate and the state minister of education. A new subject might be created by a *Privatdozent,* but it could not by virtue of that become established in the university. An old university had an established allocation of resources, and its beneficiaries would not readily allow it to be changed in favor of a subject lacking the legitimacy of age and accomplishment. New universities were more likely to allow new subjects; thus sociology was given the dignity of a professorial chair first in Cologne, founded (for the second time) in 1919, and in the University of Frankfurt, founded as a private university in 1914. It became established as a teaching subject more easily in technical colleges than in universities. It was more sympathetically viewed by ministers of culture after the republican regime was established at the end of the First World War.

In France, as in Germany, universities were immobile. The oligarchy of established professors of a faculty and the high degree of centralization of control over the total university system in the hands of the national ministry of education hindered the creation of new chairs for new subjects. In British universities, although a chair was created at the London School of Economics in 1902 (on a private endowment), no additional chair was created until after the Second World War.[15] At Oxford and Cambridge, the matter was not even canvassed until after the Second World War and there the democratic oligarchy of the representatives of established subjects prevented the diversion of resources to a subject of questioned legitimacy.

In the United States, in contrast to the European academic system, the universities were independent of central control and professors of established subjects did not rule the universities. Sociology first became institutionalized in the era of the autocratic university president. Such a president could create a new department if he could persuade the board of trustees to agree and could raise the financial resources to pay for it. The availability of private financial support and the practice of its active solicitation gave a flexibility to university budgets which the European universities did not have.

It was not, however, only the structure of government of the universities which facilitated the earlier establishment in the United States. There were in the midwestern American culture of the period of the establishment of sociology a number of features which made a large difference. (It was in the Midwest that sociology first became academically established—in Chicago, Wisconsin, Michigan, primarily, but also in Indiana, Iowa, Nebraska, and Illinois.) The intellectual leadership of the Midwest was antagonistic toward the hegemony of the older eastern universities; they were distrustful of what they thought was the excessive respect of those universities for the past. They thought that knowledge was not degraded by being about contemporary things. There was, in short, no hard, thick incrustation of genteel, traditional, humanistic, Christian, patrician culture such as prevailed in the eastern universities. The hierarchy of deference was weaker in the Midwest; there was more equalitarianism, greater sympathy for the common life, more understanding for ordinary people, and, therefore, more readiness to be intellectually concerned about them.

Among midwestern intellectuals—publicistic and academic social science intellectuals—there was a more critical attitude toward the activities of the business class—industrialists, railway magnates, and bankers of the Eastern seaboard—and a greater skepticism about the adequacy of the classical economic theory which was adduced to explain their actions. It was in the Midwest that "institutional economics," developed most notably by John R. Commons, emphasized that there was more to society than what was accounted for by classical economic theory.[16] Sociological jurisprudence was part of the same atmosphere; in the United States it was largely the work of Roscoe Pound from Nebraska and Brandeis, an outsider by origin to the dominant culture of the educated classes of the Northeast.

Quite apart from the cultural relations of the Midwest and the Eastern seaboard, it should be pointed out that the midwestern universities, even where they were not established after the Civil War, became intellectually expansive only after that war. Harvard, Princeton, Columbia, Yale, Pennsylvania had developed strong cultures of their own before the Civil War and they were, therefore, less receptive to the German academic culture which was being brought back by an increasing number of young men with scientific and scholarly ambitions in the post-Civil War period. German historicism, which was one of the main sources of sociology, and the conception of the university as a scene of teaching *and* research were more readily received in the Midwest—at Chicago, Wisconsin, and Michigan especially—than in the older universities of the East. It might, indeed, be said that the culture of the German universities from which Max Weber's sociology grew but which could not produce an academically institutionalized sociology in

Germany because of the structure of German university was able, upon transfer into an academic structure of greater flexibility, to realize its sociological potentiality.

5. *Sociology and the Larger Cultural Tradition*

In Great Britain empirical sociology grew up in the setting of a cognitively oriented tradition of the discussion of the exercise of authority. From Bacon to Bentham, a tradition of belief developed in Great Britain which asserted that systematically gathered empirical knowledge could be an instrument for the improvement of man's estate. The great civil servants who encouraged the system of social reporting to accompany the legislation of Benthamite inspiration were themselves the heirs of this tradition. Field surveys and the acquisition of information from experts and experienced persons through questionnaires and depositions before commissions of enquiry—in these ways knowledge was to be acquired and assimilated into the process of passing laws and verifying their efficacy. The ancient English universities and the Scottish universities did not share this attitude toward knowledge, and public opinion outside the universities did not share it sufficiently to pervade the universities and cause them to change their minds. It was confined to limited circles of politicians, businessmen, and administrators.

The situation was otherwise in the United States. Not only was there a fairly widespread belief in the value of being in possession of *les lumières,* of the better *quality* of a life illuminated by knowledge of nature and of man, but there was also a widespread belief in the value of knowledge as an integral part of the action of betterment. The federal establishment of the land grant colleges and generous support by midwestern state legislatures were inspired by this belief. In Wisconsin, academic social scientists, perhaps earlier than anywhere else, were summoned by politicians to aid them in the drafting and execution of legislation. Robert Park probably never read Bentham but he believed that social surveys were an important part of reform movements, and in his analysis of "collective behavior" he used to point to the role of the "survey movement" as a stage in the mobilization of belief about what had to be done to improve conditions. This cognitive orientation toward the organization of political action, coupled with an untraditional belief in the legitimacy of knowledge of contemporary things, contributed to the willingness of wealthy individuals and organizations to support sociological research. It gave sociological research a ground for self-respect at a time when its intellectual achievements were insufficient for that purpose.

This is not to say that the substantive content of the sociology which came to be established with the support of this extra-academic attitude was determined by this attitude. It does mean, however, that the prevailing extra-academic culture contributed to the institutionalization of sociology in the United States. It also means that the subject matter of sociology and, to some extent, its techniques—in the United States as in England—were selected by criteria which derived from this extra-academic culture. The principles of interpretation—the substance of the central tradition—derived, however, from more purely intellectual (and intra-academic) sources.

Ecology: International Movement

The development of an intellectual field or discipline may be viewed as including an ecological process as well as a process of institutionalization. Intellectual constructions have a spatial aspect. Their production and cultivation take place in space; institutionalization takes place at particular points and it expands or contracts territorially as well as varying in the direction of less or more institutionalization. Institutionalization is one of the factors which affects the direction of spatial movement of ideas, just as it is a mechanism of the elaboration, promotion, or suffocation of ideas.

(1) *From Germany:* Ideas are likely to move with more momentum if they are institutionalized at their place of origin—just as they are more likely to be taken up, elaborated, and "used" (at least up to a certain point) the more institutionalized they are at their place of reception.[17]

Max Weber was located in Heidelberg from 1896 to 1918. It was in Heidelberg that he developed his most significant sociological ideas. He did not teach during most of this time, and his influence was confined to personal friends and his "circle." There was no established academic except Ernst Troeltsch, who worked out any of Max Weber's ideas. Ernst Troeltsch was professor in Heidelberg from 1894 to 1914 and while there he published his *Die Bedeutung des Protestantismus für die Entstehung der modernen Welt* and *Die Soziallehren der christlichen Kirchen und Secten*, the crucial ideas of which Max Weber in his turn elaborated, amplified, and set afloat in the world.

Before the First World War, Weber seems to have been quite unknown in the rest of Europe and in the United States, despite his journey to the St. Louis Exposition and his publication of one paper in English in the proceedings of the conference held there. Troeltsch's *Bedeutung* was translated into English in the Crown Theological Library and this too passed unnoticed by English-speaking sociologists.

In France, Weber seems to have caused not even a ripple. Durkheim reviewed Marianne Weber's book on the legal status of women—rather slightingly—but he did not discuss Max Weber. I do not know of any reference to Weber in France aside from those contained in the French translation of Sombart's *Der Bourgeois*. The two leading Germanists in France at this time were Charles Andler and Lucien Herr. Neither of them seems to have paid any attention to Weber. In Italy, too, Weber was unknown.[18] Pareto never referred to him, nor did Mondolfo or Antonio Labriola. Only his protégé, Roberto Michels, who in 1907 had become a *Privatdozent (libero docento)* in Turin referred to his writings, but there was no echo.

Simmel was better known. In France the *Année Sociologique* published a translation of his monograph *Über Soziale Differenzierung*. Durkheim's early approval changed markedly: both *Soziologie* and *Die Philosophie des Geldes* were censured by Durkheim, and that was the end of Simmel as far as French sociology was concerned. In the United States various essays of Simmel appeared in English translation in the *American Journal of Sociology*, and this before the appearance in Chicago of Robert Park. It was undoubtedly Small who was responsible, but there is no trace of Simmel's ideas in Small's work. Only when Robert Park came to Chicago did Simmel's ideas take up residence in Chicago sociological research. Simmel's influence received its definitive expression in Wirth's "Urbanism as a Way of Life," whence it became a common possession of American sociology.

Tönnies, despite his sojourn in Great Britain, his major contribution to Hobbes's studies, his close affinity to Henry Sumner Maine, and his great emphasis on quantitative work, found no response in British sociology. Durkheim knew *Gemeinschaft und Gesellschaft* and, despite a certain community of outlook, criticized it sharply; his adherents made no use of Tönnies' ideas. In the United States, on the other hand, his ideas were resonant in Park's writing and teaching. Like Simmel, his main theorems entered into American sociology.

Weber had close relations with Heinrich Rickert and Karl Jaspers in Heidelberg but they were in philosophy and psychiatry and had little to do with sociology. There was no professorship in sociology in Heidelberg and the *Archiv für Sozialwissenschaft und Sozialpolitik* covered all the social sciences, devoting relatively little space to sociological writings. Weber shared the editorship of the *Archiv* with other scholars whose interests were less definitely sociological. His intellectual-convivial life was spent in a diverse company which included philosophers, jurists, and literary critics, and Weber neither gave it a focus nor shaped its output in the manner in which Durkheim had done with his entourage.

In the next generation, Karl Mannheim was Weber's main sociological continuator. Mannheim was in Heidelberg through the 1920's. His main personal connection was with Alfred Weber; from him he adapted the idea of *freischwebende Intelligenz*. He attended Marianne Weber's colloquia. Max Weber's ideas were a living presence, but only a very general one at Heidelberg at that time. None of the numerous professors and *Privatdozenten* who professed sympathy and admiration for Weber's ideas undertook to expound them systematically or to apply them. Mannheim was the exception.[19]

It was Weber's ideas on bureaucracy which Mannheim absorbed and which he tried to elaborate in his essay on *Erfolgsstreben* and which later became the central idea in *Mensch und Gesellschaft im Zeitalter des Umbaus*, which he wrote and published while in exile. It was also an idea of Weber, the idea of the charismatic revolutionary directly inspired by a believed contact with ultimate things, which Mannheim adapted in the essay on the utopian mentality (*Ideologie et Utopie*) and which he took up again toward the end of his life in his development of the notion of "paradigmatic experiences" (*The Diagnosis of Our Time*).

Mannheim's relations with Weber's ideas were affected by the fact that he came to Heidelberg only after Weber had died, that his mind was confused by an appreciation of Lukacs' type of Marxism and his awareness of its untenability, and that he worked in Heidelberg in a situation in which there was no systematic analysis or discriminating application of Weber's substantative sociological ideas.[20]

When Mannheim went to Frankfurt he replaced Franz Oppenheimer, who had been a *Privatdozent* and finally a professor after nearly twenty years in private medical practice and whose sociological ideas, written down in an immense eight-volume treatise, had no intellectual after-life at all. Mannheim brought Weber's ideas to Frankfurt.

The image of mass society was developed in the 1930's by Mannheim, Horkheimer, and Lederer. One source was the patrician tradition in Germany of the German *Bildungsschicht* and its contempt for bourgeois culture; another was the experience of the National Socialist movement in the last years of the Weimar Republic. But much of it came from an interpretation of Weber's ideas of bureaucracy and the charismatic type of authority, through Mannheim's *Mensch und Gesellschaft im Zeitalter des Umbaus*, which was one of the very few German works to bear the impress of Weber's ideas. Lederer's treatment of mass society also had its intellectual roots in the Weberian atmosphere of Heidelberg where Lederer had been a *Privatdozent* before going to Berlin.

For the rest there was scarcely any impact of Weber's ideas on German sociology. Some of the work on white-collar employees owed something to the stimulus of Weber's writing on bureaucracy, although I think that only Hans Speier's work is clear-cut in this respect. (Speier was a pupil of Mannheim.) The professors of sociology in Germany in the Weimar period—Vierkandt, Rumpf, von Wiese, Geiger, Freyer, Meusel, Dunckmann et al.—were proper German professors in the sense that they were rivalrous toward their peers. Most of them wrote books of "principles" which were largely variations on the theme of *Gemeinschaft* and *Gesellschaft*. They made little use of Weber's work. They were, however, unlike the classical type of German professor in that they created nothing important of their own and they trained no *Nachwuchs*. Albert Salomon, who was a professor in a teacher-training college, was a devotee of Weber but he had no products and no heirs. Paul Honigsheim was another in a position which had no influence.

Thus Max Weber faded out of German sociology in a decade when a number of professorships in sociology were created in universities and technical colleges. It was an outsider to Heidelberg and Germany who did more for the animation of Weber's sociological ideas than anyone else. Talcott Parsons came to Heidelberg in the middle of the 1920's. Having studied "institutional economics" under Walton Hamilton at Amherst and being endowed with greater analytical powers than his older contemporaries, he was quick to see that Sombart gave a more sophisticated form to the antitheoretical attitude contained in American "institutionalism" and that Weber succeeded in doing justice to what was valid in the intentions of institutionalism while avoiding its errors. Parsons' Heidelberg dissertation on capitalism in the writings of Sombart and Weber was the first in a long series of publications which sustained and extended the intellectual accomplishment of Max Weber.

On his return to the United States, Parsons became a tutor in economics at Harvard; there was no sociology there at that time. The institutionalized obligations of the tutorial role forced him to study economics proper in a department which was already very eminent, thanks largely to the presence of Joseph Schumpeter and F. W. Taussig. The study of Marshallian economics led Parsons to put Weber and Marshall into a common set of categories and thus enabled him to see what was distinctive in each. Pareto was pressed upon his attention by Lawrence J. Henderson, the distinguished physiologist who was a powerful figure at Harvard, a man of wide-ranging curiosity and a great deal of intellectual self-confidence. Henderson had been giving a seminar on Pareto for some time. Parsons' preoccupation with Durkheim was reinforced by the interest which Elton Mayo, T. N. Whitehead, and the Western Electric Research Group at the Harvard Business School had in *anomie* as one of the "human problems of an industrial civilization."[21]

By no means all contacts with an intellectual tradition are instigated through personal contacts provided by membership in a large, differentiated institution. Traditions can be sought out and found with only a minimal degree of institutionalization. Parsons' bringing in Durkheim and Pareto heightened contrasts, disclosed uniqueness more sharply, and revealed common elements.

What emerged from this synthesis were a number of themes which came to be taken up in various ways in the quarter of a century which followed the publication of *The Structure of Social Action*. One of these themes was the assertion of the importance of value-consensus as a major variable alongside individual and corporate ends (including definitions of "interest"). This in turn was connected with another theme regarding the importance of beliefs about the legitimacy of authority in society as regulators of perpetually present potentialities of conflict between parties of conflicting interests. Another set of themes, centering about categories for the description of action, was promoted through Parsons' study of the medical profession; this was further developed into an analysis of the professions in modern society, which, a few years after the publication of *The Structure of Social Action*, specified some of the properties of the moral ethos of the learned professions.

To do this, Parsons drew on Tönnies' conception of *Gemeinschaft*, which most writers had interpreted as a total antithesis of *Gesellschaft* (which was equated with modern capitalist society). The Tönnies dichotomy had been assimilated but never systematically analyzed into its component elements by Weber. Parsons went further and began the process of an analytical breakdown of the conceptions of *Gemeinschaft* and *Gesellschaft* into their components. This was subsequently "completed" in *Towards a General Theory of Action*[22] where the pattern-variables were systematically elaborated as affectivity-affective neutrality; diffuseness-specificity; particularism-universalism; collectivity orientation-ego orientation; and ascription-achievement orientation. All of this development was an extension of Weber's substantative ideas.

(2) *From France:* The sociological ideas of Durkheim largely died away in France not very long after Durkheim himself died, as far as their application to modern society was concerned. Davy, who published an important Durkheimian treatise in the early 1920's, entered educational administration and became a professor of sociology only a quarter of a century later. (Had there been some sociology in the universities in the 1920's and 1930's, Davy as a senior official would have been in a crucial position to promote its institutionalization. Unfortunately, there was practically nothing for him to promote.) Bouglé, too, became an administrator and ceased to do sociology. Simiand worked as a government

official and only later in his life became a professor—of labor history. Granet continued to be a faithful and creative Durkheimian and a great sociologist, but his teaching was entirely sinological and he had no students of sociology. Mauss became professor of the "religion of the *non-civilisées*" early on and later became director of the *Institut d'Ethnologie* at the Sorbonne as well as professor at the Collège de France; in none of these capacities did he have students of sociology understood either as a general theory of society or as the study of modern societies. Durkheim left behind an extraordinary band of disciples in specialized fields which were at the periphery of sociology, but since sociology was not institutionalized at the center, their accomplishments remained dispersed. There was no consolidation and no sociological *Nachwuchs.*

Only Halbwachs continued the Durkheimian tradition as a sociologist studying modern societies, but aside from his own work, there was little extension. Why was Halbwachs unable to reestablish a school extending and deepening the traditions flowing from Durkheim? Halbwachs had a vast range of knowledge—perhaps wider than Durkheim's—great working capacity, and high technical skill, and his was a more supple and sinuous mind. But nothing came of his work in the sense of its being taken up by younger French sociologists and entering into the course of sociological development. (Of course his own books still stand as monuments.)

The answer does not seem too difficult. Halbwachs was a professor at Strasbourg, a distinguished university, but not one to which the best students flocked from all over France. At Strasbourg, Halbwachs as a sociologist was alone. He did not, as far as I know, supervise doctoral dissertations, this privilege being reserved for Parisian professors. He had no younger colleagues to associate themselves with him. Thus, this powerfully and sensitively intelligent man who carried forward the Durkheimian tradition, but in no uncritically submissive way, had no juniors of sufficient quality and in sufficient numbers on whom to leave a mark. The intellectual society around him was not dense enough; the necessary complement of institutions—specializing students, post-graduate training, research projects and assistants, and journals—were not available to him.

Durkheim's influence could, of course, have survived in a free-floating form. Not all intellectual filiations need be transmitted by an apostolic succession within the framework of a corporate body. Influences need not inevitably be institutionally transmitted and impressed by reenactment in research and publication under supervision and in collaboration. Durkheim's tradition failed, however, to operate in the way in which literary traditions operate—works set loose in the world to list whither they will, finding reception here or there at the discretion of or in accordance with the disposition of the recipient but without the reinforcement by the discipline of systematic and supervised study under an authority and with the support of peers whose critical acceptance facilitates ones own critical acceptance.[23]

For founding ideas to travel across national boundaries in subjects which have no universally apprehensible symbolism, language barriers must be absent or there must be many translations. There must also be, as in the case of the Weber–Parsons relationship, a strong intellectual personality who had immersed himself deeply in the founding ideas and who can sustain the newly acquired ideas either with the aid of institutions in the new setting or through sheer force of intellectual character.

Durkheim's ideas traveled little because there was no one to carry them. Ziya Gökalp took them to Turkey but that was a dead end because Gökalp, although he taught sociology at Salonika and became professor of the subject in Istanbul, was not a strong intellectual personality; he was more interested in legitimating Turkish nationalism and modernization than in sociological research. Nor did he have research students. If he had any, there would have been no careers for them. There was nothing in British sociology of the period before the Second World War which showed Durkheim's influence. Durkheim visited England once, I think, at the beginning of the century and delivered a lecture before the Sociological Society justifying the existence of sociology. The brevity of the visit, the inconsequentiality of the subject, and the absence of receiving institutions formed no bridge over which Durkheim's ideas could travel. Sociology did not exist as a university subject in England at the time. Sociology was a subject of the laity which was interested in eugenics, evolution, and the condition of the poor. There was nothing in Durkheim which made an immediate sympathetic connection with any of these.[24]

Germans took no interest in Durkheim's ideas, which were available to the German public through the surveys of sociological theories by Squillace and Sorokin and in a dissertation of a Roumanian named Georges Marica. Since German sociologists paid so little attention to Weber, it was scarcely likely that they would clasp Durkheim to their bosoms. Germany was really out of the question as a place for Durkheim to be influential. Before the First World War Germany was the academic center of the world. German scholars of that period could, according to their fuliginous lights, be courtly, but they really did not regard foreign science or scholarship—except perhaps in Islamic studies or mathematics—as worthy of serious consideration. And sociology?! German scholars were not acknowledged academically in that field—although Weber, Simmel, Troeltsch, Tönnies, and Sombart were men of very great merit and Germans to boot—so why take a Frenchman seriously? All this in addition to the deteriorating political relations between France and Germany at that time.

(3) *From the United States:* In the United States before 1914, sociology was either under the dominion of German-inspired ethnographers and *Volkskündler*—for example, W. G. Sumner—or it was conscientiously making its first contacts—unprecedented except for Charles Booth, Seebohm Rowntree, the forgotten French surveys of *les classes dangereuses,* and the British poor law and sanitary inspectors—with the urban masses. It had no use for theory. (The American sociologists were—and still remain—ignorant of Halbwachs' prewar researches on the development of needs in the working classes.) There were only two American sociologists capable of receiving the Durkheimian influence before the First World War and to whom it would have made a difference had they experienced it. One was William I. Thomas whose study of the European sociological literature ceased as he immersed himself in his great inquiry into the Polish immigrant in America.[25] As far as I know, Thomas was quite ignorant of Durkheim; he certainly did not regard him as a source of his ideas, although as a matter of fact his views in many matters were very close to Durkheim's.[26]

The other was Robert Park, who had studied in Strassburg, Heidelberg, and Berlin with Windelband and Simmel, but whose knowledge of Durkheim—and Weber for that matter—was slight or indifferent.[27]

A Columbia Ph.D. named Gehlke, later professor at Ohio State University, wrote a dissertation on Durkheim, mainly dealing with the definition of a social fact, but scarcely touching substantive matters. Durkheim had been given a bad name by Alexander Goldenweiser in 1916 and this impeded his acceptance in the United States.

It was, in fact, Professor Parsons who became the chief protagonist of Durkheim in the United States. He was reinforced by Elton Mayo.[28]

Robert Merton, who first studied Durkheim under Professor Parsons and Sorokin, has perhaps given the greatest impetus to the diffusion of one element of Durkheim's analysis of modern society, namely, to the analysis of *anomie.* His elucidation of the components of the phenomenon of *anomie* has been taken up by a number of investigators—one of them his Harvard contemporary, Leo Srole. It has now become, like charisma, part of the stock in trade of American sociology.

(4) *From Great Britain:* The relatively uninstitutionalized structure of British sociology up to 1914 permitted a great miscellany of sociological tendencies to have a stunted and erratic existence. Some of them, such as the eugenics current, came subsequently to have a great impact on American and then on world sociology, through the development

of refined statistical measures and a less weighty but still considerable influence on the study of social mobility. (Francis Galton was in both cases the great progenitor.) They did not, however, come to America at once. Much more important was the social survey method in the study of the urban working classes. The Booth and Rowntree surveys had a marked influence on American surveys, although American academic sociology did not go in for sample surveys until after the First World War. The Middletown survey was the first major American survey and the technique received a tremendous impetus from the development of public opinion polls from the 1930's onward. Robert Park assimilated their lessons and adapted them to the financial position of Ph.D. research at the University of Chicago.

The evolutionary sociology deriving in part from Spencer did not find a response in the United States, except in the work of W. G. Sumner and the early works of W. I. Thomas. Family studies, which were mainly study of family disorganization, were not influenced by Westermarck. They derived rather from the British social survey tradition, with its interest in the conditions of life of the poor. They were also to some extent influenced by the German standard of living studies which had been initiated by Engel.

(5) *A Note on Center and Periphery:* At the end of the nineteenth century, the three leading countries of the world were Great Britain, France, and Germany. In the academic sector of intellectual life, Germany was preeminent. Sociology, being academically the least institutionalized of the scientific disciplines, was not in a position to offer *actively* much to any other country. Internationally, its traditions were freely floating and it is not surprising, therefore, that each of these countries acted as if the others scarcely existed sociologically.[29]

In contrast, the United States went to school in Europe and in Germany above all. The relationship turned round in the course of time. American sociology became central. European sociology became peripheral. The far greater degree of institutionalization of sociology in the United States, the large scale of its output, the ascent of the United States to a condition of academic centrality (and the greater power and prominence of the United States outside the intellectual sphere), and the formation of an, to some extent, international sociological culture have all contributed to change the direction of the ecological process.

This process has been much aided by the increased institutionalization of sociology in teaching and research in Europe. Institutionalization increases receptive power just as it increases radiative power between countries as well as within countries. Once sociology reduced the freedom of its exponents to believe what they wished and to call sociology whatever they liked to think sociology ought to be, American influence was bound to increase. The mere organization of regular courses of study required syllabi and textbooks, and this increased the likelihood that the voluminous American literature would be brought in. Of course, the cessation of sociology in Germany from 1933 until the last years of the 1940's was responsible for a vacuum in the literature of sociology in the German language, and American literature was available to fill the vacuum. (The American occupation of a large part of Western Germany and the easy physical availability of American sociological publications also helped to bring about this condition.) But whereas the teaching of sociology in Germany has been extensively institutionalized, sociological research has not had the same experience. As a result, the proportion of empirical research in the total body of sociological publication is fairly small. One is struck with the very considerable German knowledge of American research and the contrast with the small amount of research done in Germany. German sociological monographic series, for example, *Göttinger Abhandlungen zur Soziologie, Frankfurter Beiträge, Bonner Beiträge zur Soziologie, Soziologische Gegenwartsfragen,* although containing more reports on field research than was the case with comparable series during the Weimar Republic, still contain many definitional, programmatic, and *dogmengeschichtliche* works. The same is true of the articles in the *Kölnische Zeitschrift für Soziologie.*

In France, on the other hand, there has been much more production of empirical research, thanks to the generosity of the *Conseil National pour la Recherche Scientifique* (CNRS). The point of reference, however, is, as in Germany, American research.

The institutionalization of sociology has made for greater intellectual interdependence. It has become necessary to master "the literature" if one is to teach or do research nowadays. A sociologist cannot live from himself nowadays in the way in which Simmel did or even Durkheim or Tarde or Fritz Sander or Haserot. This link with the outside is not random in direction or terminus.

Ecology: Within National Sociological Communities

I shall confine my observations of the ecology of sociology to the United States and, more particularly, to the relations between dominance and institutionalization. In the 1920's and 1930's the University of Chicago was the center par excellence of sociological studies, although in the 1930's it was existing on the momentum of the 1920's. As the leading place for postgraduate studies in sociology, the University had continuously present a group of mature and concentrated students, each of whom was engaged in a piece of research—almost always field research—under the direction of two men who were in a close and assymetrical consensus, Professors Park and Burgess. What they taught had been in principle codified—rather primitively by present-day standards—in "Park and Burgess." Intensive instruction was amply provided in seminars and informal lectures, as well as in numerous individual consultations between teachers and students. The institutions of the term paper and the seminar paper were in full operation. Each student regarded it as his obligation to produce such a paper, and the teachers regarded it as theirs to supervise production and scrutinize results. Relations were established with various local civic organizations to facilitate the students' entry into the field for direct observation, interviewing, and the collection of documents. There were certain adjunct institutions to which students could be attached for their research (and sometimes employment), such as the Institute for Juvenile Research, the Chicago Crime Commission, and the Chicago Real Estate Board. For the most successful there were the rewards of publication in the journal conducted by the department, the *American Journal of Sociology* or, even higher, inclusion of a finished work in the Chicago Sociological Series. The journal was the official organ of the American Sociological Society and the only important national journal on the subject. Relatively few attained publication in the monograph series, but its existence provided a number of standard-exemplifying works, and it added to the awareness of membership in a central institution. For many years the seat of the American Sociological Society was the University of Chicago.

In nearly every important state university of the Midwest and Far West, sociology was Chicago sociology. The departments were headed by Chicago graduates. It was to Chicago that they turned for new members, and it was to Chicago that they sent their best graduates for further study. There were other departments of sociology in the country, most notably at Columbia, Michigan, North Carolina, Brown, Pennsylvania, and Yale, and they had their small areas of hegemony to which their graduates were appointed. There were distinguished sociologists at some of these institutions—Ogburn, Lynd, and MacIver at one time or another at Columbia, Cooley at Michigan—and some valuable monographs based on dissertations were produced. None of these lesser centers had the common standpoint, continuity of effort, and liveliness and intensity of an intellectual community which Chicago had, and that made the great difference.

Chicago maintained its position as center of sociological studies not only because of the intellectual power of some of its staff members, but also because it was more institutionalized; it produced more work with a common stamp and the quantity as well as the quality aroused attention and respect. This, in turn, enabled it to draw outstanding graduate students whom it trained and then sent out to maintain the system of Chicago dominance.

Although Chicago sociology was embedded in a more ramified corporate network, its institutionalization was in one important respect like that of Durkheim's school. It depended very much on one major intellectual personality at a time. For a while, mainly over the years of the First World War, there were two intellectual leaders, but when Thomas' academic career came to an inglorious end, there was only one. When Robert Park withdrew from Chicago toward the end of his life, preferring to spend the last years at Fisk University with Charles Johnson, his old pupil and then president of the university, Chicago sociology faltered. It lost its focus on urban and ethnic studies; new persons, changes of interest, and a diminution of intellectual authority within the institution in consequence of intellectual and then personal disagreements weakened the sense of centrality.

Perhaps more important was the fact that the fundamental ideas of Chicago sociology were not clarified, extended, and deepened. With Park's departure, the analysis of ecological processes diminished; the studies of communities also declined, so that there was no development of the concepts which would have permitted more refined studies of the same kinds of situations which had once been so productive. Burgess' family adjustment studies with a predictive interest, Ogburn's statistical time series of various social phenomena and concentration on the improvement of techniques of quantitative analysis, without a persistent substantive interest reduced the radiative and attractive power of Chicago as a center. Ogburn's interest in the quantitative description of trends and his simplistic and undifferentiated concept of "cultural lag" was not fitted into microsociological analyses of situations which could be studied by methods of participant-observation. The inchoate global, macrosociological interests of Park found no forceful reformulation in a way which could give coherence to the work of the department. Sociology persisted because it was well institutionalized—unlike its Parisian counterpart—but the orientation provided by Thomas and then by Park became powerless. There was no successor to Park at Chicago who was strong and expansive enough intellectually and temperamentally to continue the Chicago tradition and to assimilate into it the new problems and modes of thought and inquiry which began to emerge in the years just before the outbreak of the Second World War. Intellectual incapacity to amalgamate new and diverse problems and knowledge furthered disaggregation from temperamental uncongeniality. The loss of a center within the department accompanied the loss of centrality in the national sociological system. The relationship was a circular one.

Not only internal developments reduced the centrality of Chicago. New centers were emerging in the East, places which had never been subcenters of Chicago. Columbia and Harvard were of special prominence. Columbia, by the early 1950's, markedly surpassed Chicago as a center, for several important reasons. One of the most obvious is that Columbia possessed two major intellectual personalities, Robert Merton and Paul Lazarsfeld, who combined what was most "needed" in sociology—ingeniously contrived techniques of survey research with interesting, quite specific, substantive hypotheses. A second factor, closely connected with the first, was the formation of a superior form of institutionalization at Columbia in the Bureau of Applied Social Research. The scale of operation, the interests of the moving spirit (Lazarsfeld), and the shape of development of sociology at the time required and permitted a higher degree of formalization of procedures—intellectual and organizational—than temperament, capacity, the primitive state of the subject, and the small scale of the operation had fostered at Chicago. The routinization of training facilitated the routinization of research organization and research procedures, at the very moment when new techniques of observation—sample surveys—and new techniques of analyzing data were brought into sociology from market research and statistics. The research for which Columbia graduate students were trained could be done without personal inspiration: it was easily reproduced and multiplied. This increased the proliferation of the Columbia center, at a relatively high level of technical competence, and encouraged the diffusion of its procedures and mode of thought to many nonacademic institutions which sociology had not penetrated before. It also made Columbia a national and international center for sociology. (Partial subcenters were formed at Chicago and at Berkeley —wherever, in fact, a "research facility" was created. The Survey Research Center at the University of Michigan was an exception to this; it did not derive from Columbia.)

Another reason for the emergence of Columbia as a center was the intellectual style of Robert Merton. A product of Harvard, where he had studied under Parsons and Sorokin, Merton had a very wide sociological culture. Unlike Parsons—who possesses a profound vision of society which he repeatedly struggles to articulate—Merton's view of society is more orderly and more adequately articulated, but his exposition is more fragmentary and eclectic. Merton's espousal of "middle range theories" is less penetrating than Parsons' efforts at a "general theory," but they are easier to apprehend and therefore more effectively taught. They are easier to formulate in a testable way, they seem to involve no *weltan-*

schauliche commitment, and they are more congenial to institutionalized routines. All these factors together resulted in a larger production of works bearing a common stamp and of persons capable of producing more such works in the future. This led to a wide diffusion of Columbia sociology and it generated—more on the level of procedure than of substance—a consensual element in sociology which will probably endure.

Harvard's centrality, based on traditions transplanted from Heidelberg (Max Weber) and Vienna (Freud), was attributable to a new pattern of institutionalization. There, under the leadership of Talcott Parsons, Henry Murray, and Clyde Kluckhohn, a deliberate attempt was made to integrate the theories of social structure, culture, and personality. The teaching program was adapted to this conception of the subject, but the research training program did not keep up with it. The Laboratory of Social Relations never became the intellectual factory and drill ground which the Bureau of Applied Social Research became shortly after the end of the Second World War. It became the administrative sponsor and home of a variety of investigations which, to a greater extent than at Columbia, were small-scale projects which required neither an intricate division of labor nor a thoroughgoing routinization and uniformization of procedure. It never became the solidary collectivity with an identity of its own which the Columbia Bureau succeeded in becoming and, even before the death of the gifted Samuel Stouffer, it became more of an administrative name and rather less of a corporate intellectual reality. Although valiant efforts were made to create it, Harvard lacked a high degree of consensus among its central personalities such as was possessed to such a degree at Columbia. As long as the center at Harvard was a triumvirate with an outward appearance of unity, it presented a powerful force to the outer world of sociology. Nonetheless, it was in fact a center consisting internally of several relatively noncommunicating segments—and this reduced its capacity to impose itself effectively on the subject as a whole. At the same time, it should be stressed that each of the major segments was a powerful intellectual personality—Parsons, Murray, Kluckhohn, Bruner, Stouffer, and Homans—each of them in one way or another a forceful generator of ideas and works. A high degree of consensus among them might have swept the field. (It did pretty well as it was!) It also lacked what Chicago in the 1920's and early 1930's possessed, namely organs of publication and stable extra-academic institutional links with the local community for research and training purposes.

On the other hand, it had in Professor Parsons a motor force of great power; his continuous and pervasive productivity spread his influence over the country into the by-this-time numerous subcenters with highly institutionalized training provision for postgraduate students. The role of one particular publishing enterprise—The Free Press—which became the main source of sociological nutriment for the growing number of graduate students and young teachers from the end of the Second World War until about 1960—provided Professor Parsons with a surrogate for his own journal. His numerous, widely scattered essays were concentrated into a few easily available volumes. These essays enabled a new generation to acquire the underlying disposition of Parsons' theory as it was manifested in confrontation with a wide variety of particular problems. The enhancement of the prestige of Harvard University as a whole coincided chronologically with the prestige of the Department of Social Relations, which was largely the prestige of Talcott Parsons—although not exclusively so. This attracted to Harvard a group of gifted students.

The sociology department of the University of California at Berkeley also should be mentioned in these observations about the emergence of new centers in American sociology. The University of California at Berkeley was the last great American university to establish a department of sociology. It came into existence after Chicago, Harvard, and Columbia were well established with dominant traditions of their own. Berkeley could not recruit its staff from within, except at the level of its youngest members; it had to recruit from other universities. For example, Lipset trained in Columbia under Merton and Lazarsfeld, Blumer at Chicago, Bendix at Chicago, Selznick at Columbia, Kornhauser at Chicago, Kingsley Davis at Harvard, Lowenthal at Frankfurt. The size of the staff, the size of the student body, and the amplitude of resources all made for centrifugality. The high degree of institutionalization coexisted with a diversity of standpoints, so that the department gained eminence as an aggregation of outstanding and productive sociologists whose works drew the attention of the country and the world, but it lacked the coherence which Durkheim, Park, and Parsons gave to their respective circles of colleagues, collaborators, and students. We know from Chicago in its earlier years and from Harvard and Columbia in the several decades after the Second World War how effective a unitary center created by one dominant or two strong consensual figures is. It remains to be seen whether a pluralistic assemblage of eminent figures can be equally effective. Thus far, Berkeley has not shown its capacity to create a *Nachwuchs* which will renew the department while unifying it or to produce a body of graduates who will diffuse its "line" more widely in the United States.

Sociology at the University of Michigan had been gently influential through the writings of Charles Cooley ever since the turn of the century. Both Thomas and Park had assimilated Cooley's ideas about the primary group into their own dominant traditions. There had not, however, been an independent department of sociology at Michigan until 1930, when Roderick McKenzie, a Chicago product and a close collaborator of Park, came there from the University of Washington. The Department of Sociology at the University of Michigan became one of the major departments of the country, partly through the extension of ecological studies and partly through the establishment of the survey research center there. The latter was an important step in the institutionalization of training and research at Michigan; in its ecological and related studies, Michigan continued and developed the Chicago tradition with the aid of many recruits from Chicago.

The Department of Sociology at the University of California at Berkeley was the last of the five major centers to emerge. The Berkeley department is not associated in the minds of those who live in the international sociological system with any particular ascendant individuals. Can a center be effective if it consists of an un-unified plurality? The question about the University of California at Berkeley might be academic in the pejorative sense because the disorders in American universities, of which Berkeley is at the forefront, might so disaggregate and demoralize an institution that it loses its capacity to operate effectively on its students. The functions of a center might still, however, continue nationally and internationally to be performed through the other channels of communication which are integral to institutionalization, namely, learned journals, monograph series, and books. But one essential organ of institutionalization is the face-to-face encounter of teachers with students. It is through these encounters that the *Nachwuchs* is formed and that ideas are generated, tried out, and selected from among a variety of alternatives.

The absence of a systematically trained *Nachwuchs* does not mean that an intellectual subject cannot develop: it did so in the heroic age of science prior to the formation of the modern university. The absence of a systematically trained *Nachwuchs* means that the pressure of tradition is lightened. The development of an intellectual subject under such circumstances was more a matter of individual genius than it became with the transformation of universities into institutions of research and training. Such a situation might slow down the rate of development of the subject in question by diminishing the density of work in it.

Sociology has now had its organizational, economic, and demographic revolutions; it has not yet had its intellectual revolution. The question may be asked whether a unified center is possible under present circumstances in the world of learning.

The proliferation of sociologists and of sociological works in the past quarter of a century resulting in great specialization in a subject still deficient in fundamental theory might mean that it will disintegrate, losing the small coherence which it has gained over the past three-quarters of a century. Sociology might break up into specialisms such as those which exist now in the physical and biological sciences, without being able to draw on a basic or fundamental science like chemistry or genetics of physiology.

The constituent elements of the traditions of present-day sociology have been in a process of selection and coalescence as a result of the work of Weber, Durkheim, Parsons, et al. This coalescence is still very imperfect. (It can never be perfect and if it did become perfect it would come to a complete halt.) Greater coalescence than it possesses at present is, however, a necessity, not just out of aesthetic or architectonic considerations, but also because until there is such an improvement the interrelations of the different phenomena in society will be too poorly understood. Thus the multiplication of sociological works and persons might render more difficult the entry into ascendancy of a more complex and comprehensive theory because it renders more difficult the ascendancy of a theorist whose mastery of the results of specialized study would be great enough to call forth the respect and adherence of the practitioners of the diverse and numerous specialisms.

Sociological Traditions: Selection, Rejection, Coalescence

If one takes up the compendious work of Sorokin, *Contemporary Sociological Theories,* which presented a gallery of the stock of sociology of the fifty years which lay across the turn of the century, one is struck with how few of the names cited are known today by any but sociological antiquarians. Weber, Durkheim, Pareto, Park, and Thomas are there; so are Floyd Allport, Otto Ammon, Emory Bogardus, Stuart Chapin, Filipo Carli, E. de Roberty, Charles Ellwood, Franklin Giddings, Maxim Kovalevsky, E. A. Ross, Gabriel Tarde, William Graham Sumner, Alfred Vierkandt, Leopold von Wiese, L. Winiarsky, and Sorokin himself, who is cited more than any other two authors. What has become of these authors and of the ideas they espoused? They have fallen by the wayside.

Some of them were weak intellectually and their ideas did not recommend themselves to critical contemporaries. Some of them were in the wrong places; their books were the only ways in which they spoke to their contemporaries; they had no students or no students who had the ability, received the inspiration, and learned how to press further with what they had received. They might have been in the wrong countries and written in the wrong language. Some of them were known and respected figures in their time, which was contemporary with Durkheim, Weber, and Park, but the world turned away from them.

There were many ideas floating about in sociology in the latter part of the nineteenth century. Practically all of them were influenced in one way or another by Hegel, Comte, and Darwin. Evolutionary and biological ideas were common. The major idea was one of historical stages of human society from simple to differentiated, from traditional to rational, from unenlightened to enlightened, from societies with tool-based technology to societies with a machine-based technology using artificially-generated power, from loosely integrated societies to closely integrated societies. The Darwinian evolutionary viewpoint fitted into and rendered more plausible this theory of stages of development; the prestige of Darwinism also activated a tendency to analyze social structures and their functions in society in the categories appropriate to the study of biological organisms. (This type of sociological analysis was submerged rather early and left little trace.) The Darwinian influence was also present among those sociologists who, impressed by the competition of species, stressed the importance of the competition and the conflict of races as a central phenomenon in society. Others, also impressed by the idea of natural selection, were concerned with social selection and hence with the social origins and the presumed biological heredity of the members of various strata, above all, the leading strata in contemporary western societies. Many studies were intended to demonstrate the biological transmission of socially relevant characteristics. Much of the substance of these propositions no longer has any place in the effective tradition of sociology.

Certain variables in this set of beliefs have survived—for example, the study of elite-composition and the social origins of particular professions—but the matrix of propositions has fallen away. And what has survived has not been comfortably assimilated into the presently effective tradition of sociology.[30]

A few words should be said at this point about Marxian sociology, which has so many self-alleged proponents at present. "Marxian sociology," such as it is, is an amalgam of Hegelian evolutionism into which has been inserted a conception of conflict drawn from Darwinism[31] and a conception of interest generalized from the eighteenth-century conservative conception of the "landed interest." This results in something different from the mere substitution of "class" for "group"—ethnic, national, territorial—which was developed by Ratzenhofer and Gumplowicz, and it is also different from that type of conflict which arises from contention for the possession of scarce objects. The prevailing conception of society derived from Weber and Durkheim allows, despite the assertions of the critics of the "consensus-model,"[32] ample place for conflict, but it can make no place for transcendent interests which are independent of what is desired by actual human beings in situations of scarcity. This historical metaphysics of transcendent interests could obviously not find a place in a tradition which received one of its possible promulgations in *Towards a General Theory of Action*. The same may be said about the distinction between the superstructure and the infrastructure and the causal primacy of the latter. The distinction derives from a conception of human action which is alien to the more realistic, synthetic tradition contained in much of contemporary sociology.

Thus a considerable amount of the Darwinian or biologistic and the metaphysical evolutionary—Hegelian in fundamental pattern but rendered "scientific" and therefore acceptable by Darwinian idiom—could not be fused with the traditions which prevail in contemporary sociology. They have either become extinct or they lead a restless uncomfortable existence recurring intermittently at the margins of sociological culture, functioning to criticize but incapable of positive development in a culture which is committed to empirical theory—however abstract—and empirical research—however imprecise.

Of the Comtian and the Hegelian concepts of social evolution, certain very important elements have survived. The conception of society as a moral order in which discrete individuals are bound together in collective actions of various sorts by common images of themselves as parts of that order and by common beliefs defining themselves and their obligations to the collectivity is one of the chief precipitates of the coalescence of the various intellectual traditions of the nineteenth and eighteenth centuries. From Rousseau's conception of the collective will, through Hegel and the German theorists of the *Volksgeist*, from Comte's portrayal of the morally disintegrated condition of the then contemporary society in its "critical," skeptical state, and from Marx's conception of the dissolution of the moral bonds of modern bourgeois society emerged a variety of conceptions of the significance of beliefs—about the ultimately valuable—in the regulation of individual conduct and in the determination of the degree of integration of society. Max Weber's typology of legitimate authority was developed from the presupposition of the significance of beliefs and of a consensus of beliefs about ultimate things; the moral state of society which engaged Durkheim's interest, the conception of *Gemeinschaft* and *Gesellschaft* in Tönnies, the conception of the moral order in Park—all these beliefs about the scale and significance of a consensus of moral beliefs in the working of society have been the products of a confluence of traditions into which Hobbes, Rousseau, Hegel, Marx, Lazarus and Steinthal, H. S. Maine, et al. have contributed. The Marxian concept of bourgeois society unbridgeably divided into competing and conflicting classes; the Simmelian conception of urban society differentiated by division of labor, sectional interests, and individuation; Tönnies' conception of modern bourgeois society—*Gesellschaft*—which resembled Hobbes's picture of the state of nature where a relationship of *homo homini lupus* prevailed all implied a great concern with the extent and the significance of a consensus of moral beliefs. The problem was: to what extent do individual men act on the basis of conceptions of themselves, their rights and obligations, in ways which minimize or aggravate the conflicts between individuals and between groups? Consensus became the main variable of analytical (or theoretical) sociology. Even those authors who claimed that modern societies were atomized had this point of departure in an image of a consensual society: they regarded the dissensual character of modern societies as the phenomenon to be accounted for. The very definition of modern society involved placing the category "consensus/dissensus" at the center of attention.

The struggle for existence which was one variable of the sociological Darwinism and the basic theorem of the *Rassenkampf*-sociology could be assimilated into the emerging sociological tradition because it constantly raised the problem of "consensus-deficit." In Ratzenhofer and Gumplowicz, the conflict of races was given; it was not problematic. In the analytical sociology, which became ascendant in the work of the European and American founders, conflict was not given; it became a variable to be explained. Its converse—consensus—was likewise not given. It, too, had to be explained. Research on the adjustment of immigrants or family disorganization, ethnic conflict and segregation, suicide, *anomie*, sect-formation, the role of charismatic authority and of rational-legal authority were all attempts to cope with this fundamental problem. Features of other ideas which could not be assimilated were discarded.

Certain other important ideas of contemporary sociology were possible only within the context of a concern about the moral beliefs of a society. The idea of the charismatic or the sacred was an extension of this conception. The delineation of the properties of bureaucratic authority and the assessment of the role of bureaucracy in modern societies were elaborations of the conceptions of *Gesellschaft* in Tönnies and of capitalistic society in Marx. The same obtains for the conceptions of "mass society," "modernization," and "organizations" which assumed such prominence in subsequent sociological studies; they were developed out of the fundamental conceptions of the sociology formed by the continental founders of the subject.

Even before it became anywhere nearly as institutionalized as it has been for twenty-five years, this tradition was in the process of gaining ascendancy. It was, as Professor Parsons points out in *The Structure of Social Action,* contained implicitly in the three main European traditions of idealism, positivism, and utilitarianism, and it was elicited from potentiality by the powerful intelligences of the major European theorists. The American theorists of the founding generation were not of equally powerful theoretical intelligence, but they too—Thomas and Park—came upon the essential. Neither Thomas nor Park ever succeeded—nor did they try in any large-scale effort—in discarding the elements of incompatible traditions. In Park's case, fragments of Darwinism remained in his ecological ideas. Thomas, indeed, after his exclusion from the University of Chicago, turned away from not only his sociological work but also the framework which he had developed in it and which brought him near to Weber and Durkheim; he was attracted by an eclectic behavioristic psychology which was alien to what he had believed previously. There is some reason to believe, however, that after this digression he began to find his way back again, but no published work ever announced his arrival.

Somehow, despite its very scanty institutionalization in Germany, despite its submergence in France, despite its numerous competitors in Great Britain, the now dominant tradition grew and penetrated. With institutionalization and the emergence of new centers in which a more stringent formulation of the tradition was presented, it finally triumphed over competitors—not totally—in the United States. However, after the United States became the center, the reinterpreted and modified tradition then spread back to Europe, where, despite resistances, it established itself in all the main European countries in a way in which it had never done before when Europe was itself the center from which the tradition emanated.

To what extent is its wide diffusion and dominance attributable to institutionalization?[33]

The turning of direction of a tradition, the amalgamation of elements of several traditions, is a creative action. It is not an institutional action. It is the work of an individual mind and it can and has been done under conditions of very rudimentary institutionalization. As a matter of fact, it might even be said that institutionalization, as important as it is for consolidation, multiplication, and diffusion, can be something of a hindrance to the turning and amalgamation of traditions. Max Weber had the advantage of an uninstitutionalized sociology in his time. He had studied law and economics; he came on the study of religion without institutional supervision. Durkheim had not studied sociology as an academic subject. Park studied it under conditions of very scanty institutionalization, and Thomas studied it in Chicago when the subject still scarcely existed even though an institutional structure had been created for it.

A tradition can give birth to a product at one stage of its existence which it could not produce at an earlier time. Its potentialities might be presumed to have a sequentiality and it is quite possible that this stage had been reached at the end of the nineteenth century, so that in a number of countries persons who had assimilated positivism, utilitarianism, and idealism began to produce results which had marked affinities with each other. Those who produced these results entered into practically no contact with each other. It was for the next generation of those born after the beginning of the new century, with Parsons in the forefront, to draw these separately generated approximations to a tradition together into a more explicit and coherent pattern. The subsequent history of the consolidation and diffusion of the tradition is very much a history in which the institutional system of sociology has played a crucial role.

Exogenous Traditions, Endogenous Traditions, and Exogenous Stimuli

An intellectual discipline exists when a number of persons believe themselves to possess an identity defined by the common subject of their intellectual concern and when many or all of the problems which they study are raised by or derived from the tradition—that is, the body of literature and oral interpretation produced by those who regard themselves as practitioners of the discipline. An intellectual discipline is an academic discipline when it is taught—that is, discussed and investigated—through academic institutions which bear the name of the discipline or something akin to it. When its members publish works in organs bearing the name of the discipline, a discipline has an intellectual and social structure.

In its early years, while its proponents were still struggling to establish their intellectual and academic dignity, a relatively common activity of sociologists in Germany, France, Great Britain, and the United States was the delineation of the field of sociology, its demarcation from other already existing academic disciplines, and the laying out of its internal subdivisions. Sociologists at that time had no social structure for their discipline in most instances, and so they sought to define it by its intellectual properties. This was extremely difficult to do because the ancestry to which they had claim was not universally acknowledged by nonsociologists and because there was little in the way of substance which they could invoke in self-legitimation. Therefore, they argued in principle and made claims for the future. This activity has now practically vanished, partly because most sociologists have come to accept a body of problems, ideas, and procedures as sociological, despite unclear boundaries and heterogeneity of subject matters, problems, and approaches. The large body of literature containing these problems, ideas, and procedures, which appears in sociological journals, is reviewed there, and is produced by persons called sociologists, fortifies this sense of sociological identity. Sociology is also now recognized as a subject by the educated public inside and outside the universities, and this too helps to define the boundaries of the subject and of the profession which cultivates it or at least allays apprehensions about its legitimacy.

The definition of sociology by the social structure of sociology and by its corpus of works has a retroactive effect on the construction of the sociological tradition. While traditions work forward in time, the construction of a legitimatory, inspirational tradition is a temporal movement in the reverse direction. Tradition is not, however, a mythological construction, although it has some of the functions of mythology. The works and ideas which have been admitted to a tradition—that is, those which are regarded as part of it—set the problems. They can set problems in a situation in which the ostensibly valid part of the tradition is the work which has most recently been published, although the problems of the recently published works were set earlier by a major work which might go back many years. Alternatively—although the situation is not so different—current work might constantly refer explicitly to a major work done many years earlier. Sociology is at present a mixture of these two relationships to "its own past." The significant fact is that sociology now has a fairly consolidated central tradition and a number of less dominant traditions. Some of these less dominant traditions are in a sense countertraditions. These countertraditions are mainly polemical rather than substantive. Behaviorism as exemplified in the work of George Lundberg was one of these; the self-designated "critical sociology"—an evasive name for highbrow Marxism—is another. Both of these countertraditions, insofar as their adherents produce substantive works, turn out to have little nutritive value intellectually and their adherents turn out to be dominated by the prevailing central tradition in their selection of problems and in their selection of the major variables with which they operate. Allowing for the ambiguity and lack of rigor of practically all sociological work and the large degree of freedom of interpretation permitted by the poor data and the disjunctiveness of detailed empirical research, even the interpretations offered by adherents of the countertraditions turn out to be little different from the interpretations offered by the adherents of the dominant tradition.

The subsidiary traditions, on the other hand, are not polemical. They have a more positive relationship to the dominant tradition. They are viewed sympathetically by adherents of the dominant tradition. They have become tributaries into the dominant tradition, even though they arose outside it. Comparative religious studies from Spencer onward are of this sort. Thanks to the work of Weber and Durkheim, the results of comparative religion have found a place in the dominant tradition, although comparative religious studies still constitute an autonomous field of work with a strong central tradition of its own. Modern economic analysis is another. At one time, in the writings of the European founders, economics up to the time of Mill and Senior found its way into sociology. Talcott Parsons succeeded in bringing in Marshallian economics too, but his later efforts to assimilate the Keynesian development from neoclassical economics have been less successful.

Animal and plant ecology and the economic theory of location (von Thünen) also coalesced with the central sociological tradition as it was emerging into dominance; it faded as the center shifted from Chicago. Since these subsidiary traditions were never persuasively and explicitly integrated in an authoritative work by Park or his associates and protégés, they became submerged or disassociated as far as sociology was concerned.

The psychoanalytic theory of personality is another tradition which coalesced with the central sociological tradition. The particular propositions of psychoanalysis might not be true; the mechanism of the Oedipus complex might not be as psychoanalysis has described it; the same can be said about the genesis and mechanisms of aggression, conscience, anxiety, repression. Nonetheless, the indication of these variables in the coherent pattern in which psychoanalysis set them forth has been assimilated into the sociological tradition. The naming and description of these phenomena has made sociologists more aware of them, more realistic in their perception of them, more able to assess their magnitude and to estimate the probability of their occurrence under certain conditions.

There have been other coalescences of the sociological tradition and intellectual traditions external to sociology. Traditions from various currents of philosophy—for example, from phenomenology, from legal studies, from anthropological study of social structure and cultural patterns—have flowed into the sociological tradition, sometimes becoming well integrated conceptually, sometimes less so. Sometimes the traditions which reach toward sociology are contemporaneously active. Sometimes they have been dormant even outside of sociology for a long time. The most striking instance of the latter is the revival of Tocqueville's ideas about the consequences of equality and the delineation of the structure of an equalitarian society. The revival of Tocqueville occured first in political science after the Second World War and it moved into sociology with the increased attention to the structure of postwar society. The account of Tocqueville as a sociologist in Raymond Aron's *Les Étapes de la Pensée sociologique* has now made Tocqueville's ideas retroactively into a constitutive element in the contemporaneously effective tradition of sociology.

Sociology and comparative religious studies, sociology and economics, sociology and ecology, sociology and psychoanalysis, sociology and cybernetics, sociology and the theory of administration—all these coalescences with their simultaneous modifications of the content of sociology today and of the image which sociologists have of their own past are all coalescences of substantive traditions, of traditions which have grown out of the sociological tradition endogenously and of intellectual traditions which have developed outside of sociology. But they are all substantive traditions. Sociology is not, however, merely a body of substantive assertions: it exists within the more broadly embracing tradition of discourse which requires proof of its assertions. In consequence of this sense of intellectual obligation, the substantive traditions of sociology have increasingly, although unequally, affiliated themselves to traditions of observational organization and technique.

There is a long and many-streamed tradition of sociological procedure which has become more and more coalescent, in practice and in retrospect, with the substantive traditions. Paul Lazarsfeld has created a retrospective map which goes back to Petty and the English demographers of the seventeenth century. More recently he has established Quetelet as a precursor. Philip Abrams has now presented a coherent account of the course of the "statistical" tradition. Every new step in sociology toward a more rigorous procedure of empirical research adds a new set of ancestral deities to the pantheon of the sociological tradition. These, in turn, add their force to the movement of coalescence.

The initiative for the coalescences of endogenous and exogenous intellectual traditions has always come from within sociology. In that sense sociology has become a realm of its own with its own center of gravity, its own discriminatory powers. This is tantamount to a continuing expansion of sociology in its subject-matters, in the differentiation of its analytical scheme, and in its corresponding interpretative hypotheses; it also entails the increasing stringency of its procedural standards.

The tradition which is regarded as sociological is heterogeneous, not just in its substantive focus, but also in its implicit assessment of the relative intellectual dignity of different strands. It is heterogeneous because of the imperfections in the assimilation or integration of the many coalescent subsidiary or tributary traditions. The various strands of the sociological tradition are not shared equally by all sociologists. Not only are there the limitations imposed by the specialization of knowledge, but there are also differences in interest and esteem which mean that any particular sociologist will be more sympathetic to one strand or family of traditions and more hostile toward certain others.

Sociology does not, however, live from its intellectual inheritance alone, and its exertions are not confined to the refinement and enrichment of that inheritance by the study of the problems which that inheritance offers and by the incorporation of new elements from exogenous intellectual traditions. Sociology does not just live within its primary institutional system, it also belongs to the larger world. Its subject matter is that larger world. Sociologists are members of social classes, ethnic groups, religious communities, and political organizations, and they share to some extent the culture of these social systems. They belong to their society and their generation and they share to some extent in the cognitive beliefs and evaluative attitudes of their society and generation. Sociologists are citizens in their own societies and they respond to the problems of their societies. The institutionalization of sociology has entailed training students for roles in that world. The institutionalization itself has been greatly furthered by the belief of those in positions of authority and influence in that larger world that sociology "has something to contribute" beyond its contribution to the improvement of the sociological tradition. Part of the institutionalization of sociology in recent decades has involved either the subsidy (or contracting) of research by university research workers on particular subjects designated by the patron as well as the conduct of research within the corporate framework of public or private bodies concerned in a practical way with certain problems.

It has been claimed that sociology often chooses its problems and its subject-matters according to criteria of value-relevance and not just in terms of the problem-complex presented by the intellectual tradition. For example, the study of the Negro, as developed by the Chicago sociologists, has been asserted to be a response to the problems of urbanization and migration. But it was at least as important as a study of the processes of competition, conflict, accommodation, and assimilation. These categories had been developed by Park from his study of Simmel, Kistiakowski, and Gumplowicz, from his study of the work of a Danish ecologist, and from his own observations, both direct and historical. These seem to me to be still more or less the right categories in which to study the relations between groups, including those between whites and Negroes in the United States. Before the Second World War, the study of the Negro in America was further enriched by the introduction of certain approximately psychoanalytical variables, and a number of valuable monographs were produced along these lines. But at a time when the Negro question was not less urgent than it had been before, and for no good reason other than the relocation of the center of the sociological institutional system to a place where the sociological tradition espoused had not dealt with ethnic relations as a particular concrete subject-matter, the Negro ceased to be an object of study. This occurred practically at the time that Myrdal's vast synthesis of previous American—to a large extent, Chicago—research appeared. Why? Had the analytical problems involving competition, conflict, accommodation, and assimilation been so exhaustively solved that the subject no longer had anything interesting to offer? Or did the Negroes themselves cease to be a "value-problem" in the United States? Neither of these answers seems to me to be acceptable.

The decline in intellectual interest in the Negroes in the United States was a result of several factors. The first is that the new centers of sociological study—Harvard and Columbia in particular—had their own substantive intellectual traditions which did not include the Negro. In Max Weber, primordial things received little explicit attention: in *Wirtschaft und Gesellschaft* only one chapter of eleven pages is devoted to *ethnische Gemeinschaften,* and in the writings of Professor Parsons they received no attention until the 1960's when the Negro problem became urgent in the United States. I think that until relatively recently not more than two Harvard Ph.D. dissertations were written on Negroes. At Columbia, where several interesting dissertations had been written on Negroes, for example, *Sea Island to City,* in the 1930's, the subject nearly disappeared in the Merton-Lazarsfeld epoch. The eclectic analytical outlook of Professor Merton could certainly have been accommodated to almost any subject-matter and the technical virtuosity of Professor Lazarsfeld was likewise applicable to the study of race relations and to the study of Negro society. At Chicago, in the disorientation referred to earlier and under the impact of the Harvard and Columbia styles of sociological theory and research, interest in the Negroes was allowed to lapse. The old center lost its confidence in its own substantive tradition.

In the 1960's, when the "Negro problem" came forward with an unprecedented urgency, the study of the Negro was revived, but within a narrower framework and, in any case, a different one from that which guided the earlier studies. New methods of measurement of occupational discrimination, of the share of the Negro communities in the national income, and so forth, have been undertaken, but the social structure of Negro-white relationships as developed in the tradition of Thomas, Park, and Burgess, by Franklin, Frazier, Bertram Doyle, and Charles Johnson has not been taken up again. The "Negro subject-matter" of sociology might have survived the period from 1940 to 1960 if the Chicago tradition had received an authoritative theoretical formulation capable of demanding explicit incorporation into a new analytical scheme developing at Harvard. Since, however, it had not gone beyond fragmentary, only implicitly coherent, statements, its competitive power was not sufficient. Thus, despite its analytical affinity with the newest developments in the sociological tradition, its idiom was submerged and, with that, its particular subject-matter. Had its idiom survived in the new analytical scheme, its associated subject-matter would likewise have survived.

The vicissitudes of the study of the Negro in the United States would indicate that the substantive analytical or theoretical tradition was not affected by exogenous, nonintellectual occurrences. The change in the substantive or subject-matter tradition was, in my view, to a large extent the function of the adhesion of particular subject-matters and traditions to particular analytical traditions and of changes in the institutional system of sociology (for example, the relocation of centers). The re-emergence of the Negro as a subject-matter of sociological research has indeed been a consequence of exogenous nonintellectual events, certainly of nonsociological events. But has the scheme of analysis been affected by extra-intellectual events? Rather it remains very much a product of the sociological tradition and of an expansion in the direction of sociology emanating from the tradition of economic analysis (see especially the work of Professors Gary Becker and Kenneth Arrow).

The study of "popular" or "mass culture" is another illustration of the interplay of endogenous and exogenous intellectual traditions and exogenous nonintellectual events. Tönnies, Simmel, Weber, Mannheim, and Horkheimer created the tradition, but it was the response to the National Socialist movement and its triumph in Germany and the great expansion of the study of the content of and "exposure" to mass communications which led to the increased attention to mass society in American sociology in the 1950's. Changes in the political culture of literary and publicistic intellectuals in the United States—a disillusionment with Marxism attendant on the identification of Marxism and Stalinism and disillusionment with the working classes as the agents of revolution and the "heirs of German classical philosophy"—made for a more active attention to certain elements of the central tradition of sociology. The nonintellectual events—National Socialism and mass culture—did not divert the tradition of sociology; they were simply assimilated into it.

The career of industrial sociology and the sociology of work is another instance of a subject which was developed in the framework of the central tradition of sociology—much influenced by Durkheim's ideas about *anomie* and Max Weber's ideas about bureaucracy—and supported by private business management. In the United States, but not in Great Britain, the subject has lost its glamour—although obviously not its intellectual or practical significance. American sociologists have gone off that subject.[34] They have gone on to organizational analyses, under the inspiration of Max Weber and Herbert Simon, and on to studies of the professions of medicine, law, and science. The theoretical or analytical scheme remains what is offered by the cultural tradition, enriched and differentiated by the new data. The change in particular subject-matters is perhaps more a product of the availability of financial support for the study of these subjects, the much greater prominence in public opinion of the learned professions and especially of the scientific profession, and the decline among intellectuals of the prestige of the working classes. Latterly in the United States, sociologists have returned to the proletariat. But not so much to the employed "respectable" proletariat as to the unemployed, the "poor," or what was called by Marx the *Lumpenproletariat*. This change is partly a function of changes in the focus of attention in public opinion and in intellectual circles and of the opportunities afforded by the availability of financial support for such research. The relocation of subject-matter interest is relatively recent, but thus far there is no evidence that it has affected the intellectual pattern of the central tradition.

In Great Britain, a large proportion of the much-increased volume of sociological research is concerned with the working classes and the poor. This testifies to the strength of the British intellectual tradition—largely pre-academic, deriving from the statistical surveys of the nineteenth and early twentieth centuries—as well as the availability of financial support for inquiries into such subject-matters. The power of this tradition in Great Britain has been so great that an analytical tradition such as has been preponderant on the continent and the United States has scarcely developed. Even there, however, the chief analytical achievements—T. H. Marshall's *Citizenship and Social Class* and the writings of Michael Young and David Lockwood—remain within the framework of the dominant central tradition.

What we see from these few sketchily presented illustrations is that exogenous nonintellectual stimuli have indeed played a significant part in bringing forward subject-matters into the center of attention of sociological research. They cannot, however, be said to have made a comparable impact on the analytical framework of sociology which is contained in the central tradition. This framework has extraordinary continuity and stability. At the same time, this framework remains vague and ambiguous in its major categories and has experienced relatively little differentiation or deepening over the past quarter century. The main innovations have come from the intrusion of exogenous intellectual traditions and these have not yet been systematically integrated into the dominant tradition.

The Form of a Sociological Work

A tradition can be influential in an academic discipline in several ways. In one way, the most recent period's output provides the immediate point of departure, the prior output having set the point of departure for the former, and so on, while the works of the relatively remote past—the past of the middle distance—are influential either through a similar point of departure function or through the constitution of a framework of concepts or variables and fundamental problems within which the works which constitute a sequence of subsequent points of departure are performed. The latter are the monuments of the subject, not merely honorific monuments, but rather effective ones. They are what Thomas Kuhn calls "paradigms." But they are not referred back to except on honorific occasions. What is living in them? What is relevant for the development of knowledge in the field is contained in the latest stage of the tradition, the detailed research which provides the immediate point of departure for the next future stage of the subject. A third way in which tradition can be effective is through constant recourse to the monument.

These three patterns of relationship of present work in a given subject to the tradition of that subject correspond to the forms of work in both the natural sciences and sociology.

The most common form of scientific work is the short research report or paper which states the hypothesis, locates the hypothesis in relation to previous research, describes the research arrangements or experimental procedure, reports the observations or experimental results, and proposes the modification of previous hypotheses or beliefs required by the results of the reported investigation. There are also synthetic, compendious summaries of existing theoretical and experimental work which intend not to establish new knowledge, but to consolidate and order what is known in a broad field. These might be monographs with an element of originality of interpretation of what has been demonstrated by research or they might be textbooks which summarize what students and research workers at a certain level of their scientific development are expected to know.

In sociology—parallel to the differences in procedures and intellectual structure between sociology and the sciences—works tend to take the following forms. Monographic treatises describe a particular subject-matter, for example, a territorially delineated sector of society, a group, a stratum, or a profession. These sometimes take the physical form of a monograph of 20,000 words or, more frequently, a full-length book of 100,000 words. (The latter often contains the former.) They give much descriptive data with analytical remarks and usually a general interpretation of the phenomenon in relation to its larger setting at the end. These monographs or monographs-within-books are usually directed at a problem (relationships between variables), although their ostensible point of departure is a particular subject—for example, the medical profession in a given area; a particular village; a process, such as social mobility within a given country. There are also monographs which seek to deal with problems. These begin with an attempt to explain a major phenomenon—for example, specifically capitalistic acquisitive behavior as in Max Weber's *Protestant Ethic*, the causes of various kinds of suicide as in Durkheim's *Suicide*, or the effect of political campaigns on voting choice as in Lazarsfeld's *The People's Choice*.

Then there is the journal article (corresponding to the scientific paper), which summarizes the existing state of knowledge on a particular problem, formulates a hypothesis, presents the data which report observations made by the author or by others, and attempts to harmonize the resultant interpretation of the data with previous interpretations.

There are also essays which clarify, refine, and differentiate particular ideas (concepts, categories of processes, variables, and so on) without close reference to observations except for illustrative purposes.

There are theoretical treatises which present "all" the major concepts or variables which the author believes relevant to sociology as a whole and hypotheses about the behavior of these variables under different conditions. One might mention here works as various as Weber's *Wirtschaft und Gesellschaft*, Parsons' *Social System*, and Homan's *Social Behavior*. These theoretical treatises verge at their lower levels into textbooks which cover all the concepts of variables and present illustrative data. Such works as Kingsley Davis' and Wilbert Moore's *Society: An Analysis*, H. M. Johnson's *Sociology*, and Park's and Burgess' *Introduction to the Science of Sociology* are representative. Sociological textbooks usually contain much descriptive material from monographs and articles, but very little rigorously established, widely accepted knowledge—unlike textbooks in the sciences.

Finally, there are collections of essays of single authors, a form which has become relatively widespread since the end of the Second World War. Indeed, collections of essays have been among the most influential books of the past two decades. The essays have usually been analyses of concepts and elaborations of themes, rather than scientific articles.[35] Such collections are also found in the sciences, but for the most part they are monumental in intention, being the works of very distinguished figures, living or dead, and many of the papers published in them have already played a great part in the development of these subjects and have outlived their immediate usefulness. They have a different function from the collection of essays in sociology because of the longer life of the sociological essay.

The function of the books composed of sociological essays discloses a major difference between the characteristic patterns of the growth of knowledge in sociology and in the sciences. In the latter, the most important means of speedy, non-oral communication are the research paper, and more recently the preprint. The research paper in its original form is important in science; when reprinted in book form, it is likely to have only historical interest. The emergence of the preprint in science is the very opposite of the reprint in sociology of an essay in book form. The scientific preprint is required because of the speed with which the tradition is being modified or supplemented; the reprint of a sociological essay is a function of the continued dominance of the tradition, particularly the dominance of the monuments.

The sociological essay can be reprinted in a book because it has a long life, whereas a scientific paper must perform its function within a relatively short time after its moment of publication.[36] The difference might be accounted for by the much sharper focus of research in the sciences, by the more consensual perception and interpretation of the results of any particular piece of research, and by the more immediate response in the form of further investigation of the problems it raises.

The completion of a piece of research on science entails its publication because science is a collective possession—the possession of the scientific community—and knowledge which is not in the possession of the scientific community is not knowledge. Discovery, therefore, integrally entails publication. To launch on a campaign of discovery is to launch on a course which requires by its internal structure the publication of the end state of the act of discovery. A work of research which does not contribute to the pool of knowledge is a wasted effort and failure to publish is such a waste. If the delay in publication is long, then the results of research will usually be rendered out of date. It will have been superseded by the results of the research of someone else. This high probability of supersession is a function of the greater precision in formulation of problems and greater specificity and concentration of the focus on problems in the sciences.

There are relatively few instances of the very specific articulation of successive research papers in sociology as compared with any field of comparable scope in the sciences. This is partly a result of the much lower degree of specialization, and the much less differentiated division of labor in sociology. The less differentiated division of labor in sociology is in turn a result of (a) the much greater quantity of literature produced in any field of the sciences and (b) a closely related phenomenon, the larger number of persons working in any special topic in the sciences. But without the specificity of focus and the greater consensus about the crucial problem—which exists in the sciences as compared with sociology—the larger body of workers on a given problem would be of little consequence for the life span of a research paper or report. This greater concentration of minds on particular, specifically delineated problems results in the previously mentioned more rapid obsolescence of works in the sciences.

The form of the work, then, has an intimate connection with the style of thought and procedure of analysis of a subject, as well as with the scale and the structure of its institutional system.

The Professionalization of Sociology

Whereas at the turn of the century there were very few persons who called themselves sociologists anywhere in the world and even fewer who made a livelihood from it, there are now many thousands who teach what is called sociology in university course lists and thousands who are employed by governmental bodies, independent research institutions, market research organizations, public opinion surveying organizations, industrial firms, hospitals, and military organizations to do research which they classify as sociological. Most of the world's colleges and universities now provide teaching in sociology and permit specialization in it as qualification for a degree. There are nearly one hundred journals in all languages which have the name sociology in their titles or subtitles. There are about thirty national professional sociological associations, and in some countries there are also more specialized professional associations which regard themselves as covering a section of sociology. Sociology has become a subject of a learned profession.

One consequence of this professionalization is a movement toward homogeneity. National differences still exist among the dominant types of sociology—theoretical inclinations, subject-matters of research, preferred techniques,—but there is also a large common sociological culture, just as there is, despite specialization, within each country which has many sociologists. It is true that sociology in the English language enjoys the advantages of an asymmetrical accessibility. Sociologists everywhere read what is written in English by sociologists, but sociologists in the English-speaking world read less of what is written in most other countries, even if it is published in English (as are *Acta Sociologica* and *Sociologica Neerlandica*).

Does professionalization—the triumph of institutionalization—entail the isolation of the sociological tradition from other intellectual traditions, academic and nonacademic? Not necessarily. Linguistically, sociology in the English language has shown the potentiality of isolating itself through ponderous and inexact terminology and stylistic barbarism, but the isolation has not in fact occurred. The curiosity of the educated classes outside of sociology has grown as their stylistic intolerance has diminished. Perhaps they do not know any better, perhaps they do not mind, but whatever the explanation, sociology has not been rendered inaccessible to nonsociologists by reason of its execrable literary qualities. It is not even likely that the mathematization of sociology which is in prospect—in some measure at least—will isolate sociology from the increasingly mathematized educated classes.

The accessibility of sociology to nonsociologists is, however, of less interest to my concern with the growth of sociology than is the accessibility of intellectual traditions outside of sociology to sociologists. If the sociological tradition in its present protoscientific condition were to become isolated from other intellectual traditions, much harm would result, but I think that the danger is not great. For one thing, the pattern of the dominant sociological tradition is open; its toughness and persistence is not a function of its logically rigorous coherence. Its openness is partly a product of the particularity of the central problems with which it began and its excessive generality. It stands in much need of, among other things, a frequent confluence from other intellectual traditions. The large number of sociologists and the diversity of substantive things studied by sociologists guarantees that at least a small proportion will include within their knowledge developments of intellectual traditions other than their own. A small proportion of sociologists, coming from other intellectual fields or curious about what goes on in them, will bring into sociology new subsidiary traditions. There is, furthermore, despite the self-congratulatory attitude of sociologists toward their constitution as a profession, enough discontent with what they have inherited to render them receptive to innovations. The institutional structure of departments of sociology, the competitive relations between centers, and the emulative relations between centers and peripheries are likely to forestall the growth of Byzantinism in sociology.

The Dominance of Tradition and the Growth of Sociology

As sociology has become more institutionalized, it has become more productive. The institutionalization of training has produced a human product with a mastery of some techniques of research, a modicum of a sociological culture, and an ethos which praises research. The institutionalization of research—the institutional provision of research roles, resources, and facilities; the standardization of research procedures; the general approbation of research within intellectual institutions and the larger society—contributes to an increased volume of research. This is true in the United States, above all, but it is also true of countries where the contemporary type of research is a much newer growth, whether it is either a new implantation or a renewed practice of an older tradition.

This change has taken place under the auspices of "science." Sociologists increasingly believe that their work belongs within the family of science: they wish to be scientific. The old discussions regarding the differences between the natural and social sciences have evaporated with the arguments which demonstrated that sociology had a rightful place in "the classification of the sciences." The discussion as to whether sociology is a natural science is cold mutton. Part of the scientific self-conception of sociology is expressed in its preferred form of publication, the journal article. With the multiplication of the literature, the size of the subcommunity of sociologists is becoming large in almost every subfield of sociology; a sociologist can now spend his whole life or most of his sociological career as a specialist on military organization, the aged, the mass media, urbanization, the police, or narcotics addicts. Many of the specialized fields acquire and establish an identity of their own with their own elites and counterelites, their own culture, sometimes their own journals. The literature of the other specialized fields becomes more and more remote from them. As sociologists become "professionals," as so many proudly aver, specialization is accepted sometimes with the stoicism counseled in *Wissenschaft als Beruf,* sometimes with an air of self-congratulation. The new professionals of sociology are a tough-minded lot. They shun armchairs, they call questionnaires "instruments," they speak of "research technology," they are conversant with the language of computers, they are condescending toward "grand theory," negotiations for grants for research projects are frequently on their minds and lips. They are building a science, and as a science under present-day conditions, it must be specialized and it must be cumulative.

But is it cumulative? Many years ago I wrote about the discontinuity of sociological development; I had in mind among other things the flightiness of sociologists who take up a subject, cultivate it for a time, and then drop it before the subject has been brought to intellectual fruition. I had in mind, as instances, human ecology and primary group studies. I also had in mind not only the dropping of subjects or themes before their fulfillment but also the more specific unconnectedness of exact investigations of the same variables. This latter type of discontinuity between specialized investigations of what are ostensibly the same variables persists relatively undiminished, despite the greatly increased "scientificity" of sociology.

Specialization has its limits. The limits are not those given by the sterility of an uninhibited narrowing of the focus of attention—of "knowing more and more about less and less." They are limits engendered by the same institutionalization which has made specialization possible. There forms within each country (and increasingly between countries) an intellectual community, the members of which read each other's works, some of them even before publication. They read more or less the same body of literature: many of them are in contact with each other not only through reading each other's writings but also though correspondence and through personal meetings at conferences and on visits. There are, of course, disagreements within these specialized communities regarding techniques, findings, and interpretations, but they try to iron out differences and to avoid them in the future through the standardization of categories and techniques. Nonetheless, these specialized fields do not wholly isolate their members from those of other specialized fields.[37] No field fails to impinge on another equally specialized field. Thus, for example, a sociologist studying the aged cannot avoid the necessity for studying the literature on the family, on urbanization and demography, on geriatric medicine and medical sociology, and so on. As a result, the consensus on techniques of research, findings, and lines of interpretation extends beyond the boundaries of any community of specialists into adjacent communities of specialists. In this way a series of overlaps links the entire profession of sociology. Specialization in sociology has not yet progressed to the point, nor is it likely to do so in the foreseeable future, where one field cuts itself off entirely from its intellectual neighbors or from the sociological family as a whole.

There are, furthermore, limits to specialization imposed by the dominance within the institutional system of the tradition which is embodied in the essay, and less frequently in the treatise. After all, the institutionalization has occurred in universities, and in universities abstract ideas have a pride of place. If a postgraduate student has done sociology as an undergraduate, he has acquired a culture which is common to sociologists; even if he has done a different subject as an undergraduate and begins his sociological studies in a situation committed to specialized research, he must nonetheless absorb a certain amount of the common culture which is made up largely of the dominant tradition and some of the subsidiary traditions.

Sociological training is much less specialized than sociological research. A student must cover a much wider body of literature than that directly pertinent to the field in which he will do his research. Sociology now being an academic subject, the leading research workers tend also to be teachers of sociology; this means that they have often to teach subjects which are somewhat outside their specialized field of research. This, too, has a unifying effect, although it does not create or maintain a completely common sociological culture. The fact that sociology is no longer the work of amateurs, but is taught and largely practiced within universities in which "theory" has prestige, means that the prestige-conferring "classics" are taught in practically all universities for better or for worse. There is a *tronc commun* from which sociologists derive some of their intellectual dignity. Durkheim, Weber, Mannheim, Parsons, Merton are repeatedly quoted and invoked over a very wide range of specialized fields of research.

These famous sociologists are so generally drawn in, not just because those who draw them wish to avoid the appearance of being no better than intellectual hewers of wood and drawers of water. They are drawn in because there is a deeply felt need among sociologists to be part of an ongoing intellectual tradition and because there is a no less deeply felt need to explain one's particular findings by reference to certain major variables, such as solidarity, equality, deference, authority, charisma, bureaucracy. Contemporary sociologists are as much the creators of their tradition as they are its beneficiaries and prisoners.[38]

It might be said that the unifying classics of sociology are important to its contemporary practitioners because sociology is not at present a real science. If sociology were a science, its master-works would have become so assimilated into the flow of sociological work that their accomplishments would have been taken for granted. They remain important because sociological analysis is so much devoted to the elucidation of certain major themes which continue to dominate the attention of sociologists, partly because the great figures drew attention to them, and partly because experience and the concern which they have molded have shown them to be the proper subjects of study.

An effective intellectual tradition is not merely a timeless gallery of works, nor is it merely a sequence in time of works arranged by later scholars. An effective intellectual tradition is a linkage of works through influence exercised over time. This is the kind of tradition which operates in the real sciences and it also operates in sociology. In the real sciences, intellectual influence is usually fairly immediate—there are exceptions—and the works which are constitutive of tradition function through entering into the continuing flow; they lose their identity in so doing. Such tradition-setting works provide a framework of concepts which sets research tasks—and they provide the next stage of research, particular hypotheses to be revised, rejected, or confirmed, and observations which may be accepted as given or which require reinterpretation. In sociology the situation is both similar and different.

It is similar in the sense that major works provide a framework of concepts which set research tasks. It is different in that particular investigations are only loosely and vaguely linked with their guiding and legitimating ideas and are also only very loosely linked with those which have gone just before, although treating presumably the same problem at a more advanced stage. This situation in which the "subject—[is] presented [as] a library of the works of deceased giants, to be admired and discussed but most unlikely to be extended save by the rare appearance of new genius,"[39] must be superseded by a more continuous process in which one investigator takes up where his predecessor left off.

It is at just this point that the great difficulty of present-day sociology lies. To take up where one's predecessor left off entails the acceptance of the delineation of the variables studied by the predecessor in exactly the form in which they were put by the predecessor. The point of leaving off in sociology is difficult to locate because the vague interpretations of data always exceed the data. The categories involved in interpretation, that is, the theoretical construction, are always broader and vaguer than the categories in which the data were collected or the observations made. The latter vary too much from one investigator to another, even though the investigators believe themselves to be working on the same problem. For good reasons or for poor, each subsequent investigator "improves" on his predecessor's working delineation of the relevant variables. What in fact happens is that a slightly different variable replaces the one previously studied; they are both thought to represent the same thing when in fact they do not. As a result, the cumulative and progressive development of sociological knowledge is rendered more difficult. The source of the fault lies in the "great ideas" of sociology. They obviously refer to very important things—otherwise they would not hold the attention of sociologists as they have for so many decades—but they are extremely ambiguous and they defy authoritative clarification.

The unity which transcends specialization in sociology rests on this common devotion to a relatively small number of "key words." That unity is very expensively purchased. The key words and the ideas which they evoke have become inexpungibly enmeshed in the sociological tradition, so much so that they can never be merely an honorific decoration. They have become constitutive of sociological analysis. But the fact remains that they weigh like an Alp on the sociological mind. Theory is recognized as such by the presence of those alpine key words in all their misty and simple grandeur. This is not good enough. Sociology needs a much more differentiated set of categories, a much more differentiated set of names for distinguishable things. It must name many more things and name them in agreed and recognizable ways. The "slippage" between "concepts" and "indicators" must be reduced by increasing and refining the variety of "concepts."

This situation has arisen from peculiar features of the institutionalization of sociology and the flow of its traditions. Sociology arose both in France and in Germany informally, in a form which did not find at the time of its formation incorporation into the academic system. (In France the immediate or proximate institutionalization which Durkheim formed around himself was not incorporated into the ongoing university-institutional system.) The two great founders of sociology worked within humanistic faculties and themselves drew on traditions in which the technique of evidence was mainly those current in philosophy and historiography. Unlike the natural sciences in which theory and empirical investigation developed in close and reciprocal connection with each other, the dominant tradition of sociology developed without such a relationship. What Durkheim did on suicide and Weber on the selection and adaptation of industrial workers and on the East Prussian agricultural laborers showed that the founders were aware of the need for exact evidence for their theories; the fact that most of their work did not employ such evidence showed the strength of the traditions from which they had come. The same disproportion occurred in the United States, even though both the central and subsidiary traditions became institutionalized simultaneously. Despite their simultaneous institutionalization, they could not be integrated. At Columbia University, under the inspiration of Lazarsfeld and Merton, from the late forties into the sixties they came closer to integration than elsewhere, but there too the integration was loose and discontinuous. The gap persists and it is difficult to see how it is to be closed. But closed it must be.

References

1. Observation includes second- and third-hand observation through informants, documents, and printed works.

2. The situation was little different in political science. The intellectual sources of present-day academic political science were much more heterogeneous than the forerunners of sociology. Bentham, Constant, James and John Stuart Mill, Tocqueville, Ostrogorski, Bagehot, Bryce, et al. were not engaged professionally in university teaching and research. Hegel at the beginning of the century and T. H. Green, A. L. Lowell, J. W. Burgess, Jellinek, Gaetano Mosca, Woodrow Wilson, and Léon Duguit at the end were university teachers.

3. Spencer's *Descriptive Sociology* played no part whatsoever in his installation as a source of sociology—which in its subsequent form departed very far from his ideas and has only recently showed a partial turn toward them.

4. Because the EPHE was not dominated by professors, it was possible for Durkheim to attract to his circle young men who were preparing to enter other disciplines. This would have been impossible in Germany. There, a professor would not hold commerce with a research student who did not give promise of becoming a disciple, and a student would not dare to risk his career by serving a professor outside his own subject, the subject in which he wished to make his career. French professors were also jealous and students also depended on them for patronage.

Nonetheless, Durkheim did attract young men who could not make a career in sociology and who would ultimately have to make a career in some other academic discipline. Perhaps this is to be accounted for by the power of Durkheim's intellectual personality and by the fact that in the fields from which Durkheim drew his adherents, there were so few opportunities that even subservience to the incumbent professor was professionally fruitless.

5. Von Wiese's pupils did do a small amount of field work; see, *Das Dorf als Soziales,* Kölner Vierteljahreshefte für Soziologie, Beiheft I (1928). The result was simply the discovery of illustrations of the concepts of *allgemeine Beziehungslehre.* Dr. Willy Gierlichs, a pupil of von Wiese, taught in a police training college. Again this first step in the institutionalization of sociology in an applied form led to nothing because von Wiese chose the most sterile part of Simmel's sociology to try to develop. Neither the conception of formal sociology nor of *Zu- und Auseinander* was a very fertile point of departure, and von Wiese did not have the imaginative powers to draw out of them what little they contained.

6. In the index of the *International Encyclopaedia of the Social Sciences,* there are two references to Hobhouse, fifty-eight to Durkheim, and sixty-four to Weber.

7. Westermarck, whose own work was in the tradition of German *Völkerkunde,* had one pupil who transformed the subject—that was Malinowski and what he did became anthropology.

8. Institutionalization includes incorporation into the academic system, not just within a single university; other universities increase the density of self-consciousness, and they provide *debouchés* for students produced by the training system.

9. Professor Talcott Parsons spent one year at the London School of Economics before going to Heidelberg for a year. Hobhouse, whose seminar he attended, made no intellectual impact on him. All he could say of it years later was, quoting Crane Brinton: "Who now reads Herbert Spencer?" Weber, who had been dead about five years when Parsons came to Heidelberg, made a profound impression of life-long duration. It is obvious that institutionalization is not the only factor which determines intellectual impact. In both Germany and England, in their different ways, sociology was very slightly institutionalized academically. The power of the intellectual message, the pregnancy of the problem, the scope of its ramifications, the persuasiveness of the argument (insight, logic, evidence), and the magnetism of the rhetoric are surely of importance. Yet the fact that Weber had no impact in Germany despite his greatness and his ubiquitous fame shows, I think, that the factor of institutionalization is an independent variable of some importance.

10. The greatest British survey was *The Life and Labour of the People of London,* in seventeen volumes, the culmination of a tradition begun in the *Reports* of the poor law commissioners and the work of several great officials interested in public health, housing, and "sanitary conditions." Practically none of this work was done under academic auspices. The surveys of the middle of the century were done by government officials. Mayhews' *London Labour and the London Poor* was the work of a journalist, and Booth was a wealthy shipping merchant who financed and directed his survey as a private inquiry. Rowntree, too, was a wealthy businessman without university connections.

    When in the 1920's a *New Survey of London Life and Labour* was undertaken, it was directed from the London School of Economics, but it had no connection with the Department of Sociology at the school. Furthermore, it was a transient enterprise. When the survey was finished, the organization through which it had been conducted was disbanded. It might be pointed out here that when the Population Investigation Committee functioned at the London School of Economics and produced important works in demography and social selection, it worked wholly independently of the Department of Sociology. It did not serve to train and recruit professional sociologists, except for Professor David Glass, who was at that time a geographer and came into the Department of Sociology only after the Second World War. The Department of Social Research, which was created after the war, was an entirely separate entity from the Department of Sociology, but it at least represented the institutionalization of social research.

11. As did William Graham Summer at Yale. Summer never went beyond this, and as most of his students were undergraduates who had no interest in pursuing sociology further, Yale did not become an important center of sociology until many decades later.

12. Thomas spent the academic session of 1888-1889 at the Universities of Berlin and Göttingen. This was before he enrolled as a Ph.D student at the University of Chicago, which was, in fact, nonexistent during Thomas' German years.

13. He undoubtedly encountered the ideas of Tönnies about *Gemeinschaft* and *Gesellschaft,* but there is no reference to Tönnies in his dissertation, *Masse und Publikum,* or in his autobiographical reminiscences. He did reproduce a section from Tönnies' *Die Sitte* (1913) in the *Introduction to the Science of Sociology* (Chicago: University of Chicago Press, 1921), pp. 103-105.

14. Norbert Elias, *Uber den Prozess der Zivilisation,* 2 vols. (Basel: Verlag zum Falken, 1938); Hans Gerth, *Die sozialgeschichtliche Lage der burgerlichen Intelligenz in Deutschland um die Wende des 18ten Jahrhunderts;* Wilhelm Carlé, *Weltanschauung und Presse;* Jakob Katz, *Die Judenassimilation im Deutschland.* The latter three were printed only as dissertations.

15. Westermarck's chair was filled only during the Easter term of each year from 1907 to 1930.

16. Walton Hamilton, prior to going to Amherst, where his pupils included Talcott Parsons, was at the University of Texas, away from the pressure of conventional economics. It should, however, be acknowledged that Johns Hopkins University in Baltimore played an early part in this introduction of German historicism into the United States. But Johns Hopkins was *the* new university of the country until Chicago was founded. (Simon N. Patten does not fit into the picture presented above.)

17. Science in the heroic age should be reexamined in the light of the approach taken in this paper. Aside from the genius of its main actors, it owes some of its growth to its informal organization as a society of correspondence. The directly addressed audience of known persons remains an important feature of contemporary science. The publication of books and papers apparently addressed to an almost entirely anonymous audience does not mean that their contents move in all directions randomly and indifferently to their authors' intentions.

18. It is unlikely that his "Agrarverhältnisse im Altertum" and *Romische Agrargerchichte . . .* were unknown to Italian classical historians.

19. Alexander von Schelting understood Max Weber's ideas with a precision and elaborateness which was unique in Germany. He was not, however, *habilitated* until after the end of the Weimar Republic, when he went into exile and taught briefly at Columbia, before the renaissance created by Merton and Lazarsfeld. Then, after years of poverty in Switzerland and a brief period as an *Extraordinarius* in Zurich, he died without influence and with the sociological part of his study of the Russian intelligentsia unfinished.

20. Weber's methodological views and the Protestant Ethic thesis were the only aspect of his work with which German scholars were concerned in the 1920's. Jaspers and Rickert, who were closest to Weber's outlook, were interested exclusively in his methodological writings, as was von Schelting before his exile. The Protestant Ethic thesis was seriously discussed by historians but little of this flowed into sociology.

    It should also be said that the fading of the imprint of Weber's ideas was fostered not just by the absence of any sustaining intellectual-institutional structure around Mannheim, but also by Mannheim's hidden and ambivalent struggle to establish a distinctive line of his own different from Scheler's, different from Marx's, and different from that of German idealism.

21. The diffusion of Max Weber's ideas in America also owed something to Frank Knight who was the first translator of any work of Weber into English. His translation of *Wirtschaftgeschichte* bore the English title, *General Economic History* (1927). Knight was an extraordinarily acute and subtle economic theorist who saw the limits of neoclassical theory and who was also aware of the fundamental intellectual weakness of "institutional economics." In the middle 1930's, Knight conducted a seminar on the first chapters of *Wirtschaft und Gesellschaft,* in which I participated, where we studied the text line by line. Melchior Palyi was at that time on the staff at Chicago; he had been Weber's assistant at Munich and after Weber's death had helped to establish in the text of *Wirtschaftgeschichte* with Professor Hellmann and of *Wirtschaft und Gesellschaft* with Marianne Weber. He encouraged my Weber studies, although I had first discovered Weber through Tawney, Troeltsch, and Burkharin.

22. The subsequent career of the Weberian sociology through Parsons' role in the Department of Social Relations at Harvard is dealt with below.

23. The one form in which Durkheimian tradition was institutionalized was in the textbooks of civil morals in the *lycées.* But rather than recruit sociologists, it alienated persons who might have become sociologists. Raymond Aron was one of those who was alienated from Durkheim's tradition by this mode of presentation. He was recovered only by encountering Max Weber when he went to Germany at the beginning of the 1930's.

24. Durkheim exerted through Radcliffe-Brown a marked influence on British social anthropology, but until very recently the social anthropology deriving from Radcliffe-Brown and Malinowski had very little connection with British sociology.

25. It is interesting that at a not very distant time, Max Weber had studied the German side of the same problem—the impact of the immigration of Polish agricultural laborers into the East Elban landed estates—while Thomas was studying the consequences for the Polish immigrants of their movement into the United States. There is no evidence that they were aware of each other's work.

26. Thomas' *Source Book for Social Origins* contains nothing by Durkheim and there are only two references to articles by Durkheim in a bibliography of forty-two pages.

27. I knew him well in the period just after his retirement from the University of Chicago and often used to talk with him, or rather listen to him, but he never mentioned Durkheim or Weber, even though he knew I was more or less conversant with European sociology. I attended his last lectures at the University of Chicago in the spring of 1934, and, although Bagehot, Tarde, Sighele, and all sorts of books including Theodore Geiger's *Die Masse und ihre Aktion* were mentioned, Durkheim was passed over without a word. There were, however, two excerpts from Durkheim in *Introduction to the Science of Sociology*.

28. I am also not certain about how Elton Mayo came to be interested in Durkheim. Perhaps via Radcliffe-Brown or Lloyd Warner.

29. Although Durkheim disregarded foreign writers who called themselves sociologists, just as he disregarded French *soi-disant* sociologists, he did make much use of ethnographic work from Germany, Great Britain, and to some extent, the United States. Spencer and Gillen on Australia was the factual foundation of *Les Formes élémentaires de la religion*.

30. I refer here to the awkwardness of present "elite studies." Attempts to legitimate such studies are sometimes made by associating them with the study of power and authority—very unconvincingly—or with the "circulation of elites," which likewise has not found thus far a comfortable home in contemporary sociology.

31. The variant of the Darwinian tradition elaborated by Ratzenhofer and Gumplowicz had an enduring after-life via Albion Small, W. I. Thomas, and Robert Park. The affinity of this conception of conflict to competition in the animal and plant worlds had a powerful effect on Chicago sociology. It also had a pronounced influence on Arthur Bentley and through him on American political sociology.

32. A characteristic allegation of this sort is combined in the essay by Professor Ralf Dahrendorf, "Out of Utopia," *American Journal of Sociology*, 61 (1958).

33. It must be remembered that what is institutionalized is an intellectual process and that an account of the process is not exhausted by an account of the institutional setting in which it occurs. There is the sheer power of intelligence in confrontation with problems. The problems can be perceived through a number of mechanisms. They can be contained in the intellectual works which a searching, questing intelligence has discovered—either by its own curiosity or by moving along paths which it has made for itself in only very lightly chartered territory. Intellectual works themselves are linked with other intellectual works contained in them by references to other books and authors. Even without the aid of the contemporary institutional elaboration of teaching, bibliographical services, libraries, and learned societies, a seeking intelligence with the sensitivity of an Indian tracker could find like-minded persons across the stretches of a relatively unorganized Europe. Nonetheless, the task was more difficult before institutionalization through universities and learned journals.

34. During the depression years of the 1930's, interesting investigations were made into the effects of unemployment on individual morale, family life, and so forth.

Such studies were dropped from the agenda of sociological research when nearly full employment replaced massive unemployment.

35. The collection of essays is different from the "reader," which in sociology is a more "scientific" type of textbook than the conventional compendium of discursive expositions of concepts and illustrative data. The "reader" as a textbook is usually a collection of research reports which have appeared as articles in journals. The editors, unlike the authors of textbooks in scientific subjects, do not consolidate existing knowledge as it bears on particular problems; they simply present the sources—journal articles—from which in scientific textbooks an interpretative consolidation is made.

36. The tendency, now widespread among social scientists, to distribute their writings in mimeographed, dittoed, or xeroxed form of publication is not an effort to inform the rest of the social science community of results achieved and verified, but rather the opposite. The distributed paper is a preliminary version, and the distribution is an effort to elicit criticisms; it is an acknowledgment of the tentative and uncertain status of what is presented in the papers circulated.

37. One of the reasons for the sterility of much of experimental social psychology is that it has become an isolated "specialism" which is content to remain within the limits fixed by the identities of men and pigeons.

38. One of the difficulties is that we cannot imagine anything beyond variations on the themes set by the great figures of nineteenth- and twentieth-century sociology. The fact that the conception of "post-industrial society" is an amalgam of what St. Simon, Comte, Tocqueville, and Weber furnished to our imaginations is evidence that we are confined to an ambiguously defined circle which is more impermeable than it ought to be.

39. A. H. Halsey, *Encounter* (April 1970).

TALCOTT PARSONS

On Building Social System Theory: A Personal History

THE EDITORS have urged contributors to be autobiographical and to write informally. In this vein, I may start by saying that the main subject matter will be the evolution of a contribution to the generalized theoretical analysis of the phenomena of human action, with special concern for its social aspects—that is, the theory of the social system. This has entailed combining elements from a variety of sources, combinations that are not frequently tapped by persons more inclined to disciplinary specialization.

I was perhaps predisposed in this direction by a highly unorthodox education. As an undergraduate at Amherst, influenced by an older brother who had gone into medicine, I had intended to concentrate in biology with a view either to graduate work in that field or to a medical career. But in 1923, my junior year, I was converted to social science under the influence especially of the unorthodox "institutional economist," Walton Hamilton. Then all concentration plans were disorganized by the dismissal, at the end of my junior year, of Alexander Meiklejohn as president of the college. None of the teachers whose courses I had elected was on hand the next fall. I made do with more courses in biology, some in philosophy, including one on Kant's *Critique of Pure Reason,* and some in English literature.

From the beginning I assumed I would pursue graduate study. But although sociology in a broad and vague sense attracted me, the regular American graduate programs, though I knew relatively little about them, did not. When an uncle offered to finance a year of study abroad, I elected to go to the London School of Economics. There the names of L. T. Hobhouse, R. H. Tawney, and Harold Laski particularly attracted me. Only after arriving did I discover the man who proved the most important to me intellectually—Bronislaw Malinowski, the social anthropologist.

I was not a degree candidate at London and my plans were unformed, so I was ripe to seize the offer of an exchange fellowship to Germany, for which I was recommended by Otto Manthey-Zorn, with whom

I had had a seminar in German philosophy at Amherst and who also had much to do with my appointment, a year later, as instructor in Economics at Amherst. I was assigned to Heidelberg with no personal voice in the matter, but this located me where the influence of Max Weber, who had died five years before, was strongest. It is significant that I do not remember having heard his name either at Amherst or in London.

Weber's work, especially *The Protestant Ethic and the Spirit of Capitalism* (which a few years later I translated into English[1]), immediately made a strong impression on me. On going to Heidelberg, I had had no intention of taking a degree, but I subsequently learned that this could be accomplished with only three semesters' credit, oral examinations, and a dissertation. I decided to write a dissertation under the direction of Edgar Salin (later of Basel, Switzerland) on "The Concept of Capitalism in Recent German Literature." I began with a discussion of Karl Marx, said something about some less central figures such as Lujo Brentano, and then concentrated on Werner Sombart (author of the huge work *Der moderne Kapitalismus*[2]) and Max Weber. This work crystallized two primary foci of my future intellectual interests: first, the nature of capitalism as a socioeconomic system and, second, the work of Weber as a social theorist.

During the year of teaching at Amherst, which allowed time for a good deal of work on my dissertation, it gradually became clear to me that I wanted to go thoroughly into the relations between economic and sociological theory. I became particularly indebted to discussions with Richard Meriam, who had come to Amherst as chairman of the Department of Economics after my graduation. Meriam persuaded me that, although economic theory was one of my examination subjects at Heidelberg, I needed to know much more about it, and I decided to make acquiring this knowledge my next step. Although a German Dr. Phil. was not the equivalent of a good American Ph.D., I decided not to become a candidate for the latter. Meriam recommended that I go to Harvard and arranged an instructorship appointment for me there for the fall of 1927.

Allyn Young, perhaps the most important man for my interests, had just gone to England, but I was fortunate in having contact with Harvard economists F. W. Taussig, T. N. Carver, W. Z. Ripley, and Joseph Schumpeter (who had a visiting appointment, though later he came permanently to Harvard). Edwin F. Gay, the economic historian, was thoroughly familiar with the German background of my Heidelberg training and sympathetic to my interests.

Meriam was quite right in maintaining that the knowledge of economic theory I could acquire at Harvard was far superior to that I had learned at Heidelberg. It gradually became clear to me that economic theory should be conceived as standing within some sort of theoretical matrix in which sociological theory also was included. I tried my hand at statements of this idea in some articles that Taussig kindly published in the *Quarterly Journal of Economics*,[3] of which he was then editor. Most important, however, was a decision to explore this theme in the work of Alfred Marshall, who at that time was the dominant influence in "orthodox" or "neoclassical" economic theory, with a view to extracting Marshall's "sociology" and its articulation with his strictly economic theory. The results, published in 1931-1932,[4] crystallized the first stage of a theoretical orientation that seemed to me to go beyond the levels attained by my teachers in articulating the theoretical structures of the two disciplines.

Contact with Schumpeter was particularly helpful in laying the groundwork for this development because, on the problem of the scope of economic theory, he was a strict constructionist in contrast to Marshall's refusal to draw any sharp boundaries. Knowledge of Vilfredo Pareto, gained both on my own and through contact with L. J. Henderson, was also especially important because Pareto had been an eminent economic theorist, a good deal in the same tradition as Schumpeter, but at the same time he had attempted the formulation of a more comprehensive system of sociological theory which, in his view, included a rather strictly defined economic theory.[5] Thus both Schumpeter and Pareto served as critical reference points from which to attempt to distinguish between the economic and the sociological components of Marshall's thought.

From this starting point there gradually evolved the project of including not only Marshall and Pareto—on whom I had written a monograph-length analytical paper shortly after completing the Marshall study[6]—but also Weber and Durkheim, in a comprehensive study of this group of "recent European writers." Weber's general ideas on the nature of modern capitalism, which had been the principal focus of my dissertation, and more specifically his conception of the role of the ethic of ascetic Protestantism, provided sufficient ground for hope that a Marshall-Pareto-Weber "convergence" could be worked out.

Increasingly, it seemed to be desirable to include Durkheim in the picture, but this presented greater difficulties. Of the four, Durkheim had had by far the least to do with economics as a technical discipline. Moreover, I never had a teacher on Durkheim in the same sense that my Heidelberg teachers were such on Weber, Taussig and Schumpeter were on Marshall, and Schumpeter and Henderson were on Pareto. Furthermore the exposure to Durkheim I had been given, especially by Ginsberg and Malinowski in London, was not only not very helpful, but also positively misleading: much unlearning of what was not true about Durkheim became necessary. The clue, however, was available. It lay in Durkheim's first major work, *De la division du travail social* (1893),[7] which interestingly was seldom mentioned in the English language secondary literature of that period. Careful study of this book showed that its analysis could indeed be very directly articulated with Weber's analysis of capitalism, and that in turn with Marshall's conception of free enterprise. The theory as such then articulated with the sociological, rather than the strictly economic, components of the work of Pareto and Weber and, more indirectly, Marshall. The key conceptual complex concerned the institutional framework of property and especially contract as distinguished from the "dynamics" of economic activity as such, and as constituting, for its understanding in the theoretical sense, a field for sociological rather than economic investigation.

A First Major Synthesis

The outcome of this complex series of studies was *The Structure of Social Action*, published in 1937 but completed in first draft—though substantially revised afterward—nearly two years before.[8] The book was meant as a study of the writers' ideas about the modern socioeconomic order, capitalism, free enterprise, and so forth, and, at the same time, of the theoretical framework in terms of which these ideas and interpretations had been formulated. In this respect the most immediate interpretive thesis was that the four—and they did not stand alone—had *converged* on what was essentially a single conceptual scheme. In the intellectual milieu of the time this was by no means simple common sense, but as it unfolded it surprised even me a great deal.[9]

To arrive at this conclusion required three resources in particular. One, of course, was careful critical study of the rather large body of relevant texts, as well as secondary literature, though most of the latter was worse than useless. The second was the development of a theoretical scheme that would be adequate to the interpretation of these materials. The third, finally, in a sense lay back of that. It included a kind of philosophy of science orientation about which a few things ought to be said.

All persons aspiring to some degree of sophistication in intellectual fields, since long before the 1920's when these problems became salient to me, had to develop some conception of the nature and conditions of empirical knowledge, and in particular of the nature and role of theory in that knowledge. I was introduced to such problems partly through courses in empirical science, especially biology, and partly through philosophy, including, as I have noted, an intensive course in Kant's *Critique of Pure Reason*.[10] The Heidelberg experience carried me substantially farther, particularly in studies of the issues involved with Weber's *Wissenschaftslehre.* Notable among these were, first, the problems centering around the German historical traditions and hence the status of generalized theoretical conceptualization in the social and cultural disciplines, and, second, those of the status of the interpretation of subjective meanings and motive in the analysis of human action, what the Germans called the problem of *Verstehen.*

Returning to this country I found behaviorism so rampant that anyone who believed in the scientific validity of the interpretation of subjective states of mind was often held to be fatuously naïve. Also rampant was what I called "empiricism," namely the idea that scientific knowledge was a total reflection of the "reality out there" and even selection was alleged to be illegitimate.

Weber had insisted on the inevitability and cognitive validity of selection among available factual information. The importance of analytical abstraction was further strongly emphasized by Henderson in his formula: "A fact is a statement about experience in terms of a conceptual scheme."[11] The culmination of this conception came for me in A. N. Whitehead's work, especially his *Science and the Modern World,* including his illuminating discussion of the "fallacy of misplaced concreteness."[12] Through channels such as these I arrived at a conception which I called "analytical realism," which treated the kind of theory I was interested in as inherently abstract but by no means as "fictitious" in the sense of Hans Vaihiger.[13] This seemed to me in particular to fit the treatment of the status of economic theory by Schumpeter and Pareto. I also found various of James B. Conant's writings on the nature of science, especially the role of theory, very helpful.

Closely related to this was the conception of "system." Schumpeter and Whitehead were important in forming the background of this concept, but I think it crystallized above all in the influence of Pareto and Henderson. As Henderson was never tired of pointing out, Pareto's model in this respect was the idea of system as used in the theory of mechanics, but he attempted to apply it both to economics and to sociology. Hence Henderson's statement that perhaps Pareto's most important contribution to sociology was his conception of the "social system," a dictum which I myself took so seriously that I used the phrase as the title of a book some years later.

Henderson's own primary model, which he explicated at some length in *Pareto's General Sociology,* was the physico-chemical system.[14] He related this, however, to biological systems. He was a great admirer of Claude Bernard and wrote a foreword to the English translation of the latter's *Experimental Medicine.*[15] The central idea here was that of the "internal environment" and its stability. This connected closely with the idea of W. B. Cannon of homeostatic stabilization of physiological processes and also with the residues of my own exposure to biology.[16]

Thus a certain foundation for the transition from the concept of system as used in mechanics and the physico-chemical system as explicated by Henderson to that of the special character of "living systems" was laid in these early years. This was essential to the later phase of my thinking which is usually referred to as "structural-functional" theory and culminated in my book *The Social System.*[17] Further steps were greatly influenced by a continuing Conference on System Theory held from about 1952 to 1957 under the chairmanship of Dr. Roy Grinker in Chicago. Among the several participants whose ideas were important to me, the social insect biologist Alfred Emerson stands out. What he had to say, including some of his writings, strongly reinforced my predilection for the homeostatic point of view of Cannon. He spoke, however, in such a way as strongly to predispose me, and I think others, in favor of the then just emerging conceptions of cybernetic control, not only in living systems but also in many other kinds of systems. This later became a dominant theme in my thinking.

Finally, Emerson put forth what turned out for me to be an especially fruitful conception which did much to seal my conviction of the fundamental continuity between the living systems of the organic world and those of the human sociocultural world.[18] This was the conception of the functional equivalence of the gene and, as he put it, the "symbol." We can perhaps restate these as the genetic constitution of species and organism and the cultural heritage of social systems. In more recent years this perspective has come to be of fundamental theoretical importance to me.

Personal and Professional Concerns

The Structure of Social Action marked a major turning point in my professional career. Its major accomplishment, the demonstration of the convergence among the four authors with which it dealt, was accompanied by a clarification and development of my own thought about the problems of the state of Western society with which the authors were concerned. The state of Western society which might be designated as either capitalism or free enterprise—and on the political side as democracy—was clearly then in some kind of state of crisis. The Russian Revolution and the emergence of the first socialist state as controlled by the Communist party had been crucial to my thinking since undergraduate days. The Fascist movements affected friendships in Germany. Less than two years after the publication of the book the Second World War was to begin, and, finally, came the Great Depression with its ramifications throughout the world.

My personal concerns involved a growing family[19]—three children born from 1930 to 1936—and difficulties of status at the professional level. Though the situation was not strictly comparable to what it would be now, it was even then anomalous that I remained at the instructor level for nine years, the first four in the Department of Economics, the last five in the newly constituted Department of Sociology. I had the misfortune of serving under unsympathetic chairmen—in economics the late H. H. Burbank, in sociology P. A. Sorokin. My promotion to an assistant professorship occurred in 1936 and was pushed not by Sorokin but, notably, by E. F. Gay, E. B. Wilson, and Henderson, all of whom were "outside members" of the Department of Sociology. The first draft of *The Structure of Social Action* was in existence by then, and known to all the principals.

Even with the assistant professorship, however, I was by no means certain I wanted to stay at Harvard. In the (to me) critical year of 1937 I received a very good outside offer. Since Gay had retired the previous year and gone to California, I went to Henderson—not Sorokin. In those days before the ad hoc committee system, Henderson took the matter directly to President Conant who, with Sorokin's consent, to be sure, offered to advance me immediately to the then extant "second term" assistant professorship, with a definite promise of permanency as associate professor two years later. On those terms I decided to stay at Harvard.

On the intellectual side I have noted that I had very good relations with Taussig, Schumpeter, and Gay. The above crisis occurred shortly after the completion of an extraordinary relationship with Henderson. I had known him through his Pareto seminar and in other respects before the manuscript of my book was referred to him for critical comment in connection with my appointment status. Instead of the usual limited response, he got in touch with me (stimulated I think mainly by my discussion of Pareto) and started a long series of personal sessions at his house, something like two hours, twice a week for nearly three months. In these sessions he went through the manuscript with me paragraph by paragraph, mainly the sections dealing with Pareto and Durkheim; he passed over Marshall rather quickly and did not enter into my treatment of Weber at all.

This was an extraordinary experience, both personally and intellectually. Those who knew him will remember that Henderson was a formidable person, who could be dogmatic, both on political matters, in which he was a pronounced conservative,[20] and on many intellectual matters, as exemplified by his unfairness—as I judged it—to all but one or two sociologists. He had, however, an immense knowledge of science, especially on the level of the philosophy of science and the nature of theory, and if one stood up to him and did not allow oneself to be overridden, he was an exceedingly insightful critic and very helpful with precisely *my* intellectual problems. I benefitted enormously from this special contact, and took about a year to complete the revisions which the discussions convinced me should be undertaken.[21]

In these early Harvard years, I also had important experiences with age-peers and, in course of time, students. A group of junior faculty members that met fairly regularly included Edward S. Mason, Seymour Harris, Edward Chamberlin, and for a time Karl Bigelow in economics, Carl J. Friedrich in government, and Crane Brinton in history. With the move to sociology, which brought me closer to psychology and anthropology, I began to get to know Gordon Allport, who had recently returned to Harvard from Dartmouth, and Henry Murray. In anthropology, there were two particularly significant contemporaries. The first was W. Lloyd Warner, brought to Harvard mainly by Elton Mayo, who, under Henderson, directed the Western Electric Study, to make a community study that eventually became the well-known "Yankee City Series." When Warner left Harvard to go to Chicago he was replaced by Clyde Kluckhohn, a young social anthropologist who was quite independent of the Henderson group but associated with them. He became a close friend of Murray, and with Allport, Murray, and Kluckhohn the nucleus of the eventual promoters of the Social Relations experiment was present.

Also by the mid-thirties significant relations with graduate students, a few of whom received teaching appointments, began. The most important single one was with Robert Merton, who was in the first contingent of graduate students in sociology, but after him came Kingsley Davis, John Riley, and—not as a student—Mathilda Riley, Robin Williams, Edward C. Devereux, Logan Wilson, Wilbert Moore, Florence Kluckhohn, and Bernard Barber. An informal discussion group on problems of sociological theory composed largely of these people met in the evenings in my study in Adams House, while I was still an instructor.

Theoretical Interests after The Structure of Social Action

Completion of *The Structure of Social Action* constituted a gratifying accomplishment, though at that time I was not aware how large a reputation it would achieve.[22] The theoretical framework which had enabled me to demonstrate the thesis of convergence clearly had the potential of further use and development, but there were several alternatives of what to attempt next with its help. In the Bellagio conference, where this essay was first presented, there was considerable discussion of why I did not identify with economics as a discipline. By the time

The Structure of Social Action neared completion that decision had already been made on one level by my transfer from economics to the new Department of Sociology. In spite of the friendliness of Taussig, Gay, and Schumpeter, I am quite sure I could not have counted on a future in economics at Harvard. But basically I did not want to do so, and in retrospect the focus of the reasons lay in my permeation with Weber's thinking, then Durkheim's: Freud was yet to come. Though I wanted to keep my contact with economic theory, and in fact have done so in various ways, I saw clearly that I did not want to be primarily an economist, any more than Weber turned out to be.

There was one interesting episode which might, at a relatively late time, have turned me at least farther in the direction of economics. After my formal transfer to sociology, Schumpeter organized a small discussion group with younger people, mostly graduate students, on problems of the nature of rationality. After a few meetings he proposed to me that the group should aim at producing a volume, of which he and I should be at least coeditors, if not coauthors. Though not specifically rejecting the proposal, at least immediately, I remember having reacted rather coolly, and in fact I let it die. I am not wholly clear about my motives, but I think they had to do with the feeling that I needed a relatively complete formal break with economics.[23]

The Professions and the Two Aspects of the Rationality Problem

The actual choice I made was to undertake a study of some aspect of the professions as social phenomena. This interest grew logically out of the combination of my concern with the nature of modern industrial society and the conceptual framework in which I had approached it. It was empirically nearly obvious that the "learned professions" had come to occupy a salient position in modern society, whereas in the ideological statement of the alternatives, capitalism versus socialism, they did not figure at all. Indeed, what is now habitually called the "private, nonprofit" sector of organization and activity which is occupationally organized, as distinct, for example, from kinship, did not figure ideologically. In retrospect it can be said that both ideological positions stated versions of the "rational pursuit of self-interest"—the capitalist version, grounded in utilitarian thought, the interest of the individual in the satisfaction of his wants, the socialist version, the interest of the collectivity (on lines deriving from Hobbes and Austin) in maximization of satisfaction of the public interest.

Within this field I chose to study some aspects of medical practice. This choice had, I think, good technical grounds, but it also goes back to personal motives. Certainly I could say that my previous renunciation of biological-medical interests played a part: the role of a sociological student of medical practice was an important way of satisfying both motives. The Henderson-Mayo group, however, also had a bearing. Henderson was himself medically trained, though he never practiced, and his early Harvard appointments were in the Medical School. He had combined his medical and sociological interests in a famous paper, "Physician and Patient as a Social System,"[24] which stated an approach highly congenial to me; hence it was natural that I should consult Henderson and Mayo—but also W. B. Cannon—about my plan of work. All three strongly reinforced my own feeling of the probable fruitfulness of such a study. I proposed, in addition to canvassing the literature, to approach it through the methods of participant observation and interview. The semi-public character of medical practice in the modern hospital made a good deal of the former possible, such as—equipped with a white coat and the (albeit nonmedical) legitimate title of doctor—making ward rounds, observing operations, going on the home visit service of the Tufts medical center and the like. (Perhaps, with current concern over ethical aspects of research, this mild deception of patients would now be regarded as unethical.) The other source of data was a series of interviews with a rather large number of physicians, chosen to represent different types of practice.

One other aspect of the situation involved in this choice turned out to be of primary significance. This was a time in which ideas about psychosomatic relations were beginning to take hold among the intellectual elite of medicine—typified perhaps by internists at the Massachusetts General Hospital, where I spent a good deal of time. In the background was psychoanalysis and the fact that the professor of psychiatry at Massachusetts General Hospital, Stanley Cobb, had recently been the principal founder of the Psychoanalytic Institute in Boston. The Henderson-Mayo group had also been very much concerned with this and related movements of thought—they were devotees of Pierre Janet, but also of Jean Piaget.

For me the decisive event was a talk with Elton Mayo about my interests in medical practice, in which he asked point blank how well I knew the work of Freud. My reply had to be, only very fragmentarily. He then earnestly advised me to read Freud seriously and comprehensively. This was fortunately a time when I had a good deal of free time, thanks to an assistant professor's term leave, and I followed his advice. It was too late to build the implications of Freud's ideas into *The Structure of Social Action*, but this proved to be one of the few crucial intellectual experiences of my life. This, of course, prepared the way for formal psychoanalytic training—at the level permitted—about a decade later.

I used the economic paradigm of the "rational pursuit of self-interest" as the major point of reference, but in this case negatively, to throw light on the *differences* between the classical economic model of market orientation and the professional case with which I was concerned. Major differences were not far to seek. In the individual practitioner they were, first, fee-for-service relation to patients' situations—the so-called "sliding scale" or higher charges for well-to-do patients and lower or remission of charges for less well-to-do patients. And, second, the objection to "shopping around"—namely patients testing out various physicians with respect to their judgment of the value of the proferred medical service, financially and otherwise. In later years I was to modify substantially this close assimilation—with all the differences—of the professional relation to the ideal type commercial one.

The most important theoretical implications concerned the problem of the nature of rationality, the very question with which I had been involved, not only in my own work, but also in my work with Schumpeter. What opened up was a distinction, not only between economic and noneconomic aspects of rationality, but also, within the latter category, between two different modes or directions of considering the rationality problem. The first concerned a very old problem, even for me, namely the relation between rational (mainly scientific) knowledge and action in the sense of "application." Medicine, especially at the time I was studying it, was a kind of prototype of the possibilities of generating potentially useful knowledge and applying it to the solution of critical human problems. What was called "scientific medicine" then enjoyed a sort of heyday, and its significance was strongly impressed upon me by my brother, who had been trained at Johns Hopkins. There were, of course, very definite connections between this aspect of medicine and the more general setting of the rationality problem in *The Structure of Social Action*, notably in Pareto's concept of "logical action."

The concern with psychosomatic problems, and eventually those of mental illness, noted above, raised a different set of questions. These concerned the significance of the scientific modes of rational investigation and analysis for the understanding, and in some eventual sense control, of the non- and irrational factors in the determination of human action, in the first instance individual, but clearly also social. Concern with these issues permeated the thought of all of my authors except Marshall, but intensive contact with Freud rounded out the pattern and gave new dimensions to it, notably concerning the relevance of nonrational factors and mechanisms in the more intimate microsocial processes of interaction. I suspect that concern with this complex of problems had a good deal to do with my coolness toward Schumpeter's overture, flattering as that was for a young and still insecure scholar.

In this type of interest I was clearly influenced by my growing association with such colleagues as Clyde Kluckhohn and Henry Murray; Gordon Allport, by contrast, was a psychologist of pronouncedly rationalistic bent. At any rate, *both* facets of the "rationality complex" happened to be combined in my study of medical practice and both stood in sharp contrast to my earlier concentration on the economic-political aspects of rationality. In the background stood concern for the status of religion in any general analysis of social action—a concern which I had come by honestly as a result of my family background and which was brought into focus by Weber's analysis of the Protestant Ethic and his more general studies on the comparative sociology of religion.

As I see it now, these three (or, including religion, four) foci of the "rationality problem" have been nearly dominant in the structuring of my theoretical interests since this crossroads situation of the late 1930's. The first main swing was away from the economic-political complex toward the socio-psychological, that is, the problem of nonrationality seen more from the perspective of Freud than that of Weber or Pareto, different as these two were from each other. In the course of this move I was fully conscious of the importance of "cognitive rationality" in particular as constituting the cultural basis of the scientific component of medicine. Intensive concern with this, however, was to await my coming around to concern with the anchorage of the professions in the system of higher education and research, a concern that has become central to me in recent years. In a certain sense it is a case of the "return of the repressed."

In the context of the social-emotional aspect of medical practice, I came to analyze some of the phenomena which were then commonly called the "art of medicine" (as contrasted with the "science") in terms of Freud's conception of the relation between analyst and analysand, notably the phenomenon of transference, which I regard as one of Freud's great discoveries. Freud clearly had not invented the physician-patient relationship, which in Western tradition goes back at least to Hippocrates (vide Henderson), but he had accepted it as his primary social framework for psychoanalytic practice and immensely deepened the basis of its understanding. It became clear that the psychoanalytic relation presented the extreme, and, hence, in a kind of limiting sense paradigmatic, example of this relationship, and that the vast and vague range of psychosomatic relationships fitted into this. There was, of course, a basic link between the two primary aspects of the rationality problem of interest here, namely via the aspiration of psychoanalysis to scientific status, which in my opinion in spite of much controversy has been broadly validated and is perhaps in the context of application best symbolized for the process of therapy by Freud's aphorism "where Id was, there shall Ego be."

It was mainly by this path that I came to think of illness as in some sense a form of social "deviance" and of therapy as belonging in a very broad band of types of "social control"—a view for which I have paid dearly in the accusation of being an agent of Establishment interest in maintenance of the status quo. There is, nevertheless, a certain element of truth in this position, truth which I think to be basically independent of the particular form of social order. The important theoretical point is the shift from considering the application of medical science as a biophysical technology only, to considering it as a field of social interaction as well. Put in more technical terms which were crystallized later, the traditional view of medical practice conceived it as a relation between cultural systems (scientific knowledge) and organisms, with social agents only implementing the obvious implications of knowledge, whereas the other perspective involved treating medical relationships, at least in part, as cases of the subtle interplay between unconscious motives at the personality level and particularities of the structure of social systems. There are further sociological correlates of the difference which involve a duality of levels—in the medical case between physician as competent agent of social control and patient as recipient of these important services—which cannot be gone into in detail here. This turns out, however, to be a major focus of social structure which does not have a place either in the predominantly economic or the political models (capitalism and socialism).

From Medical Practice to the Theory of Socialization

In a period of my career so much under the influence of Freud, it was perhaps natural to shift interest from the analysis of the social situation of psychoanalysis, and more generally of medical practice, to that of the origins of the problems which analyst and analysand confronted. These clearly were problems in the first instance in the personalities of analysands—the bases of "countertransference" were analyzed somewhat later—and for the sociologist led to the consideration of conditions of child development in the family as a social system. Freud himself had increasingly laid stress on "object relations" but could scarcely be said to have developed an adequate sociology of the family. In this connection a major conception on which there had been impressive convergence came to play a central role. In my reading of Freud, I gradually realized the importance of what I and others came to call the phenomenon of "internalization" (Freud's own term was "introjection")—both of sociocultural norms and of the personalities of the others with whom an individual has been interacting, above all a "socializing agent" (the latter case sometimes being called "identification").

This idea first emerged clearly in Freud's thought in the conception of the superego, though one may say it was present from an early stage, especially in the conception of transference (for example, the treatment of the analyst as if he were the analysand's father). Freud came to consider moral standards, particularly as implemented by the father, integral parts of the child's personality, through some phases of the learning process. Gradually the scope of this aspect of Freud's theory of "object relations" widened in the course of his later work to include not only the superego, but also the ego and even the id.[25] At about the same time, it became clear to me that a very similar conception, developed from a quite different point of view, was essential to Durkheim, especially in his conception of social control through moral authority. The idea was also at least implicit in Weber's treatment of the role of religious values in determining behavior, and appeared with great clarity in the work of a group of American social psychologists, notably G. H. Mead and W. I. Thomas. The conception of the internalization, in successive series, of sets of cultural norms and concrete social objects has become a major axis of the whole theory of socialization, figuring in new forms even in the most recent treatments of the problems of higher education.

Internalization is a feature of the structure of the personality as a system. I have called the parallel phenomenon for social systems institutionalization, especially the shaping of social relationships by components of normative culture, which come to constitute direct structural parts of the social system of reference. Perhaps the preeminent theorist of this line of analysis was Weber, especially in his comparative sociology of religion, but Durkheim made contributions of similar importance. Both these conceptions, furthermore, could make sense only if the primary subsystems of a general system of action were conceived to be *interpenetrating* as well as interdependent. Thus, certain components of the cultural system came to be *at the same time* components of certain social and personality systems. This very central conception, in turn, depended heavily on the conception of the abstractness of the subsystems under consideration. Thus a social system—for example, a "society"—is not a concrete entity, but a way of establishing certain relations among components of "action" that are distinctive relative to the manifold of concrete reality.

Concern with this set of problems of the nonrational was certainly reinforced by the circumstances of the times. Discussions of the problem of German character, in connection with which Erik Erikson first became a salient figure to me, were important at this time.[26] So were personal matters, including my medical brother's premature death (1940) and the aging and eventual deaths of my parents (1943 and 1944).

This set of circumstances seems to me to provide the principal explanation of a major failure of my career: not to carry through my intention to publish a major monographic study of medical practice. I had, I think, gained immensely from the venture, but had been diverted into more general concerns with the subtler aspects of social control and the genesis of the problems in socialization processes outside the professional context.[27] In any case, I did stall on completion of this enterprise and, in addition to a good many discussions of aspects of the problems in various papers, settled for a longish chapter, "The Case of Modern Medical Practice," in *The Social System*.

Starting, perhaps, in the later 1940's there was a shift back from concern with psychological and microsociological problems to the more macrosociological, including the economic, but also perhaps a renewed sense of participation in the European scene, which was inaugurated by teaching at the Salzburg Seminar in the summer of 1948. In a sense this process culminated in a revisit to the problems of the relation between economic and sociological theory which began during my year as a visiting professor at Cambridge, England (1953-1954).

In 1946, however, I entered into formal psychoanalytic training as a "Class C" candidate at the Boston Psychoanalytic Institute. The more general intellectual grounds for my interest are perhaps clear from the above discussion, though I also had some personal reasons for seeking psychotherapeutic help. I count myself exceedingly fortunate in having had as my training analyst Dr. Grete Bibring, who had been a member of the original Freud circle in Vienna until forced into exile by the Nazi takeover of Austria. Of course, without a medical degree I could not have aspired to practice psychoanalysis, and according to the rules in force at that time I was not permitted to take control cases; indeed, I was admitted to the clinical seminars only as a special concession. However, I never had any intention of practicing.

In addition to deepening my understanding of psychoanalytic theory and the phenomena with which it dealt, this experience had the effect of contributing to my "weaning" from an overconcern with the psychoanalytic level of dealing with human problems and was hence a kind of corrective to the effect of the original reading of Freud and the early phases of my study of medical practice. Both an efflorescence of concern with more abstract and analytical problems of theory and empirical concern with less psychological areas—for example, economic and political again, and later educational—began to grow in relative strength.

I was not one of the many Harvard faculty members who was called to war service away from Cambridge. I did, however, do some teaching in the School for Overseas Administration, of which my friend Carl Friedrich was director, which administered Area and Language programs and programs for Civil Affairs officers. I taught in the fields of

European and East Asian societies. During the latter part of the war period I served as a consultant to the Enemy Branch of the Foreign Economic Administration, dealing with the postwar treatment of Germany. I wrote several memoranda in opposition to the so-called Morgenthau Plan.

In 1944, in response in part to a very good offer from outside Harvard, I was appointed chairman of the Department of Sociology, with something of an understanding that important reorganization would take place soon. Allport, Kluckhohn, Murray and I had had important discussions on the possibility of reorganization. In 1945 there were two openings for permanent appointments in sociology. One of them went to George C. Homans, who had been in the department before leaving to serve in the Navy. The other went to Samuel A. Stouffer, who was just winding up his eminent service as director of research in the Information and Education Branch of the War Department. These events, late in 1945, made possible the faculty's action in establishing the Department of Social Relations, which opened in the fall of 1946. Stouffer became director of the Laboratory of Social Relations, an associated research organization, and I became chairman of the department, which included, besides sociology, social anthropology and social and clinical psychology. I continued as chairman for ten years until 1956. The role of Paul H. Buck, dean of the Faculty of Arts and Sciences and provost, was of outstanding importance in the establishment and development of the Department of Social Relations.

During this period I also became more active in professional affairs outside Harvard. I had served in 1942 as president of the Eastern Sociological Society, but as it was a war year the post did not entail much activity. I was elected president of the American Sociological Association for 1949 and this, of course, proved to be a much bigger job. The association was then undergoing a major organizational crisis occasioned mainly by growth in membership and activities. During my term a new office was set up, the first paid executive officer was appointed, and the association's constitution was substantially revised. After an interval of a few years I again became active in the affairs of the association, serving as chairman of the Committee on the Profession, then as secretary for five years, and finally for two years as the first editor of *The American Sociologist,* a house organ of the association dealing with matters of concern to the profession rather than "contributions to knowledge." In the 1950's I was also active in the American Association of University Professors, serving on the special committee on loyalty-security cases and for one term each on the Council and the Committee on Academic Freedom and Tenure.

Theoretical Development, 1937-1951

For my role as teacher, especially at the graduate level, the early years of the Department of Social Relations came to be a true golden age. Opening just a year after the end of the war, the department attracted, aided by the G. I. bill, an unusually able sample of the backlog of young men whose training had been interrupted by the war. Among those who had already been at Harvard were Bernard Barber, Albert Cohen, Marion Levy, Henry Riecken, and Francis Sutton, with Robert Bales staying through the war. New arrivals included David Schneider, Harold Garfinkel, David Aberle, and Gardner Lindsey. A little later came James Olds, Morris Zelditch, Joseph Berger, Renée Fox, Clifford Geertz, François Bourricaud (from France on a Rockefeller Fellowship), Robert Bellah, Neil Smelser, Jackson Toby, Kaspar Naegele, Theodore Mills, Joseph Elder, Ezra Vogel, William Mitchell (in the Department of Government), Odd Ramsoy (from Norway), and Bengt Rundblatt (from Sweden).

In the late 1950's and early 1960's came a third wave of especially important graduate students, including Winston White, Leon Mayhew, Jan Loubser, Edward Laumann, Charles Ackerman, Enno Schwanenberg, Victor Lidz, Andrew Effrat, Rainer Baum, Mark Gould, John Akula, and, in a special collaborative relationship after completion of his formal studies, Gerald Platt. Close association with advanced students of such caliber has been one of the most rewarding features of my academic career. Such young minds cannot fail, it seems to me, to have a most stimulating effect on their teachers. My own experience has strongly reinforced my belief in the importance of *combining* the teaching and the research functions in the same organizations and roles.

A few of these direct collaborative relationships—with such people as Robert Bales, James Olds, Neil Smelser, Winston White, Victor Lidz, and Gerald Platt—resulted in collaborative publications. Other important relationships have been with David Schneider, Clifford Geertz, Leon Mayhew, and (though not as a formal student) my daughter, the late Anne Parsons.

In spite of the shifts noted above, there seems to be a certain unity in the period of intellectual interests and theoretical development running from completion of *The Structure of Social Action* to the two major books published in 1951, *Toward a General Theory of Action*, a collaborative work coedited with Edward Shils,[28] and my own *The Social System*. The most important thread of continuity lies, I think, in what came to be called the "pattern-variable" scheme.

This scheme originated as an attempt to formulate a theoretical approach to the interpretation of the professions. Clearly nothing in the capitalism-socialism dichotomy would do, so I turned to a famous distinction crystallized for German sociology by Toennies and used by Weber: that between *Gemeinschaft* and *Gesellschaft* as types of social organization.[29] The problem of self-interest was the initial point of reference, in a sense which posed an alternative far short of the total public interest in the socialistic sense. The professional orientation was, as I initially put it, "disinterested" (later "collectivity-oriented"), in the sense in which the physician professes to be above all concerned with the welfare of the patient. This criterion put the professions in the *Gemeinschaft* category.

The scientific component of medicine, the universalistic character of the knowledge applied to problems of illness, fitted with an extensive complex of features of modern society which Toennies and his numerous followers would have to classify as *Gesellschaft*. The obvious inference was that the Toennies dichotomy should not be treated only as variation in terms of a single variable, but also as a resultant of a plurality of independent variables. If these were indeed independent, there should be not just two fundamental types of social relationship, but a substantially larger family of such types. My suggestion was that the professional type belonged in this family but neither as *Gemeinschaft* nor as *Gesellschaft*. Important as the problem of self-interest was, it seems to me now that even more so was that of how to bring the universalism especially characteristic of cognitive rationality, and the problem of the status of nonrational emotion or affect, into the *same* analytical scheme. Quite early a dichotomous variable which I called "affectivity—affective neutrality" was formulated and incorporated in the same system which also included "Universalism—particularism."

The pattern-variable scheme underwent over the years rather complex vicissitudes which need not be detailed here. The project that produced the volume, *Toward a General Theory of Action*, however, produced a first genuine synthesis. It was initiated as a kind of theoretical stocktaking of what underlay the social relations experiment, and for this project Edward Shils and the psychologist E. C. Tolman were brought to Harvard on a visiting basis. Shils and I developed a particularly close collaboration from which came our joint monograph, *Values, Motives and Systems of Action,* which is, in a sense, the theoretical core of *Toward a General Theory of Action*. In it we developed the pattern-variable scheme as a theoretical framework not only, as I had initially assumed, for the theoretical analysis of social systems, but also for analysis of action in general, and, especially within our purview at the time, of personalities and of cultural systems. As such it was no longer a catalogue of dichotomous distinctions, but became very definitely a "system," which, however, contained seeds of much further complication than we were aware of at the time.

This generalization and systematization seemed to constitute a real theoretical breakthrough which emboldened me to attempt a general statement on my own responsibility of the nature of a social system, including most definitely the macrosocial levels. The outcome, my book *The Social System,* apart from its codification of received sociological wisdom, rested above all on two features which might be considered original. The first of these was the clarification of the relations between social systems, on the one hand, and psychological—or personality—and cultural systems, on the other. The second was the consciously systematic use of the pattern-variable scheme as the main theoretical framework for the analysis of the social system.

Economics and Sociology Reconsidered

The two 1951 books were in some ways a culmination, but even more important they were a foundation for a major new phase. Shils's and my contribution was organized about the pattern-variable scheme, which our collaborative work generalized from the social system level to that of action generally. Robert Bales[30] and I meanwhile had had many discussions of the relations of this scheme to his scheme for the analysis of small-group interaction. These discussions became so important that we invited Shils to work with us for the summer of 1952, and the three of us produced the *Working Papers in the Theory of Action* (1953).[31]

The crucial outcome, in light of subsequent developments, was the emergence of what we now call the "four-function paradigm." Its genesis lay in a convergence between the system comprising the four elementary pattern variables and a classification that Bales had set forth in his *Interaction Process Analysis.*[32] We concluded that systems of action generally could be exhaustively analyzed in terms of processes and structures referable to the solution—simultaneously or in sequence—of the four functional problems that we called "adaptation," "system (not unit) goal-attainment," "integration," and "pattern-maintenance and latent tension-management." Though there were many defects in our formulations at that time, this basic classification has remained with me for the more than fifteen years since it first emerged and has constituted a primary reference point of all my theoretical work.[33]

A consequence, closely related to Bales's work on small groups, was an extension of the analysis to the socialization process, harking back to my study of medical practice. This extension eventuated in the book, *Family, Socialization, and Interaction Process,*[34] written in collaboration with Bales, James Olds, and others. Its main theme was that the nuclear family, emergent in modern industrial societies, could be treated as a small group and differentiated according to the four-function paradigm, on the axes of generation and sex, in ways closely analogous to the pattern of differentiation of many of the small experimental groups with which Bales and his associates had been working. This was also, perhaps, the point at which I began to be intensely interested in the phenomenon of differentiation in living systems more generally. This interest was linked to my previous biological concerns and stressed the importance of "binary fission."[35]

This line of theorizing, a continuation of the concern with problems of nonrationality discussed above, was soon overshadowed by another which led me back to the old problem of the relations of economic and sociological theory. For the academic year 1953-1954 I was invited to serve as visiting professor of social theory at the University of Cambridge. While there I was invited to give the Marshall Lectures, sponsored by the Cambridge Department of Economics in memory of Alfred Marshall. The subject assigned to me was, specifically, the relation between economic and sociological theory.

For some years I had not been intensively concerned with problems of the status of economic theory. And, on accepting the assignment, I was not sure I could go much beyond the level attained in *The Structure of Social Action.* But the theoretical development in the interim, notably the four-function paradigm, turned out to have laid a basis for a quite new phase of analysis.

In preparation I studied thoroughly, for the first time, Keynes's *General Theory of Employment, Interest, and Money* and carefully reread large parts of Marshall's *Principles of Economics.*[36] It suddenly struck me that Marshall's extension of the basic classification of the factors of production and the shares of income from land, labor, and capital could, with the addition of Marshall's fourth factor, which he called "organization," to the three classical categories, be regarded as a classification respectively of inputs and outputs of the economy as a social system, analyzed in terms of the four-function paradigm.

This insight proved to be the starting point of a fundamental reconsideration of the problem, which was only partially carried out by the time of the delivery of the three lectures in November 1953. By unusual good luck, however, Neil Smelser, whom I had known as an undergraduate at Harvard, was then in the second year of his Rhodes scholarship at Oxford, studying economics. I sent him the manuscript of my lectures, and he responded with such detailed and pertinent comments that we arranged a series of discussions during the academic year in England. Then, the following fall, we were both back at Harvard and completed the collaboration which produced the book *Economy and Society.*[37]

We did, I think, succeed in working out a new and more generalized analysis of the relations of economic and sociological theory This analysis extended to the relation of the economy, conceived as a subsystem of a society, to the society as a whole. This theoretical restructuring, in addition, proved to be generalizable to a reconsideration of the other primary functional subsystems of societies. Hence it has opened up a whole new view of the structure and functioning of total social systems, of which a society is one particularly important type.

The key connection was that what economic theorists called the economy should be regarded as one of four primary functional subsystems of the society, with a primarily *adaptive* function—that is, as agent of the generation of generalized resources. Three of the factors of production and shares of income then should be regarded as, respectively, inputs from and outputs to each of the other three primary subsystems. The fourth pair—land and rent—should be treated as a special case, as it long had been in the theoretical tradition of economics. The key is the famous doctrine that the supply of land, unlike that of the other factors, is not a function of its price. This property fitted the logical requirements of the pattern-maintenance function, which we had been treating as the stabilized reference base of an action system. In the process we considerably revised the traditional economic conception of land to include not only natural resources, but also any economically significant resources that were unconditionally committed to the function of production in the economic sense, which included commitments at the value level to production. In our aspect, then, economic rationality became a category of value, not of psychological motivation.

Given that our assignments of sources of input and destinations of output among the other three subsystems were correct, and that we were able to work out comparably correct classifications and categorizations for the inputs and outputs of the other three primary subsystems, it has eventually proved possible to work out a complete "interchange paradigm" for the social system as a whole.[38] This has taken several years and has required a long series of conferences between Smelser and myself as well as with others.

This line of thought both introduced a new complication and opened up a new set of opportunities. The primary reference model of interchange for us became the Keynesian focus on the interchange between households and firms: the former are placed in the pattern-maintenance system, which made good sociological sense, the latter in the economy. Two, not four, categories, however, were involved: what economists have called "real" inputs and outputs and the monetary categories of wages and consumers' spending. This naturally raised questions on the status of money as a medium of exchange and on its other functions—for example, as measure and store of economic value.

Monetary theory has, of course, become increasingly central in the discipline of economics, but economists and others have tended to treat money as a unique phenomenon. If the idea of a generalized interchange paradigm for the social system as a whole made sense, however, it would seem to follow that money should be one member of a family of comparable generalized media; indeed there should be four of them for the social system.

It was not too difficult to work out some of the necessary aspects of the sociology of money for it to be treated in this way; but the other media presented greater difficulties. The first success came with the attempt to treat power in the political sense as such a medium, different from but comparable to money.[39] This entailed a far more substantial reordering of conceptualizations used by political theorists than those used by economists working in the monetary context. It necessitated introducing the conception of the "polity" as defined in abstract analytical terms parallel to that of the economy—hence not as government—and concerned with collective goal-attainment to the exclusion of integration as its primary societal function. Above all, the conception of power as a symbolic medium—which is parallel to the property of money as having value in exchange but not in use—was almost totally missing in political thought, where "intrinsic effectiveness" in the Hobbesian tradition has clearly been paramount as a reference point. Nevertheless, I think a coherent paradigm of power as such a medium has been worked out (see note 38). Once this step was taken, it was much easier to extend the analysis to include two other media: "influence" and "value-commitments," these terms being used in a specifically technical sense.[40]

Media of Interchange and Social Process

By this path the stimulus of the Marshall Lectures eventually opened up an approach to not only the structural, but also the processual analysis of social systems that promises to bring treatment of their noneconomic aspects to a level of theoretical sophistication more nearly comparable to that attained for the economy and that includes the dynamics of interrelation between these other subsystems and the economy. For example, the conceptions of inflation and deflation, as these have been used by economists, now seem to be generalizable to the other three societal media and their interrelations not only with money, but also with one another. One example of the many difficulties in working this out may be cited. The monetary dynamics referred to is clearly incompatible with the idea that money is a "zero-sum" phenomenon. Credit expansion and contraction are, of course, central features of monetary inflation and deflation. Political theorists, however, have been predominantly of the opinion that power should be treated in zero-sum terms. Hence, to make money and power comparable in this vital respect, it has been necessary to investigate the basis of this contention and to show that it was untenable for my purposes.

The conception of action systems and their relation to subsystems, which had crystallized in the four-function and the interchange paradigms, strongly suggested the desirability and importance of extending the analysis. In one direction this has been pursued in considerable detail—namely, to the "general action" level, as we have called it. A first stage was central to the two 1951 books and may be regarded as a development from the two facets of the problem of rationality that emerged in my study of medical practice. This stage treated the social system as being flanked, as it were, by the psychological or personality system on the one hand, and the cultural system on the other, and interdependent with and interpenetrating with both. The logic of the four-function paradigm gradually made it clear how the "behavioral organism"—not the total concrete organism—could and should be fitted in. This extension was facilitated by the revival and expansion of biological interests, particularly through contact with Alfred Emerson and by close association with James Olds, who had gone from social psychology into the field of brain research. The functional locations of the four subsystems of action are clear and stable. They are, namely, in the adaptive position, behavioral organism; in that of goal-attainment, personality; in the integrative position, the social system; and in that of pattern-maintenance, the cultural system.

This perspective has been prominent now for some years, and attempts at a first-order breakdown of the systems other than the social have been made (for example, the paper on psychological theory for the Koch symposium and, on the cultural system, the Introduction to Part 4 of *Theories of Society*).[41] Only recently, and so far only very tentatively, has it proved possible to work out a general interchange paradigm for the general-action level.[42] An interesting convergent outcome has emerged. The categories of generalized media that have tentatively been settled on can be identified with those introduced as the "four wishes and the definition of the situation" by the social psychologist W. I. Thomas more than a generation ago. They are, respectively: the adaptive medium, parallel to money at the social system level; intelligence, which can include, in its positive form, Thomas' "wish for new experience" and, in its negative aspect, his "wish for security"; the goal-attainment medium, performance capacity, rewarded by Thomas' "recognition"; the integrative medium, affect—roughly in the psychoanalytic sense—rewarded by Thomas' "response"; and, finally, as the medium of pattern-maintenance process, the "definition of the situation," which, as in the other pattern-maintenance phenomena, should be and is treated by Thomas as a special case.[43]

"Structural Functional Theory?"

The concept of system, in the action field as in others, has been central to my thinking from a very early stage. With it is associated an extensive complex of empirical-theoretical problems that have entered prominently into the critical discussions of this type of theory. These concern such conceptions as equilibrium and its relations to conditions of stability and possibilities and processes of change; the status of the concept function itself; problems of "consensus vs. conflict" as characteristics of social systems; and the relation between what may be called "maintenance processes" in systems and processes of structural change, extending all the way to the conception of evolution or its opposite.

Perhaps, if I may repeat a little, my first introduction to the equilibrium problem was in the form of the Henderson-Pareto version, reinforced as it applied to the economy by Schumpeter. This form used the concept system in the sense of mechanics, with the physico-chemical systems as the model. It emphasized the conditions of stability, though Henderson was careful to point out that Pareto's conception of equilibrium was by no means necessarily static. Quite early, however, I came to be influenced by the more physiological conception of equilibrium, especially as formulated by Cannon around the concept of homeostasis.

This physiological conception articulated more directly with the functional perspective then prevalent in the thinking of social anthropologists, especially A. R. Radcliffe-Brown and his followers. Malinowski, though also known as a functionalist, was off on a different theoretical tack.[44] Radcliffe-Brown was much influenced by Durkheim and came into my orbit by that route. For a considerable time, Merton and I came to be known as the leaders of a structural-functional school among American sociologists.

Developments since the emergence of the four-function paradigm and the analysis of generalized media, in particular, have made the designation "structural-functional" increasingly less appropriate. First, it gradually became clear that structure and function were not correlative concepts on the same level—as, for example, universalism and particularism are in the pattern-variable formulation. It became evident that function was a more general concept defining certain exigencies of a system maintaining an independent existence within an environment, while the cognate of structure, as a general aspect of such a system, was process. Concern both with the maintenance of boundaries and other aspects of the functioning of an action system increasingly focused attention on problems of control. Money could thus be regarded as a mechanism through the circulation of which economic activities are controlled, in a manner analogous to that in which the circulation of hormones in the blood controls certain physiological processes. Such ideas articulated, in turn, with the strong emphasis in modern biological thinking that living systems are open systems, engaged in continual interchange with their environments.

Clarification of the problem of control, however, was immensely promoted by the emergence, at a most strategic time for me, of a new development in general science—namely, cybernetics in its close relation to information theory. It could now be plausibly argued that the basic form of control in action systems was of the cybernetic type and not primarily, as had been generally argued, the analogy of the coercive-compulsive aspects of the processes in which political power is involved. Furthermore, it could be argued that functions in systems of action were not necessarily "born free and equal," but had, along with the structures and processes implementing functional needs of the system, differential hierarchical relations on the axis of control.

Here developments of the cybernetic aspects of biological theory, especially the "new genetics," proved to be extremely suggestive for action theory. Particularly important was Emerson's idea that the "system of cultural symbolic meaning" played a role analogous (in the proper biological sense of analogy) to that played by genes in biological heredity. There was a notable theoretical fit between this conception and the role that had been assigned, in the theory of action, to the pattern-maintenance function and the structures and processes involved with it generally and to cultural systems particularly.

This perspective offered a way out of the endless circles of argument about the relative predominance of classes of factors in the determination of social processes and developments. For example: Was, in the last analysis, Marxian economic determinism more correct than cultural determinism? In general, such a question is meaningless, being of the same order as the old biological argument over heredity versus environment. The alternative, of course, is that action process involves combinations of factors that have different functions for the system in which they are combined, and that one major aspect of these functions is control in the cybernetic sense.

The cybernetic perspective also helped to open new possibilities for dealing with the vexing problems of stability and change in action systems. In this connection, it was possible to articulate new perspectives with my previous interest in the socialization of personality and similar themes. Insistence on a radical theoretical distinction between the processes by which a system-pattern is maintained—including for societies the socialization of new members—and those by which its major structure is itself altered seemed to be vindicated, as broadly analogous to the basic biological distinction between physiological processes by which the state of the individual organism is maintained or changed, and evolutionary processes involving change in the genetic constitution of the species.

Social Change and Evolution

The latter problem context has emerged into a strongly revived interest in the theory of social and cultural evolution and its continuities with organic evolution. This interest was particularly stimulated in a 1963 seminar in problems of social evolution given jointly with S. N. Eisenstadt and Robert Bellah. It has been followed up in various writings.[45] This line of interest, of course, constitutes a continuation of my concern with Weber's comparative and historical perspective, particularly in its relation to the interpretation of the nature and problems of modern society. It is also linked to more detailed problems in a series of studies of higher education on which I have recently been engaged.

A considerable part in the theoretical analysis of processes of structural change in social systems has recently been played by a paradigm that is another derivative of the general four-function paradigm. This has been labeled a paradigm of a stage in progressive structural change in a system of action, especially a social system.[46] The point of departure for this is the conception of differentiation, a process which seems on good grounds to come to focus on the binary case—that is, the division of a previous structural unit into two functionally and hence qualitatively different units. The paradigmatic case for the social system is the differentiation of the peasant type of household into a residential household and a productive agency from which income for support of the former can be derived.

For very long—for example, in the work of Herbert Spencer—differentiation has been understood to be complemented, on functional grounds, by new integrative structures or mechanisms. Partly for this reason, the newly differentiated system is also involved in new adaptive problems, very much in line with the general biological concept of adaptation as crystallized in the Darwinian tradition, but with a new emphasis on active as distinguished from passive adaptation. Finally, however, there are components of such a system that are relatively insulated from these overt processes of structural change. These components belong organically in the genetic category; action-wise, in that of pattern-maintenance. Hence a four-fold paradigm becomes appropriate at this level also. We have spoken of differentiation as focusing on the goal-attainment function, integration in the obvious place, but with a special emphasis on what we call "inclusion," upgrading of adaptive capacity as the focal adaptive category and "value-generalization" as the mode of change required to complete such a phase for the system, if it is to have the prospect of future viability.

This paradigm has played a prominent part in the spelling out of the interest in societal evolution mentioned above. The work on social evolution has been documented in a number of articles and in two small volumes written for the Prentice-Hall Foundations of Sociology series (Alex Inkeles, editor): *Societies: Evolutionary and Comparative Perspectives* (1966) and *The System of Modern Societies* (forthcoming).[47] The basis of this interest goes back to my dissertation on the nature of capitalism as a social system, which could now be more broadly redefined as an interest in the nature and main trends of modern society. This time, however, it has been approached in a broad evolutionary perspective, very much in the spirit of Max Weber, but with certain important differences from his views.

Comparative perspective has, of course, been very much in my mind, but at the same time I have been frankly concerned with the conditions and processes of modern Western development. With this in mind, and against the background of a survey of primitive and intermediate societies, I have been especially concerned with the way in which Christianity (in the context of Judaism and the culture and society of classical antiquity) set the stage for modern development. In this connection, a double set of considerations seemed to be particularly important. One was the idea that in two special cases small-scale societies—namely, ancient Israel and Greece—were able to make special cultural contributions because they became differentiated from their environments *as total societies,* on a basis which, however, did not permit them to survive for long as independent entities. Their cultures could, though, be differentiated from their societal bases and exert a profound influence on subsequent civilizations. I have called these "seed-bed" societies. In a broad sense the contributions of Israel and Greece to the modern world—especially, though not exclusively, via Christianity—are well known, but perhaps the sociology of the process is not so familiar.

The second ruling conception was that of the Christian church as a partially separated subsystem of the whole society of late Mediterranean antiquity, politically unified under Rome, which, with its strategic cybernetic position could eventually have a decisive influence on the whole process of modern development. It can be seen that the seed-bed society and the differentiated religious collectivity could be understood to serve similar functions for the long future—functions that in certain senses are parallel to the functions of investment for the process of economic development. I have tried to put forward this view for Christianity in two articles on its general development and significance.[48] Of course this line of analysis is, in certain respects, an extension and revision of Weber's famous interpretation of the significance of the ethic of ascetic Protestantism.

In this connection I have conceived, as so many others have, of Israel and Greece as contributing principal foundation components of what may be called the "constitutive" culture of modern civilization. These components were synthesized and changed in essential respects in Christianity. Beyond this rather commonplace position I have been concerned with elucidating the social processes by which the connection has taken place over time and with linking these interpretations with emerging interpretations of the essential elements of the modern system.

The Nature of Modern Societies

Weber's view of capitalism, like Marx's clearly rested on capitalism's relation to the industrial revolution. This view was consonant, on the whole, with the assumption that this basic change in economic organization—in close relation to technology, of course—was the most essential feature of the new society. Marx and Weber were in agreement on this main point, though they differed profoundly in their accounts of the factors involved in its genesis and in their analysis of the internal dynamics of the industrial structure.

In their different ways, both were also passionately interested in the political developments that first came to a head in the French Revolution and had immense and complex repercussions thereafter. But Marx, in his focus on class struggle, and Weber, in his focus on the process of bureaucratization, tended to relegate the democratic revolution to a place secondary to the industrial. Increasingly, it has seemed to me that the two should be placed in positions of coordinate importance. From the broad point of view of my paradigm, no matter how highly interdependent they were, one should be interpreted as being of primarily economic and the other of political significance in the analytical senses. In this sense, they can be seen as independent in focus but resting on a common base.[49]

From my interest in the professions, I came gradually to the conclusion that what may be called the educational revolution is of an order of significance for modern society at least commensurate with that of the other two. The educational revolution, of course, began substantially later, toward the middle of the nineteenth century. But with the proliferation of mass higher education in the last generation, this revolution has reached a kind of culmination. As a result, the occupational structure has altered profoundly—having its primary focus in the professions themselves, as specially articulated in and with the system of higher education, rather than in "line" bureaucracy.

The conception of the three revolutions—industrial, democratic, and educational—fits with the paradigm of progressive change because all three involved major processes of differentiation relative to the previous state of modern society. Moreover, all were major agencies of upgrading through the immense increase of generalized and mobile resources. All three have also clearly posed major problems of integration for the societies in which they have appeared and have necessitated major shifts of what we call value-generalization.

In accord with the logic of the four-function paradigm, it has seemed to be meaningful to seek a more fully constitutive base in some sense back of these three major transformative processes, a sense which might well include the temporal dimension. In this context, I have increasingly come to consider the possibility that the principal beginning of the modern phase of societal development lay *before* the emergence of the three revolutions—in a cultural and societal milieu that could be conceived to lay foundations common to all three. Once this possibility—as distinguished from the more common tendency to date modernity from either the industrial or the democratic revolution (or both)—was open, it seemed entirely clear that the primary locus lay in what I have called the northwest corner of the seventeenth-century European system: England, France, and Holland. In an important sense, of course, England and Holland were more closely related because both were predominantly Protestant powers with a strong economic emphasis, at that time mainly commercial. It should not be forgotten, however, that France narrowly escaped a Protestant victory and that Calvinism left a lasting influence there.

Not only was ascetic Protestantism especially strong in this whole area, but so was the aftermath of the Renaissance. The two coalesced, in a sense, in the great English and Dutch development of science in that century. The same century saw in England the first culmination in the development of common law and the establishment of the first important parliamentary regime. The France of Louis XIV, on the other hand, developed the most powerful centralized state yet seen, which served as the primary foil to the democratic revolution. The latter questioned, not the conception of the state as such—Rousseau emphasized this concept maximally—but the structure of authority by which it was controlled. Thus in terms of cultural base (especially religion and science), of legal order, and of political organization, these three countries established in the seventeenth century several of the major components of modernity. Nor should it be forgotten that these countries were at that time in the forefront of the institutionalization of the concept of nationality, which served as a basis for some of the conflict among them. The Dutch emancipation from Spanish rule was, of course, a major reference point.

There is an important sense in which all three revolutions presupposed this common base. The first was a kind of extension of economic differentiation and upgrading from the commercial level to the industrial—that is, the mobilization of the deeper-lying factors of production for economic upgrading. The second, intimately involved with nationalism, was a mobilization of deeper-lying factors of political effectiveness, notably the active support of the citizenry, who were no longer subjects of the monarch. The third constituted a mobilization of cultural resources in the societal interest, through the immensely complicated process of internalizing commitment to the major cultural patterns and the implementation of the attendant commitments.

In this connection, a major problem of interpretation of the trend of the modern system arose. The problem may perhaps be stated in terms of the contrast with a view which, with all their differences, may be said to be common to Marx and Weber. In a certain sense, they were agreed that the core problems of the modern system lay in power relationships. Marx located the core of this relationship in the dichotomous structure of the industrial firm, with its owner-manager versus worker structure, and then generalized this to the all-society basis of class. Weber, more realistically in the light of developments, located it in the conception of a much more differentiated firm as a bureaucratic system *not* bifurcated on a power basis, but in a diffuse and general sense controlling the actions of its participants.

The principal point of reference for a different view in my case has been the work of Durkheim, notably his conception of organic solidarity. The simplest ways to formulate the difference seem to be, first, in terms of the contrast between a predominantly associational and a hierarchical bureaucratic pattern of social structure and, second, between a more monolithic and a more pluralistic type of structure. The relation of associational structure to the problem of concentration versus dispersion of power is relatively obvious and has, of course, been very much involved in the democratic revolution. In discussing this context, the tendency has been to concentrate on the organization of government as such, and above all on central government. But in many modern societies, perhaps especially the United States, there has been a vast proliferation of voluntary associations of many different sorts. For my own purposes, a particularly important case of associationalism has been the professions, precisely because of their increasingly strategic significance to the occupational structure. They provide a central focus for the differences of opinion between capitalists and socialists and among the theorists who have concerned themselves with the impact of the industrial revolution. The professions clearly do not tend toward a bureaucratic type of organization, but, so far as they are involved in collective decision-making, claim considerable autonomy relative to agencies not belonging to the profession in question and act mainly as associational groups. Since the professional role is typically that of a full-time job, I have—following, indeed, Weber's and other usage—been calling this pattern "collegial." Not the least important case is the academic profession, which has over several centuries preserved a predominantly collegial pattern of organization, even though it has had to articulate with more bureaucratic ones especially in the realm of academic administration.

The problem of pluralism is somewhat more subtle. Over a wide range, differentiation of a social structure does not lead to the allocation of the personnel of the previous structure exclusively to one or the other of the resulting ones. Thus when the older peasant type of household became differentiated, the adult males continued to be members of residential households, but *also* became members of employing organizations—that is, factories and offices. The same principle applies frequently to collectivities that are units in more extensive social systems; thus various disciplinary associations are, in their corporate capacity, members of such a body as the American Council of Learned Societies. Indeed, departments as well as individuals are members of a university faculty.

Since I have attributed such importance to the process of differentiation in societal development generally, but especially in its modern phase, the phenomenon of pluralization with its distinctive features, conditions, and consequences becomes of substantive importance. In his concept of organic solidarity, Durkheim made a centrally important beginning in the conceptual analysis of this range of phenomena, a beginning on which I have increasingly been attempting to build.[50] It has become more and more evident that the phenomena in this range are of critical significance to modern society, not only in the economic sphere, but also in the articulation of occupation with kinship, ethnicity, religious structures, and various aspects of the category of community. At the same time, for a combination of reasons of ideology and intellectual history, the focus of attention and the development of conceptual tools appropriate to this area have seriously lagged behind. The tendency to focus on the two great figures of Marx and Weber and to employ either a class analysis or a bureaucratic one is indicative of this intellectual situation.

In terms of my own intellectual experience, one particularly important point of reference could be exploited in this direction. Durkheim's approach to the analysis of the modern economy in *Division of Labor* emphasized its *institutional* regulation in the relatively informal senses, but on formal levels it stressed law more than governmental administration. The central focus was on the institution of contract and secondarily on that of property. In terms of later theory, this view directly linked the economy, as the adaptive subsystem of a society, with the integrative system that I have recently begun calling the "societal community." This linkage with Durkheim and its various subsequent ramifications lead to an emphasis on differentiatedness and pluralization of structure in direct contrast to the hierarchical emphasis on power relations that is common to both Marx and Weber.

In developmental terms, it became clear that, with the exception of the organization of the French state, the fundamental structural contributions of seventeenth-century society were of the associational-pluralistic character, notably ascetic Protestantism, common law, and parliamentarism as well as science and the rapid development of a market economy. In its capitalistic form, the industrial revolution certainly moved society farther in that direction, as did the democratic revolution. From this perspective it became clear, perhaps especially under the influence of Tocqueville, that the modern society emerging in North America was beginning to play a role in the total modern system of the twentieth century somewhat parallel to that of the European northwest corner in the seventeenth. This society owed its primary distinctive features to an associational-pluralistic emphasis and not to the sharpness and rigidity of its class discriminations nor to its especially high level of bureaucratization. In addition to decentralized governmental democracy, examples would be its federalism and separation of powers, the religious constitution of separation of church and state, denominational pluralism, and the capacity to absorb, in the sense of integration by inclusion, large immigrant religious and ethnic groups, though this absorption is far from being complete.

Spelling out the ramifications of these structural trends—assuming that they indeed exist—is a complex empirical-theoretical task, but it has increasingly become my main concern in recent years. One scholar, even when he enjoys collaboration and can connect with the work done by many others, can at best work through such complex problems only in partial and fragmentary fashion.

A substantial part of the hierarchy-power preoccupation of so much generalized social thought of the last century or so I attribute to ideological factors. Thus, classically, the socialist reaction to the capitalist conception of an economy governed by the rational pursuit of unit self-interest substituted rigidly centralized control by government in the public interest. In the dilemma stated by these alternatives, attention was diverted from the actual extent to which the new industrial economy was *neither* purely rationally individualistic in the sense of the utilitarian economists nor collectivistic in the socialistic sense. As Durkheim made clear, it was governed, in considerable part, by other factors. Among such factors is a normative structure, legitimized in terms of values grounded in cultural bases, notably at religious levels. In the other direction lies the affective grounding of solidarity, precisely in Durkheim's sense, in the motivational attachments of individuals to roles, to the collectivities in which they participate, and to their fellow members.

Solidarity and the Societal Community

Parallel to the rigid capitalistic-socialistic dichotomy between individual self-interest and public interest, there is a more recent dichotomy between alienation of the individual and various subcollectivities from collective solidarities, on the one hand, and expectations and demands for total absorption of the individual or relevant subcollectivity in some macroscopically conceived community, on the other. Here also a third alternative, not simply an intermediate state, is almost certainly more important than the recent and current ideological assertions would have us believe. Such alternatives presumably fit in the broad associational-pluralistic range.

Adequate theoretical resources to define these alternatives, to diagnose and further analyze the existent phenomena where they appear as well as the features of many existent structures which block them, are probably in relative terms more inadequate than were the theoretical resources for analyzing pluralistic normative components in Durkheim's time. Above all, it is necessary to establish adequate theoretical links between the psychology of the individual, the functioning of social systems in many different respects, and the grounding of the normative factors in the cultural system. One problem is to avoid simple dichotomization of the *Gemeinschaft-Gesellschaft* type, which is so strikingly parallel to that of socialism-capitalism. There is a distressing tendency among today's intellectuals to posit return to a relatively primitive level of *Gemeinschaft* as the only remedy for what are so widely held to be the malaises and the moral evils of contemporary society.[51]

The approach to this complex problem area that has been most congenial to me has been a continuation of the analysis of the process of socialization of the individual, with special reference to the interrelations between the motivational dynamics involved and the structural setting of the process seen in terms of both the social and the cultural systems. Psychoanalytic psychology at the strict personality level has provided a solid theoretical base for getting a purchase on this problem area. For understandable reasons, however, it has tended to concentrate its attention on the earlier phases of the process—especially in its classical form, the Oedipal. Even here it has needed appreciable correction and modification in the light of sociological analysis of family and kinship systems.

Psychoanalytic theory, however, has with a few exceptions (for example, Erikson on adolescence)[52] notably neglected the progressive stages of the socialization process through the various stages of formal education. It has often been content with therapy and with the ambiguous dictum that character structure has been fully laid down by the end of the first—or more usually the sixth—year of life and that what comes after that can safely be relegated to pathology or its absence.

In earlier phases of my thinking there were some beginnings of successful analysis in this area (as in the early paper on "Age and Sex" as categories of social structure,[53] later in the *Family, Socialization, and Interaction Process,* and including perhaps the social paradigm of the conditions of psychotherapy advanced in *The Social System* and elsewhere). It has, however, not proved possible to achieve a level of analytical generality in this area comparable to that obtainable in the main areas of the political, economic, legal, and even religious structures of societies.

Important progress in this direction was made by returning after some years to a consideration of problems of kinship and the incest taboo, with special focus on the significance for the associational type of society of the taboo among siblings. This consideration led in turn to an interest in the significance of "symbolic" kinship patterns in Western institutional history, especially religious orders labeled as "brotherhoods and sisterhoods." Indeed, religious celibacy could readily be conceived to constitute a case of the "investment" pattern similar to the seed-bed societies and early Christianity.[54] My views in these areas have been immensely clarified by the work of my old student and friend David Schneider on American kinship.[55]

I had previously also been engaged in a study of the relations of secondary education to social mobility, in collaboration with Florence Kluckhohn and the late Samuel Stouffer. An extension of the thought of this study backward to the elementary-school level produced important clarification of some structural reference points.[56] My empirical interests had also, as noted, extended forward on this continuum to studies of the social structure and dynamics of higher education, especially in terms of its intimate involvement with the professions.

Substantial further theoretical progress has proved to be contingent on the development of another generalized analytical paradigm—the classification of generalized media of interchange and the spelling out of categories of interchange among the four primary functional subsystems at the level of the general system of action, involving cultural, social, psychological, and behaviorally organic systems. As outlined above, these categories converged with the scheme of W. I. Thomas.

It had long been evident that "affect"—in something like the psychoanalytic sense, which should be clearly distinguished from erotic pleasure—should be treated as a generalized medium operating at the general action level. The problem was where it should be placed; the prominence of its psychological associations made its anchorage in the personality system especially plausible. The breakthrough came in exploring the possibility and finally making the decision to place it primarily in the social system and, of course, the latter's interchanges with the other primary systems.[57] This decision treats affect as the direct parallel, at the general action level, of influence in the social system—namely, as the primarily integrative medium. Furthermore, its status as a generalized medium makes it possible to outline a series of steps in its differentiation similar to the historic steps in the evolution of monetary exchange from barter to advanced credit systems—with the marketability of the fundamental factors of production (labor, in particular) developing at an especially crucial stage (essentially that of the industrial revolution).

The solidarity of a social system may then be thought of as a state of solvency of its "affective economy," conditioned both on the flow of instrumentally significant contributions from its members and on their motivational states of gratification, which can be thought of as the positive of which alienation is the negative. In simple social systems, these factors can be conceived to be ascribed; this is true both of primitive societies and of the socializing agencies in which, in a more differentiated society, the child is placed in the earlier stages of his socialization.

How highly differentiated his personality system must become in order to achieve high levels of gratification by full participation depends, of course, on the structure of the social and cultural milieu in which he acts. What we have called pluralization of the structure of the society is thus a major aspect of the much discussed complexity of modern life. Seen in this context, it can be suggested that the development of mass higher education, such a conspicuous and to some extent disturbing phenomenon of our time, may be regarded as a response to the societal need for personalities in sufficient numbers capable of coping with this complexity both in terms of many forms of instrumental competence and of personality integration at affective levels. The new modes of inclusion of individuals and subgroups in social solidarities constitute the primary focus of this set of problems of the stability and other aspects of integration of modern societies.

It is tempting to draw parallels with the states of disturbance that followed the high development of both the industrial and the democratic revolutions. In the former case, two can be distinguished: labor disturbance and that of the business cycle. Smelser has shown convincingly that a new kind of labor disturbance appeared among classes of workers who had not been injured, as had for example the hand-loom weavers, by the transition. These disturbances underlay the development of trade unionism and of the socialist movement, so far as they involved labor movements.[58] Depressions, on the other hand, raised questions of the stability of the new system at over-all systemic levels. It is also notable that attempts to deal with both categories of primarily economic disturbance were couched mainly in terms of self-interest—wages, hours, and conditions of work and the expectations of profit on the part of firms. At the same time there was a tendency, most evident in the socialist aspect and in Marxian theory, to combine these economic considerations with those of political power.

The equivalents for the democratic revolution may be said to be, on the one hand, the struggles over power and authority internal to particular political units (for example, the waves of revolution in Europe of 1789 and its aftermath, 1830, 1848, and indeed 1917-1918) and, on the other hand, the systemic waves of disturbance concerning disequilibrium in the relations among national units. The equivalent of depression here is surely war or, short of it, severely strained international relations, with a tendency for these disturbances to become increasingly generalized. Here the place of the individual self-interest of workers and entrepreneurs has clearly been taken by the collective self-interest of the Powers in their respective positions of power. At the same time, the focus on power has been modified by a highly significant integrative reference, the most prominent manifestation of which has been nationalism. Just as power struggles over economic interests have often become economically irrational, struggles over national prestige have often become politically irrational. Thus a *Realpolitiker* like Bismarck could be more rational than a nationalistic political romanticist like Napoleon III.

I suggest that what I have called the educational revolution may be interpreted as the most salient manifestation of a new phase in the development of modern society where integrative problems rather than economic or, in the analytical sense, political problems are paramount. Student disturbances would then be parallel to labor and authority disturbances because students constitute the category of persons exposed to the most massive problem of adjustment to structurally changed conditions. The focus of their problem is, on the one hand, not realistically in power, but in the mode of their inclusion in the course of the educational process itself—a new phase of the socialization process—and in the more general societal world after completion of formal education.

From this point of view, the radicalism of the New Left is parallel to the socialism of the labor movements and to the Jacobinism of the radical democrats. The more systemic disturbance, then, is the propagation of waves of alienation and related forms of malaise, especially among the more sensitive components of modern populations, notably the intellectuals. This concerns the stability of fulfillment of expectations of improved social solidarity in the first instance, and hence the evidences of lack of such solidarity—such as poverty, racial discrimination, crime, and war—are seen to be particularly distressing. Although it may be argued that the recent manifestations center on the integrative problem, this series of disturbances resembles the earlier ones in that there is prominent invocation of the next higher level of concern or control—in this case, that of values. This is very evident in the special prominence, especially in the more radical circles, of moral concerns.[59]

Thus, over a period of more than thirty years, an empirical interest in the problem of capitalism came to a special focus on the nature and significance of the professions. A continuing interest evolved in the broadest categorization of the nature of modern society, but was no longer put in terms either of capitalism, as such, or of the capitalist-socialist dilemma. Indeed, I am sympathetic to talk of the post-industrial society, but wonder whether it should not also be called in some sense the "post-democratic" society, a suggestion to which there would probably be considerable resistance. (I would not suggest by this term that democracy no longer counted, any more than that "post-industrial" implies that industry is obsolete.) By the same token, special attention should be paid to the system of higher education—and within it the academic profession as the structural core of the system of higher education. For the medical and academic professions I have attempted a close approximation to a standard empirical study, more so than in other aspects of my work. But in both cases I have wanted to understand the professional groups in question in the context of the wider system of which they have come to constitute particularly important parts.[60]

Higher Education as Focus

It is perhaps evident that these concerns with the main trends of development of modern society would naturally bring into sharp focus problems of the nature and current status of the system of higher education in modern, especially American, society. In evolutionary terms, as the culmination of the educational revolution, it had an especially salient place; special study devoted to it seemed to be far from trivial. Second, this field was particularly important in view of my long-standing interest in the modern professions, since it became increasingly clear that university-level formal training was one of the hallmarks of the professions. Training for the most prestigious of the so-called "applied" professions had developed graduate-level professional schools, which in turn became increasingly closely drawn into the universities.

The main guardian and developer of the great tradition of knowledge had become the central academic profession, mainly that institutionalized in faculties of arts and sciences. This profession of "learning itself" could be felt to be the "keystone of the professional arch" and it was to this group that I turned my primary attention. At the same time, the study of higher education offered an opportunity to continue and develop further my long-standing interest in the processes of socialization, carrying this into much later phases of the process than had occupied the primary attention of most psychoanalytically oriented students.

This interest crystallized before the Berkeley outbreak, in a project to study academic professionals in the United States, first on a pilot basis with samples from the faculties of 8 institutions and then, starting in 1967, a nation-wide sample from 116 institutions all offering four-year liberal arts programs, with or without graduate schools. This study has been generously supported by the National Science Foundation, and the main direction of research has been carried out by my collaborator Dr. Gerald Platt. As of this writing it is nearing completion.

A more autobiographical note may be added here. It is perhaps understandable that a social scientist like myself, who had become so much absorbed in matters of general theoretical concern, should be under a certain amount of tension in relation to the strong American emphasis on the importance of solid *empirical* research. Response to this pressure was surely one factor in my decision to study medical practice. This study was conceived mainly in the anthropological tradition of participant observation and interview.

With the end of the war and the entrance of Stouffer on the Harvard sociological scene survey research had achieved a position of salience in the social science world. Soon after that he and I decided to collaborate and, associating Florence Kluckhohn with us, undertook a study of social mobility among high school boys, which centered about a graduate seminar that the three of us jointly conducted. This project assembled a substantial body of data, mainly from questionnaires administered in a sample of public high schools in the metropolitan Boston area.

A crisis supervened with the premature death of Stouffer in 1960—in the same summer in which Clyde Kluckhohn died, even more prematurely. Florence Kluckhohn and I had plans to bring out a volume and indeed the body of statistical material for which Stouffer had been primarily responsible had been carefully reworked with that in mind by the late Stuart Cleveland, but the vicissitudes of assembling the other desired contributions defeated the project.

There is, hence, a certain psychological continuity in my attempting a second round at involvement with survey research. Indeed, the use of this method had been decided upon before Platt joined the enterprise, and his training in these methods constituted one of his principal qualifications. I have considerable hope that this time it will come off, though my personal contribution will have been that of senior faculty sponsor, theoretical contributor, and critic, rather than operative survey researcher, which has been the job of Platt and the staff working with him. Especially since coming to know Stouffer well I have had a high intellectual respect for empirical social research, and very much hope that still closer alliances between these techniques and the kind of theory I have been concerned with can be worked out.

Another rather obvious continuity between the two ventures in survey research lies in the fact that they both dealt with phases of the sociology of education and both in some sense in relation to the socialization process. The mobility project was especially concerned with linking socialization, in a focus on formal education, with the occupational structure which has been so crucial to modern society since the industrial revolution. In a sense our study of higher education and the academic profession has led into a new sphere where the relations of education to the cultural tradition, coming to focus in the problem of the status of the intellectual disciplines, has taken a kind of precedence over the problems of the allocation of manpower within the occupational system, important as these are.

This is perhaps an appropriate place to take note of another context of significance for the development of my thinking, namely the American Academy of Arts and Sciences. I was elected a Fellow of the Academy in 1945 when, for a variety of reasons, including the war then just ending, it was not very active. I had attended a few "stated meetings" of the Academy, but I think a more active interest was first enlisted through *Dædalus,* my first independent contribution to which was participation in a symposium on youth in 1961. I also became a member of the Research Funds Committee of the Academy and of the Committee on the Social Science Monograph prize.

I became progressively more involved in such *Dædalus* enterprises as those dealing with the New Europe, with science and culture, and with the color problem[61]—and the Poverty Seminar that followed them, independent of *Dædalus*.[62] I also became for a time chairman of the Academy's Committee on Research Funds and a member of the Commission on the Future of the Academy. Finally, in 1967, I was elected president of the Academy, the first social scientist to serve in this capacity.

It was particularly congenial to me that interests of various groups within the Academy, especially through *Dædalus*, turned in the direction of the study of higher education. This has been evident in a number of ways: the Danforth Project on the governance of universities, the study of the ethical problems of experimentation with human subjects, the discussion of international problems of higher education in industrial societies, the recent studies of the status of the humanities, and the volume in which the present essay is published.[63] I have participated in the Academy's Assembly on University Goals and Governance, initiated in September 1969, as a central staff member, assigned to producing a set of generalized analyses of the nature of the current system of higher education, its place in modern society, and its possibilities of change. I have hence been in the interesting position of functioning in my Academy role mostly as a generalist student of higher education, but—in the role of research supervisor of the faculty study—I am also attempting to pin down some rather specific empirical generalizations about what academic people are really like and what makes them tick.

Participation in the work of the Academy has been particularly rewarding to me in a double sense. As an explicitly interdisciplinary organization, with active participation of members ranging across the whole spectrum of the intellectual disciplines and beyond, it has seemed to me one of the few best antidotes, organizationally speaking, to the alleged and in part actually existent trend to overspecialization in our culture, particularly its academic sector. Hence, for a scholar committed to the importance of highly generalized orientations, active participation in what has been going on in the Academy in the last decade or so has seemed to present an unusual personal opportunity for the kind of interdisciplinary action which is difficult in the local university setting. This has included exposure to stimuli which I would not otherwise have had to take account of—such as the presentations of the biologists in this volume—and an opportunity to act more positively in relation to the cognitive interests and sentiments of people involved in the immense range of different disciplines and intellectual interests represented in the Academy and various of its activities.[64]

From the more objective point of view of understanding the system of higher education and related phenomena, the Academy has increasingly come to represent a potential of generalization in the cultural field, in relation to the social organization of research, teaching, and application, which is not totally unique, but still probably preeminent. The very fact that such an organization can, for the time being at least, flourish in an age of allegedly rampant specialization seems to be some sort of index of the deeper concerns which are, often silently, guiding our main cultural development.

Cognitive Style and Summary of Themes

More than one commentator on the first draft of this essay has raised the question of the relation between a kind of "intellectual opportunism" and a pattern of consistency and continuity in the developments which I have outlined in the preceding pages. An attempted formulation seems more appropriate near the end of this account than near the beginning.

It is quite clear that neither in the occupational sense nor in the sense of intellectual content has mine been a meticulously planned career. The furor over the dismissal of Meiklejohn at Amherst was not foreseen when I went there, nor was the shift from biomedical to social science interests planned. Within a limited range the year at London was, but the German venture, including being assigned to Heidelberg, very definitely was not. Similarly, though going farther with economic theory was planned, involvement in sociology at Harvard clearly was not, nor was the life-long career anchorage at Harvard. Just as, when I went to Heidelberg, I had never heard of Weber, when I decided to come to Harvard I had never heard of Gay or Henderson. I was early predisposed to treat Pareto as rather indifferent and was conditioned to consider Durkheim unsound. I also had no special attraction to or knowledge of Freud until well into my thirties, and I had no special interest in the professions until nearly the same period.

There is a relation between the serendipitous element in these various career decisions at both the occupational status and the intellectual commitment level and another pattern that has continued down to the present. This is the pattern of responding to intellectual stimuli: challenges to organize association meetings and attend conferences, or, most important, to write articles on a wide variety of topics. Two early examples were a request to set up, at a meeting of the American Sociological Association (1941), a session on Age and Sex as coordinates of the role-structure of societies. It is out of this that the most widely reprinted paper I ever wrote, "Age and Sex in the Social Structure of the United States," came. The second was a request from the editor of the *American Anthropologist*, Ralph Linton, to attempt a synthesis of anthropological method in the analysis of kinship with sociological perspective on American society. The result was the paper, "The Kinship System of the Contemporary United States" (1943), which also received rather wide attention.

It is perhaps largely in response to this kind of thing that two related images of my role in American social science have emerged. One is that—largely, I presume, by contrast with "solid" empirical research contributors—I have been held to be primarily a talented and "stimulating" essayist, writing on a variety of topics but without any genuine continuity or solidity of any kind—one might suggest an "esoterically academic" kind of journalist. The second is the attribution to me of a kind of schizophrenic dual professional personality—on the one hand this kind of journalism, on the other hand a wholly unrealistic abstract kind of formalized theorizing, with the strong implication if not assertion that the two personalities had nothing substantive to do with each other.

Professor Renée Fox has especially stimulated me (in a detailed personal communication) to reconsider the problem of continuity of development, especially at theoretical levels. I hope that my conviction that there has in fact, for over forty years, been a basic continuity has come through in the course of the preceding exposition.

In attempting to understand the nature of the psychosocial process by which this continuity has developed, I have found one parallel to be particularly suggestive. In two recent academic years I collaborated with Professor Lon L. Fuller of the Harvard Law School in a seminar under the highly permissive title Law and Sociology. In the course of it I have learned a good deal about law, and not least about the common-law tradition. From the point of view of Continental European systematists of law—an especially prominent example being Hans Kelsen—the state of common law is clearly intellectually scandalous. It allegedly consists of nothing but an aggregate of particular cases and seems almost completely devoid of principles.

Fuller, more than any other person,[65] has helped me to see that, far from the "case system" being inherently antithetical to "systematization," under the proper conditions it can be a positive vehicle of systematization. The essential point is that, since in common-law terms, with a few qualifications, courts must adjudicate *any* case put before them in a procedurally acceptable form, they have the problem, not only of rendering decisions, but also of justifying them. An appellate judicial system ensures that dubious justifications are likely to be challenged on appeal and by an intellectually critical profession, for example, in law review articles. Justification in this sense involves subsuming the particular decision, not only under specific precedents, but also under more general legal principles.

There are those in my immediate relational nexus, most notably my colleague George Homans,[66] who hold that the only legitimate use of the term "theory" is to designate a logical deductive system, with explicitly and formally stated axiomatic premises and, combined with appropriate minor premises, a set of deductions from them which fit empirically verifiable statements of fact. From Homans' point of view, all I have produced is a conceptual scheme which is not theory at all. A semantic issue is surely involved here but I, along with many others, have never confined the use of the term "theory" to this narrow type. I do regard it as a legitimate goal for a course of development of theory, but to say that anything short of it is not theory at all is another matter.

However that may be, two things may be said about the development outlined in this essay. First, what is at present available in my more abstract writings is not a mature system of theory in Homans' sense. Second, the process by which it, such as it is, has been arrived at has most emphatically not been one of having sat down and formulated the basic axiomatic principles and then deduced their logical implications and checked these against the known facts.

It has, on the contrary, been a process much more like that of many developments within the common law. The work of which *The Structure of Social Action* was the outcome certainly established a theoretical orientation—in my sense of theoretical—which was not a congeries of random opinions within the areas of relevance. From this conceptual scheme, if you will, as a reference base the process has been one of exploring a rather wide variety of highways and byways of empirical-theoretical problems, not, however, in wholly random succession. In this process, along with my serendipitous encounters with intellectually significant persons and influences, I have indeed reacted to quite a number of externally presented stimuli of the sort that I have characterized, especially requests to write on topics suggested by others.

In a sufficient proportion of such cases, I hope I have reacted somewhat in the manner of a competent common-law appellate judge: namely, that I have considered the submitted topics and problems in relation to a theoretical scheme, which—though its premises were not defined with complete precision and henceforth assumed as fully given in a logically complete sense—has had considerable clarity, consistency, and continuity. In a sufficient proportion of cases, it seems to me that this kind of procedure has yielded empirical insight and rounding out, extension, and revision and generalization of the theoretical scheme. At certain points this has meant intensive concern with formally defined theoretical problems, but at other points primary concern with much more empirical issues. In any case this is essentially what I have meant by the phrase "building social system theory" as used in the title of this essay.[67]

If the above considerations throw some light on the process by which some serious continuity has been maintained, I may now say a few words, in a summary fashion about the themes which, in retrospect, seem to me to have been most important in the continuous theoretical development and the patterns of their succession.

Though a number of primary themes were involved in the theoretical patterning of *The Structure of Social Action*, notably that of the nature of the historic conceptions of economic self-interest and of economic rationality, one came to be particularly salient and has continued to be so with many variants ever since. This was what I have called the "problem of order," with reference to the human condition generally and the social system in particular. The classic early modern formulation lay in Hobbes's concept of the "state of nature" and the problem why human societies, with all their troubles, had not by and large become states of the "war of all against all." (Even with the many wars of history, the combatting units have been social systems, not isolated individuals.)[68]

My assumption throughout has been congruent with that of Hobbes in the sense that even such order as human societies have enjoyed should be treated as problematical, not assumed as obviously "in the nature of things"; in this regard perhaps I have inherited some element of Christian pessimism. Hobbes's personal solution, the "social contract" to set up an absolute sovereign who would coercively enforce order, was by the 1930's obviously unsatisfactory. But the problem remained. One of my most important reasons for linking Weber and Durkheim and, by teasing out essentially latent conceptions, Pareto was the growing insight that they had in common the recognition of the intellectual seriousness of the problem and the conviction that, in one way or another, normative factors in human action, which were analytically independent both of economic interests in the usual sense and of interests in political power, were of decisive importance.[65] Durkheim's insight about the normative components in the structure and regulation of systems of contractual relations was a truly clinching contribution to my own conceptualization; Durkheim made specific reference to Hobbes in this connection. I very much stand by the view that order in this sense is genuinely problematical, and that the nature of its precariousness and the conditions on which such order as has existed and may exist is not adequately presented in any of the views of human society which are popularly current, regardless of political coloring. There is a fundamental distinction between an intellectually competent analysis and understanding of this kind of problem and a popularly appealing ideological definition of it. They are not always at sharp variance with each other, but very generally so.

The "problem of order" is quite clearly a central focus in problem-formulation—in the German formulation *Problematik*—of the relations among, and the balances of factors involved in, states of stability, of tendencies to disorganization and dissolution of systems, and trends of change.[70]

The connection between the theme of order and that of convergence must be evident from the above and from earlier discussions. Insight into the problem of the theoretical significance of accounting for order and its further potentials, as well as its failures, could be considered to be a theoretical achievement. My thesis that an otherwise diverse group of theorists had converged on a common "direction of solution" of this problem, which was not obvious to the academic common sense of the time, could, on the other hand, be held to constitute a "finding."

The nature of this convergence has been sketched here and is spelled out at great length in various of my writings. It concerned the component of normative control as distinguished above all from coercive enforcement in ways which linked both with homeostatic conceptions in psychology and cybernetic conceptions over a much wider range.

As noted by Clifford Geertz in discussion of this essay, the theme of convergence did not stop with the cases intensively examined in *The Structure of Social Action*, but has continued to be a major theme of my whole intellectual career. The conviction of some kind of convergence between socioeconomic and biological thinking played an early part. Perhaps above all my concern with Freud made salient both the problem of convergence between social system theory and personality theory, and, gradually, the extent to which this was actually present. Of course such convergence often had to be teased out from what at face value were incompatible positions. The analytical distinction between personality and organism has been indistinct in most psychological thinking—indeed many psychologists today would totally deny its relevance—but, especially because of association with James Olds in the early phases of his work on the brain and with Karl Pribram, it seemed to me to be a case of convergent patterns within a framework of analytical distinctness. Similar considerations have operated in the field of the relation between social and cultural systems, for which I was primed above all by Weber, but also by my many associations with cultural anthropologists. In a sense, perhaps the most extensive convergence of all has seemed to occur under the umbrella of the cybernetic conception, with its many associations and ramifications.

The "problem of rationality" has constituted another very major thematic complex. Phrasing it as "the problem" I hope makes clear that I have not been a naïve rationalist, either in the sense of holding that virtually all human action *is* essentially rational, or that the intrusion of non- or even irrational elements should be condemned. The position, rather, is that to attempt to analyze the role and nature of rational components in relation to those which should not be designated as such has constituted a major focus of theoretical concerns.

It seems clear that my initial focus on problems of economic and, secondarily, political rationality—for example, in the capitalism-socialism debate—was a legitimate but limited focus. To a considerable extent I have built the above account around the relation between this focus and the two others, which in a sense are located—in the spectrum of cognitive concerns with human action—on each side of this middle one. *The Structure of Social Action,* reinforced by Freud and related influences, opened up both. For example, Pareto's conception of "logical action," strictly bound by the canons of scientific validity, opened a door into the scientific basis of the professions, of the functions of higher education, and, more generally, of "cognitive rationality" as a value pattern—as well as opening another door into the elucidation of the "psychological" nonrational.

As I have tried to make clear, for a number of years I was rather more concerned with the other alternative to economic-political rationality, namely that which linked the social system with personality in a way that proved to have highly complex relations, on the one hand to the organic complex and on the other hand to the cultural. Thus in the former context the problem of the significance of the erotic complex was salient, in the latter the problem of the role of internalized values was uppermost, starting with Freud's conception of the super-ego.

The "problem of rationality" in this context has two, or possibly three, facets. One is the question of the roles, in the determination of action, of rational and nonrational forces—for example, for Freud the ego and the "reality principle" and the id in relation to the "instinctual needs" governed by the "pleasure principle." It should be clear to the reader that my views in this area have been much less antirational than those of many other students of the problems, but I hope not naïvely rationalistic either.

The second, very vital, context concerns the accessibility of non- and sometimes irrational forces to rational understanding, that is, in cognitive terms. The intellectual movements into which I came were deeply involved in this situation. This included Freud most conspicuously, but very clearly all of my major authors, with the partial exception of Marshall. Perhaps Freud's most heroic endeavor was to set out a program for the "rational understanding of the unconscious," an entity which was, by his definition, in its very nature nonrational. This is indeed a far cry from either the rational understanding of the "rational pursuit of self-interest," or, indeed, the rational understanding of the pursuit of rationally cognitive knowledge.

The third facet, if there is one, is the link between these two. The classical aphorism is Freud's "where Id was, there shall Ego be." We might even go back to August Comte and his slogan, *savoir, c'est pouvoir.* In what senses and within what limits does rational understanding of the nonrational—which clearly includes the physical world—open the doors to control? In the most general sense of course the answer is that it does open such doors. But this remains one of the most seriously controversial areas of the rationality complex, various aspects of which have been very central to me.

The rational component of psychotherapy shares with economic and political rationality its instrumental character. But two problems beyond this arise. The more obvious one concerns the sources of legitimacy and justification of the ends or goals in the interest of which such instrumental rationality is brought to bear. The utilitarians, and still for the most part, economists, treated consumption "wants" as given, that is, as not constituting the locus of intellectual problems for their purposes. Similarly for Freud and psychiatry mental health was an aspect of general health, and its attainment or restoration almost by definition desirable. But in both contexts, and a variety of others, it is reasonable to raise the question, paraphrasing a famous book title: "Rationality for what?" (Robert Lynd, *Knowledge for What?*)

A seductively simple solution is to say that the goals of instrumentally rational action are *basically* nonrational. But as so often, this is too simple. Weber made a major contribution to further sophistication here with his concept of "value rationality" (*Wertrationalität*), which he conceived as constituting one of the primary types of action. The essential implication, which cannot be grounded here,[71] is that the "universe of values" is not devoid of rational organization and that decisions of "commitment to" values, including their more or less direct implementation, have a rational component which is independent of instrumentality.

It has turned out that the relevance of this position, which was only most explicit in Weber, has operated in two directions, not one, in a sense somewhat analogous to the "fork in the path" encountered when following out the theoretical problems implicit in the conception of economic rationality. The more obvious of these concerns religion. In a variety of ways, problems of religion have been prominent for me almost from the beginning. It was Weber's Protestant Ethic essay which set off a major development for me, and the *common* concern of Weber, Pareto, Durkheim, and later Freud with the intellectual problems posed by religion as a human phenomenon became a major reference point in the earlier phases of my career. In the circumstances this had to involve the problem of the relation between rational and nonrational components of religion.[72]

This concern with religion—in the role not of a *dis-* but more of an *un*believer, in the terminology of a recent Vatican conference—has been a major orientation point in my intellectual career. It was already a major aspect of my early rejection of "positivism," but at the same time has been a focus of a continuing attempt to understand the balance of the roles of rational and nonrational components in human action. Clearly, however, such a focus of intellectual concern leads one beyond the more purely cognitive problems of religion into those of moral commitment, affective engagement, and practical action.

The other ramification of the concept of "value-rationality" is in some ways more surprising. This concerns the status of the value-component in defining the relation of cognitive structures not to the clearly nonrational characteristics of the phenomena "cognized," such as the Unconscious or the "grounds of meaning" at the religious level, but to cognitive structures themselves. This has come to a head in recent years on the conception of "cognitive rationality" precisely as a *value*-pattern, not simply as a maxim of expedient "want satisfaction." The generalization of this conception was foreshadowed in the interpretation which Smelser and I put forward of the economic category of land as including a *value*-commitment to economic rationality.

The relevance of this perspective to many recent concerns, both of my own as a theorist and of society, is almost patent. It is clearly very central to the complex of higher education and its relation to the intellectual disciplines. Since the role of empirical cognitive knowledge was so central to the original formulation of the rationality problem, this concern with the value aspect of empirical cognition, both in the grounds of the cognitive validity of the knowledge mobilized in the instrumental aspects of rational action and in the cognitive problems of the justification of commitment of instrumental potentialities among goals, in a sense brings consideration of the problem of rationality full circle in that the considerations involved in the grounding of value-choices, including their more or less religious bases, are seen to be of the same order as those involved in the grounding of the validity of empirical knowledge.[73]

Perhaps I may conclude with a few words about my conception of the significance to me of the most important intellectual role models: clearly, Weber, Durkheim, and Freud, none of whom, it is important to note, I ever knew personally, though they all lived into the period of my attainment of some kind of personal awareness or perhaps of "identity," to use Erikson's term. In terms of substantive influence in the shaping of problems and of the many elements of empirical and conceptual structure which has been central to my thinking, it is quite clear that all three have been crucial. Others have of course been exceedingly important, first perhaps Pareto and Marshall, in that order, but again

Schumpeter, Henderson, Cannon, Taussig, Piaget, and many others.

One factor in this attribution of significance is, of course, the intellectual location of relevance. Equally or more towering figures in more remote fields have naturally not had the same significance in my development, even though they may have been very important on the periphery. This would have been true of Cannon, of the biological background figures like Darwin, of Whitehead, of Piaget, of Norbert Wiener, and various others. The other factor is, from such a locational perspective, the stature of these figures relative to those of directly comparable relevance.[74]

The sense in which the ideas of Weber, Durkheim, and Freud have permeated my own thinking should be clear from the above account. There remains the question of the sense in which they have been role models mainly in terms of what I have called "cognitive style." Here there emerges an important distinction between Weber and the other two. Substantively, Weber has been at least as important to me as any of the three. Stylewise, however, he was very different from the other two. He was much more, in Erikson's format, a Luther type, who, with all his immense preparation, underwent a single major transformative crisis—in Weber's case involving serious mental illness—from which he emerged as a new Weber, who, with truly dazzling "virtuosity" (a term he was fond of) produced within two or three years the great methodological essays (*Wissenschaftslehre*) and the Protestant Ethic as the opening step in a major reinterpretation of the nature of modern society seen in the broadest comparative-evolutionary perspective. It seems to me significant that, in many of his subsequent writings, Weber particularly stressed the centrality of the "charismatic breakthrough" as the most important process of religious, and more generally of sociocultural innovation and change. In suggesting that this is not the only way—connecting with the idea of the role of genius—I do not in the least mean to derogate the importance of what I regard as Weber's superlative intellectual achievement.

The cognitive style of Durkheim and Freud was quite different. I do not for a moment think that either was less intellectually ambitious than was Weber. Their method, however, was to settle on and thereby become committed to attempting a radical solution of certain definable problems in their respective spheres. For Durkheim it was a special version of the "problem of order" in the sense in which this has been outlined above. For Freud, it was the problem of rational understanding of the nonrational, with special references to the role of what he came to conceive as the "unconscious."

Clearly, in both cases, the process by which these intellectual commitments were entered into was motivationally highly complex—for example, in Freud's case his liberation by the death of his father. But in neither case was the maturing of his commitment a highly dramatic event, though for each it resulted in a truly seminal book, namely, Durkheim's *Division of Labor* and Freud's *Interpretation of Dreams.*

It is not quite fair, but still broadly accurate, to say that from his great breakthrough on, Weber's contribution consisted far more in a truly monumental spelling out and empirical validation of the basic insights of the critical reorientation. In the other two cases, it was a process of step-by-step development of theoretical thinking from the original problem-formulation base. In this sense there is a Weberian theory which dates from his new orientation following his recovery from his psychological disturbance—that is, about 1904-1905. In a comparable sense there is no Durkheimian or Freudian theory, but there is the documentation of an impressive process of theoretical development.

I see no reason to suggest that either of these alternative cognitive styles on the part of intellectual innovators is in any general sense superior to the other; both are critically important, but each is effective in different times and situations. Speaking personally, however, Durkheim and Freud have been my paramount role models as theoretical analysts of human action. Perhaps this has some bearing on the question of the balance between continuity and opportunism in my own intellectual history.

References

1. London: Allen and Unwin, 1930.

2. 3 vols., 2d ed., Leipzig: Duncker and Humblot, 1916.

3. "Sociological Elements in Economic Thought, I," 49 (1935), 414-453; "Sociological Elements in Economic Thought, II," 49 (1935), 645-667.

4. "Wants and Activities and Marshall," *Quarterly Journal of Economics,* 46 (1931), 101-140; "Economics and Sociology: Marshall in Relation to the Thought of His Time," *Quarterly Journal of Economics,* 46 (1932), 316, 347.

5. Pareto gave the French edition of his book the title *Traité de sociologie générale.* (The Italian was *Trattato . . .*) It has always seemed to me unfortunate that the English translation, published some years later than my work, bore the title *The Mind and Society.*

6. This paper was never published as such, but its main substance, after considerable revision, was published as the three chapters on Pareto in *The Structure of Social Action* (New York: McGraw-Hill, 1937).

7. George Simpson, trans., New York: Free Press, 1964.

8. The first part of this book that I actually committed to paper concerned Durkheim's early empirical work and dealt with the Division of Labor and his (subsequently) very famous study of suicide (Chapter VIII).

9. Thus Sorokin, whose *Contemporary Sociological Theories* was the most widely used compendium in the field in the 1930's, treated Pareto, Durkheim, and Weber as belonging to entirely different schools, and did not once mention a relation between any pair of them. Marshall, of course, he did not discuss at all because he was labeled an economist, not a sociologist.

10. In retrospect it seems to me that this experience was, even apart from the substantive importance of Kant for my problems, especially important training for my later work. It was reinforced by a seminar and oral exam on the same book under Karl Jaspers at Heidelberg in 1926. The importance lay in the fact that I undertook the detailed and repeated study of a great book, the product of a great mind, to a point of reaching a certain level of appreciation of the nature of its contribution, and not being satisfied with the myriad of current rather superficial comments about it. This experience stood me in good stead in working with the contributions of my own authors and coming through to what I felt to be a high level of understanding of them in the face of many distorted interpretations current in the secondary literature, some of which were widely accepted.

    After the Pugwash Conference of 1967, I. I. Rabi commented to me on the importance of contact with the operation of what he explicitly called "really great minds." He spoke of his own good fortune in his student or early post-student days of being able to work in the laboratories of Pauli and Bohr. I did not have anyone of that caliber among my teachers in the flesh—though possibly Jaspers was—but I had a good many teachers who had that appreciation and I was able to develop it for myself through intensive study of the writings of great minds. The course on Kant was my first introduction to that experience.

11. Bernard Barber, ed., *L. J. Henderson on the Social System* (Chicago: University of Chicago Press, 1970).

12. New York: Macmillan, 1926.

13. Hans Vaihiger, *The Philosophy of "As if,"* trans. C. K. Ogden (New York: Barnes & Noble, 1952).

14. L. J. Henderson, *Pareto's General Sociology: A Physiologist's Interpretation* (Cambridge, Mass.: Harvard University Press, 1935).

15. Claude Bernard, *An Introduction to the Study of Experimental Medicine,* trans. H. C. Green, repub. ed. (New York: Daler, 1957).

16. W. B. Cannon, *The Wisdom of the Body* (New York: Norton, 1932). In the discussion at the Bellagio conference, when this essay was first presented, a question was raised about the seriousness of this exposure to biology. I was greatly pleased when Professor Curt Stern said: "May I make one very short point in regard to Amherst that not everybody might know. At Amherst, biology was taught at a very advanced, even a graduate level, although Amherst did not give doctors' degrees. These were highly distinguished people, and probably their influence was greater than it would have been had Professor Parsons gone to another college with good but less distinguished professors."

17. New York: Free Press, 1951.

18. It should be remembered here that social scientists had been forced to expend much ingenuity and energy on fighting off pressures toward illegitimate and premature biological "reductionism."

19. I was married in 1927 to Helen B. Walker, whom I had met as a fellow student at the London School of Economics. She has worked at Harvard for many years, most recently in administration of the Russian Research Center.

20. Thus he was extremely hostile to President Roosevelt, whose general policies I personally supported.

21. A note on the then, as now, prevalence of the belief that the word was "publish or perish" and that the criteria of conformity were to publish as soon as possible and as much as possible. I can testify that the responsible advice which I received from senior faculty members at Harvard did not fit the formula. It was uniformly to take as much time as necessary to do the best job of which I was capable. Of course, it helped immensely that such people as Henderson, Gay, and Wilson knew what I was doing and had seen samples of it. Another senior critic who should also be mentioned in the same vein was the late A. D. Nock.

    As in the case of most formidable personalities, there was an underground reaction to Henderson. He wore a reddish beard and was, behind his back, widely known as "Pinkwhiskers." My sessions with him took place in the late afternoon and I generally went directly home and told my wife what Pinkwhiskers had said. I remember worrying that my children would pick up the name and on some occasion say directly to Henderson: "Are you the Pinkwhiskers my daddy talks about?"

    It is fortunate that, between the first and final drafts of this paper, there has appeared *L. J. Henderson on the Social System,* edited by Bernard Barber. It contains most of Henderson's sociological writings, with a long and informative introduction by Barber. Barber's introduction tells in considerable detail the story of Henderson's involvement in the social science scene at Harvard and the various people in it with him.

22. Perhaps I may be pardoned for noting that, though I felt I was exceedingly fortunate to get the book published in the first place without subsidy and to have it republished in 1949 by The Free Press, it has maintained a steady substantial sale for over thirty years; indeed, just thirty years after the original publication a paperback edition was brought out which has sold well. In this matter I owe a particular debt to Jeremiah Kaplan, who practically "was" The Free Press, for his imagination in bringing out the 1949 reprint and sticking with this and a considerable series of my other publications. Without Kaplan, and his principal adviser Edward Shils, the postwar efflorescence in sociological publication, a wave on which I rode, probably could not have occurred, at least not so soon.

23. I have barely noted above that in my undergraduate days a comparable basic decision was made, namely to go into social rather than biological science. My switch to social science—a qualified economics initially—was associated with my father, who, during my development, was a college teacher and administrator. When I was a student in college my father was president of Marietta College in Ohio. He had begun his career as a Congregational minister and was very much involved with the then important "social gospel" movement, which, it is now clear, had much to do with the origins of sociology in this country.

24. Reprinted in Barber, ed., *Henderson on the Social System.*

25. See my paper, "Social Structure and the Development of Personality: Freud's Contribution to the Integration of Psychology and Sociology," in *Social Structure and Personality* (New York: Free Press, 1964).

26. See my paper, "The Problem of Controlled Institutional Change," reprinted in *Essays in Sociological Theory,* rev. ed. (New York: Free Press, 1954).

27. Possibly one precipitating factor in this diversion was the death, in 1938, of my father-in-law, Dr. W. D. Walker of Andover, Massachusetts, at the age of sixty. Dr. Walker was a particularly fine type of general practitioner of medicine and had been exceedingly helpful to me in the working out of the fieldwork phase of my medical study and in general discussions of the medical scene. At the same time, he was sufficiently old-fashioned not to "take much stock" in my more esoteric psychological interests.

28. Talcott Parsons and Edward Shils, eds., *Toward a General Theory of Action* (Cambridge, Mass.: Harvard University Press, 1951).

29. Ferdinand Toennies, *Community and Society* [*Gemeinschaft und Gesellschaft*], trans. and ed. Charles P. Loomis (New York: Harper & Row, 1963).

30. Bales had been one of the few graduate students in residence during the lean war years. As a junior faculty member he began his notable program of the experimental study of small human groups.

31. New York: Free Press, 1953.

32. Cambridge, Mass.: Addison-Wesley, 1950.

33. This scheme was explicated in Chapter III of *Working Papers in the Theory of Action.* Shorthand for it became the "A,G,I,L" scheme.

34. Talcott Parsons, Robert Freed Bales, James Olds, Morris Zelditch, and Philip Slater, *Family, Socialization, and Interaction Process* (New York: Free Press, 1955).

35. This interest was, as noted above, strongly stimulated by participation in the continuing conference on theories of systems, organized by Drs. Roy Grinker and John Spiegel. I was particularly influenced, in that conference, by the contributions of the Chicago biologist Alfred Emerson. See Appendix to *Family, Socialization, and Interaction Process.*

36. New York: Harcourt, Brace, & World, 1936 (Keynes); 8th ed., London: Macmillan & Co., Ltd., 1925 (Marshall).

37. New York: Free Press, 1956.

38. The interchange paradigm appears as an appendix to my article "On the Concept of Political Power," *Proceedings of the American Philosophical Society,* 107 (June 1963), reprinted in *Sociological Theory and Modern Society* (New York: Free Press, 1967).

39. *Ibid.*

40. This analysis is in two of my articles: "On the Concept of Influence," *Public Opinion Quarterly* (Spring 1963) and "On the Concept of Value-Commitments," *Sociological Inquiry,* 38 (Spring 1968), both reprinted in *Politics and Social Structure* (New York: Free Press, 1969).

41. See the Introduction to Part IV of Talcott Parsons, Edward Shils, Kasper D. Naegele, and Jesse R. Pitts, eds., *Theories of Society* (New York: Free Press,

1961) and my essay, "An Approach to Psychological Theory in Terms of the Theory of Action," in Sigmund Koch, ed., *Psychology: A Science,* III (New York: McGraw-Hill, 1959).

42. See the Appendix to "Some Problems of General Theory in Sociology," in John C. McKinney and Edward Tiryakian, eds., *Theoretical Sociology: Perspectives and Developments* (New York: Appleton-Century-Crofts, 1970).

43. W. I. Thomas, *The Unadjusted Girl* (Boston: Little, Brown, 1923).

44. See my article, "Malinowski and the Theory of Social Systems," in Raymond Firth, ed., *Man and Culture* (London: Routledge and Kegan Paul, 1957).

45. "Evolutionary Universals in Society," *American Sociological Review,* 29 (June 4, 1964), reprinted in *Sociological Theory and Modern Society;* "Christianity," in David Sills, ed., *International Encyclopedia of the Social Sciences* (New York: Macmillan and Free Press, 1968); *Societies: Evolutionary and Comparative Perspectives* (Englewood Cliffs, N. J.: Prentice-Hall, 1966); *The System of Modern Societies* (Englewood Cliffs, N. J.: Prentice-Hall, forthcoming).

46. See *Economy and Society,* Chapter 5.

47. See note 45, above.

48. See "Christianity," in Sills, ed., *International Encyclopedia,* and "Christianity and Modern Industrial Society," in Edward Tiryakian, ed., *Sociological Theory, Values, and Sociocultural Change: Essays in Honor of Pitirim A. Sorokin* (New York: Free Press, 1963).

49. There are complexities of "crossover" involved. Thus bureaucracy in economic production is a harnessing of *analytically* political components in an economic interest, and the constituency aspect of political democracy is a comparable harnessing of integrative components in the interest of government.

50. See my essay, "Durkheim's Contribution to the Integration of Social Systems," in Kurt Wolff, ed., *Emile Durkheim, 1858-1917: A Collection of Essays with Translations and a Biography* (Columbus: Ohio State University Press, 1960), reprinted in *Sociological Theory and Modern Society.*

51. See Robert A. Nisbet, *The Sociological Tradition* (New York: Basic Books, 1966).

52. Erik Erikson, "Youth: Fidelity and Diversity," *Dædalus* (Winter 1962), pp. 5-27.

53. "Age and Sex in the Social Structure of the United States," *American Sociological Review,* 7 (October 1942), 604-616, reprinted in *Essays in Sociological Theory.*

54. See "Kinship and the Associational Aspects of Social Structure," in Francis L. K. Hsu, ed., *Kinship and Culture* (Chicago: Aldine Press, forthcoming).

55. David M. Schneider, *American Kinship: A Cultural Approach* (Englewood Cliffs, N. J.: Prentice-Hall, 1968).

56. See my articles, "The School Class as a Social System: Some of Its Functions in American Society," *Harvard Educational Review,* 29 (1959), 297-318, reprinted in *Social Structure and Personality,* and, with Gerald M. Platt, "Higher Education, Changing Socialization, and Contemporary Student Dissent," in Matilda Riley, ed., *A Sociology of Age Stratification* (New York: Russell Sage Foundation, forthcoming).

57. Professor Renée Fox, in a commentary on the original draft of this essay, suggested that the decision to place "affect" primarily in the social system was effectively made much earlier, namely in formulating the pattern-variable, "affectivity vs. affective neutrality." The decision, new or old, has been hotly contested by two of my ablest young collaborators, Victor Lidz and Mark Gould.

58. See Neil Smelser, *Social Change in the Industrial Revolution* (Chicago: University of Chicago Press, 1959) and *Essays in Sociological Explanation* (Englewood Cliffs, N. J.: Prentice-Hall, 1968).

59. See Talcott Parsons and Gerald Platt, "The American Academic Profession: A Pilot Study," multilith (Cambridge, Mass., 1968) and my "Some Problems of General Theory in Sociology."

60. Preliminary findings are reported in Parsons and Platt, "The American Academic Profession."

61. *Dædalus* issues on "A New Europe?" (Winter 1964), "Science and Culture" (Winter 1965), "The Negro American" (Fall 1965 and Winter 1966), "Color and Race" (Spring 1967).

62. Daniel P. Moynihan, ed., *On Understanding Poverty* (New York: Basic Books, 1969); James L. Sundquest, ed., *On Fighting Poverty* (New York: Basic Books, 1969).

63. *Dædalus* issues on "The Embattled University" (Winter 1970), "Ethical Aspects of Experimentation with Human Subjects" (Spring 1969), and "Theory in Humanistic Studies" (Spring 1970).

64. A broad survey of some of these interests and activities can be gleaned from the annually published *Records* of the Academy. My first two annual reports as president are in the 1968 and 1969 volumes.

65. Lon L. Fuller, *The Anatomy of the Law* (New York: Praeger, 1968).

66. George C. Homans, *Social Behavior: Its Elementary Forms* (New York: Harcourt, Brace & World, 1961); *The Nature of Social Science* (New York: Harcourt, Brace & World, a Harbinger Book, 1967); "Contemporary Theory in Sociology," in Robert E. L. Faris, ed., *Handbook for Modern Sociology* (Chicago: Rand McNally, 1964).

67. Another important focus of this process lies in the teaching role. This is analagous to the common-law courts in that the teacher, especially at the advanced undergraduate and graduate level, is obligated, within rather wide limits of definition of his sphere of competence, to try to deal with questions raised by his students —in class, perhaps especially during seminar discussions; in term papers, honors papers, or other theses; and in personal conferences. Within limits the students, not the teacher, formulate the problems. If he deals with the students with competence and integrity, he must continually refer them to the generalized theoretical structure of the relevant bodies of knowledge. I have long thought that the enormous stimulus of these interchanges, both to good theoretical thinking and to incentive to be empirically informed and sound, constitutes one of the main reasons why too great a separation of the functions of research and teaching is unlikely to be healthy for the academic enterprise.

Certainly, in my own case, interaction with a succession of generations of intellectually able and curious students has constituted a major stimulus to the development of my theoretical thinking and acquisitions of empirical knowledge. The exceptional graduate students named earlier in this essay have played an especially important role in this respect—in a gratifying proportion of cases extending well beyond their student days.

68. It may be of interest to note that I took a Kantian approach to the problem of order. Very broadly, with respect to the epistemology of empirical knowledge, Hume asked "*is* valid knowledge of the external world possible?" and came out with, by and large, a negative answer. Kant, on the other hand, posed the question in a more complex way. He first asserted that "we in fact *have* valid knowledge of the external world" then proceeded to ask "*how* is it possible?", that is, under what assumptions? Similarly some social theorists have wondered whether social order was possible at all, and often denied its possibility. I, on the other hand, have always assumed that social order in fact *existed,* however imperfectly, and proceeded to ask under what conditions this fact of its existence could be explained.

69. That the same was true of Marshall, in the sense that he assumed it but did not let it obtrude too much into his technical economics, goes almost without saying. One may say that if there ever was a late Victorian, "evangelical" Englishman, Marshall was one.

70. I have been widely accused by critics of being a last-ditch defender of order at any price, the ultimate price usually being interpreted to be fascism. Fortunately the more preceptive of the critics have seen order as a problem, not as an imperative.

71. See my article, "The Sociology of Knowledge and the History of Ideas," in Philip Wiener, ed., *Dictionary of the History of Ideas* (New York: Scribner's, forthcoming).

72. Fairly early in my Harvard teaching career I introduced a course in the Sociology of Religion which continued for at least two decades, in the later years in collaboration with Robert Bellah.

73. If, without being accused of being a "racist," I may venture to quote an old Negro spiritual, I think the phrase "there's no hiding place down there" sums up the situation admirably. By "down there" in the present context I mean the positivistic view of the total cultural self-sufficiency of science, which has been alleged to have no "deeper" connections with any components or problems of human orientation outside itself.

74. I am reminded here of a geographical case which I have personally experienced a number of times. Seen from the valley of Chamonix, the Mont Blanc massif is clearly the most important mountain mass in that region of the Alps. If, however, one moves from Chamonix and its neighborhood, not only to Geneva, but beyond to the slopes of the Jura, the preeminence of the massif becomes overwhelmingly salient—providing weather permits visual perception. It is in an analogous sense that I think my three major figures constitute the highest "peaks" of the "range" of intellectual achievement in the field most relevant to me and in the time. This is by no means to deny the importance of the rest of the range.

PAUL A. SAMUELSON

Economics in a Golden Age: A Personal Memoir

ON ALFRED NOBEL's last birthday, in Europe's most beautiful building of this century, Professor Arne Tiselius spoke of a visit he once made to a Scottish chemistry laboratory. Through some failure of communications he did not realize until ushered into a large auditorium that a lecture was expected from him. When he asked what they would like to hear him discuss, someone called out, "How does one go about getting a Nobel Prize?" Professor Tiselius confessed he could not be responsive to the question.

When my turn came for the usual Stockholm remarks of gratitude and humility, I departed from my polished text to say, "I can tell you how to get a Nobel Prize. One condition is to have great teachers." And I enumerated the many great economists whom I had been able to study under both at Chicago and Harvard.

Although necessary, this condition is not by itself sufficient. "One must also have great collaborators." And, again, from Robert Solow down, I was able to count my own blessings. "Of course, one must have great students," following which went a recital of famous names.

Finally, in a crescendo of humility, I said, "And more important than all of these, one must have LUCK."

I stand by all these impromptu remarks. But with the shade of Samuel Johnson hovering over me and warning against all cant, I must add some afterthoughts. In the dark of the late October night, when a reporter phoned my home to ask my reaction to having received the Alfred Nobel Memorial Prize in Economic Science for 1970—and I spell out the title to emphasize that economics is a latecomer at the festive table with an award that is not quite a proper Nobel Prize—my first response was, "It's nice to have hard work rewarded." My children told me later it was a conceited remark.[1] Nonetheless one must tell the truth and shame the devil. It has been one of the sad empirical findings of my life: other things equal—with initial endowments and abilities held constant—the man who works the hardest tends to get the most done, and the one who saves the most does, alas, end up richest. To find these copybook maxims valid is as vexing as to learn that the horror stories told in my youth against cigarette smoking are after all true.

One must also tell the truth and shame the angels. So I must add what I might not have felt it necessary to add twenty years ago, namely that one must have been blessed with analytical ability. Great intellects—Newton, Lagrange, Gauss, and Mill will do as examples—have often been accused of false modesty. Newton said that however he may have appeared to others, to himself he appeared like a child playing with pretty pebbles on the beach. Lagrange explained his success in solving difficult problems by "always thinking about them." In similar words, Gauss discounted his superiority over other great mathematicians, saying in effect that, if you had thought as hard as he had about these matters, you too would be a Gauss. Finally, John Stuart Mill, who seems to have had the highest IQ ever observed, tells us in his autobiography, in words that are as charming as they are naïve, "Aw, shucks, anyone could have learned Greek at three and written a history of Rome at five"—and, I may add, have a nervous breakdown by nineteen—"if only he had the advantage of a teacher like James Mill."

What are we to make of these absurd disclaimers? Certainly Newton, who anonymously led the battle against Leibniz' claims to the calculus and who declared he would go to his grave without writing up the universal law of gravitation if he were required to make acknowledgment that Hooke had also some notions about attraction according to the inverse square of the distance, was anything but a truly modest man. And the record for Gauss is not that of a generous person. (When an old friend wrote to tell that his son had discovered non-Euclidean geometry, Gauss could not forebear from saying that he himself had already done that in unpublished work decades earlier. Worse than to kiss and tell is to not publish and claim.) A truly generous scholar, Euler—who made sense of Maupertuis' mystical principle of least action but refused to take the credit for it, and who delayed his own publications in the calculus of variations so that the youthful Lagrange could publish his novelties first—did not go around belittling himself or his work. Only in the case of Mill was pathological modesty a characteristic feature. (Actually when we read his remarks about (1) his father, (2) that paragon of all intellectual and other virtues, Harriet Taylor, and finally (3) his step-daughter, with whom he lived after Harriet's death, we do not have to be a Freud to recognize that we are in the presence of neurotic pathology. Besides, Mill's notions of radical philosophy—utilitarianism, feminism, and much else—required him to have a belief in the environment as the prime determiner of all abilities.)

Properly understood, there is much truth and not merely cant in the disclaimers of these men. A fish has no reason to be aware of the water he has never left. As we live inside our own skulls, our findings become transparently clear. We not only see our discoveries; but, particularly among profound minds, we also see through them. The excitement is in the chase. Once we have conquered the theorem, there is, as Mach has stated, an inevitable feeling of letdown, almost I would say a postorgasmic relaxation. When Newton was asked how he knew that gravitational attraction would lead to Keplerian ellipses for planetary motion, he could say simply, "I calculated it." When one heard the late John von Neumann lecture spontaneously at breakneck speed, one's transcendental wonder was reduced to mere admiration upon realizing that his mind was grinding out the conclusions at rates only twice as fast as what could be done by his average listener. Is it so remarkable that a few leading scholars will again and again lead the pack in the conquering of new territory? If you think of a marathon race in which, for whatever reason, one clique gets ahead, then you will realize that they need subsequently run no faster than the pack in order to cross each milestone first.

But let us make no mistake about it. Although there is much truth in the quoted disclaimers, there is also much nonsense. Mere work will not make a bookkeeper into a Gauss: it will not even make a Jacobi into a Gauss. And John D. Rockefeller did not get that rich by saving more dimes than other people. Talent, natural talent, is a necessary even if not sufficient condition for success in these realms. My old colleague in the Society of Fellows, Stanislaus Ulam, used to tell the story of a mathematician friend who had worked out a wonderful formula for success. Success in life turned out to be a many variable function, which depended *inter alia* on how handsome you were, how wellborn, and a great variety of other matters. "Ability," said Ulam, "did enter into the formula, but after much manipulation it was found to enter both in the numerator and the denominator and could be neatly cancelled out of the final answer." That is a good story to tell at a cocktail party. But, outside the fields of college administration, it is utter poppycock. Ulam himself provides an excellent counterexample: it was not his brown eyes that explain the invention of the hydrogen bomb, the development of the Monte Carlo method, and numerous advances in the area of topology.

As mentioned, twenty years ago I would not have insisted on these trite assertions. However, by pure chance, I was for a time given some medication that excellently treated the symptoms for which it was prescribed. But during that period, I felt that it took the fine edge off my mind. Suddenly I realized how the other half lives! It was not that my performance suffered so visibly to the outer world. During that very period I wrote one of my best articles, but to myself it was clear that I was living on capital. (Paderewski used to say that if he quit practicing for one day, *he* noticed it; if he quit for two days, the critics noticed it;

if he quit for three days, the whole world noticed. Luckily I was able to change my diet before the third day.) It was not merely that my ability to discover new truth was lowered; in addition the zest to discover new truth was diminished by a more passive participation in the struggle against ignorance. In sum, there is a chemical element in intellectual achievement and one is a fool to take great pride in the chance circumstance that one's chemistry happens to be a favorable one. I may add as a corollary that it would be nice to give Newton and Gauss a potion to show *them* what they were really like.[2]

The Time and the Place: The Midway

The year 1932 was a good time to come to the study of economics. "May you live in interesting times" may be a curse against happiness, but it is surely a benison for any scientist. Louis Pasteur in Eden would have become merely a brewer of beer.

The University of Chicago was a great place to study economics then. Frank Knight and Jacob Viner were at the top of their form. Henry Schultz, with energy and passion, was introducing the new mysteries of econometrics. Paul Douglas and Henry Simons catered to the needs of the young for relevance and commitment. Outside of the Economics Department, the rest of the university was in its finest hour. The new broom of Hutchins swept in exciting innovations of undergraduate curriculum and had not yet become the wand that paralyzes and destroys.

When I appeared on January 2, 1932, at 8:00 A.M. to hear Louis Wirth lecture on Malthus' theories of population, my mind was literally the tabula rasa of John Locke's psychology. Not yet graduated officially from high school, I had never heard of Adam Smith. (I later discovered that *The Wealth of Nations* had always been in our family library, disguised as a few inches of the five-foot Harvard Classics. But having sampled to my displeasure *Two Years Before the Mast* and being conditioned against the volume on the *Aeneid* which I had used as a trot—or, as we called it in those days, a "pony"—I forewent the pleasures of a liberal education in favor of research on sex in the eleventh edition of the *Encyclopaedia Britannica* and of browsing through the debates on Christianity and socialism that my father had picked up in the second-hand bookstores. For my money, Clarence Darrow and Robert Ingersoll always won the arguments: at fourteen turning the other cheek seemed merely stupid.)

I was prepared to find college difficult. True, I had always been a bright student. Although it was fashionable then to say you hated school, I always secretly liked it as a child and looked forward to the coming of September. Before the sociologists at the university showed me that all differences in performance rest on environmental opportunities, I was a naïve Francis Galton.[3] To ourselves, my brothers and I seemed perceptibly "smarter" than our cousins, who in turn were definitely smarter than the general run. It was with incredulity that I discovered in the second grade a boy who could add faster than I, but I could rationalize this by the fact that I was always being "skipped" a term, something that was particularly easy to do under the unconventional semester plan which characterized the innovative Gary school system. However, I reached my finest hour just before my hormones changed and turned me into an underachiever. Coming into college a term behind my class, I therefore thought that hard work would be necessary to survive, an agreeable error that launched me into scholarly orbit. As Wirth, himself a distinguished sociologist and excellent lecturer, expounded the 1-2-4- . . . and 1-2-3- . . . arithmetic of Malthus, it was all so simple that I was sure I must be missing the essential point. Since then I have come to realize that Malthus was just as simple as it seemed to me at sixteen, and hence I have never been surprised by the popularity of Malthusian doctrines with the man in the street and Ph.D.'s in biology.

Although there was a minute in my sophomore year when I toyed with the notion of becoming a sociologist, my real stimulus came from an old-fashioned course in elementary economics that remained in the curriculum as a fossil from the pre-Hutchins "old plan" days, and which I was put in by virtue of my late arrival. Having missed the cosmic aspects of economics, as expounded by Harry Gideonse and the really excellent staff in Social Science Survey I, I was expected to be able to catch up by learning about marginal cost and elasticity of demand. By luck, my teacher was Aaron Director, a strong libertarian of the Knight-Hayek school. (A local joke was that, later, he used to refer to Milton Friedman as "my radical brother-in-law.") Director was also an analyst and an iconoclast, whose cold stare terrorized the coeds in the class but captivated me. In any case, even if I had had Mr. Squeers for a teacher, the first drink from the economic textbooks of Slichter and Ely would have been like the Prince's kiss to Sleeping Beauty. It was as if I were made for economics.[4] I could not believe that the rest of the class were making heavy weather of such problems as to what would be the effect upon the price of kidneys of Minot's discovery that liver cured anemia, or the effect of mutton price if orlon were invented.

Chicago was a good place to learn economics at that time precisely because it was a stronghold of classical economics, a subject which had reached its culmination thirty years earlier in the work of Cambridge's Alfred Marshall. Economics itself was a sleeping princess waiting for the invigorating kiss of Maynard Keynes, and if one had to spend one's undergraduate days marking time before that event, Chicago was a better place to do so than would have been Harvard, Columbia, or the

London School. Cambridge University was never within my ken, but since economics was also waiting for the invigorating kiss of mathematical methods, it would have been a personal tragedy if I had become merely a clever First in the Economics Tripos there. (I like to think I might have risen above the tragedy, but as Wellington said of Waterloo, it would have been a "damned close-run thing.")

I have written elsewhere that, for an economic theorist, the last half of the nineteenth century was a bad time to be born. The really great work in neoclassical economics was all done in the years 1865 to 1910. Jevons, Menger, Walras, Böhm-Bawerk, Marshall, Wicksteed, Wicksell, and Pareto had gone beyond the classical synthesis of Smith, Ricardo, and Mill. Even the Marxian branch of the classical tree, save for one brief period of Indian Summer, shows clear sign of degeneracy after the turn of the century and the demise of Marx and Engels.

I do not say that 1915 was the perfect year to be born. Much as every family thinks that it would be happier with 20 per cent more income, every scientist thinks that if he had turned up just a bit earlier many of the delays in his subject might have been avoided and many more of the victories would have been his alone. Right before my time came the wonder generation of Frisch, Hotelling, Harrod, Myrdal, Tinbergen, Ohlin, Haberler, Hicks, Joan Robinson, Lerner, Leontief, Kaldor, and others too numerous to mention. Still to a person of analytical ability, perceptive enough to realize that mathematical equipment was a powerful sword in economics, the world of economics was his oyster in 1935. The terrain was strewn with beautiful theorems begging to be picked up and arranged in unified order. Only the other day I read about the accidental importation into South America of the African honey bee, with a resulting decimation of the local varieties. Precisely this happened in the field of theoretical economics: the people with analytical equipment came to dominate in every dimension of the vector the practitioners of literary economics.[5]

Elsewhere in this volume, Talcott Parsons tells how he moved out of economics and into sociology more or less by chance and the necessity to make a career. It seems to me that this was a great stroke of luck for him. Although his genius might have turned economics in the direction of methodological system building, it would have been the Lord's own work and definitely against the tides of change. What were the major tides of economics in the decades after my 1935 graduation?

Back around 1950, at a Princeton Inn meeting of the American Economic Association's executive committee, Frank Knight once announced in his cracker-barrel Socratic manner: "If there is anything I can't stand it's a Keynesian and a believer in monopolistic competition." Being not much more than half his age at the time, but definitely old enough to behave myself better, I asked, "What about believers in the use of mathematics in economic analysis, Frank?" When told he couldn't stand them either, I realized that the indictment fitted me to a "T." And I thanked my lucky stars once again that by chance and necessity, I, like Parsons, had been forced to leave the womb. Having been the usual A student and local bright boy at Chicago, I naturally thought it to be Mecca. Why leave Mecca?

Aside from its genuine excellences as a leading center for economics, Chicago had an additional attraction for me.[6] Although still an undergraduate, I had the opportunity to take Jacob Viner's celebrated course in graduate economic theory—celebrated both for its profundity in analysis and history of thought, but also celebrated for Viner's ferocious manhandling of students, in which he not only reduced women to tears but on his good days drove returned paratroopers into hysteria and paralysis. I, nineteen-year-old innocent, walked unscathed through the inferno and naïvely pointed out errors in his blackboard diagramming. These acts of Christian kindness endeared me to the boys in the backroom of the graduate school: George Stigler, Allan Wallis, Albert Gaylord Hart, Milton Friedman, and the rest of the Knight Swiss guards. As I performed various make-work tasks for the department—dusting off the pictures of Böhm-Bawerk, Menger, and Mill in the departmental storage room which Stigler and Wallis had squatted in—we would gossip for hours over the inadequacies of our betters and the follies of princes who try to set right the evils of the marketplace.

Karl Marx, though, was right in his insistence on the economic determinism of history, including the history of economists. For a few pieces of silver I left Olympus. The Social Science Research Council tried the experiment of picking the eight most promising economic graduates by competitive exam and, in effect, subsidizing their whole graduate training. My comfortable fellowship carried only one stipulation: leave home. As a scholar, I sighed; as an opportunist, I obeyed.

But where to go? Foreign study was frowned on. The choice, and it tells us as much about the state of economic institutions as would a Carnegie or Rosenwald Report, was either Columbia or Harvard. Most of my teachers and friends advised Columbia. Harry Gideonse, whose influence along with that of Eugene Staley on my choice of economics as a major was insufficiently stressed in my earlier remarks, said: "How could anyone give up Morningside Heights in preference for New England?" (How times change!) Wallis, Friedman, and many of the Chicago students of that day and since tended to have Columbia ties as well. They advised me, only too correctly, that I would not learn any modern statistics at Harvard if I passed up the chance to attend Hotelling's Columbia lectures.

The decision was made, as so many important ones are, by nonrational processes and miscalculation. I picked Harvard. Why? It was not because of the great Schumpeter. Actually I was warned that he was kind of a brilliant nut who believed that the rate of interest was zero in a stationary state, an impossibility that had been demonstrated at length by our local sage. I had never heard of Leontief or the mathematical physicist Edwin Bidwell Wilson (not to be confused with my Society of Fellows contemporary, E. Bright Wilson, Jr., the physical chemist also at Harvard), two great reasons for studying economics at Harvard. My teacher in money, Lloyd Mints, told me that John Williams was a good man but to watch out for the inflationist Seymour Harris. (Since Harris was still an orthodox protégé of Harold Hitchings Burbank, this judgment says much for Mints's prophetic powers to smell out evil.) I must confess that glances into Edward Chamberlin's recent and great book on *Monopolistic Competition* carried some small weight in the balance in favor of Harvard.

But in the end my decision was made on quite nonscholarly grounds. Gideonse did not know his man. I went to Harvard in search of green ivy. Never having been east (I don't count Florida as east) before the age of twenty, I picked Cambridge over New York in the expectation that the Harvard Yard would look like Dartmouth's Hanover common. Expecting white churches and spacious groves, I almost returned home after my first view of Harvard Square, approached by bad chance from the direction of Central Square. No wonder I annoyed Chairman Burbank at first encounter: I told him that I (1) would not take E. F. Gay's famous (and sterile and dull) course in economic history, but instead would take Chamberlin's course given to second-year graduate students, (2) intended to "skim the cream" of Harvard since it was by no means certain that I would choose to stay more than a year, and (3) had not made advanced application to do graduate study at Harvard for the simple reason that in those days any paying customer, to say nothing of an anointed Social Science Research Council Predoctoral-Training Fellow, could get in anywhere. It was not love at first sight. But it really would not have mattered since Burbank stood for everything in scholarly life for which I had utter contempt and abhorrence.[7]

The Golden Days of the Harvard Yard

"In 1935 a brash young student from the University of Chicago appeared at Harvard."[8] Luck was with me. Harvard was precisely the right place to be in the next half dozen years. When I once told Edward Mason that I proposed to write a memoir on the golden days for economics in the Harvard Yard, he suggested I wait "until we are up again." Well, there will never be a better time than now to sing of the Age of Hansen—and of Schumpeter, and of Leontief, and (for me) of E. B. Wilson. And I can sing for all my comrades at arms—Alan and Paul Sweezy, Kenneth Galbraith, Aaron [R. A.] Gordon, Abram Bergson, Shigeto Tsuru, Richard Musgrave, Wolfgang Stolper, and others who were already established when I arrived in the graduate school, September 1935; and for many yet to come—Lloyd Metzler, Robert Triffin, Joe Bain, James Tobin, Robert Bishop, John Lintner, Richard Goodwin, Henry Wallich, Cary [E. C.] Brown, Emile Despres, Walter and William Salant, Sidney Alexander, Benjamin Higgins, to say nothing of postwar diamonds such as James Duesenberry, Robert Solow, Carl Kaysen, . . .

Harvard made us. Yes, but we made Harvard. By the time of World War II the dominance in economics of Harvard had become almost a scandal. The old-school tie counted for much, too much no doubt—but it does save time to have familiar faces around. (When I came to do mathematical work at the Radiation Lab during the war, I found that the old Princeton tie had a similar role in moving you up the mathematics queue.) When the American Economic Association commissioned an official survey volume on the state of modern economics, George Stigler, in a scholarly review, drew up a statistical matrix of index references and mutual citing of the Harvard constellation in order to document the dominance I speak of. I do not suppose he would have gone to the trouble if he had not thought it a distorted reflection of the intrinsic worth of the barbarians outside the walls: Stigler has always insisted on the old-fashioned distinction between value and market price!

And yet, that this involved more than mutual backscratching was brought home to me by a later event. In 1948 when Alvin Hansen turned sixty, some of his fond students banded together to present him with a *Festschrift.** Since he had at that time spent no more years teaching at Harvard than he had teaching earlier at Minnesota, our committee naturally thought to invite students from his Minnesota days to contribute. From the Harvard vintages, there was almost an embarrassment of riches to choose from. Yet despite an extra effort the final disproportion was striking. Years later I happened to ask Alvin why that should have been so. He said that he had indeed had some very good students at Minnesota but that, by and large, the very best students had tended to bubble off (with his blessing) to the few largest centers. This confirmed the advice that I used to give young students: If you have any reason to think you are good, beg, borrow, or steal the money to come to the top place. If you go to Rome, it is your classmates who will be the next cardinals and who will be picking popes. I presume that these casual observations from experience confirm the views about "the institutionalization of a discipline" that Edward Shils discusses elsewhere in this book.

* Published as *Income, Employment, and Public Policy* (Norton, 1958).

Readers who are noneconomists will not wish me to elucidate the many features of graduate study at Harvard in those days. Let me summarize it all by saying that my transfer from Chicago to Harvard put me right in the forefront of the three great waves of modern economics: the Keynesian revolution, of which I shall give an account presently, the monopolistic or imperfect-competition revolution, and finally, the fruitful clarification of the analysis of economic reality resulting from the mathematical and econometric handling of the subject—including an elucidation for the first time of the welfare economics issues that had concerned economists from the days of Adam Smith and Karl Marx to the present.

Much of Harvard analysis was crude and unrigorous, as I discovered to my intense surprise. But it had life and it lacked closure. The mistakes one's teachers made and the gaps they left in their reasonings were there for you to rectify. You were part of the advancing army of science. There was an extra advantage: in that pluralistic environment there was plenty of opposition to every one of the new doctrines. You were not an Uncle Tom if you enlisted in the wave of the future. Schumpeter repudiated Keynes. Williams and Hansen conducted a famous dialogue, the courtesy of which in the face of fundamental disagreement does both men credit. (The greatest capital for undercutting a colleague in asides with students would have accrued to Hansen, and all honor must go to him that never once did he indulge in such personal criticisms. I was one of his favorites and intimates and the farthest I could push him to go was to say, when I praised a colleague long dead, that the man had in effect been a closet-Keynesian who, to curry favor with businessmen, had camouflaged his wisdoms and insights.)

All autobiography loses interest when success arrives. So let me simply say that the years 1935-1940 were good to me. As a junior fellow I was completely happy, turning out paper after paper. I once had a student who told me that he could have spent his whole life hitting fungoes to his brother on the farm back in Illinois but he had a suspicion it couldn't last. I could have remained a junior fellow all my life. Indeed it was fortunate that no one offered me a permanent fellowship at Harvard—lecturer was the name given for those with second-class membership in the club—because I would certainly have happily accepted it.

In 1940 I was offered the princely post of instructor in the Department of Economics and tutor in the Division of History, Government and Economics. I gladly took it and continued in my same Leverett House office. When a month later a better offer came from MIT and when I learned that my departure would not cause irreparable grief, I took the offer.[9] My last qualm in doing so was overcome as a result of a long handwritten letter from my revered teacher of mathematical economics and statistics, E. B. Wilson, about his Hegira from Yale to MIT early in the century. Wilson was the last of the universal mathematicians. He was

Willard Gibbs's favorite student, and one of the first to do work in a variety of fields: vector calculus, functionals, mathematical physics, aeronautical engineering, vital statistics, psychometrics, and most important for me, mathematical economics. He was the only intelligent man I have ever known who loved committee meetings. He also loved to talk and we had hours of conversations following his lectures. What a teacher's pet I have always been! I was also a favorite protégé of his. In the letter he told how his agonizing decision to leave Yale for MIT which friends had warned him against as a barbarian outpost, had been the beginning of a fruitful and happy epoch. Until you leave home, he said in effect, you are a boy and not the master of your own house. (In Virginia Woolf's diary there is a striking entry, which I quote from imperfect memory: "Father [Leslie Stephen, the biographer, historian, and mountain climber] would have been ninety today. Thank God he died. I could never have realized myself otherwise.")

Since gossip always likes a good story, and since McGeorge Bundy and the cited Breit-Ransom book have commented in print on the fact of my leaving Harvard, a few words on the event may not be out of order. I left Harvard in 1940 for the same reason that James Tobin left it in 1950: I got a better offer. Just as Lord Melbourne said he liked getting the Order of the Garter because there was "no damned merit to it," my parting was eased by the fact that no one, least of all me, thought that it was lack of merit that kept me from a chair in economic theory. Those were the depression days, the days of the Walsh-Sweezy hearings, in which President Conant was rationalizing Harvard's budget and tenure procedures after the benevolent despotism of President Lowell. Even my beloved mentor, Wassily Leontief, did not yet quite have tenure in my time. It was not for another decade that, under the so-called Graustein formula for spacing departmental appointments, a theorist was appointed at Harvard. It is frustrated expectations that make for disappointment and bitterness: from the first day I appeared in Boylston Hall in 1935, I never received any letter, oral promise, or clairvoyant message even hinting that a permanent appointment awaited me. When last year a Harvard Crimson reporter asked why I had not responded favorably to a couple of calls from Harvard in the later days of Bundy and Pusey, I heard myself stating the simple, prosaic truth: "After a cost-benefit analysis, I decided to stay put."

In any case, on a fine October day in 1940 an *enfant terrible emeritus*[10] packed up his pencil and moved three miles down the Charles River, where he lived happily ever after.[11]

What Economics Is About

It is easy for the artist to write about himself as a young man, but hard for him to write about his art. Has there ever been even one good novel about a scholar or artist that conveyed any notion of his work? It was Hans Zinsser who said that when any genuine biologist reads Arrowsmith's prayer in the Sinclair Lewis novel, he wants to throw up. (Yet I have seen elsewhere the biography of at least one distinguished scientist who was led into the field by reading that same book.) When my wife was a girl, she thought of her father, a small-town banker, as doing nothing down at work except sitting with his feet upon the desk. That I suspect is every child's view of what his parent does during that hiatus between morning and dusk. If James Joyce were to write the account of a single day in the life of a scientist, that would be unutterably dull except to another scientist.

G. H. Hardy claimed that the one romance of his life as a Cambridge mathematician was his collaboration with Ramanujan, the self-taught Indian genius. The great romance in the life of any economist of my generation must necessarily have been the Keynesian revolution. Perhaps the best impression I can convey of its impact is given by the following quoted passage, taken from a eulogy I was invited to write on the occasion of Keynes's death in 1946, less than a decade after the period I have been writing about:

> I have always considered it a priceless advantage to have been born as an economist prior to 1936 and to have received a thorough grounding in classical economics. It is quite impossible for modern students to realize the full effect of what has been advisably called "The Keynesian Revolution" upon those of us brought up in the orthodox tradition. What beginners today often regard as trite and obvious was to us puzzling, novel, and heretical.
>
> To have been born as an economist before 1936 was a boon—yes. But not to have been born too long before!
>
> "Bliss was it in that dawn to be alive,
> But to be young was very heaven!"
>
> The *General Theory* caught most economists under the age of 35 with the unexpected virulence of a disease first attacking and decimating an isolated tribe of South Sea islanders. Economists beyond 50 turned out to be quite immune to the ailment. With time, most economists in-between began to run the fever, often without knowing or admitting their condition.
>
> I must confess that my own first reaction to the *General Theory* was not at all like that of Keats on first looking into Chapman's Homer. No silent watcher, I, upon a peak in Darien. My rebellion against its pretensions would have been complete except for an uneasy realization that I did not at all understand what it was about. And I think I am giving away no secrets when I solemnly aver—upon the basis of vivid personal recollection—that no one else in Cambridge, Massachusetts, really knew what it was about for some 12 to 18 months after its publication. Indeed, until the appearance of the mathematical models of [it] there is reason to believe that Keynes himself did not truly understand his own analysis.
>
> Fashion always plays an important role in economic science; new concepts become the *mode* and then are *passé*. A cynic might even be tempted to

speculate as to whether academic discussion is itself equilibrating: whether assertion, reply, and rejoinder do not represent an oscillating divergent series, in which—to quote Frank Knight's characterization of sociology—"bad talk drives out good."

In this case, gradually and against heavy resistance, the realization grew that the new analysis of *effective demand* associated with the *General Theory* was not to prove such a passing fad, that here indeed was part of "the wave of the future." This impression was confirmed by the rapidity with which English economists, other than those at Cambridge, took up the new Gospel; . . . and still more surprisingly, the young blades at the *London School* . . . who threw off their Hayekian garments and joined in the swim.

In this country it was pretty much the same story. Obviously, exactly the same words cannot be used to describe the analysis of income determination of, say, Lange, Hart, Harris, Ellis, Hansen, Bissell, Haberler, Slichter, J. M. Clark, or myself. And yet the Keynesian taint is unmistakably there upon every one of us . . .

Instead of burning out like a fad, today ten years after its birth the *General Theory* is still gaining adherents and appears to be in business to stay. Many economists who are most vehement in criticism of the specific Keynesian policies—which must always be carefully distinguished from the scientific analysis associated with his name—will never again be the same after passing through his hands.

It has been wisely said that only in terms of a modern theory of effective demand can one understand and defend the so-called "classical" theory of unemployment . . .

Thus far, I have been discussing the new doctrines without regard to their content or merits, as if they were a religion and nothing else . . .

The modern saving-investment theory of income determination did not directly displace the old latent belief in Say's Law of Markets (according to which only "frictions" could give rise to unemployment and overproduction). Events of the years following 1929 destroyed the previous economic synthesis. The economists' belief in the orthodox synthesis was not overthrown, but had simply atrophied . . .

Of course, the Great Depression of the Thirties was not the first to reveal the untenability of the classical synthesis. The classical philosophy always had its ups and downs along with the great swings of business activity. Each time it had come back. But now for the first time, it was confronted by a competing system—a well-reasoned body of thought containing among other things as many equations as unknowns. In short, like itself, a synthesis; and one which could swallow the classical system as a special case.

A new *system*, that is what requires emphasis. Classical economics could withstand isolated criticism. Theorists can always resist facts; for facts are hard to establish and are always changing anyway, and *ceteris paribus* can be made to absorb a good deal of punishment. Inevitably, at the earliest opportunity, the mind slips back into the old grooves of thought since analysis is utterly impossible without a frame of reference, a way of thinking about things, or in short a theory.[12]

The final lines of this quoted passage contain certain notions that have brought fame to Thomas Kuhn, who argues in *The Structure of Scientific Revolutions* (University of Chicago Press, 1962) the importance of scientific paradigms in conditioning the thought of each school of science. This is a seminal idea, but, as expressed in the first edition of that work, one which fails to do justice to the degree to which "better" theories, and I mean intrinsically better theories, come to dominate and replace earlier theories. Now if Dr. Kuhn had been talking about the softer science of economics, I could provide him with much grist for his mill. And yet I must confess to the belief that truth is not merely in the eye of the beholder, and that certain regularities of economic life are as valid for a Marxist as for a classicist, for a post-Keynesian as for a monetarist.

In short, economics is neither astrology nor theology.

REFERENCES

1. They were apparently not alone. In a recent biographical work, W. Breit and R. L. Ransom, *The Academic Scribblers: American Economists in Collision* (New York: Holt, Rinehart and Winston, 1971), the authors quote the remark, with the prefacing words: "He was characteristically brash about his own achievements and prize." Despite some acknowledgments of help from me in personal correspondence on some fine points of doctrine, the authors' account does not jibe with my views on a number of matters.

2. Reminiscing one night at the Society of Fellows, I. A. Richards told me that the great Cambridge philosopher, Frank Ramsey, whose death at twenty-six was such a tragedy to economics as well as philosophy, found it hard to believe that others could not solve complicated syllogisms of mathematical logic in their heads. "Can't you *see* it?" he would ask. No they couldn't. (A good joke about von Neumann is in order. According to legend, when asked the old problem of how much distance a fly traverses in running back and forth between two approaching locomotives, von Neumann is supposed to have given the right answer—by *summing* the series. The other night, instead of counting sheep, I tried to do the same in my head. Although I dislike puzzles, it turned out to be fairly easy—given *all* the time one needs.)

3. A few years ago on a plane from Chicago to Washington, conversation with a fellow passenger enabled me to infer that he must be a Chicago professor of sociology. Without identifying myself, I chanced to remark: "At the University of Chicago I was taught to believe that differences in abilities and achievements are almost solely a function of differences in the environment. But now that I have become the father of six children, three of them triplets, I have had to modify my views." This drew down upon my head the following reproof: "Actually, you were right the first time. It *is* environment which is all important. Just to illustrate, take my own case: you may not realize it, but I am a distinguished professor and scholar. Yet I had two uncles who I can assure you were of the caliber to hold a chair in any great university in the world; but being poor immigrant boys, they were deprived of environmental opportunity." I could not resist replying: "Isn't it remarkable that a man who doesn't believe in the importance of heredity could report that no less than three of his immediate family were capable of being great scholars." Fortunately, he never turned his head.

4. Years ago I heard Boris Goldovsy interviewed on radio. He was asked whether his son had talent, and replied: "Of course both his mother and father are musical, but still it is uncanny. Even at the age of five, when he goes to the piano it is as if his fingers have a wisdom that the little boy himself does not know." Or to vary the analogy, no cat ever took to catnip the way our first-born, Jane, reacted at the age of several months to her first taste of ice cream: her eyes rolled to know that such delights existed, and if she could have climbed a tree or rassled a bear she would have. So with me, on looking into economic books. Possibly, I would have done well in any field of applied science or as a writer, but certainly the blend in economics of analytical hardness and humane relevance was tailor made for me or I for it.

5. That the apparatus of marginal revenue and imperfect competition should have had to be painfully rediscovered at the beginning of the 1930's, almost a century after Cournot's definitive 1838 work, testifies to the decadence of economics at that date. Or to illustrate with a harder problem that my economist readers will understand: in 1936, after I had already taken graduate courses in economic theory from Jacob Viner at Chicago and Joseph Schumpeter at Harvard, I still had to go around the Harvard Yard like Diogenes asking, "Why is it necessary for optimality that price should have to equal marginal cost?" And I was to receive no definitive understanding on the matter until my classmate Abram Bergson gave his 1938 definitive reformulation of modern welfare economics.

6. It is fashionable to recall how much one hated school days at Eton, and it is subtly self-flattering to state how badly one was educated at college. (Henry Adams even complained that he was never assigned *Das Kapital* in Harvard College, which demonstrates that the faculty there was already up to its tricks of not assigning books not yet written.) I had a great education at the University of Chicago from 1932 to 1935, and one of my resentments against Robert Hutchins is that, by the time of his retirement, he had reduced a thriving college of 5,000 undergraduates into a few hundred under-age neurotics. To be sure the withdrawal of Rockefeller's princely patronage also contributed to the decline. Among the great noneconomist lecturers I heard, and got to know outside of class, were the mathematician Gilbert Bliss, the biologist Anton Carlson, the anthropologists Fay-Cooper Cole and Robert Redfield, the paleontologist Alfred Romer, and nearer to my own field, Frederick Schuman and Harold Lasswell in political science, Louis Wirth in sociology, and W. T. Hutchinson in American history (the single best course lecturer I have heard anywhere). My education was more in width than depth, not a bad way to spend the years sixteen to nineteen prior to specialized research. In a 1972 obituary article of Jacob Viner, to appear in the March-April *Journal of Political Economy,* I include some further recollection of the Chicago scene.

7. It tells much about the Harvard of President Lowell's days that such men received office space and held power. But to admit the rich texture of life, I repeat a conversation between the late Alfred Conrad, Richard M. Goodwin (Hoosier, Rhodes Scholar, former Harvard don, and now a fellow in Peterhouse College at Cambridge) and me, as we were driving back to Boston from Joseph Schumpeter's funeral in Connecticut. Conrad: "Say what you will about Burbie, knowing that I had tuberculosis at Saranac, never a week went by that he did not ask me about my health. No one else in the huge Harvard environment seemed to know whether I was dead or alive. You were in a human relation with Burbank." Goodwin: "Indeed you were. He once said to me, 'Dick, I'd like to do with you

what my father once did to me. Tie you to the wheel of a wagon and apply a long black whip to your back.' " One should add that he was genuinely fond of Goodwin. During the war years a dear friend of ours was Burbank's secretary, but we avoided all strain by not talking about the man she worshipped and I despised.

8. Breit and Random, *Academic Scribblers,* p. 111.

9. A couple of months later, Ed Chamberlin, serving as an interregnum Harvard chairman, phoned to ask whether MIT was paying me a full-year's salary despite my late-October recruitment. When I said yes, he asked me whether I would not in that case return my September Harvard check, saying "you wouldn't want to deprive a young scholar of a full-year's income." I meekly returned the check, although it occurred to my wife that perhaps there might have been found a different solution for that poor man's problem.

10. A phrase of Provost Peter Kenen of Columbia, not mine.

11. A distinguished foreign economist who read this memoir commented to me: "You know, the popular explanation for your leaving Harvard is that you were a victim of antisemitism." I replied, "I suppose that would be the *simplest* explanation." "As a scientist," he asked, "aren't you required to accept the simplest explanation?" I do use Occam's Razor to slice away the reasons when they are otiose; but if a simple hypothesis cannot explain all the facts, I don't think it mandatory to embrace it. Lest I be misunderstood, let me state that before World War II American university life was antisemitic in a way that would hardly seem possible to the present generation. And Harvard, along with Yale and Princeton, was a flagrant case of this. So if anyone wants to understand why Jews, in relation to their scholarly abilities, were underrepresented on the Harvard faculty in those days, he can legitimately invoke the factor of antisemitism. (In many of the humanities faculties, bigots genuinely believed that Jews were no good; in science and mathematics, the belief was that they were too good; one could, so to speak, have one's cake and eat it too by believing—as did the eminent mathematician George D. Brikhoff and, to a degree, the eminent economist Joseph Schumpeter —that Jews were "early bloomers" who would unfairly receive more rewards than they deserved in free competition. Again, lest I be misunderstood, let me hasten to add the usual qualification that these two men were among my best friends and, I believe, both had a genuine high regard for my abilities and promise.) To illustrate, though, the failure of any one factor to account for the richness of reality, the following questions can be asked of those who know economists of that period well. (1) Why was Lloyd Metzler not given a tenure post at Harvard, since he suffered only from the disability of being from Kansas? (2) If you contemplate the academic careers in America of three men—Oskar Lange, Jacob Marshak, and Abba Lerner, not all Jewish—how can you cover the facts with the simple theory of antisemitism?

12. Quoted, with grateful acknowledgment to the original journal, from "Lord Keynes and the General Theory," *Econometrica,* 14 (1946), 187-199, and reprinted in P. A. Samuelson, *Collected Scientific Papers,* II (Cambridge, Mass.: MIT Press, 1966), pp. 1517-1533.

CURT STERN

The Continuity of Genetics

The Dawn of Classical Genetics: Mendel's Laws

When in 1865 Gregor Mendel presented his paper "Experiments on Plant Hybrids" he reported on various types of pea plants and gave ratios such as tall to short plants or yellow to green seeds in the offspring of hybrids. When only a single pair of contrasting traits was considered, ratios like 3:1 or 1:1 among the progeny of crosses were presented, and when two or more pairs of traits were followed simultaneously, ratios like 9:3:3:1 or 1:1:1:1 and even more complex ones were described. These ratios were expressions of empirical data and as such were open to discussion and interpretation. Mendel himself went beyond the empirical realm and provided a rational framework from which the facts could be deduced. He distinguished between the traits which he observed and the conceptual *Elemente* which were in the cells of his specimens and were responsible for the formation of the traits. He recognized the existence of pairs of these elements which corresponded to the pairs of contrasting traits, and he conceived of the existence of many pairs of elements in the cells of his plants corresponding to the many pairs of observable contrasting traits.

The ratios which Mendel had determined appeared as the consequences of two simple processes postulated by him: segregation and independent recombination of the elements. The first of these consisted in the separation of the partners of each pair of elements during formation of germ cells, or, more accurately stated, during formation of the ovules and pollen. Though the two contrasting elements of a pair are present together in the vegetative cells of a hybrid, only one of them is assigned to any one ovule or pollen produced by the hybrid. Specifically, Mendel stated, one half of the germ cells receive that element of a pair which came to the hybrid by way of the ovule from which it developed after fertilization and the other half of the germ cells receive that element which came to the hybrid by way of the pollen grain which had fertilized the ovule.

The second process which Mendel postulated took into account the fate of more than a single pair of elements. If A^1 and A^2 represent the elements of one pair and B^1 and B^2 those of another, a hybrid may be symbolized as $A^1A^2B^1B^2$. Segregation demands that all germ cells of the hybrid receive either A^1 or A^2 and either B^1 or B^2. But which of the A^1A^2 elements will be associated with B^1, which one with B^2? Mendel found that A^1 and B^1 are as frequently together as A^1 and B^2, and A^2 and B^2 as frequently as A^2 and B^1. Thus, independent recombination occurs, based on a process which gives no preference to any one association. In 1909, more than forty years after Mendel's exposition, his elements were named "genes" by the Danish geneticist Wilhelm Johannsen, who derived the term from William Bateson's 1906 designation of the science of heredity and variation as "genetics." Mendel had postulated the existence of genes without being able to assign them a physical basis. Johannsen could have done so but insisted that they should be regarded as no more than operational units.

Four years after Mendel's report in Brünn, Friedrich Miescher in Tübingen discovered a new chemical substance in the nuclei of pus cells and later in the nuclei of fish sperm. This substance, which he called nuclein, was deoxyribonucleic acid, DNA. It is nearly certain that Mendel never knew of Miescher's finding and that Miescher never heard of Mendel. If they had, it would not have helped them much in their diverse endeavors. Even though it was speculated early that DNA was the heredity substance, for a long time no connection was established between the specific genes of Mendel and Miescher's chemical analysis of nuclei en masse. Only slowly did the cellular role and the biochemistry of DNA become clarified, and not until three quarters of a century after Miescher's initial work was a belief in the genetic significance of DNA based on more than general arguments. We shall return to this late union of genes and DNA.

It is well known that Mendel's paper did not make an impression on his contemporaries and remained virtually unknown until the turn of the century, when its findings were rediscovered. Three investigators, without knowing of Mendel, experimented with crosses of varieties of plants and subsequently found references to Mendel's paper. With the publication of their data together with their acknowledged "Mendelian" interpretation the field of genetics became permanently established. Although the gap in hybridization studies which lasted from 1865 to 1900 seems to belie the theme of this essay, "The Continuity of Genetics," there was no gap in the wider pursuit of genetic tasks. During this period the fundamental facts of fertilization and of cell and nuclear division were established and incorporated into theories of heredity and development. Soon after 1900 the stream of efforts in cell studies was joined by that in hybrid analysis, and in the following decades the river of genetics sent branches to all areas of biology and received tributaries from them.

It might seem a strange coincidence that three separate individuals succeeded in the same year of 1900 to rediscover the Mendelian laws. In reality the part played by each of these men was different. Of the three, Hugo De Vries, in Holland, was the first to publish an account of his findings. This prompted Carl Correns, in Germany, to write up an account of his own work within a day of receipt of a reprint of De Vries' paper. Erich von Tschermak, in Austria, composed his account after seeing the papers of his two predecessors. All three investigators independently retrieved Mendel's forgotten paper, references to it being easily found in a well-known book by the botanist Wilhelm Focke, published in 1881 (this compilation surveyed all previous work in plant hybridization but did not grasp the significance of Mendel's work).

With the rediscovery of Mendel's publication the common merit of De Vries, Correns, and Tschermak ends. De Vries and Correns not only claimed that they had discovered the essential facts and developed their interpretation before they found Mendel's article, but they demonstrated in their own publications that they fully understood the essential aspects of Mendel's theory. On the other hand, Tschermak's papers of 1900 were inadequate in the analysis of their own data and were lacking an interpretation. I remember well a conversation with T. H. Morgan in the middle twenties. Morgan judged that Correns had shown clearly by his later work that he could have founded "Mendelism" without seeing Mendel's paper, and that Tschermak had shown that he was unlikely to have been an independent discoverer. Morgan was inclined to judge De Vries' independence as not proven by the later work of this undoubtedly great scientist.

The rediscovery of the Mendelian laws in 1900 was a far lesser accomplishment than Mendel's original feat. This has been stressed by Correns. In a later account of his part in the "confirmation of that which had been discovered more than 30 years before" he wrote: "And through all that in the meantime had been discovered and thought (I think above all by Weismann), the intellectual labor of finding out the laws anew for oneself was so lightened, that it stands far behind the work of Mendel."[1]

The process of fertilization of the egg which initiates the development of all sexually reproduced organisms had remained obscure from the dawn of biology until 1875. In that year Oskar Hertwig at the age of twenty-six observed for the first time the essentials of fertilization—in the sea urchin. In this organism the small transparent eggs are spawned into sea water, where a sperm enters an egg. After entrance the sperm's head, its nucleus, swells up and assumes an appearance similar to the nucleus of the egg. The two nuclei then coalesce into a single unit as fertilization occurs. It was not long until the essence of Hertwig's observations was confirmed in other animals as well as in plants: fertilization consists of the complete union of two and only two sexual cells, and it involves the union of the two parental nuclei.

Mendel himself had been concerned with the nature of fertilization. In his historic paper on plant hybridization he wrote that "according to the opinion of famous physiologists" reproduction in higher plants involves the pairwise union of ovule and pollen cell to form a single cell. He found support for this view in his experiments with peas, where among the offspring of hybrids both original parental types recur in equal numbers. He specifically rejected the idea that the ovule plays only the role of a foster mother of the pollen cell, since in that case the hybrid would resemble the pollen plant exclusively or very closely. "A thorough proof," he wrote, "for complete union of the content of both cells presumably lies in the universally confirmed experience that it is immaterial to the form of the hybrid which of the parental types was the seed or pollen plant."

Notwithstanding these valid considerations and the opinion of famous physiologists, Mendel felt the need to reexamine the view that only a single pollen grain suffices for fertilization. In 1869 and 1870, although he was troubled by an eye ailment and his time for experiments was greatly curtailed by his duties as abbot of his monastery, he could not bring himself to postpone a crucial experiment. He placed single pollen grains on the stigma of flowers of *Mirabilis Jalapa,* the four-o'clock, and obtained well-developed seeds and from them plants which could not be distinguished in any way from those produced by ordinary pollination. He thus refuted the opinion of Charles Darwin and of the French plant hybridizer Charles Naudin, who had come close to being an independent discoverer of Mendel's laws. Naudin and Darwin had maintained that a single pollen grain does not suffice for fertilization of the ovule. (A detailed analysis of Naudin's and Darwin's views will appear in a forthcoming publication by Robert Olby.)

In his experiments Mendel anticipated Hertwig's observations with sea urchin eggs, yet Mendel's fundamental findings of the sufficiency of a single pollen grain remained even less accessible to other investigators than his pea experiments. The latter, after all, had been published in a scientific journal (the *Transactions of the Brünn Natural History Society*), while the former were reported solely in letters to the Swiss botanist Carl Nägeli, letters which remained unknown to everyone else until their publication early in the twentieth century. In one of these letters Mendel refers to another experiment under way, ingeniously designed to find out whether two pollen grains may simultaneously participate in fertilization. No report on the results of the experiment has come down to us.

The volumes of egg and sperm of a given species are nearly always very disparate. The egg is a large, nucleated cell rich in cytoplasm and laden with reserve materials, the sperm a very small cell consisting mostly of a condensed nucleus and possessing only a small amount of cytoplasm. Clearly the quantitative contributions of egg and sperm are unequal. Why then does not the offspring of two parents resemble its mother more than its father, instead of, by and large, resembling both parents to the same degree?

The answer was provided by Hertwig's microscopic observations and those of his followers and contemporaries in the 1880's. Since the two uniting nuclei involved in fertilization are more or less alike in size and general content, *they* must be the bearers of the hereditary material which the new individual obtains in equal amounts from his two parents. The extranuclear cytoplasm of egg and sperm and its content apparently is neutral as far as inherited traits are concerned.

The visible equivalence of egg and sperm nucleus was resolved further. During division the otherwise diffuse nuclei of cells were shown to form well-defined threads or loops, which were later called "chromosomes" since these bodies preferentially take up certain stains. The Belgian zoologist and parasitologist Edouard van Beneden discovered that *Ascaris,* the large intestinal round worm of the horse, had only four chromosomes in its fertilized eggs, one of the very few species with so small a number. In 1883 van Beneden showed that in fertilization the egg nucleus and the sperm nucleus alike contain two chromosomes; not only are the nuclei in bulk contributed equally by the two parents, but specifically the countable, formed elements—the chromosomes. It was reasonable to conclude that it is not the diffuse content of the nuclei but their chromosomes which are the bearers of the hereditary material.

Independently of van Beneden's publication the presumably genetic role of the chromosomes had been pointed out in a brilliant, brief pamphlet, "On the Significance of the Nuclear Division Configurations" (1883). Wilhelm Roux, the author of this "hypothetical discussion" (as the title page termed it), asked himself what the meaning was of the complicated processes of nuclear division—mitosis. The work of a galaxy of students on the fine structure of the cell—work made possible by great advances in optical theory and the engineering of optical instruments and by the introduction of fixation and staining procedures for the study of cells—had culminated in the finding that each chromosome is split lengthwise during mitosis and that the two resulting daughter chromosomes subsequently move to opposite poles and so become included in two different cells. The cytoplasm of a dividing cell is usually simply constricted into two portions, each of which becomes a new cell. Why does the nucleus not divide itself up in a similar rough fashion, instead of arranging its contents in long threads—the chromosomes—and then cleaving them lengthwise?

Roux's approach was frankly teleological. The founder of *Entwicklungsmechanik*, the causal analysis of developmental processes, he was willing to ask questions about the purpose of a biological phenomenon which were complementary to questions about its causal antecedents. "Let us consider the task," he wrote, "of dividing into two halves a mixture of different substances in such a way that in each half is found one half of each of the different substances." If the number of different substances is very large, and the amount of each very small, then division of each "granule" representing one of the substances and separation of the sister granules into opposite locations would be a suitable method. Moreover, the task of dividing and distributing the sister granules would be eased greatly if many granules were arranged together longitudinally, forming one or several threads, dividing lengthwise whole threads and distributing their daughter threads into two separate assemblies. "Our purpose was the halving of a mixture of substances not just in terms of its total mass but also in terms of each separate quality . . . If the purpose discussed by us is likewise that of mitosis, then all those marvelous processes of mitosis would be recognized as fully purposeful. Conversely, since we cannot assume . . . that a process can be useless that is so generally present, that requires so much time and energy and that is so complicated and difficult to acquire, we must admit a certain probability that our postulated purpose is also the purpose of mitosis."[2]

For Roux the function of the extraordinary multiplicity of nuclear qualities consisted in their importance for physiological, behavioral, and developmental processes. Explicitly the word "heredity" itself cannot be found in the pamphlet, but its essence is implied. In the final pages Roux stresses that the apparent homogeneity of the chromosomal substance of the nucleus should not deceive a critical observer. "One must be aware," he wrote (translated somewhat freely), "that we look at the molecular processes of the cell in a way similar to looking at a factory from a balloon that floats suspended in the highest regions of the atmosphere: the diameters of the parts are a millionfold smaller than the distance from which we see them and the most diversified objects therefore appear homogeneous."[3]

At the same time that Roux recognized the chromosomes as the bearers of the submicroscopic units which are transmitted from cell to cell and which control the fundamental life processes of the organism, August Weismann came to a similar conclusion and explicitly attributed to the chromosomes the role of bearers of hereditary units. His elaborate theoretical framework included a point of view which was diametrically opposite to the one held by most biologists from antiquity up to and beyond Charles Darwin. The old view derived the hereditary constitution of the germ cells from messengers sent out by the body parts of the organism. Darwin, for instance, in his "provisional hypothesis of pangenesis" conceived of each organ sending out particles into the blood stream which then were assembled in egg and sperm. This hypothesis implied the possibility of the inheritance of acquired characteristics. The experiences of an organ would mold the nature of its "gemmules," and the hereditary properties of the germ cells would reflect the acquired properties of the organ. Weismann's new view looked at the changeable individual as only a passing expression of the invariant content of the generations of germ cells. Instead of the egg being considered the hen's way of making more hens, the hen was being regarded as the egg's way of making more eggs. The principle of conservation of the hereditary material through the chain of life was applicable only to the contents of eggs and sperm—the germ plasm—not to individuals. There was no inheritance of acquired characteristics.

When Weismann erected his theoretical edifice he had not heard of Mendel, but when Mendel's work became known, Weismann was jubilant. "The Mendelian laws are a confirmation of the basic tenets of the theory of the germ plasm."[4] Indeed, Mendel's "elements" were observed to segregate unchanged in their properties by the nature of the individual which had carried them from fertilization to maturity. Crosses of tall by dwarf pea plants yielded tall hybrids. When these produced pollen and ovules, half of them carried the unchanged gene for tallness, which had been present in the tall hybrids in the same cells which also held the gene for dwarfness; the other half of the pollen and ovules carried the unchanged gene for dwarfism, which had been present—unexpressed—in the same cells which also held the gene for tallness.

The Confluence of Classical Genetics with Cell Biology: The Genetic Role of Chromosomes

While Roux and Weismann argued at a biological level in favor of the chromosomes as the hereditary material, other investigators contributed chemical evidence. In 1881 the botanist E. Zacharias showed that Miescher's nuclein was located not indiscriminately in the nucleus but specifically in its chromosomes. On the basis of this fact, Hertwig and others who had recognized the nature of chromosomes as bearers of the hereditary material postulated that it is the nucleic acid moiety of the chromosomes which carries the genetic information. A proponent of this hypothesis was E. B. Wilson. In the first edition of his *The Cell in Development and Inheritance* (1896), one of the most scholarly books written by an American biologist, he summarized the findings which suggested that inheritance depends on a specific chemical substance in the chromosomes, the nucleic acid of Miescher. Wilson's synthesis undoubtedly was read widely, yet it remained without consequences: awareness of the existence of a substance in an organism obviously does not carry with it knowledge of its function. Altogether the observations and speculations which pointed first to the cell nucleus, then to its chromosomes, and further to their nucleic acid as the bearers of the hereditary material were inadequate to establish a complete "chromosome theory of inheritance." Such a theory demanded knowledge in two areas, cell biology and genetics, and the second of these did not yet exist.

When Mendelism came into being at the turn of the century there was waiting the accumulated evidence and speculations on the significance of the chromosomes in development and hereditary transmission. It took only a few years until the facts of hybridization of plants and animals and the microscopic studies of cells were coordinated (1903, 1904). In independent accounts by Wilson's American graduate student William B. Sutton and the German zoologist Theodor Boveri it was pointed out that the deduced type of transmission of Mendel's elements mirrored the observable type of transmission of chromosomes from one generation to the next. Sutton and Boveri reasoned as follows. Elements—that is, genes—occur in pairs; so do chromosomes. Members of each pair of genes segregate at the time of formation of germ cells so that each germ cell receives only one of the two genes. Chromosomes likewise occur in pairs and segregate at the time of germ cell formation, assigning only one of each pair to any egg or sperm. The members of different pairs of genes recombine at random at the time of segregation. Different chromosome pairs may well do the same: although observations at that time could not prove this inference to be true, they did not contradict it. The parallelisms of genetic and cytological facts found their explanation in the assumption that genes are located in the chromosomes—one of a pair of genes in one chromosome, the other in the partner chromosome. Different pairs of genes were assumed to lie in different pairs of chromosomes.

If this Sutton-Boveri theory of Mendelian inheritance was indeed not valid, one would have to discover other cellular particles that behave exactly like chromosomes. If it was valid, a tremendous advance had been made in the understanding of the basic organization of living systems. Instead of vague speculations about inheritance, whose prime rule was, it had been said, that there was no rule, the recognition of genes as separable entities and their placement in the chromosomes provided insight into the structural makeup of the self-replicating genetic material.

The theory, however, did not conquer the minds of all geneticists. William Bateson, the foremost Mendelian in the early years of the century, continued to see his task as the abstract analysis of complex crosses. It led to the discovery of ever more gene pairs, without reference to their physical basis in the chromosomes. Others too were reluctant to see in the parallelism between the behavior of genes and chromosomes any more than a chance coincidence. What was needed was further, independent evidence for an intrinsic nature of the parallelism. The first fully convincing proof of the genetic importance of chromosomes came from Boveri in a preliminary report in 1902 and five years later in a volume of such exhaustive and penetrating critical ingenuity that it should be mandatory study for every biologist—were it not, alas, written in German.

Boveri was concerned with the results of abnormal fertilizations. When it had been established that fertilization normally involves the fusion of one egg nucleus with one sperm nucleus, it was soon seen that the occasional fertilizations of sea urchin eggs by two spermatozoa resulted in early death of the developing egg. Boveri set himself the task of finding the cause of the ill effect of double fertilization. He observed a great variability between different dispermic eggs in the stage at which they died and in the type of embryonic organ whose disintegration led to death. He also observed that within the same embryo or larva different quadrants developed in often strikingly different ways, one quadrant perhaps nearly normally, another very abnormally, and the remaining two quadrants in two different intermediately abnormal ways. In normal, monospermic sea urchin egg fertilization the sperm brings into the egg not only a nucleus but also a special organelle, the centrosome. This divides into two and forms a spindle between its two parts (Figure 1, A). There the chromosomes of egg and sperm nucleus arrange themselves in an equatorial plate, and after each chromosome has divided to form two sister chromosomes the spindle separates and assembles them in two sister nuclei. These come to lie in the two cells which are formed by cleavage of the original single egg cell.

The sea urchin species with which Boveri worked has 18 chromosomes in each of the fusing egg and sperm nuclei—that is, 36 chromosomes jointly. The longitudinal division of these 36 chromosomes results in the presence of 72 chromosomes, 36 of which go to one cleavage cell and 36 to the other. In dispermic fertilized eggs two centrosomes are introduced, which after division form four centrosomes. In many cases the four centrosomes come to lie in the egg like the four corners of a square, and four spindles are formed between them like swollen sides of the square (Figure 1, B). The first cleavage division of the egg results in four cells instead of the normal two. Each of the four ultimately forms

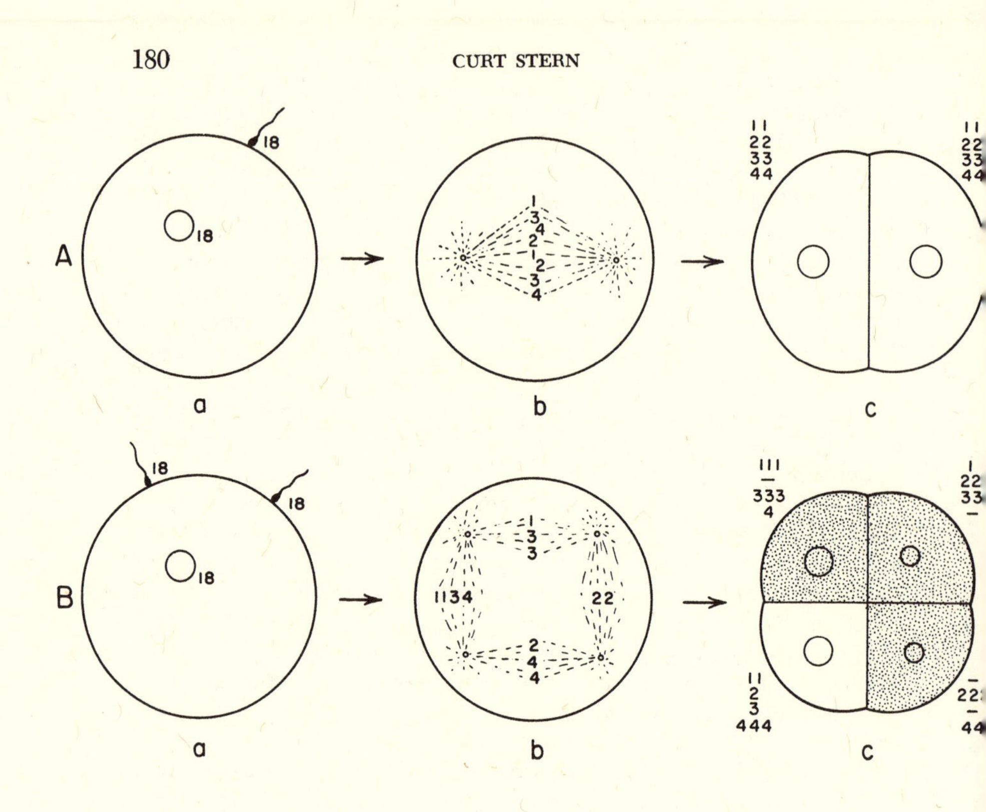

Figure 1.

A. (a) Normal fertilization of a sea urchin egg. The sperm head and the egg nucleus each carry 18 different chromosomes. Four pairs of the chromosomes, one of each pair from each parent, are shown in (b). They are designated 1, 2, 3, and 4.

(b) The first cleavage division with chromosomes 1, 1; 2, 2; 3, 3; and 4, 4. The chromosomes have not yet divided.

(c) After chromosome division and movement to the two poles of the division spindle, two cells are formed whose nuclei are identical in chromosomal content.

B. (a) Fertilization by two spermatozoa.

(b) Four cleavage spindles are formed with chromosomes 1, 1, 1; 2, 2, 2; 3, 3, 3; and 4, 4, 4. The chromosomes are irregularly distributed over the four spindles.

(c) Four cells are formed whose nuclei vary in size according to number of chromosomes as well as in kinds of chromosomes. The three stippled cells lack one or two kinds of chromosomes and have totals of 7, 5 and 5 chromosomes, respectively. The descendants of these cells die. The fourth cell, not stippled, has 7 chromosomes including at least one of each type. Its descendants will form a surviving quadrant of the embryo.

one quadrant of the developing embryo or larva, just as the four cleavage cells formed by two successive divisions of a normally fertilized egg. The latter, however, form a normal larva; the former, a very abnormal one. In various ways Boveri showed that the difference in development cannot be attributed to double fertilization per se or to a number of other, secondary circumstances, but must be connected with the constitutions of the chromosomes. In a dispermic egg, egg and sperm nuclei together furnish $3 \times 18 = 54$ chromosomes at fertilization. These become distributed over four spindles. Observation showed that chance plays a large role with respect to which chromosomes go on which spindle. There is great variation from egg to egg; sometimes all spindles obtain approximately equal numbers of chromosomes, sometimes extremely unequal numbers.

The 54 chromosomes assembled on the four spindles split lengthwise thus result in 108 chromosomes. When the sister chromosomes move to opposite poles they are assembled in four nuclei. The size of these nuclei and their descendants in the four quadrants of the embryo depends on the number of chromosomes contained in them. By comparing the size of the diameters of the nuclei with the degree of developmental success in each quadrant of an embryo, as well as the developmental success in the quadrants of different embryos which have nuclei of similar size, Boveri came to the conclusion that variable numbers of chromosomes cannot account for the variability in development. Further considerations then made it clear that this variability was caused by qualitative differences between the nuclei.

When chromosomes were first investigated in such organisms as salamanders and onions they seemed all alike in size, and Weismann had speculated that they were alike too in genetic content. When Boveri studied the chromosomes of his sea urchins he was aware that they showed some size differences and that even more striking differences in size and shape had been found in other organisms. From there it was not a far cry to think of qualitative differences between the various chromosomes of a cell and to attribute the specific types of cellular degeneration of the embryonic quadrants to lack of specific chromosomes with their special qualitatively unique content. This theory could be tested in detail. According to it the 54 chromosomes of the just-fertilized dispermic egg should consist of three sets of 18 different chromosomes. Chance would assign any one chromosome to one of the four simultaneously formed spindles. One could calculate how often a resulting nucleus would fail to obtain all three of one of the 18 kinds of chromosomes, and then one could compare this theoretical expectation with the frequency of abnormal quadrants.

Boveri discussed the statistical problem with his colleague W. Wien in the physics department of Würzburg University. Wien, who knew that Boveri did not like even simple arithmetic computations, advised against an abstract treatment and instead proposed an experimental copying of the natural processes involved. Accordingly Boveri prepared a circular wooden tray on which he poured the well-mixed contents of a beaker —54 wooden spheres painted, in groups of three, with successive integers: 1, 1, 1; 2, 2, 2; . . . 18, 18, 18. He then placed onto the tray a wooden cross with arms the length of the diameter of the tray, thus separating the 54 spheres into four random groups that corresponded to the four chromosome groups present on the four spindles. In the living egg the chromosomes of each of the four groups would divide and the sister chromosomes would be distributed to opposite poles, thus forming four nuclei. In the model experiment the distribution of the chromosomes over the four nuclei could be determined from their distribution over the spindles. Whenever each "chromosome" was present in a nucleus at least once, the situation would lead to normal development. Whenever one or more "chromosomes" were not represented at all, the situation would lead to pathological development. Repeating the model procedure a hundred times yielded percentage estimates of expected normal and pathological development of quadrants which could be compared with actual results from dispermic eggs. The agreement between expectation and observation in this as well as in variants of the experiments was reasonably good, and the theory of qualitative differences between chromosomes became generally accepted. Worded differently, the theory stated that different chromosomes carry different genes.

Independent support for the theory came from consideration of a special chromosome found in grasshoppers and plant bugs. In 1891, in a descriptive study, the German investigator, H. Henking had found an unusual nuclear body of unknown function which he called X; this was present in one half of all sperm and absent in the other half. Eleven years later the American zoologist Clarence McClung wrote an article entitled "The Accessory Chromosome—Sex Determinant?" He had recognized the chromosomal nature of the odd body, and finding that all males of certain species had 11 pairs of ordinary chromosomes plus one "X chromosome," he proposed that this chromosome was the determiner of male sex. The suggestion that the X chromosome is sex-determining proved to be correct, although the specific conclusion that it was male-determining turned out to be wrong. Since his reasoning was based on an erroneously reported count of 22 chromosomes in females, McClung's assignment of male-determining properties to the accessory chromosome had been logical. But the female chromosome count should have been 24, consisting of 11 pairs of ordinary chromosomes and one pair of X chromosomes. It follows that the role of the X chromosome in the species concerned is female-determining: while one X is insufficient to direct development to femaleness, two X's do accomplish this goal.

McClung's fusion of the cytology of the X chromosome with the ancient riddle of sex determination was the first step in an analysis of the genic content of the X chromosome. Typical Mendelian experiments had shown that it made no difference whether a specific inherited trait was present in the mother or the father of a cross. Also, daughters and sons of such crosses were alike in either showing or not showing the trait. The chromosome theory of inheritance readily accounts for this "symmetry" in transmission, since most chromosomes occur in pairs, alike in females and males. The X chromosome, however, is an exception to the rule of alikeness: in many insects, as we have seen, and in many other species including man the X chromosomes occur in an "asymmetrical" fashion—twice in females and once in males. One could have predicted from this cytological situation that genes located in the X chromosomes would be transmitted asymmetrically.

Actually, not prediction but empirical facts of unexpected types of inheritance led to an interpretation in terms of genes in the X chromosome. The classical example of "sex-linked" inheritance is that of the white-eye trait in the fruit fly *Drosophila.* Crossing a red-eyed female with a white-eyed male produces daughters red-eyed like the father and sons white-eyed like the mother ("criss-cross" inheritance). This experiment was Thomas Hunt Morgan's, and his interpretation (in 1910) was the following. As a result of segregation of chromosomes the eggs produced by a female fly are all alike in having one X chromosome, but the sperm are of two kinds, with and without an X (the latter carrying a Y chromosome). X sperm fertilizing an X egg leads to development of a female; non-X sperm to that of a male. Every female, therefore, derives its two X chromosomes from both parents, one chromosome from each. Every male derives its single X chromosome from its mother. Now, if one assumes that the X chromosome carries one or the other of the contrasting genes for red (+) or white (w) eyes, one can symbolize the cross (1) red female by white male as $X^+X^+ \times X^w$ and the reciprocal cross (2) white female by red male as $X^wX^w \times X^+$. The daughters of both crosses will be alike, X^+X^w with the dominant w^+ resulting in redness. The sons of the two crosses will be different, being X^+, red, in cross (1) and X^w, white, in cross (2).

The chromosomal interpretation of sex-linked inheritance in *Drosophila* was easily extended to man. Red-green colorblindness, hemophilia, and other traits whose transmission was known to follow certain rules with an obscure rational basis became the first examples of human genes to be assigned to a specific chromosome, the X.

The success of the chromosomal interpretation greatly strengthened the evidence for the chromosome theory in general. If the parallelism between the usual transmission of genes and the usual chromosomal situation could have been a chance coincidence, the additional parallelism between sex-linked inheritance and X-chromosomal behavior made a causal connection nearly obvious. Moreover, it was found that sex-linked inheritance occurred in two different ways—the *Drosophila* and man way, and the moth and chicken way. The results of genetic experiments with moths and chickens required that the females form two kinds of eggs, those with and those without an X chromosome, and that the males form only one kind of sperm, with an X chromosome. In other words, the females should be X and the males XX, which ran counter to the many findings on X chromosomes of various species in which the females always had been XX and the males X. No one, however, had thoroughly studied the chromosomes of moths and chickens. Cytological investigation was stimulated by the genetic experiments, and studies carried out by Jacob Seiler (1913) showed that the females indeed were X and the males XX. Here was a genetic exception from normal transmission of traits which was opposite in kind from the exception represented by white eyes in *Drosophila.* And the new type of exception found its parallel in a new kind of X-chromosome inheritance.

One more study in *Drosophila* provided the clinching evidence for the chromosome theory. If X-linked inheritance was an exception to the normal type, rare exceptions from X-linked inheritance were also encountered. Thus, in crosses of white-eyed females to red-eyed males an occasional daughter was white-eyed like the mother and an occasional son red-eyed like the father, contrary to typical criss-cross inheritance. Could the chromosome theory account for these exceptions from exceptions? It could. As Morgan's student Calvin Bridges reasoned in his doctoral thesis (1916), the exceptional white-eyed daughters must have received two X chromosomes from the mother instead of one, and the fertilizing sperm must have been of the non-X kind—that is, containing the Y chromosome. According to this interpretation the exceptional females must be of the sex-chromosomal constitution XXY. No such constitution had ever been seen. The exceptional red-eyed sons must have received no X chromosome from the mother, and the fertilizing sperm must have been the X-chromosome kind. Accordingly, the exceptional males must be of the constitution XO, where O stands for neither X nor Y—again a constitution which had never been seen. But when the sex-chromosomal constitutions of the exceptional offspring were studied under the microscope the cells of the females showed two X chromosomes and one Y chromosome and those of the males a single X chromosome and no Y. The prediction of cytological constitutions based on the aberrant offspring from crossing experiments seemed an achievement of the human mind comparable to the prediction in 1846 of the existence of a new planet, Neptune, from deviations in the expected path of Uranus.

Chromosome Maps

It had been realized early that organisms have more genes than chromosomes. Each chromosome therefore must carry many genes. This is obvious for the X chromosome since many genes show X-linked inheritance, and it was also proven to be true for other chromosomes. A priori it was not known whether genes in the same chromosome always remain together or whether their physical "linkage" permits new combinations. Experiments proved the latter to be true. If in an individual one chromosome of a pair is marked by genes A^1 and B^1 and the other by A^2 and B^2, then his germ cells consist not only of the original classes A^1B^1 and A^2B^2 but also of two additional classes A^1B^2 and A^2B^1. In 1911 Morgan explained this phenomenon which he had encountered in his *Drosophila* crosses by assuming physical exchange of corresponding parts between the chromosomes of a pair during germ cell formation. He postulated that the genes in a chromosome are arranged in a linear sequence—prophecy of Roux!—and that each gene occupies a fixed locus in the sequence. If in an A^1B^1/A^2B^2 hybrid no exchange point falls in the interval between the loci of A and B, then the "noncrossover" combinations A^1B^1 and A^2B^2 are recovered among the offspring. If the two chromosomes break between the two loci and reunite in a crosswise way, then the "crossover" combinations A^1B^2 and A^2B^1 are produced. To these postulates Morgan added another. He had found that the frequency of crossovers among all combinations was different for different pairs of linked genes: for instance, in round numbers, 10 per cent for A-B, 18 per cent for A-C, and 1 per cent for C-D. These differences, the postulate went, were the consequences of differences in distance between the genes: C and D are located close together, A and B less so, and A and C least.

Morgan was not the first to observe linkage and recombination of linked genes. In England some years earlier, William Bateson and R. C. Punnett had discovered these phenomena in crosses of sweet peas. But the explanation which these geneticists proposed was abstractly formal and did not include reference to chromosomal events. Morgan's ingenious interpretation was partly suggested to him by a recently published paper (1909) of the Belgian cytologist F. A. Janssens, who described in fine detail the behavior of the chromosomes of a salamander during the development of its germ cells. This behavior suggested to Janssens that chromosomes may exchange sections between them. Morgan's "Attempt To Analyze the Constitution of the Chromosomes," as he entitled his "transforming" paper, was carried one step further by another of his great students. A. H. Sturtevant went beyond the general statement regarding genes lying "close together" or "more distant" from each other by using the numerical proportion of crossovers as an index for the distance between any two genes. "Then by determining the distances (in the above sense) between A and B and between B and C, one should be able to predict AC. For, if proportion of crossovers really represents distance, AC must be approximately, either AB plus BC, or AB minus BC" depending on whether the sequence of loci is A-B-C or A-C-B.[5] On the basis of these considerations Sturtevant, in 1913, constructed the first genetic map of a chromosome.

Experiments with many genes confirmed the prediction of the value of AC as either the sum or the difference of the values of AB and BC. This was taken as evidence for the reality of the linear arrangement of genes along the chromosomes. Later some geneticists emphasized that the additive and subtractive properties of linkage values could be interpreted in a variety of ways of which physically linear sequence was only one possible mode. Evidence accumulated, however, that provided proof of the Morgan-Sturtevant theory. One type of proof came from Harriet Creighton's and Barbara McClintock's study of the transmission of linked genes in corn concomitantly with that of the specific chromosomes involved in the linkage (1931). They used a pair of chromosomes whose partners differed visibly in some detail at each end: one chromosome had a "knob" at one end and was abnormal at the other end; the other chromosome had no knob at one and was normal at the other end. When the noncrossover combinations of genes located in the chromosome pair were studied microscopically the original doubly marked chromosomes were recovered, but when crossover combinations of genes were studied two visibly new types of chromosomes were seen: (1) one type having a knob at one end and being normal at the other, (2) the second having no knob and being abnormal. An equivalent proof of the chromosomal theory of crossing-over for *Drosophila* was published nearly simultaneously with that for corn.

Another type of proof for the theory of linear arrangement of genes in the chromosomes came from the analysis of the chromosomes in the salivary gland cells of *Drosophila* larvae.

In the 1920's when I was a postdoctoral Rockefeller Fellow working with Morgan, Bridges, and Sturtevant in the "Fly Room" at Columbia University, some of us playfully hoped that Santa Claus would bring us a microscope a hundred times more powerful than the best instruments then available to look at chromosomes. Of course, we knew from elementary physics that our hope could not be fulfilled (this was before the idea of an electron microscope had been conceived). Unknown to us, however, nature had kept in store an equivalent gift: in the nuclei of the salivary gland cells the chromosomes are a hundred times larger than the chromosomes in other cells, and they show a microscopically visible constant differentiation of thousands of bands specifically thick or thin, widely or closely spaced, analogous to the sequence of absorption bands in a spectrum. The salivary gland chromosomes are actually "cables" made up of thousands of closely held-together uncoiled individual strands, the products of successive divisions of originally single strands. This gives them their large diameter and makes visible their great length which in the uncoiled state is not clearly recognizable since a single strand is too narrow for recognition. When analysis of abnormal genetic situations led to deductions concerning abnormal chromosome constitutions the "salivary chromosomes" proclaimed the validity of the deductions. Thus, when it was postulated that some specific genes had been lost, this was mirrored in the loss of a section of bands or a single band of a chromosome. Likewise, when it was postulated that some genes had been translocated from their normal position to a new one, the salivary chromosomes showed an absence of a chromosome section at its usual place and its presence somewhere else. And when the genetic findings led to the hypothesis that a gene sequence had become inverted so as to produce the sequence A-D-C-B-E-F from the normal A-B-C-D-E-F, the salivary chromosomes showed the presence of the genic inversion by an inversion of a sequence of bands. The giant banded salivary structures had been seen decades before, but their true chromosomal nature was only established in 1933, by Emil Heitz and Hans Bauer. The correlations between positions of genes and order of bands in the chromosomes were demonstrated in the same year by Theophilus Painter, who profited from the advice of Wilson Stone.

The assignment of specific genes to specific chromosomes and the construction of chromosome maps by means of linkage data were the crowning achievements of age-old efforts to understand the morphological composition of the human body and other organisms. The first of the systematic attempts to gain such knowledge led to the development of the science of gross anatomy. When the skill of the anatomist became inadequate for the elucidation of the finer structures of the body, further advances were achieved by means of the microscope, methods of differential staining, and microchemical methods. Then the electron microscope made its appearance and revealed cellular structures close to and at the molecular level. Even such a powerful device of direct observation, however, would have been unable to lead to recognition of the genetic function of the structures observed or to the assignment of specific activities to specific structural elements. It required the subtle tools of the mind to specify the topography of the genes within the chromosomes, the microscopically visible assemblages of genes.

Morgan's dual role in the development of "higher Mendelism" is worthy of a comment. His own contributions in *Drosophila* genetics provided an understanding of sex-linked inheritance and laid the basis for the analysis of the genetic architecture of the chromosomes. After these accomplishments he served primarily as the synthesizer of genetic advances made by the members of his "school," particularly Sturtevant, Bridges, and H. J. Muller. The mutual influences which emanated from the division of labor between "the boss" and his collaborators were strengthened by a climate of equality created by Morgan. How rare to find a mature, already famous biologist inviting his students, who hardly had earned their doctoral degrees, to be coauthors with him of the fundamental textbook of the new genetics—*The Mechanism of Mendelian Heredity* (1915)!

For Morgan himself the *Drosophila* work was only one aspect of a biologist's searching. Morgan always had a multitude of projects going simultaneously: some promising, others foolish, he said, and still others so foolish that he did not want to talk about them. He was a prominent experimental embryologist, a lifelong student of regeneration, of sex determination, of fertility and sterility, of the hormonal effects of castration in chickens, of the genetics of mice, rats, and pigeons, of evolution. For a biographical memoir of Morgan, Sturtevant made a list of over fifty kinds of animals and one plant on which Morgan had published results of his work.[6]

Even at the height of his achievements during the first years of the *Drosophila* period he experimented with other forms and wrote extensive accounts of this work. Sturtevant characterizes Morgan's strengths as "involving great skill and patience in the collecting and care of animals, insight in seeing what were the critical points to study, and ability to recognize and to follow up unexpected facts."[7] No wonder that Morgan made many important discoveries and also that he soon shifted his emphasis from the *Drosophila* work. Notwithstanding, for more than one and a half decades he remained the synthesizer of many of the findings of the "Morgan school." Gradually, however, he was left behind. I remember the "awe-full" moment when Bridges explained to him a particularly intricate new result and the initiator of it all left the room, shaking his head and saying "too much for me!"

Physical Nature of the Gene

Although the chromosome theory had lifted the concept of the gene from a solely operational unit to a material entity of known localization, it had not brought clarification of the physical nature of the gene. Chromosomes are made up of nucleoproteins—macromolecules of nucleic acids joined to protein macromolecules—and the question remained, What part of the chromosomes represent the genetic material? On the basis of inadequate information on the chemistry of nucleic acids it seemed unlikely that these molecules occur in a sufficiently large number of variants to equal the very large number of different genes.

Moreover, cytological observations seemed to show that chromosomes may at certain stages lose their nucleic acid component. E. B. Wilson, who in the first edition of his earlier-cited book on the cell (1896) had regarded nucleic acid as the hereditary substance of the chromosomes, had abandoned this viewpoint when the third edition appeared in 1925. In a section entitled "Staining-Reactions of the Cell-Substance" he wrote: "So far as the staining reactions show, therefore, it is not the basophilic element (nucleic acid) that persists, but the socalled 'achromatic' or oxyphilic substance. The nucleic acid component comes and goes in different phases of cell activity and it is the oxyphilic component that seems to form the essential structural basis of the nuclear organization."[8]

Attention was focused therefore on the "oxyphilic" protein moiety of the chromosomes. Proteins were known to be of innumerable kinds, functioning in many different ways; those that were enzymes were effective even when present in very small amounts. For these reasons genes were usually believed to be proteins, and the nucleic acid molecules were regarded as perhaps skeletal rather than genetic elements of the chromosomes.

Lacking exact information on the genic nature, how should the geneticist define a gene minimally? H. J. Muller (1890–1967) revolved this question in his mind all during his long, active life and emphasized two aspects of genes: their ability to replicate themselves accurately and their ability, if changed, to replicate themselves in the changed form. Genes, of course, do not replicate without a cellular machinery that provides the material and the energy for building two genes where only one is present to begin with. This machine itself, however, is not self-replicating; its existence and multiplication depend on genes. If one removes some of the nonnuclear material from a cell, it is able to regenerate fully the normal amount. If one removes a gene, no regeneration occurs. A gene then is a cellular unit which arises only from a preexisting gene. This minimum definition is clearly not sufficient to lead to decisions concerning the chemical nature of a gene.

Using ultraviolet radiation and exploring the mutagenic efficiency of different wavelengths of the ultraviolet spectrum, Lewis Stadler, Alexander Hollander, and others had attempted to characterize the class of compounds to which genes would belong. This work, begun in the middle thirties, gradually bared a close relation between the absorption spectrum of nucleic acid and the mutagenic effectiveness of the components of the spectrum. The investigators were on the right track, but various complications prevented definitive conclusions.

Then, chemical knowledge came from an unexpected source. In 1928 the British physician F. Griffith worked with rabbits which had been infected with a harmless strain of the bacterium pneumococcus. He also had a virulent strain of the organism. Nothing happened when he injected killed bacteria of the virulent type into noninfected rabbits. But when he injected the harmlessly infected rabbits with killed bacteria of the virulent type the nonvirulent pneumococci changed to virulence and the rabbits died. O. T. Avery at the Rockefeller Institute for Medical Research set himself the task of discovering what part of the killed bacterial material was essential in the transformation of avirulent into virulent microorganisms. In 1944, working together with M. McCarty and C. M. MacLeod, he had the answer. It was DNA. In a letter of May 13, 1943, to his brother the 66-year-old Avery wrote: "The . . . material . . . conforms *very* closely to the theoretical values of pure desoxyribose nucleic acid . . . (Who could have guessed it) . . . If we are right . . . then it means that nucleic acids are not merely structurally important but functionally active substances in determining the biochemical activities and specific characteristics of cells."[9] It was to be nearly a decade before the double helix model for the structural composition of DNA was worked out by James Watson and Francis Crick. It provided the beginning of an answer to the problem of self-replication of an original DNA gene as well as of changed forms of it.

Having stressed the chromosome theory of inheritance, it is necessary to raise the question of whether some part of inheritance must be assignable to extranuclear, nonchromosomal elements. Presumably such elements should be located in the cytoplasm of cells and would be expected to be transmitted in a non-Mendelian manner. There was a period in which the proponents of cytoplasmic transmission of certain traits regarded themselves as superior to the proponents of chromosomally located genes, an attitude which had its counterpart in an aggressive defense reaction by the latter proponents. In the course of time the two camps modified their divergent opinions and agreed that most traits follow Mendelian rules and are dependent on nuclear genes. A few traits, however, such as some inherited variations of chlorophyll bodies in plants or of respiratory enzymes in yeast cells, are transmitted by self-replicating elements in the cytoplasm—cytoplasmic genes. During the last few years it has been found that these genes are molecules of DNA just as the nuclear genes.

In 1927, Muller had succeeded in experimentally causing "artificial transmutation of the gene" in *Drosophila,* and the same was soon accomplished by Stadler in plants. The application of X rays and other ionizing radiation to these organisms caused genes to mutate. How the energy-rich radiation changed the gene molecules remained obscure—not surprisingly in view of the lack of knowledge of genic structure. During the Second World War new mutagens were discovered: certain chemicals such as mustard gas. Disappointingly, neither radiation nor chemical mutagenesis was specific in singling out individual genes for change. By and large, mutations were produced at random among the thousands of possible targets. Nevertheless, the impact of these discoveries was great: it was the first time in history that man intentionally could alter the hereditary makeup of individuals by changing the genetic material itself. All earlier manipulations of genes were of a type in which selection or chance sorted out some already existing genes for survival and others for rejection.

Specifically, as Muller recognized from the very beginning, artificial mutations were inducible in man as well. Muller warned of the ill consequences for future generations when human ovaries and testes are irradiated for diagnostic and therapeutic reasons and when mutations thus induced in the germ line are transmitted to the offspring. A fortiori such concerns have deepened ever since the genetic effects of nuclear explosions, radioactive fallout, and contamination have disturbed the thoughts of a whole generation of biologists and of mankind in general. More recently the possibility of chemical mutagenesis by man-made agents in the environment or by individual use of such agents has become a cause of warnings and of calls for intense specific research. It is implied in the foregoing that mutations are harmful to their bearers; although this is not the case for all mutations, it is so for many.

Genes have been recognized not by direct study of their structure but by their effects. Usually these effects were far removed from the primary actions of genes, actions which until recently were not accessible to observation. A wide gap lies between a chromosomal gene for seed shape and its visible effect on the seed, or between a gene for body size and the actual stature attained. It was only slowly that these gaps were bridged by tracing backwards as far as possible the development of the end effect. For a long time geneticists were primarily occupied with the Mendelian type of separation of the genic content of organisms into its components and with the exhilarating successes of gene localization. Thus it came about that an early fundamental foray into the problem of genic action remained isolated and without significant influence.

It was the English physician A. E. Garrod who early in the present century studied certain biochemical abnormalities in human individuals, found that they were due to Mendelian genes (then newly rediscovered), and traced in detail the biochemistry of what he termed "inborn errors of metabolism."[10] He recognized that these consisted in the inability of the organism to perform a step in a sequence of chemical reactions. Normally, specific enzymes permit the various steps to be made; the errors of metabolism depend on the absence of a normal enzyme, resulting in an interruption (a "block") at some specific point in the sequence of reactions. This basically new insight into genic action became well known to biochemists but had no deeper influence on the thoughts of geneticists until thirty years later.

Only when the first urgent problems of genetics had been solved did interest return to the mode by which genes exert their effects. Large-scale experiments by George W. Beadle and Edward L. Tatum on the bread mold *Neurospora* furnished an abundance of "biochemical mutations," and their effects were traced to specific blocks in the multidimensional network of reactions. As in man these blocks were traceable to deficiencies of specific enzymes, and a general hypothesis of genic action was summarized in the phrase "one gene–one enzyme," meaning that genic action consists in the production of enzymes and that each gene is responsible for a different enzyme. This fruitful hypothesis later became further generalized under the impact of the molecular genetics of the reactions controlled by the DNA genes and the unraveling of the structure of proteins. "One gene–one polypeptide chain" encompassed dependence of enzymes on genes as well as of other types of molecules such as hemoglobins. Genetics had become an integral aspect of biochemistry and cell physiology.

The Influence of Genetics on the Biological Sciences

Genic action par excellence is involved in the differentiation of the fertilized egg during development. Historically, embryology had first been dealt with as a purely observational discipline, then an experimental science. Many of the insights into the processes that determine the fate of different parts of the developing whole had been obtained at a level which did not require references to specific genes. Yet it had long been obvious that differentiation is the result of differential gene activity in various parts of the embryo. There seemed to be a paradox between the knowledge that mitosis furnishes all cells of an embryo with the same genic-chromosomal content and the evidence for inequality of differentiation in different cells. Weismann had tried to overcome the paradox by denying equality of gene content and postulating instead hereditarily unequal cell divisions during development. This artificial postulate never carried conviction, and during the 1890's other biologists such as O. Hertwig, Wilson, and Hans Driesch developed theories of differentiation according to which all genes are present in all cells, with some genes being called into action in some cells and other genes in other cells. These theories remained mainly at a general level, but here and there evidence became available for effects of specific genes on specific types of differentiation. (An example is furnished by a gene in *Drosophila* which directs the differentiation of certain cells to become sense organs. In the absence of the gene the same cells differentiate into skin cells.) How genes become turned on and off during development is now viewed in light of findings on the molecular regulation of gene activity in microorganisms, where regulator and operator genes, affected by extragenic variables, determine the functioning of other genes.

In still another biological area, that of evolution, genetics has played an outstanding role. It has redirected the emphasis of evolutionists from consideration of the individuals of a species to the abstract permanent gene pool of the population, whose mortal recombinants arise from it. Evolutionary genetics has thrown light on the permanence of species as well as on their changes, has studied differences between species in individual genes as well as in whole chromosomes. This is not the place to enter into discussions about experimental and theoretical population genetics as applied to evolutionary processes. Neither can we deal with evolutionary cytogenetics—the correlated study of cells under the microscope and of genes in hybridization experiments in their joint bearing on evolution.

One evolutionary event should be detailed, though, since it deserves much wider recognition than is usually awarded to it. This is the fact that a "good," that is, well-defined, plant species of the mint family, the hemp nettle *Galeopsis Tetrahit*—assigned specific rank by no less a systematist than Linnaeus—has been recreated out of other good species in the experimental plots of the Swedish geneticist A. Müntzing. Furthermore, several other new forms have been similarly produced which fulfill the designation species. The experiments consisted of hybridization between distant natural species. Species hybrids are usually sterile, because the chromosomes of the two parental forms do not pair well and as a result the ovules and pollen grains of the hybrids obtain abnormal combinations of the parental chromosomes. Exploitation of rare chromosomal processes, however, may lead to the doubling of all chromosomes in a hybrid, which will enable the doubled chromosomes to pair with each other and will therefore lead to fertile ovules and pollen. In different organisms evolution proceeds in different ways, and the examples of sudden "cataclysmic" origin of new species are typical for only a fraction of evolutionary events. Nevertheless, it is important to stress that the origin of species has been observed directly, and not only deduced theoretically.

Genetics as Applied to Man

Genetics as a science encompassing all living systems has been constructed out of discoveries in peas, flies, mice, fungi, bacteria, and viruses. Each kind of organism added to the generality of genetic findings at the same time that it demonstrated specific and often striking variations in the form in which the general principles were expressed. Man is an organism which physically represents a typical animal species but has psychological and sociological attributes setting him apart from all other forms. Human genetics at the physical level has shown man to be subject to the same genetic regularities as other animals. Who would have not expected it to be so? On the other hand, discussions concerning genetic determinants of man's mental characteristics and of his cultural achievements have been a battleground between extremists: those who believed in almost complete genetic determinism and those who believed in the complete nongenetic malleability of the mind and the potentially ubiquitous realization of its highest cultural achievements. Dogmatism on both sides has led to estrangement between camps which would have profited from collaboration, but the ideological warfare has also had its constructive aspects. For every finding that made one side proclaim "never underestimate the power of genes," the other side adduced a finding that justified its warning "never underestimate the power of the environment." The education of each camp by the other has made progress, and the number of die-hard extremists has greatly declined. Genetics will not cease to remain a central issue in uniquely human concerns: the preservation of the genetic endowment of mankind, its improvement by selective processes, the future application of biological methods to change genes at will, the not-so-utopian possibility of perpetuating "desirable" gene combinations in successive asexually reproduced generations, the problems of the genetics of race and of socio-economic stratifications, to name some of them.

Genetics, as applied to man, is a vulnerable science. Its ultimate usefulness in concepts and in practice is matched by its potential misuse. Such misuse has played a role throughout the twentieth century in the rash conclusions and popular exploitation of the opinions of naïve eugenicists and in their tragic consequences in Hitler's Germany. Another example of the vulnerability of genetics was the period of the dilettante movement of Lysenkoism in the Soviet Union and its physical and spiritual satellite countries (which on a dogmatic level also included China).

Conclusion

The period of genetics which has been surveyed in this essay clearly belongs to "little science." Mendel's small experimental plot in the garden of his monastery, Morgan's crowded Fly Room at Columbia University, and even Muller's thousands of glass vials in which he analyzed the mutagenic effect of X-rays all involved limited enterprises, in an external sense. When Bridges and Sturtevant had earned their doctoral degrees in 1915, Morgan obtained a grant from the Carnegie Institution of Washington to pay for their salaries as research assistants and also for the maintenance of living stocks of *Drosophila*. It would be of interest to know how large—or rather how small—the annual budget was.

The little-science attributes of the Morgan school extended to the physical state of the Fly Room itself. When rain fell on the roof of Schermerhorn Hall, a bucket had to be placed in the room to catch the water which leaked through the ceiling. The "microscope lamps" were ordinary light bulbs, shaded by metal sheets cut out of ordinary cans. The room abounded with fruit flies either escaped from milk bottles which served as culture vessels or permanently bred in the never thoroughly cleaned garbage can of the room. In addition there was other life. During the two years I worked in the inspiring spirtual atmosphere of the Fly Room I never opened my desk drawer without looking away for a while to give the cockroaches a chance to run into the darkness. Once I said breathlessly, "Dr. Morgan, if you put your foot down you'll kill a mouse." He did! Some years later, Morgan founded a well-equipped modern laboratory at the California Institute of Technology, but during the Columbia years little science was the appropriate way to make scientific progress instead of assigning time to technical perfection.

The continuity of genetics rests on internal necessity. So many pieces in the storehouse of biological facts fit together that it was unavoidable to recognize the unifying power of the discovery of the gene. In detail, however, instead of continuity there have been starts and all-too-early stops, witness Mendel and Miescher and Garrod.

The way in which genetics developed from the confluence of separate currents was hardly the only possible one. Historical accident played a large role. One can speculate that basically the transformation of the pneumococcus microbe from avirulence to virulence could have been discovered in 1870 instead of six to eight decades later. It would have led very early to identification of the genetic material with DNA. Extensions of the transformation work to other genic properties of the pneumococcus and the discovery of linkage in simultaneous transformation of several genic properties could have led to the construction of a genetic map long before Sturtevant's feat in *Drosophila*. Or, one may speculate what would have happened if chromosomes and the essentials of their behavior had not been discovered in the 1880's but fifty years later. Mendelian analysis from 1900 to 1930 would have remained without the clarification in form of the chromosome theory which actually came early. One may speculate even more brazenly. Could not one of the pioneering agriculturalists in the Middle East have counted the numbers of different plants in a segregating field? Could he have observed 3:1 ratios (perhaps helped in his arithmetic by belonging to a people who used the duodecimal rather than the decimal system)? Could his or his followers' interpretation of what we now call Mendelian ratios have led to the discovery of the binomial theorem millennia before Newton?

These are idle questions, but I raise them to confess my belief in a reality of the external world however inexhaustible in character and however incompletely represented at any stage of its recognition. Given enough time, genetics, whether born in ancient history or in the most recent centuries, would have reached a state very similar to the present. Whatever the paths taken during the rise of genetics, the purification of the concept of the gene would have been the final result. The gene at first was regarded as a *living* unit, a submicroorganism. This concept had to yield to the recognition that the gene as a molecular species is the basis of life in living *systems*, organisms. With this recognition more modest demands were made in assigning functions to the genes, and simultaneously more penetrating analyses of these functions became possible.

References

1. H. F. Roberts, *Plant Hybridization Before Mendel* (Princeton: Princeton University Press, 1929), p. 337.

2. Wilhelm Roux, *Über die Bedeutung der Kerntheilungsfiguren: Eine hypothetische Erörterung* (Leipzig: Engelmann, 1883), reprinted in W. Roux, *Gesammelte Abhandlungen über Entwicklungsmechanik der Organismen* (Leipzig: Engelmann, 1895), II, 125-143.

3. *Ibid.*, p. 19.

4. August Weismann, *Vorträge über Descendenztheorie,* 3d ed. (Jena: Fischer, 1913), II, 33.

5. A. H. Sturtevant, "The Linear Arrangement of Six Sex-Linked Factors in *Drosophila* as Shown by Their Mode of Association," *Journal of Experimental Zoology,* 14 (1913), 45.

6. A. H. Sturtevant, "Thomas Hunt Morgan, 1866-1945," *National Academy of Sciences Biographical Memoirs,* XXXIII (1959), 283-325.

7. *Ibid.*, pp. 289-290.

8. E. B. Wilson, *The Cell in Development and Heredity,* 3d ed. (New York: Macmillan, 1925), p. 653.

9. Quoted in C. Bresch, *Klassische und molekulare Genetik* (Berlin: Springer, 1964), p. 130.

10. A. E. Garrod, *Inborn Errors of Metabolism,* The Croonian Lectures delivered before the Royal College of Physicians of London, June 1908 (London: Oxford University Press, 1909), reprinted in H. Harris, *Garrod's Inborn Errors of Metabolism* (London: Oxford University Press, 1963), pp. 1-93.

GUNTHER S. STENT

DNA

The Golden Jubilee

IN SEPTEMBER of 1950 the Genetics Society of America held a symposium at Ohio State University to celebrate the fiftieth anniversary of the rediscovery of the work of Gregor Mendel, founder of the scientific study of heredity. The proceedings of that symposium were published under the title *Genetics in the 20th Century*, in the form of twenty-six essays written by some of the most eminent geneticists of the time.[1] In his introduction to these essays the editor, L. C. Dunn, of Columbia University, writes that their "primary purpose was to survey the progress of the first fifty years of genetics and to exemplify the status of some of its problems today." These problems, Dunn points out, extended beyond the confines of pure and applied biological science to the arena of the then ascendant Cold War: "no one in 1950 can be unaware of the fact that the principles upon which genetics rest have been declared politically unacceptable in Russia and that the other communist countries generally have followed suit." However, its alleged conflict with the tenets of dialectical materialism notwithstanding, "genetics has become a many-sided body of knowledge and method dealing with questions which are recognized as of central importance in all efforts to understand living matter—how it perpetuates itself through reproduction, how it changes and adapts itself to its environment. Many of its principles have turned out to have a general character, so that not only do the rules apply to plants, as Mendel first found, but to animals of all kinds, to man himself, and to the whole world of microorganisms, bacteria and viruses, revealed since Mendel's time." Finally, Dunn observes that "in spite of its evident diversification, genetics has fortunately retained the essential unity given to it by [Mendel's] discovery of a fundamental element of heredity, the gene, so that varied problems can be stated in a common language which is becoming more generally understood."

This essay is dedicated in grateful affection to my teacher André Lwoff.

The achievements which the distinguished essayists celebrated form a body of knowledge that is nowadays generally referred to as *classical genetics*. And it is precisely in the gene concept, which had provided its common language (and which had given such offense to primitive ideologues such as Trofim Lysenko), that classical genetics is to be differentiated from the *molecular genetics* that was to follow in its wake. The fundamental unit of classical genetics is an indivisible and abstract gene. In contrast, the fundamental unit of molecular genetics is a concrete chemical molecule, the nucleotide, with the gene being relegated to the role of a secondary unit aggregate comprising hundreds or thousands of such nucleotides.*

As perusal of the Golden Jubilee essays shows, the gene concept had remained largely devoid of any material content for the fifty years following the rediscovery of Mendel's work. Besides not having fathomed its physical nature, classical geneticists had been unable to explain how the gene manages to preside over specific cellular physiological processes from its nuclear throne, or how it manages to achieve its own faithful replication in the cellular reproductive cycle. Herman J. Muller of Indiana University, then one of the elder statesmen of genetics, epitomizes that condition of classical genetics in his essay "The Development of the Gene Theory" in these terms: "the real core of genetic theory still appears to lie in the deep unknown. That is, we have as yet no actual knowledge of the mechanism underlying that unique property which makes a gene a gene—its ability to cause the synthesis of another structure like itself, in which even the mutations of the original gene are

---

* The draft of this paper referred to the classical gene as being not only an indivisible and abstract unit but also a *transcendental* unit. In the Bellagio conference discussion my use of all three of these adjectives was criticized. So far as "transcendental" is concerned, I have now eliminated it, as a possible source of confusion, even though I still think that its common (rather than Kantian) meaning, namely possessing attributes so fantastic as to be beyond ordinary comprehension, *is* applicable to the classical gene. As far as the other two adjectives are concerned, Curt Stern expressed the view that the gene was "indivisible" only because geneticists did not *know* how to divide it, and that it was not really "abstract" because geneticists were perfectly aware that it had a material basis, genes having in fact been shown to reside on chromosomes. To me it appears, however, that both indivisibility and abstractness of the classical gene were fundamental epistemological qualities, which took their origin in the kind of operations by means of which the gene was then studied. From the classical purview, it would have been *meaningless* to divide the gene and *futile* to endow it with a concrete physical identity. For instance, I remember that in 1949 there appeared in *Life* magazine an electron micrograph of an ultrathin chromosome section, which purported to show the first picture of a gene. The patent speciousness of this claim at the time illustrates the basic cognitive difficulty that existed as recently as twenty years ago: how could one recognize a gene as a gene even if one happened to lay eyes on it?

copied . . . What must happen is that just that precise reaction is *selectively* caused to occur, out of a virtually infinite series of possible reactions, whereby materials taken from a common medium become synthesized into a pattern just like that of the structure which itself guides the reaction. We do not know of such things yet in chemistry." But for the classical geneticist study of the detailed nature and physical identity of the gene, though undoubtedly of great intellectual interest, is not really an essential part of his work. His theories on the mechanics of heredity and the experimental predictions to which these theories lead are largely formal, and their success does not depend on the knowledge of structures at the submicroscopic, or molecular, level where the genes lie. Application of the adjective "classical" to that first phase of genetic research is rather analogous to its well-established use in "classical physics." The fundamental and indivisible conceptual unit of nineteenth-century classical physics was the atom, which despite its unfathomed nature, had allowed very far-reaching insights into the macroscopic properties of matter. "Atomic" twentieth-century physics later succeeded in explaining the nature of the atom in terms of subatomic phenomena, just as molecular genetics was to succeed in explaining the molecular nature of the gene in terms of subgenic phenomena.*

One Gene-One Enzyme

Among the twenty-six Golden Jubilee essays there are only three which deal to any extent with matters that were of immediate relevance for the then-nascent molecular genetics. One of these essays, "Genetic Studies With Bacteria," was written by Joshua Lederberg, at the time a twenty-five-year-old professor-prodigy in the University of Wisconsin. Lederberg reviews his recent work on bacterial sexuality, which he and Edward L. Tatum had discovered just four years earlier.[2] His discussion of bacterial genetics is still entirely "classical," in that the

* At the Bellagio conference Erik Erikson wondered what is actually meant by "classical." He received a variety of responses to his query, such as that "classical" means "of no concern to me," and hence "embalmed"; "what you learn in school"; "perfect in content and form"; "simple and perfect in form"; and that it pinpoints dogmas from which young people are encouraged to make heretical departures. Léon Rosenfeld stated that for Bohr "classical physics" meant perfection in physical description, and that the later atomic physics is "nonclassical" in the sense that the language of everyday macroscopic experience can give only an imperfect description of the microscopic world of atoms. Dr. Rosenfeld's remark makes the correspondence between classical physics and classical genetics even closer than I had previously thought. For the macroscopic character differences on which the concept of the Mendelian gene was based are likewise closer to our everyday experience than the microscopic nucleotide unit of molecular genetics.

gene is treated as the fundamental unit responsible for the phenomena under study. Indeed, Lederberg takes some pains to make clear that the genetic mechanisms of the lowly bacteria can be understood from the classical viewpoint developed through study of higher forms. His essay pertains to molecular genetics only insofar as it describes what was to become one of the chief "molecular" experimental systems, the sexually fertile bacterium *Escherichia coli* K12 isolated at Stanford University "in the fall of 1922 from the stools of a diphtheria convalescent."

George W. Beadle, then Thomas Hunt Morgan's successor as chairman of the Biology Division at the California Institute of Technology, contributed an essay entitled "Chemical Genetics," which traces the development of the concept that genes preside over cellular function by controlling the chemical reactions of cell metabolism. Beadle describes how in the 1930's he and Boris Ephrussi (who also contributed an essay, concerned mainly with extranuclear inheritance) worked on the genetic control of the formation of the eye color of the *Drosophila* fruit fly and how he finally became discouraged over the difficulties encountered with that material. In 1940 he and E. L. Tatum turned their attention to a more favorable organism, the bread mold *Neurospora.* According to Beadle: "With the new organism our approach could be basically different. Through control of the constituents of the culture medium we could search for mutations in genes concerned with the synthesis of already known chemical substances of biological importance. We soon found ourselves with so many mutant strains unable to synthesize vitamins, amino acids and other essential components of protoplasm that we could not decide which ones to work on first." Beadle reports that in the intervening ten years he, Tatum, and their collaborators managed to analyze the genetic and biochemical characteristics of enough *Neurospora* mutants to lend strong support for the "one-gene–one-enzyme" hypothesis.[3] (In his essay Beadle actually prefers the name "one-gene–one-function" hypothesis, but it was the former name which became popular in the community of geneticists.) This hypothesis states that each gene has only one primary function, which in most or all cases is to direct the synthesis of one and only one enzyme, and thus to control one single chemical reaction catalyzed by that one enzyme. Though (as Beadle emphasizes) the idea that genes control single functions was not really original with him or Tatum, there can be little doubt that their clear formulation and strong experimental evidence for that hypothesis had a profound impact on subsequent thought about the gene. It must be noted, however, that the gene of Beadle's essay is still the classical, indivisible, abstract unit. But in promulgating the belief that each gene is doing only one thing, the one-gene–one-enzyme hypothesis gave hope of ultimately being able to find out how that thing is done.

The third essay touching on molecular genetics was one by Alfred E. Mirsky of the Rockefeller Institute for Medical Research, entitled "Some Chemical Aspects of the Cell Nucleus." Chemical study of the cell nucleus was begun in the 1860's by Mendel's contemporary, the Swiss chemist Friedrich Miescher,[4] before the notion had emerged that the cell nucleus is the seat of heredity. Miescher undertook an analysis of cells, such as pus cells and salmon sperm, in which he knew the nucleus to represent a large fraction of the total cell mass. These analyses revealed that the nucleus contains a hitherto unknown, phosphorous-rich, acid substance, to which a later worker gave the name *nucleic acid.*

By the turn of the century the ubiquitous presence of nucleic acid in plant and animal cells had been demonstrated, and the German biochemist Albrecht Kossel had identified the nucleic acid building blocks (Figure 1): the four nitrogenous bases *adenine, guanine, cytosine,* and *uracil* (the former two belonging to the class *purines* and the latter two to the class *pyrimidines*), a five-carbon sugar, and phosphoric acid. Further analytical work, largely by P. A. Levene[5] and by W. Jones in the 1920's, showed that there exist two fundamentally different kinds of nucleic acid, which are now called *ribonucleic acid,* or *RNA,* and *deoxyribonucleic acid,* or *DNA.* RNA contains *ribose,* whereas DNA contains *deoxyribose* as its five-carbon sugar. DNA, furthermore, does not contain the pyrimidine uracil; instead of uracil it contains 5-methyl uracil, or thymine. Nitrogenous base, sugar, and phosphoric acid were found to be linked to form a *nucleotide.* By the 1930's it had been shown that nucleic acid molecules contain several such nucleotides linked through phosphate diester bonds between their ribose or deoxyribose sugar molecules (Figure 1). Another ten years had to elapse before the enormously high chain length of nucleic acids was finally appreciated. As we now know, nucleic acid molecules represent *polynucleotide* chains of thousands, and sometimes millions of nucleotides in continuous chemical linkage.

Mirsky's essay is concerned mainly with the biological significance of the DNA type of nucleic acid. It sets forth how the invention in 1924 by R. Feulgen[6] of a specific color stain for DNA made it possible to show that the cell DNA is located almost exclusively in the *chromosomes* of the nucleus, whereas the cell RNA, despite its being a "nucleic acid," is located mainly in the cytoplasm. Since the chromosomes are the cell organelles in which the genes were known to reside, it did not seem farfetched to imagine that DNA plays some important role in hereditary processes. But as the chromosomes contain even more protein than DNA,

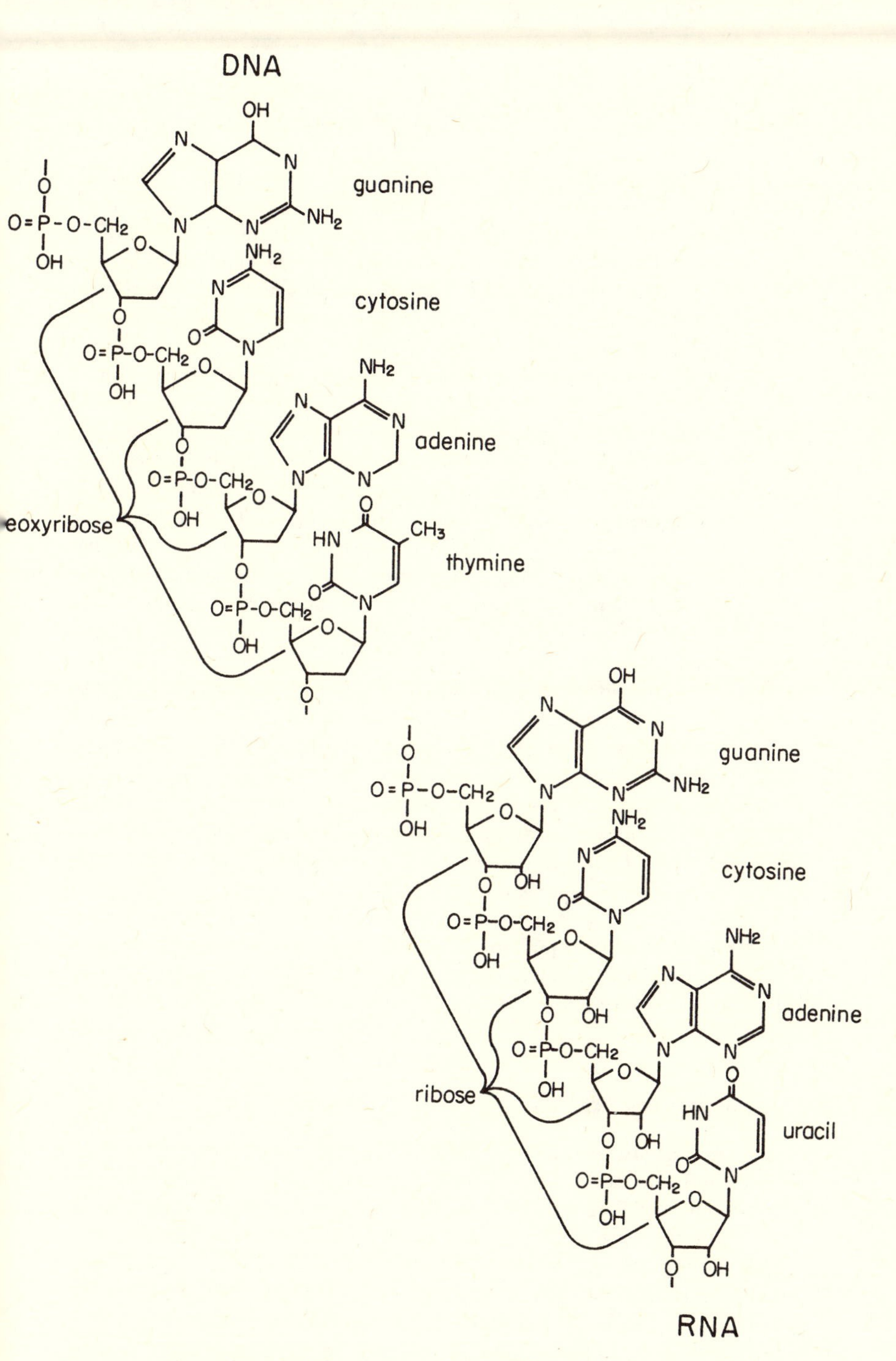

Figure 1. The polynucleotide chains of deoxyribonucleic acid, or DNA, and of ribonucleic acid, or RNA.

it was not necessary to infer that the genes actually contain any DNA. In support of the view that "DNA is part of the gene substance," however, Mirsky cites his own observations that "in the different cells of an organism the quantity of DNA for each haploid set of chromosomes is constant . . . constancy per cell is certainly an unusual characteristic for chemical components of cells . . . Even a substance, such as RNA, present in all cells, varies greatly in amount in different cells."

As far as the chemistry of DNA is concerned, Mirsky announces the recent demise of the *tetranucleotide theory*, "according to which a polynucleotide containing one of each of the four bases was considered to be a fundamental unit in DNA. Using chromatographic procedures for analysis [Erwin] Chargaff and his colleagues have shown that the four bases [adenine, guanine, cytosine, and thymine] are not present in equimolar proportions [as had been previously believed], and this has removed whatever experimental basis there was for the tetranucleotide theory." Furthermore, "Chargaff and his colleagues have analyzed the DNA's prepared from a number of different sources and have obtained results showing that in preparations from different organisms the ratios of the bases are different, although they are the same in preparations from different tissues of the same organism. It is highly probable, therefore, that the proportion of the four bases differ in DNA's of various organisms."

Mirsky does not, however, mention another seemingly less important finding which Chargaff presented in the very paper[7] cited in the essay. This finding is the DNA compositional equivalence rule, which states that although the proportion of the four bases differs in the DNA's of various organisms, it is nevertheless true that the molar proportion of adenine is always equal to that of thymine and the molar proportion of guanine is always equal to that of cytosine. Three years later, this rule was to provide a crucial clue for divining the structure of DNA.

Transforming Principle

In 1944, or six years before the Golden Jubilee, work had been published (and was certainly well known to most of the essayists) which proved that DNA is not only "part of the gene substance" but *is* the gene substance. Yet Mirsky's essay is the only one of the twenty-six in which the implications of that work are discussed. (Lederberg's Golden Jubilee essay refers to it briefly as a promising development in bacterial genetics, and Beadle's essay devotes two sentences to it, saying that it "has certainly introduced another chapter in genetics, and one that promises to be among the most exciting. It has given chemists new incentive to learn about the nucleic acids, compounds which everyone recognizes to be extremely important biologically and about which so little is known.")

This work had been carried out by Oswald T. Avery[8] and his collaborators at the Rockefeller Institute and represented the identification of the active principle of *bacterial transformation*, a phenomenon first observed in 1928 by the British bacteriologist F. Griffith.[9] Avery could show that upon addition of purified DNA extracted from normal *donor* bacteria to mutant *recipient* bacteria, which differ from the donor bacteria in one mutant gene, some of the recipient bacteria are transformed hereditarily into the donor type. Thus the normal donor gene must have entered the transformed recipient bacterium in the form of a donor DNA molecule and there displaced its homologous mutant gene. Hence it followed that the bacterial DNA represents the bacterial genes. In 1944 this conclusion seemed so radical that even Avery himself was reluctant to accept it until he had buttressed his experiments with the most rigorous controls.

Avery's controls were evidently not rigorous enough for most contemporary geneticists, including his Rockefeller Institute colleague Mirsky, who was of the opinion that "it is quite possible that DNA, and nothing else, is responsible for the transforming activity, but this has not been demonstrated conclusively. In purification of the active principle more and more of the protein attached to DNA is removed . . . It is difficult to eliminate the possibility that the minute quantities of protein that probably remain attached to DNA, though undetectable by the tests applied, are necessary for activity." But Mirsky concedes that "it can be regarded as established that DNA is at least part of the active [transforming] principle." As we shall see presently, Avery's work, just as Mendel's, had been too far ahead of its time, so that for some eight years it had very little impact on genetic research. Although, in contrast to Mendel, Avery and his discovery were well known, the genetic role of DNA had to be rediscovered in 1952 through work with bacteriophages.*

* In the Bellagio conference discussion Charles Weiner stated that he "always recognizes it as a danger signal when someone says 'so and so was too far ahead of his time,' a discovery or a man. This is a method of obscuring what went on . . . to say that it was ahead of its time is, essentially, to take a shortcut through history." Of course, pronouncing a discovery to have been ahead of its time just *because* it was not immediately appreciated is to make an empty tautology. But I am appealing here to another criterion of premature discovery that ought to illuminate rather than obscure what went on. By this criterion a discovery is "ahead of its time" if the inferences to which it leads cannot be connected by a series of simple logical steps with contemporary canonical knowledge. Thus, Curt Stern's essay shows that Mendel was "ahead of his time" because the statistical laws which Mendel found to govern

The Hershey-Chase Experiment

Bacteriophages, or *phages* as they are called in the trade, are subcellular parasites of bacteria that were discovered in 1917.[10] They occupy only about one-thousandth of the volume of their bacterial host cell and hence are so small that they cannot be seen in ordinary microscopes using visible light. Phages were first seen in 1940, in the wake of the development of the electron microscope, and they were found to be tadpole-shaped particles having a head and a tail. (It was eventually established that the tail is composed of protein and that the head represents a stuffed bag whose casing is protein and whose stuffing is DNA.)

In 1938 the physicist Max Delbrück, then a postdoctoral fellow at the California Institute of Technology, started experimenting with phages, in the expectation that study of their self-replication might throw light on what Muller's essay described as "that unique property which makes a gene a gene—its ability to cause the synthesis of another structure like itself, in which even the mutations of the original gene are copied." Delbrück's work began with designing the *one-step-growth experiment,* in which he used as his experimental material a phage active on *Escherichia coli.*[11] The one-step-growth experiment showed that each phage particle infecting an *E. coli* bacterium gives rise to some hundred phage progeny particles after a brief half-hour latent period. Thus this ex-

---

heredity could not be connected with the then known aspects of anatomy and physiology; and I am trying to show in the following just why it was difficult in 1944 to connect DNA with the gene. Lest it be concluded that the judgment of prematurity can be rendered only with hindsight, I shall provide an example of a recent discovery that falls within the purview of this essay and can be judged as being ahead of the *present* time. Three or four years ago there appeared reports purporting to have shown that the memory of a task learned by a trained donor animal can be transferred to a naïve recipient animal through the vehicle of nucleic acid molecules. Now whereas it is quite generally appreciated that the possibility of such memory transfer would, in fact, constitute a fact of capital importance for our understanding of the higher nervous system, these reports have so far remained without heuristic effect on brain research. For there is no chain of reasonable inferences by means of which our present, albeit very imperfect, view of the functional organization of the brain can be reconciled with the possibility of its acquisition, storage, and retrieval of experiential information through nucleic acid molecules. Thus for the community of neurobiologists there is no point in paying serious attention to these claims, or even to spend any time on checking whether they are true or false. This attitude may, at first glance, appear to be "unscientific," but it is, in fact, the very way in which science has to operate. Or, as Arthur Eddington advised his fellow scientists, it is not a good policy to put overmuch confidence in facts until they have been proven by theory. This appears to be also the view taken by Michael Polanyi (*Science,* 141 [1963], 1010) who showed that a premature discovery of his own was understandably ignored for forty years because during all that time it could not be connected with the then orthodox theories of physical chemistry.

periment brought clearly into focus the central problem of self-replication: how does the parental phage particle manage to produce its crop of a hundred progeny during that half hour? Two years later Delbrück met Salvador Luria, then a recently arrived refugee from war-torn Europe, and Alfred Hershey, of Washington University in St. Louis. This meeting brought into being the American Phage Group, whose collective memory has been preserved in a series of autobiographical essays.[12] The members of this group were united by a single common goal—the desire to reach what Muller referred to as "the real core of genetic theory," or to extend the frame of reference of genetics beyond the billiard ball gene.

Although the intellectual foundations were laid by the Phage Group for the edifice of molecular genetics during the next dozen years, the first real breakthrough came only in 1952 with an experiment by Hershey and his young assistant, Martha Chase.[13] Hershey and Chase showed by use of phage particles labeled with radiophosphorus ^{32}P in their DNA and with radiosulfur ^{35}S in their protein that at the outset of phage infection of the *E. coli* cell only the DNA of the phage actually enters the cell; the protein of the phage remains outside, devoid of any further function in the reproductive drama about to ensue within. Thus it could be concluded that the genes of the parent phage responsible for directing the synthesis of progeny phages reside in its DNA. This second demonstration that DNA is the hereditary material had an immediate and profound impact on genetic thought.

Why did Avery's announcement that DNA is the genetic material have in its day so much less effect on genetic thought than the Hershey-Chase experiment proving the same point eight years later (with much less compelling evidence, it might be noted)? First, it was only in the late 1940's that, thanks to the pioneering efforts of Luria and Delbrück,[14] bacteria and phages came to be accepted as genuine genetic organisms to which the gene concept could be legitimately applied (Lederberg's Golden Jubilee essay still has the flavor of a missionary effort to spread that gospel). So, in 1944 many people regarded bacterial transformation as some bizarre metamorphosis without relevance to hereditary processes in higher forms. Further, and more important, the tetranucleotide theory of DNA structure still held sway, so that it was very difficult to imagine how a DNA molecule made up of monotonously repeating units containing one each of the four bases *could* be the carrier of *genetic information.* Thus even those persons who were prepared to accept in the mid-1940's the conclusions that bacterial transformation is a truly hereditary process and that DNA is really the transforming principle were wont to consider the phenomenon as a case of gene *mutation.* That is to say, they imagined that DNA is a chemical capable of causing specific or directed mutations rather than being the gene substance itself. But the demise of the tetranucleotide theory announced in Mirsky's essay now meant that the four types of bases can follow each other in any arbitrary order in the polynucleotide chain. Since the base composition was found to be different in DNA samples obtained from different organisms, it could be supposed at the time of the Hershey-Chase experiment that any given DNA molecule harbors its genetic information in the form of a precise sequence of the four bases along its polynucleotide chain.

With the acceptance of the parental phage DNA as the genetic material and the birth of the notion that genetic information is encoded into DNA as a nucleotide base sequence, the fundamental problem of biological self-replication could now be restated in terms of two DNA functions, *autocatalytic* and *heterocatalytic.* By means of the autocatalytic function, the phage DNA replicates its own precise nucleotide base sequence a hundredfold to generate the genes with which its progeny phages are to be endowed. And by means of the heterocatalytic function, the phage DNA directs, or presides over, the synthesis of the phage-specific proteins that are to furnish the body of its progeny. The successful elucidation of these two DNA functions was to be the work of the next decade.

The Double Helix

In 1951, one year after the Golden Jubilee, Linus Pauling, of the California Institute of Technology, discovered the basic structure of protein molecules.[15] Proteins, we might recall here briefly, are long chain molecules, built up of a sequence of twenty different kinds of *amino acids.* These amino acids are joined to each other through a chemical linkage called the *peptide* bond. The length of different kinds of protein chains present in living cells varies considerably, but on the average these chains contain about three hundred amino acids linked end-to-end. Pauling found that the three-dimensional conformation of the amino acid chain is a helix, to which he gave the name *α helix.* The discovery of the α helix represented the first great achievement in the attempt to work out protein structure by the methods of X-ray crystallography.

Pauling's success inspired James D. Watson, a twenty-two-year-old member of the Phage Group and a pupil of Luria's, to abandon the genetic and physiological experiments on phage reproduction that he had been carrying out. Watson instead decided to try to work out the three-dimensional structure of the DNA molecule, which, a few months after his decision, the Hershey-Chase experiment showed to be the carrier of the phage genes. To gain the necessary skills in X-ray crystallography,

Watson joined John C. Kendrew of the University of Cambridge, who, like Pauling, was studying protein structure. In Cambridge, Watson met Francis Crick, a Ph.D. student, to whom it had also occurred that the three-dimensional structure of DNA would be likely to provide important insights into the nature of the gene. Watson and Crick then began a collaboration whose story is now so well known through Watson's famous autobiographical account[16] that there is little need to retell it here. Suffice it to say that in the spring of 1953 Watson and Crick discovered that the DNA molecule is a *double helix*, composed of two intertwined polynucleotide chains (Figure 2). The DNA double helix is self-complementary, in that to each adenine nucleotide on one chain there corresponds a thymine nucleotide on the other chain, and that to each guanine nucleotide on one chain there corresponds a cytosine nucleotide on the other chain. The specificity of this complementary relation devolves from hydrogen bonds formed between two opposite nucleotides, adenine-thymine and guanine-cytosine, at each step of the double helical molecule (Figure 3). The complementary base-pairing relation of the two chains thus accounted for the base composition equivalence rule noticed by Chargaff three years earlier.

On first sight, Watson's and Crick's discovery of the double-helical structure of DNA resembled Pauling's then two-year-old discovery of the α helix, particularly since the formation of specific hydrogen bonds also plays an important role in Pauling's structure. But, on second sight, the promulgation of the DNA double helix emerges as an event of a qualitatively different heuristic character. First, in working out the structure of the double helix Watson and Crick had for the first time introduced genetic reasoning into structural determination, by demanding that the evidently highly regular three-dimensional structure of DNA must be able to accommodate the informational aspect of an arbitrary nucleotide base sequence along the two polynucleotide strands. Second, the discovery of the DNA double helix opened up enormous vistas to the imagination. It was to provide the high road to "the real core of genetic theory" which, according to Muller, was still lying in the deep unknown at the time of the Golden Jubilee. Molecular genetics had now become a going concern.

Watson and Crick had concluded their letter to *Nature*,[17] in which they first described the DNA double helix, with a sentence that can surely lay claim to being one of the most coy statements in the literature of science: "It has not escaped our notice that the specific [nucleotide base] pairing we have postulated immediately suggests a possible copying mechanism of the genetic material." In a second letter to *Nature*[18] they soon told what it was that had not escaped their notice: the DNA molecule can achieve its autocatalytic function upon separation

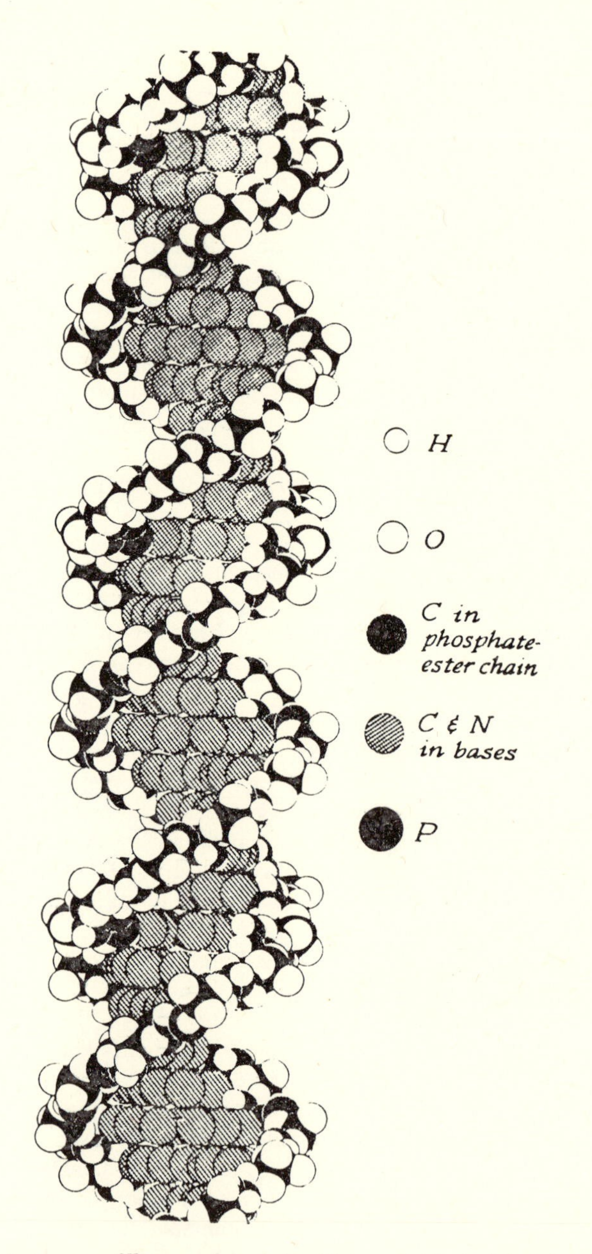

Figure 2. A space-filling model of the Watson-Crick double helix structure of DNA. (From G. S. Stent, *Molecular Biology of Bacterial Viruses* [San Francisco: W. H. Freeman, 1963].)

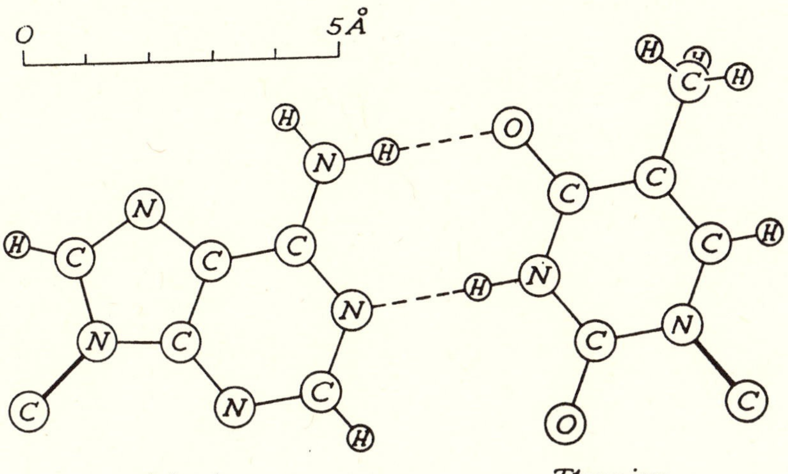

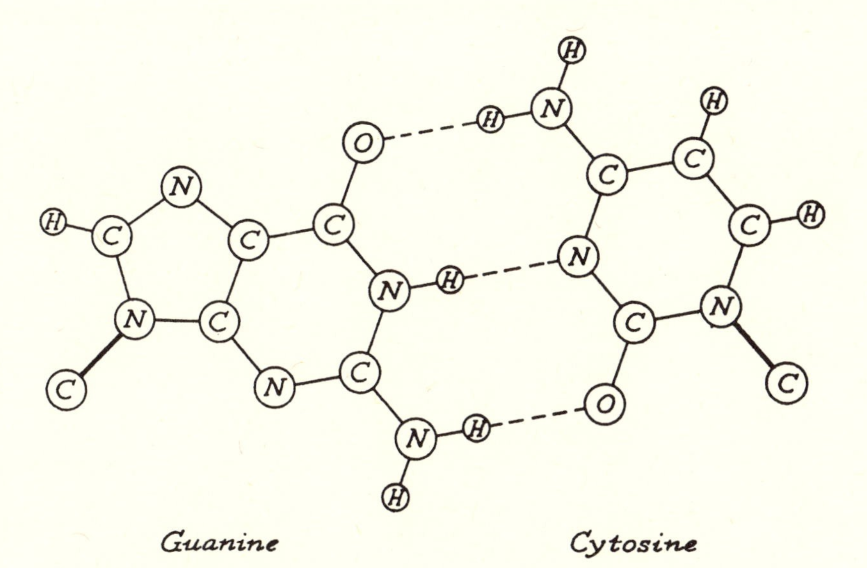

Figure 3. The complementary base pairing of adenine and thymine, and of guanine and cytosine in the double-helical DNA molecule. Hydrogen bonds are shown as dashed lines. The carbon atom of the deoxyribose to which each base is attached is also shown. (From G. S. Stent, *Molecular Biology of Bacterial Viruses* [San Francisco: W. H. Freeman, 1963].)

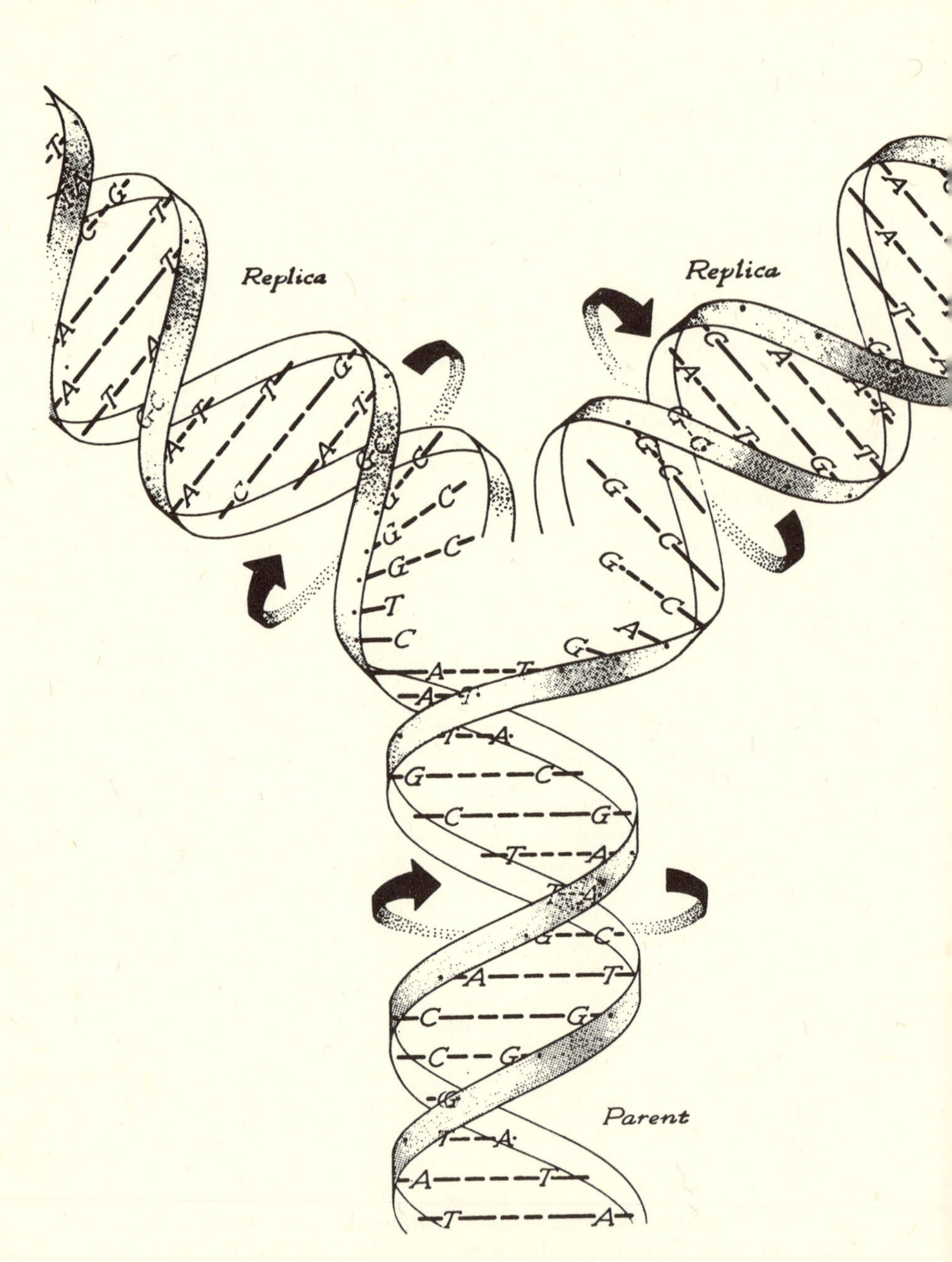

Figure 4. Replication of the DNA double helix, according to the mechanism of Watson and Crick. (From G. S. Stent, *Molecular Biology of Bacterial Viruses* [San Francisco: W. H. Freeman, 1963].)

of the two helically intertwined, complementary polynucleotide strands (Figure 4). Each of the two parent strands would then serve as a *template* for the ordered assembly of its own complementary daughter strand, by having every nucleotide on the parent strand attract and line up for polynucleotide synthesis a free nucleotide carrying the complementary purine or pyrimidine base. Thus in the case of phage reproduction, the DNA of the infecting phage particle would undergo successive rounds of unwinding and complement addition. In this way an intrabacterial pool of replica phage DNA molecules would be built up, identical in nucleotide base sequence to the DNA of the parent phage; this pool would provide the genes for the offspring phages.

It took about five more years to prove that the solution to the problem of the autocatalytic function proposed by Watson and Crick is essentially correct. The main proof depended on the prediction[19] that under that proposal the atoms of the parental double helix ought to become distributed in a *semi-conservative* manner. That is to say, each of the two daughter molecules generated by fission of the parental DNA double helix should contain one polynucleotide chain of parental provenance and one chain synthesized *de novo.* In 1958, Matthew Meselson and Frank W. Stahl, one a graduate student of Pauling's and the other a postdoctoral student in Delbrück's laboratory at Caltech, managed to demonstrate that the DNA of *E. coli* is replicated semi-conservatively.[20] For the purpose of this demonstration, Meselson had adapted the method of *equilibrium density gradient sedimentation* to the detection of minute density differences of large molecules, particularly for measuring the density increase produced by the replacement of a common atomic species, such as nitrogen ^{14}N, by its rare heavier isotope, such as ^{15}N, in the molecules. This method was to become one of the most widely used techniques in later molecular biological research.

Reform of the Gene

Understanding of the heterocatalytic function of DNA, which from the very outset of its formulation appeared as a more complex problem than the autocatalytic function, required a rather greater effort and a somewhat longer time. First of all, it proved necessary to reform the concept of the classical gene. Today it seems impossible to make an accurate reconstruction of the historical course taken by this reform and to ascertain to whom each facet can be attributed. Many of the essential ideas were first proposed in informal discussions on both sides of the Atlantic and were then quickly broadcast to the cognoscenti by private international bush-telegraph. Months and often years elapsed before a new idea was committed to print, and then very often it was not by the person who had first thought of it. Nevertheless, it can hardly be doubted that here too Watson and Crick played a dominant role, as well as Seymour Benzer, then at Purdue University.

One of the points of departure of this reform was Beadle's and Tatum's one-gene–one-enzyme theory. By the early 1950's the analytical work of Frederick Sanger,[21] of the University of Cambridge, had lent direct support to the credence that a given species of enzyme represents a homogeneous class of protein molecules in which a definite number of the twenty different kinds of amino acids are assembled in a unique sequence. From this credence developed the so-called *sequence hypothesis,* which states that the exact spatial conformation of a protein molecule, and hence the specificity of its biologic function, is *wholly determined* by that unique amino acid sequence from which it is built. At the time that this hypothesis was advanced there was no proof whatsoever of its validity. But it came to be embraced immediately as molecular-genetic dogma, because it allowed rephrasing of the one-gene-one-enzyme theory in more concrete terms: the gene directs the synthesis of one enzyme by directing the assembly of the twenty kinds of amino acids into a protein molecule of given amino acid sequence. But since it was already taken for granted that the genetic information is held in the form of a particular nucleotide base sequence in DNA, it now became clear that the *meaning* of the particular nucleotide base sequence making up a sector of DNA corresponding to a gene could be nothing other than the specification of an amino acid sequence of the corresponding protein molecule. By means of incisive *fine structure* genetic experiments on a single gene of one of the *E. coli* phages, Benzer was able to gather convincing support for the notion that the gene is, in fact, a linear array of DNA nucleotides which determines a linear array of protein amino acids. It was Benzer more than anyone else who showed that the fundamental genetic unit is the DNA nucleotide base.[22]

Genetic Code

This reform of the gene concept led directly to the belief that there must exist a *genetic code* that relates the nucleotide base sequence in the DNA polynucleotide chain to the amino acid sequence of the corresponding enzyme protein. An obvious consideration quickly revealed that this code could be no simpler than one involving the specification of each amino acid in the protein molecule by at least *three* successive nucleotide bases in the DNA. Certainly there cannot exist a *one-to-one* correspondence between nucleotide bases in the DNA and amino acids in the protein molecule, because the four kinds of nucleotide bases taken *one* at a time could specify only one out of four, not one out of twenty, kinds of amino acids. Nor would it suffice that *two* adjacent nucleotide bases specify one amino acid, since in that case only $4 \times 4 = 16$ kinds of amino acids could be coded for by the four kinds of nucleotide bases. But four kinds of nucleotide bases taken *three* at a time provide $4 \times 4 \times 4 = 64$ different code words, or *codons*, and hence each of the twenty kinds of protein amino acids could be represented by at least one such codon in the genetic code, with the extra number of codons allowing for the possibility that each kind of amino acid is represented by more than a single codon.

These a priori insights had certainly been reached within a few months of the discovery of the DNA double helix and were first published in 1954 by the physicist-cosmologist George Gamow.[23] But it was not until 1961 that it was finally proven that the genetic code does involve a language in which a triplet of successive nucleotide bases in the DNA polynucleotide chain stands for one protein amino acid. That proof came from purely formal genetic experiments carried out by Crick[24] and his collaborators on the same phage gene which had figured in Benzer's earlier reform of the gene concept. As we shall see presently, the code was unexpectedly broken that same year.

Central Dogma

It was one thing to have formulated the general principles according to which genetic information is stored and replicated in the DNA. But it was quite another to work out the molecular mechanisms of the heterocatalytic function through which that information is realized as protein molecules. Here Watson and Crick also played a dominant role by formulating in the years 1953-1955 what came to be known as the *central dogma.* According to that dogma the heterocatalytic function is a *two-stage process,* in which the other type of nucleic acid, RNA, also becomes involved. In the first stage, the DNA molecule serves as the template for the synthesis of an RNA polynucleotide chain onto which the sequence of nucleotide bases in the DNA chain is *transcribed.* In the second stage, the RNA chain is then *translated* by the cellular machinery for protein synthesis into protein molecules of amino acid sequence specified via the genetic code. It is to be noted that an essential feature of the central dogma is the one-way flow of information,

$$\text{DNA} \rightarrow \text{RNA} \rightarrow \text{protein},$$

a flow which is never reversed.

In order to study the processes envisaged by the central dogma it became necessary to employ the methods of biochemistry to open the black box containing the cellular machinery which actually effects the transcription-translation drama of the central dogma. One of the first insights then provided by the application of biochemical methods was the identification of the *ribosome* as the *site* of cellular protein synthesis.[25] The ribosome is a small particle present in vast numbers in all living cells (one *E. coli* bacterium contains about 15,000 ribosomes); its mass is composed of about one-third protein and two-thirds RNA. But how is the information for specific amino acid sequences encoded in the gene made available to the ribosome in its protein assembly process? In answer to this question it was proposed in 1961 by François Jacob and Jacques Monod,[26] of the Pasteur Institute, that the RNA onto which, according to the central dogma, the genetic nucleotide base sequence is first transcribed is a molecule of *messenger RNA* (Figure 5). This messenger RNA molecule is picked up by a ribosome, on whose surface then proceeds the translation of RNA nucleotide sequence into protein amino acid sequence, codon by codon. In this translation process, the messenger RNA chain runs through the ribosome as a tape runs through a tape recorder head. While the tail of a messenger RNA molecule is still running through one ribosome, its head may already have been picked up by another ribosome, so that a single molecule of messenger RNA can actually service several ribosomes at the same time.

How the amino acids are actually assembled into the correct predetermined sequence by the messenger RNA as it runs through the ribosome had been envisaged by Crick[27] before the concept of the messenger RNA had even been clearly formulated. Crick thought it unlikely (working from first principles, as was his wont) that the twenty different amino acids could interact in any specific way directly with the nucleotide base triplet on the RNA template chain. He therefore proposed the idea of a nucleotide *adaptor,* with which each amino acid is outfitted prior to its incorporation into the polypeptide chain. This adaptor was thought to contain a nucleotide base triplet, or *anticodon,* complementary (in the Watson-Crick nucleotide base pairing sense) to the nucleotide triplet codon that codes for the particular amino acid to which the adaptor is attached. The anticodon nucleotides of the adaptor would then form specific hydrogen bonds with their complementary codon nucleotides on the messenger RNA and thus bring the amino acids bearing the adaptor into the proper, predetermined alignment on the ribosome surface.

No sooner had the adaptor hypothesis been formulated than students of the biochemistry of protein synthesis began to encounter an ensemble of specific reactions and enzymes that gradually resembled more and more the a priori postulates of that hypothesis.[28] First, a special type of small RNA molecule, the *transfer RNA,* was discovered, which contains about 80 nucleotides in its polynucleotide chain. Each cell con-

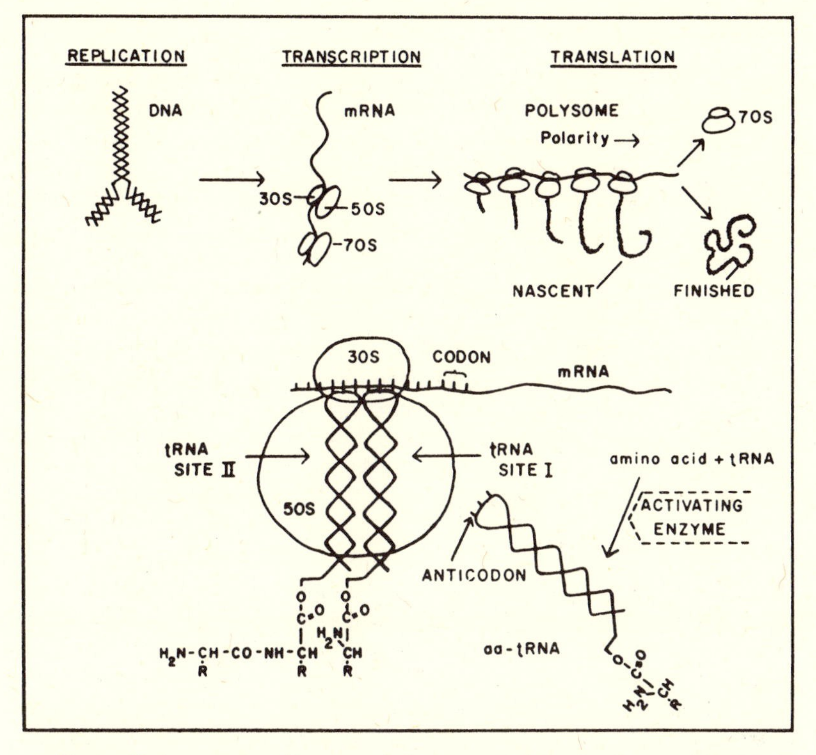

Figure 5. A summary diagram of the autocatalytic (replication) and heterocatalytic (transcription and translation) functions of DNA. *Upper left:* The DNA double helix replicates according to the semiconservative Watson-Crick mechanism. *Upper center:* The DNA nucleotide sequence has been transcribed onto a molecule of messenger RNA, or mRNA. The mRNA molecule is engaged by ribosomes, which are composed of two subunits, a smaller one called 30S and a larger one called 50S; the two subunits together constitute the intact, or 70S, ribosome. *Upper right:* Each of several ribosomes working in tandem on the same mRNA (a "polysome") translates the mRNA nucleotide sequence into the corresponding polypeptide chain. When a ribosome has translated the entire nucleotide sequence corresponding to a gene, the completed protein chain is released and the ribosome is free to attach itself to another mRNA molecule. *Lower part:* Details of the process of amino acid assembly. The nascent protein chain (here consisting merely of two amino acids) is attached to that molecule of transfer RNA, or tRNA, which figured as adaptor of the last amino acid to be added into the chain. This molecule of tRNA is in turn held to site II of the 50S ribosomal subunit. The next amino acid to be incorporated into the nascent protein chain is specified by that nucleotide triplet codon of the mRNA which faces site I of the 50S subunit. Into site I can fit only a molecule of tRNA whose anticodon matches the codon displayed by the mRNA and to which the appropriate amino acid has become attached, thanks to the recognition effected by the activating enzyme. Once the tRNA has entered site I, its amino acid is brought into juxtaposition with the last amino acid of the nascent protein chain, and the next peptide bond can be formed. When the chain has thus been elongated by one amino acid residue, mRNA and tRNA molecules move over the ribosome from right to left, to translocate the tRNA now carrying the nascent chain from site I to site II and to display the next codon at site I. (From H. K. Das, A. Goldstein, and L. C. Kanner, *Molecular Pharmacology,* 2 [1966], 158.)

tains several dozen distinct species of transfer RNA, each species being capable of combining with one and only one kind of amino acid. This transfer RNA turned out to be Crick's postulated adaptor, since that transfer RNA species which accepts any given amino acid contains the anticodon nucleotide triplet in its polynucleotide chain which is complementary to the codon representing that same amino acid in the genetic code.

Second, a set of enzymes, the *amino acid activating enzymes,* was discovered, each of whose members is capable of catalyzing the combination of one kind of amino acid with its cognate transfer RNA molecule. Thus the set of activating enzymes which matches each amino acid with its proper transfer RNA adaptor (by means of which the amino acid is recognized in protein synthesis) evidently represents the *dictionary of heredity,* the cellular agency that "knows" the genetic code.

Breaking the Code

The actual deciphering of the genetic code began with a discovery made by the young biochemist Marshall Nirenberg at the National Institutes of Health. In the spring of 1961 Nirenberg had managed to develop a "cell-free" system capable of linking amino acids into protein molecules. This system contained ribosomes, transfer RNA, and amino acid activating enzymes extracted from *E. coli.* Though Nirenberg was by no means the first to reassemble *in vitro* the cellular machinery for protein formation, his system had one very important advantage over its predecessors: here protein synthesis depended on the addition of messenger RNA to the reaction mixture. Thus it became feasible to direct the *in vitro* formation of specific proteins by introducing into this system specific types of messenger RNA. Now when Nirenberg introduced a synthetically produced *monotonous* RNA containing *only* the uracil nucleotide (instead of the four types of nucleotide bases present in natural messenger RNA), he obtained a dramatic result. Addition of the artificial, monotonous messenger RNA resulted in the *in vitro* formation of an equally monotonous "protein," namely a "protein" containing only one kind of amino acid—phenylalanine.[29] This result could have only one meaning: in the genetic code the uracil-uracil-uracil (or in the DNA, the equivalent thymine-thymine-thymine) nucleotide triplet represents the amino acid phenylalanine.

Nirenberg announced his identification of the first codon in August 1961, at the International Congress of Biochemistry in Moscow, where it caused a sensation. (Crick later wrote that he was "electrified.") Thus at one stroke the breaking of the genetic code had become accessible to direct chemical experimentation, because now the effect of introducing various synthetically produced types of messenger RNA of known composition into the cell-free protein-synthesizing system could be examined. The Moscow announcement set off a code-breaking race, sometimes called the Code-War of the U3 Incident, which culminated in deciphering the definite, or at least probable, meaning of many of the 64 codons by 1963.

In 1964 Nirenberg made a second experimental breakthrough in the deciphering of the genetic code by means of his cell-free system for protein synthesis.[30] At that time he discovered that it is possible to detect in his reaction mixture the specific *binding* to ribosomes of molecules of transfer RNA carrying their cognate amino acids. In particular, he found that addition to his reaction mixture of a very short polynucleotide chain consisting of only three nucleotides, instead of messenger RNA, promoted the *specific binding* to ribosomes of those and only those transfer RNA molecules that carry the anticodon complementary in the Watson-Crick base-pairing sense to the nucleotide triplet added to the reaction mixture. Protein formation does not, of course, occur under these conditions, since the short nucleotide triplet cannot serve as the template for directing the assembly of many amino acids. By means of this new technique Nirenberg found, in confirmation of his earlier identification of the codon representing the amino acid phenylalanine, that addition of the uracil-uracil-uracil nucleotide triplet promotes the binding of phenylalanine transfer RNA to ribosomes. And since it was relatively easy to prepare by chemical methods all 64 possible nucleotide triplets, the entire code could be worked out by means of this binding method within little more than a year.

The results of Nirenberg's work, which were presently supported or confirmed by other methods, are now generally presented in the form of a table, the arrangement of which was conceived by Crick (Figure 6). It has been suggested that this table of the code represents for biology what the periodic table of the elements represents for chemistry. The important features of the genetic code are these:

(1) The code contains synonyms, in that many amino acids are coded for by more than one kind of codon. For instance, the nucleotide base triplets uracil-uracil-uracil *and* uracil-uracil-cytosine are synonyms, in that they both code for phenylalanine.

(2) The code has a definite structure, in that synonymous codons representing the same amino acid are nearly always in the same "box" (see Figure 6) of the table. That is, the synonymous codons generally differ from each other only in the third of their three nucleotides. An explanation for this aspect of the code was provided by Crick in terms of the geometry of the hydrogen bonds involved in the recognition of the messenger RNA codon by the transfer RNA anticodon at the ribosomal site of protein formation.

THE GENETIC CODE

1st↓ 2nd→	U	C	A	G	↓3rd
U	PHE	SER	TYR	CYS	U
	PHE	SER	TYR	CYS	C
	LEU	SER			A
	LEU	SER		TRP	G
C	LEU	PRO	HIS	ARG	U
	LEU	PRO	HIS	ARG	C
	LEU	PRO	GLUN	ARG	A
	LEU	PRO	GLUN	ARG	G
A	ILEU	THR	ASPN	SER	U
	ILEU	THR	ASPN	SER	C
	ILEU	THR	LYS	ARG	A
	MET	THR	LYS	ARG	G
G	VAL	ALA	ASP	GLY	U
	VAL	ALA	ASP	GLY	C
	VAL	ALA	GLU	GLY	A
	VAL	ALA	GLU	GLY	G

Figure 6. The table of the genetic code. In this table the letters U, C, A, and G represent the four kinds of nucleotides, containing the bases uracil, cytosine, adenine, and guanine. The three- and four-letter abbreviations represent the twenty kinds of protein amino acids. The codon corresponding to any given position on this table can be read off according to the following rules: The base of the first nucleotide of the codon is given by the capital letter on the left, which defines a horizontal row containing four lines. The base of the second nucleotide is given by the capital letter on the top, which defines a vertical column containing sixteen codons. The intersection of rows and columns defines a "box" of four codons, all of which carry the same bases in their first and second nucleotides. The base of the third nucleotide is given by the capital letter on the right, which defines one line within any given horizontal row. The triplets UAA, UAG, and UGA are "nonsense" codons, to which there corresponds no amino acid.

(3) The code is very nearly universal. Though most of the code was deciphered by use of the protein-synthesizing machinery of *E. coli,* later tests showed that the results are very much the same whether the transfer RNA and amino acid activating enzymes (the agency that "knows" the code) are obtained from bacterial, plant, or animal sources (including mammals). The universality of the code among contemporary living forms shows that the code has remained unchanged over a very long period of organic evolution. One explanation which has been offered for the, at first sight surprising, evolutionary permanence of the code is that any genetic mutation engendering a change in the code—by means of which evolution of the code would have had to proceed—would necessarily be lethal to the organism in which it took place: such a mutational alteration of the code would cause a sudden change in the amino acid sequence of *all* of the protein molecules of the mutant organism. And such a wholesale change would be most unlikely to be compatible with survival of the organism. Another possible explanation for the evolutionary permanence of the code is that there exists some as yet unfathomed geometrical or stereochemical relation between the anticodon nucleotide triplet and the amino acid which it represents. Indeed, if such a relation exists, it would be bound to hold one of the keys to understanding the origin of life.

The Operon

The developments recounted so far have revealed the *qualitative* aspect of the heterocatalytic function: how the nucleotide base sequence carried by the DNA is finally translated into the predetermined amino acid sequence. However, there must pertain to the heterocatalytic function also a *quantitative* aspect: how the DNA manages to govern the synthesis of appropriate amounts of the different proteins whose structure is encoded into it. For it is an easily ascertainable fact that in *E. coli* the number of protein molecules read off one gene per generation can exceed by more than ten-thousandfold the corresponding number read off another gene in the same cell. Furthermore, the rate of translation of any given gene is not always the same, being very high under some physiological conditions and very low under others.

The framework for understanding this quantitative aspect, which was not really covered by the original formulation of the central dogma, was finally provided in 1961 by the *operon theory* of Jacob and Monod[26] (Figure 7). In order to explain the functional regulation of the genes, this theory envisages that a group of genes, or an *operon,* is subject to coordinate control. The genes belonging to the same operon occupy contiguous sectors of the DNA; that is, they are closely linked

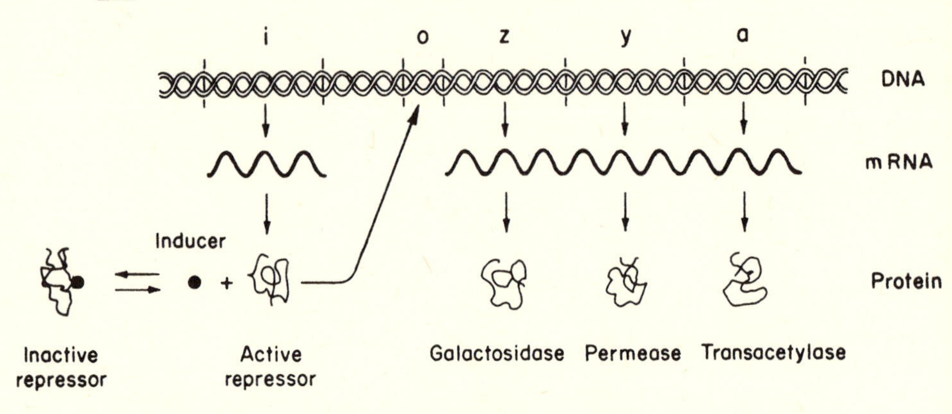

Figure 7. The operon theory of Jacob and Monod, as applied to the lactose fermentation genes of *E. coli*. Three contiguous genes, *z, y,* and *a,* code for the protein structure of three enzymes, galactosidase, permease, and transacetylase, and share a common operator gene *o*. A gene *i* codes for the protein structure of the repressor, which can attach to and thus "close" gene *o*. Combination with an inducer, such as lactose, inactivates the repressor, thus preventing it from "closing" *o*. (From G. S. Stent, in *The Neurosciences* [New York: Rockefeller University Press, 1967].)

and share a common, special regulatory segment of DNA, their *operator*. This operator, which is located at one extreme of the operon group of genes, can exist in two states: open or closed. As long as the operator is open, messenger RNA can be transcribed from all the genes of the operon, and hence synthesis of the protein species encoded into these genes may proceed on the ribosomes into which these messenger RNA molecules are fed. As soon as the operator is closed, transcription of the messenger RNA, and hence synthesis of the corresponding protein species, ceases. Thus the rate of translation of the genes belonging to any operon depends on the fraction of the total time during which their operator happens to be open. Now whether the operator is open or closed in turn depends on whether it has interacted with a *repressor* protein, itself the product of a special regulatory gene. Combination of the operator segment with its related repressor inhibits messenger RNA transcription of the genes of that operon and therefore closes the operator. Hence, in the last analysis, the rate of translation of any gene is a function of the intracellular concentration of active repressor capable of combining with the operator segment of DNA to which that gene is linked.

Thus, by the mid-1960's the general nature of both autocatalytic and heterocatalytic functions of the DNA was understood.[31] Through formation of complementary hydrogen bonds DNA achieves both functions by serving as a template for the synthesis of replica polynucleotide chains, making DNA chains for the autocatalytic and RNA chains for the heterocatalytic function. RNA, in turn, completes the heterocatalytic function by formation of complementary hydrogen bonds with the anticodons of transfer RNA molecules in the amino acid assembly processes. The real core of genetic theory was lifted from the deep unknown in which Muller had found it to lie only fifteen years earlier. We now *do* have actual knowledge of the mechanism underlying that unique property which makes a gene a gene: Formation of complementary hydrogen bonds seems to be all there is to how like begets like.

Alas, the very success of molecular genetics in explaining one of the most profound mysteries of life in terms of workaday chemical reactions altered the spiritual qualities of this field. Molecular genetics now presents an integral canon of biological knowledge which must be preserved and passed on to succeeding generations in the academies. Moreover, as a subject for scholarly research it is far from exhausted, and its practical exploitation in such biotechnological domains as agriculture and eugenics has, as yet, barely begun. But its appeal as an arena for heroic strife against the Great Unknown is gone. And in contrast to classical genetics, which at the apogee of its development still harbored the enigma of the gene as a skeleton in the closet, molecular genetics seems to have left no transcendant legacy. So the would-be explorer of uncharted territory must direct his attention elsewhere. One of the most formidable unsolved problems of biology recommending itself to such romantic types is embryology. Understanding the processes responsible for the orderly development of the fertilized egg into a complex and highly differentiated multicellular organism still seems to boggle the imagination. The recent course of embryological research, however, suggests that it may be just more of the same old molecular genetics, albeit at a much more complicated level.

There now seems to remain really only one major frontier of biological inquiry which still offers some romance of research: the higher nervous system. Its fantastic attributes continue to pose as hopelessly difficult and intractably complex a problem as did the gene a generation ago. Indeed, the higher nervous system presents the most ancient and deepest biological mystery in the history of human thought: the relation of mind to matter. Increasing numbers of veteran molecular geneticists are now turning their attention to the higher nervous system in the hope of finding relief from jejune genetic investigations along more or less clearly established lines. They have good reason to hope that, unlike the quest for fathoming the gene, the quest for understanding the brain will not soon reach a disappointingly workaday denouement. For since the mind-matter mystery is not likely to be amenable to scientific analysis, the most interesting attribute of life may never be explained.[32]

I am indebted to Curt Stern and Alfred E. Mirsky for helpful criticisms of the manuscript of this essay.

REFERENCES

1. L. C. Dunn, ed., *Genetics in the 20th Century* (New York: Macmillan, 1951).

2. E. L. Tatum and J. Lederberg, "Gene Recombination in the Bacterium *Escherichia Coli*," *Journal of Bacteriology,* 53 (1947), 673-684.

3. G. W. Beadle, "Genes and the Chemistry of the Organism," *American Scientist,* 34 (1946), 31-53.

4. F. Miescher, "Ueber die chemische Zusammensetzung der Eiterzellen," *Hoppe-Seyler Medizinisch Chemische Untersuchungen,* 4 (1871), 441.

5. P. A. Levene and L. W. Bass, *Nucleic Acids* (New York: The Chemical Catalog Company, 1931).

6. R. Feulgen and H. Rossenbeck, "Mikroskopisch-chemischer Nachweis einer Nukleinsäure vom Typus der Thymonucleinsäure und die darauf beruhende elektive Färbung von Zellkernen in mikroskopischen Präparaten," *Zeitschrift für Physiologische Chemie,* 135 (1924), 203-248.

7. E. Chargaff, "Chemical Specificity of Nucleic Acids and Mechanisms of Their Enzymatic Degredation," *Experientia,* 6 (1950), 201-209.

8. O. T. Avery, C. M. MacLeod, and M. McCarty, "Studies on the Chemical Nature of the Substance Inducing Transformation of Pneumococcal Types," *Journal of Experimental Medicine,* 79 (1944), 137-157.

9. F. Griffith, "Significance of Pneumococcal Types," *Journal of Hygiene,* 27 (Cambridge, Eng., 1928), 113.

10. Readers interested in obtaining an overview of the contribution of phage research to modern biology are referred to G. S. Stent, *Molecular Biology of Bacterial Viruses* (San Francisco: W. H. Freeman, 1963).

11. E. L. Ellis and M. Delbrück, "The Growth of Bacteriophage," *Journal of General Physiology,* 22 (1939), 365.

12. J. Cairns, G. S. Stent, and J. D. Watson, eds., *Phage and the Origins of Molecular Biology* (New York: Cold Spring Harbor Laboratory for Quantitative Biology, 1966).

13. A. D. Hershey and M. Chase, "Independent Function of Viral Protein and Nucleic Acid in Growth of Bacteriophage," *Journal of General Physiology,* 36 (1952), 39-56.

14. S. E. Luria and M. Delbrück, "Mutations of Bacteria from Virus Sensitivity to Virus Resistance," *Genetics,* 28 (1943), 491.

15. L. Pauling, R. B. Corey, and H. R. Branson, "The Structure of Proteins: Two

Hydrogen-Bonded Helical Configurations of the Polypeptide Chain," *Proceedings of the National Academy of Sciences,* 37 (Washington, D. C., 1951), 205.

16. J. D. Watson, *The Double Helix* (New York: Atheneum, 1968).

17. J. D. Watson and F. H. C. Crick, "A Structure for Deoxyribose Nucleic Acid," *Nature,* 171 (1953), 737.

18. J. D. Watson and F. H. C. Crick, "Genetic Implications of the Structure of Deoxyribonucleic Acid," *Nature,* 171 (1953), 964.

19. M. Delbrück and G. S. Stent, "On the Mechanism of DNA Replication," W. D. McElroy and B. Glass, eds., *The Chemical Basis of Heredity* (Baltimore: Johns Hopkins Press, 1957), p. 699.

20. M. Meselson and F. W. Stahl, "The Replication of DNA in *Escherichia Coli,*" *Proceedings of the National Academy of Sciences,* 44 (Washington, D. C., 1958), 671.

21. F. Sanger, in D. E. Green, ed., *Currents in Biochemical Research* (New York: Interscience, 1956).

22. S. Benzer, "The Elementary Units of Heredity," W. D. McElroy and B. Glass, eds., *The Chemical Basis of Heredity* (Baltimore: Johns Hopkins Press, 1957), p. 70.

23. G. Gamow, "Possible Relation Between Deoxyribonucleic Acid and Protein Structures," *Nature,* 173 (1954), 318.

24. F. H. C. Crick, L. Barnett, S. Brenner, and R. J. Watts-Tobin, "General Nature of the Genetic Code for Proteins," *Nature,* 192 (1961), 1227.

25. K. McQuillen, R. B. Roberts, and R. J. Britten, "Synthesis of Nascent Proteins by Ribosomes in *Escherichia Coli,*" *Proceedings of the National Academy of Sciences,* 45 (Washington, D. C., 1959), 1437.

26. F. Jacob and J. Monod, "Genetic Regulatory Mechanisms in the Synthesis of Proteins," *Journal of Molecular Biology,* 3 (1961), 318.

27. F. H. C. Crick, "On Protein Synthesis," *The Biological Replication of Macromolecules,* Symposium of the Society for Experimental Biology, XII (London: Cambridge University Press, 1958), p. 138.

28. M. B. Hoagland, P. C. Zamecnik, and M. L. Stephenson, "Intermediate Reactions in Protein Biosynthesis," *Biochimica et Biophysica Acta,* 24 (1957), 215. M. B. Hoagland, M. L. Stephenson, J. F. Scott, R. J. Hecht, and P. C. Zamecnik, "A Soluble Ribonucleic Acid Intermediate in Protein Synthesis," *Journal of Biological Chemistry,* 231 (1958), 241.

29. M. Nirenberg and J. H. Matthei, "The Dependence of Cell-Free Protein Synthesis in *Escherichia Coli* Upon Naturally Occurring or Synthetic Polyribonucleotides," *Proceedings of the National Academy of Sciences,* 47 (Washington, D. C., 1961), 1588.

30. M. Nirenberg and P. Leder, "RNA Codewords and Protein Synthesis: The Effect of Trinucleotides Upon Bringing sRNA to Ribosomes," *Science,* 145 (1964), 1399.

31. By that time the Cold War had eased, both in its strategic and its genetic aspects. Molecular genetics had become definitely an in-subject in the Soviet Union, thanks in part to the final fall from favor of Lysenko after the dismissal of N. Khrushchev. But it seems likely that in any case the "materialist" concrete DNA gene of molecular genetics (and the possibility of inheritance of "acquired" characters through phenomena such as the DNA-mediated bacterial transformation) is more palatable to dialectical-materialist thought than the "idealist" abstract gene of classical "Mendelism-Morganism." See Zh. Medvedev, *The Rise and Fall of Lysenko* (New York: Columbia University Press, 1969).

32. To readers interested in the development of genetic thought, I strongly recommend A. H. Sturtevant, *A History of Genetics* (New York: Harper and Row, 1965). Some of the philosophical issues attending the rise of molecular genetics are treated in M. Delbrück, "A Physicist Looks at Biology," *Transactions of the Connecticut Academy of Sciences,* 38 (1949), 173 (reprinted, see note 12); in G. S. Stent, "That Was the Molecular Biology That Was," *Science,* 160 (1968), 390; and in D. Fleming, "Emigré Physicists and the Biological Revolution," in D. Fleming and B. Bailyn, eds., *The Intellectual Migration: Europe and America, 1930-1960* (Cambridge, Mass.: Harvard University Press, 1969), 152. Professional introductions to molecular genetics are provided by J. D. Watson, *The Molecular Biology of the Gene* (New York: Benjamin, 1965), and G. S. Stent, *Molecular Genetics* (San Francisco: W. H. Freeman, 1971). A fuller treatment of the notions touched on in the closing paragraphs of this essay can be found in G. S. Stent, *The Coming of the Golden Age: A View of the End of Progress* (New York: Natural History Press, 1969).

ROBERT OLBY

Francis Crick, DNA, and the Central Dogma

THIS ESSAY does not present a definitive account of Francis Crick's contribution to molecular biology, but rather is an attempt at a brief description of how Crick, a British physicist caught up in the war effort, came into biology; how he collaborated with James D. Watson, the Chicago-born biologist, in discovering the structure of DNA; and how Crick went on to play a central role in the formulation of the theory underlying molecular biology today. Several of his scientific contributions have been omitted in order to focus attention on the double helix, the sequence hypothesis, the central dogma, and the genetic code.[1]

Born in June 1916, during the First World War, Francis H. C. Crick grew up in those uncertain interwar years which were marked by disillusionment, the slump, and financial ruin. Many a family business had to be closed down, including the Northampton boot and shoe factory run jointly by Francis' father, Harry, and his uncle, Walter. The shoe shops which Harry Crick continued to run failed to restore the wealth of the family, and to preserve the tradition of Mill Hill schooling for all the family, the Cricks moved close to the school so that Francis' younger brother, Tony, could become a day-boy; and Francis, who left the school in 1934, was thus able to commute daily to University College, London. Michael Hart, headmaster of Mill Hill School, recalls that Francis Crick's father was an Old Millhillian, as were three of his uncles. Of Francis he said: "We remember vividly all the characteristics (piping shrill voice and laughter) described by his contemporaries at Cambridge today. He took Higher Certificate and gained distinctions in Physics and Maths for Science, and also a pass in Chemistry. He was a highly competent VI former who was expected to do very well, but we had no real expectation of his future brilliance."[2]

The Crick family was well-to-do middle class; Francis' father and his uncles were fond of bridge and tennis parties and were good Congregationalists. But with the exception of Francis' grandfather,

Walter Drawbridge Crick, a prominent member of the Northamptonshire Natural History Society and Fellow of the Geological Society, Francis was alone in the intensity of his scientific curiosity. It was W. D. Crick who reestablished the family name in Northamptonshire when he rose from the position of clerk in the Goods Department of the London and North-Western Railways Co. to director of the boot-and-shoe firm of Latimer Crick and Co.[3] and owner of the fine house, Nine Springs Villa, with its spacious gardens, in a select part of Northampton. All this he achieved in a short life of forty-six years, during which he also found the time to collect liassic fossils, freshwater bivalves, Cruikshankiana, books, porcelain, and furniture. As a student of freshwater bivalves he had the distinction of a mention by Charles Darwin.[4]

Of his father's generation (see Table 1) Francis said that Willie (killed in the 1914-1918 war) was reckoned the cleverest, but all were bright. Uncle Walter, who had taken over the direction of the factory at the age of nineteen was deeply disturbed by the events of the slump, blaming the bankers whose high interest rates had reduced liquidity and deflated the economy with ruinous efficiency. Although he became quite fanatical on this subject, the pamphlet *Abolish Private Money, or Drown in Debt,* written with Frederick Soddy, does demonstrate considerable command of the data and the ability for forthright expression.[5] Neither Walter nor Harry was interested in making money; both would have preferred a different career. While Harry remained in England to supervise the London shoe shops, Walter emigrated to the United States, where he spent the last eighteen years of his life as a sales agent for shoe manufacturers.[6] Only Uncle Arthur had a flair for business, and he amassed a fortune as the owner of a chain of drugstores in Kent.

Table 1.

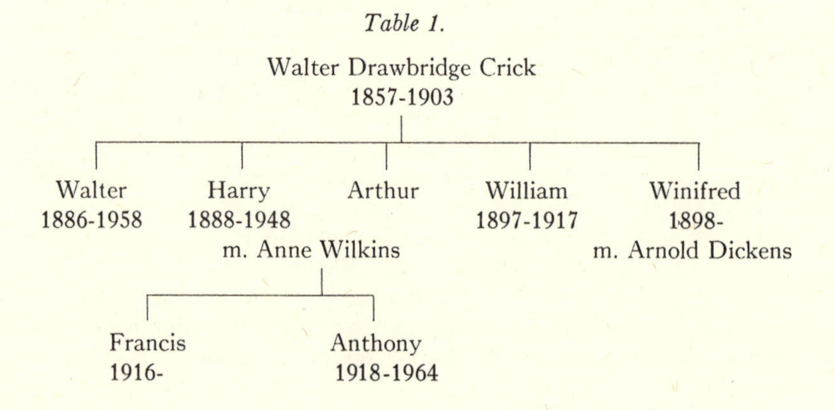

What restricted the intellectual interests of this generation of Cricks was the fact that they never went to a university, a result, in part, of the early death of their father. Their interests, too, were no doubt determined by the nonacademic, nonconformist circle in which they moved. Harry Crick had wanted to become a lawyer, but his real passion was tennis;[7] such was his zeal that he would play secretly on Sundays. As church secretary he could not have done this openly, and Francis, as the son of so respected a figure, did not find it easy to evade regular attendance at services. His resistance to this duty led him at an early date to take the antivitalist's position, a stand which was later to make the field of molecular biology particularly attractive (and, as we shall see, it was he who gave a rigorous statement of the reducibility of cell physiology to chemistry and physics in the sequence hypothesis and the central dogma).[8]

It would be a mistake to describe Crick's family as boring and humdrum. While Francis found something lacking in his family circle at Northampton, the family found something lacking in him. This was not due to conflict between radical and bourgeois ideas, for the family admired Northampton's free thinker, Charles Bradlaugh, and Uncle Walter took sides with the workers over the causes of the slump. But there was no one in the family circle with whom Francis could talk about the scientific topics which fascinated him. While some people operate best in seclusion, others need sounding boards, critics, catalysts; for them ideas are formed to be discussed, and part of their zest for life comes from the joy of communicating ideas. Crick undoubtedly belongs to the latter group. A part of what one might call his style is intense and prolonged discussion, something which his good-natured and forthright family could not provide. What interests they had were either at the natural history level or, as in the case of his college-trained mother, on the arts side. But to the young Crick science was everything; the family called him "crackers," so lopsided did his interests appear.[9]

As a schoolboy he had worked quite hard, and he took his work at University College seriously. But he cannot now recall that the physics course stimulated him in any special way: it was too early for the exciting developments in the quantum theory to permeate undergraduate physics courses. Although Professor E. N. da C. Andrade gave probably the best course on the Bohr theory of the atom in those days (he was author of the definitive book, *The Structure of the Atom*, 1923), no more than one question on this theory appeared in the physics final examinations; the remainder were on more traditional topics, such as "Describe a method suitable for measuring the viscosity of a gas," "Outline the experiments that would have to be carried out to enable the mass of the earth to be determined."[10]

Naturally a first was expected of Crick, so there was general disappointment when he was awarded only a good second in the physics finals of 1937.[11] As a graduate student he was given the task of building an apparatus with which to measure the viscosity of water at temperatures above 100°C. Since the 1930's Professor Andrade had paid special attention to viscosity because of his interest in the liquid state, and he wanted to test his assumption of 1934 that the viscosity of a liquid "is proportional to the average number of molecules possessing the minimal kinetic energy necessary for the transmission of momentum.[12] This implied a linear relationship between log η and 1/T at constant volumes. The water had to be studied at above 100°C because at lower temperatures, as a result of hydrogen bonding, its molecules depart widely from ideal conditions.

So for two years Crick busied himself building a copper sphere containing water at high temperatures. When the sphere was oscillated, the damping effect of the water was measured, from which its viscosity was to be calculated. This work was near completion when war was declared, the laboratory was closed, and Crick joined the Admiralty Research Laboratory at Teddington. In the first year of the war a mine (at that time the Germans were using naval mines as bombs) fell on University College and destroyed his viscosity apparatus. In a report dated 1946 Professor Andrade wrote of Crick: "He showed great ability as an experimental and theoretical physicist, has an ingenious and lively mind and shaped very well as a research student. He is a very able physicist with plenty of initiative."[13] Not surprisingly he had been awarded the Carey Foster Research Prize in Physics in his second year of Ph.D. work.

At Teddington, Harrie Massey (now Sir Harrie) had collected a crack team of scientists to work on the sweeping of enemy mines. They also considered an antisweep device for British mines. Here all members of the group learned to value the clear and incisive mind of Crick and his strong physical sense. When the team moved down to Havant, near Portsmouth, to join the Mine Design Department with its long-established naval staff, there were frequent tensions, and Massey found himself soothing officers hurt when Crick had told them they were talking nonsense. One incident which illustrates Crick's ability to see straight through to the heart of the matter was when Massey asked his advice about how to overcome the powerful magnets in German sweeps which activated the British mines at a safe distance from them. At once Crick replied that all you had to do was to lay some mines so insensitive to the magnets that they would only be fired right under the sweeps. Most of the naval staff could not see the sense of laying mines too weak to blow up a ship; they failed to see that the problem was a statistical one, the aim being to put out of action the maximum number of minesweepers.[14] The success of the Mine Design Department is shown by the fact that throughout the war 30 per cent of German shipping was in dry dock undergoing repairs. By the time sonic, magnetic, and electric operators were being used for British mines the German sweeps were trailing an army of devices to counter all such mechanisms—magnets, alternating circuits, cannons, and clappers.

When the war was over in 1946 Crick stayed on as an Admiralty scientist working in the Naval Intelligence Division in London, with the long-term aim of going into fundamental research either in particle physics or physics applied to biology. Seven years in the Admiralty made him competent in hydrodynamics and magnetism but not in fundamental physics, still less in chemistry or biology. Like so many scientists, he had found the first five years of his university career dull. On top of this was the financial effect of the depression; and had it not been for the generosity of Uncle Arthur, and his aunt, Miss Ethel Wilkins, his studies, like those of Pauling before him,[15] would have been interrupted. As it was, their generosity eased things for Crick before and during the war and in Cambridge afterwards.

Just how little Crick knew about the world of chemistry in those days is illustrated by his reaction to a speech of Linus Pauling's in 1946 on modern structural chemistry. Had Crick then known who Pauling was he would have expected to be impressed by his speech, but never having heard of Pauling before he was simply surprised and delighted at the view expressed by this structural chemist on the role which weak intermolecular forces will be shown to play in physiological processes. Pauling had said: "I believe that usually the specific physiological properties of substances are determined not by these strong intramolecular forces, but instead by the weak forces—van der Waals forces, hydrogen bonds—which operate between molecules."[16] Clearly Crick was already in 1946 receptive to the extension of orthodox physical chemistry into biology rather than to the search for unorthodox long-range forces and the like.

It was reading such articles and talking to his naval officer friends about antibiotics and similar subjects which convinced him that biology was what interested him. Also, he had read Erwin Schrödinger's book *What Is Life?* This noted physicist had asked "How can the events in space and time which take place within the spatial boundary of a living organism be accounted for by physics and chemistry?" and had concluded that "The obvious inability of present-day physics and chemistry to account for such events is no reason at all for doubting that they can be accounted for by those sciences."[17] As a physicist used to statistical laws in which the regularities in the behavior of particles result from the average behavior of large numbers of them, Schrödinger failed to see how so small a number of atoms as the thousand or so in the sensitive area of a gene could show such extraordinary constancy. Had they been regularly repeating crystalline polymers he could have understood it, but since the "hereditary codescript" must be "aperiodic," he concluded that "we must be prepared to find a new type of physical law prevailing in life."[18] The way out of the dilemma lay in the concept of template molecules upon which complementary aperiodic molecules can be laid down under the influence of weak intermolecular bonds. Schrödinger deliberately presented the problem from the point of view of one who doubted that bonds exist and who thus viewed the gene as a collection of unbound atoms. He went on to attribute the constancy of the gene to its crystalline-solid constitution, and there is no doubt that this concept had an impact on those physicists who read his book and turned to biology.

What impressed Crick about the book at the time was that "fundamental biological problems could be thought about in precise terms, using the concepts of physics and chemistry." One was left with the impression that "great things were just around the corner."[19] This, then, was a general impression, but the significance of Schrödinger's statement on the hereditary codescript did not strike Crick until after the DNA structure was out. Then he wrote to Schrödinger enclosing a reprint of the 1953 *Nature* paper, but received no reply. Professor Cyril Darlington also recalled meeting Schrödinger at this time and found him unwilling to talk about the problems he had earlier discussed with such enthusiasm.[20]

A more powerful stimulus than *What Is Life?* leading Crick to biology was the "religious" one of an atheist who wanted "to try to show that areas apparently too mysterious to be explained by physics and chemistry, could in fact be so explained. That is why (having decided that since I knew so little anyway I had a free choice of subject!), I selected as the two regions: (1) the borderline between the living and the non-living (molecular biology as we would call it to-day) and (2) the brain. After a great struggle I chose the former because it seemed nearer to things I knew about already."[21]

Entering Biology

It was through Massey that Crick was introduced to the physiologist, A. V. Hill, and Sir Edward Mellanby, Secretary to the Medical Research Council, to whom he went for advice on how to get into biological research. By this time J. D. Bernal's wonderful work on the

X-ray crystallography of proteins and nucleic acids had come to Crick's attention, and he was anxious to join Bernal's group at Birkbeck College, London. Now the way seemed clear to escape from magnetism and hydrodynamics into X-ray crystallography. But when Crick went to Birkbeck and asked the secretary if he could see Bernal with a view to working under him, he was cut down to size by the reply, "Don't you realize that people from all over the world want to come and work under the Professor?"[22] He retired abashed and sought the advice of A. V. Hill. What followed is described from the records of the Medical Research Council.

Crick explained to Hill that he was particularly attracted by the application of physical methods to biological products—as typified by Bernal's work on the structure of viruses—and with encouragement from Professor A. V. Hill he came to see Sir Edward Mellanby in the summer of 1947 to discuss his future. Sir Edward was obviously impressed by Crick's approach, but pointed out the difficulties of placing a man of his standing in the biological world at a commensurate salary; Crick was in no way discouraged by the situation and indicated that he would be prepared to enter the research field by means of an MRC studentship. All such studentships were fixed at £350 untaxed, regardless of age and experience.

Sir Edward assured him that an application sponsored by Professor Bernal would go through automatically, but suggested that in the latter's absence abroad, Crick should ask Professor Hill to speak on his behalf. He went on to surmise that at some future date Crick might well consider the possibility of joining Wyckoff's team of workers in Washington and Crick endorsed this view, expressing his own particular interest in their kind of inter-disciplinary approach to biophysics.

Crick's application for a research studentship was warmly supported by Professor A. V. Hill, but it had not been possible in Professor Bernal's absence to arrange supervision at University College. Professor Hill was in fact of the opinion that Cambridge would be a preferable centre, in view of the considerable amount of work in progress there which would give Crick the necessary experience in the biological field. In his report on Crick dated July 9, 1947 Hill stated that: ". . . by far the best place for him if someone would father him and his work there would be Cambridge. There he could form the necessary biological knowledge while beginning to learn about and apply the various techniques necessary for the purpose he has in mind." As the application was to be considered by Council within a few days, there was no time to explore the question of a supervisor at Cambridge and, quite exceptionally, the Council awarded him a research studentship without any place of work being specified. They formed a most favourable impression of his originality and potential which is clearly expressed in his application; he wrote:

> "The particular field which excites my interest is the division between the living and the non-living, as typified by, say, proteins, viruses, bacteria and the structure of chromosomes. The eventual goal, which is somewhat remote, is the description of these activities in terms of their structure, i.e., the spacial distribution of their constituent atoms, in so far as this may prove possible. This might be called the chemical physics of biology."

After a further meeting with Sir Edward, it was arranged that Crick should visit Cambridge and discuss the whole matter of his future in the first place with Dr. (now Dame) Honor Fell at the Strangeways Research Laboratory; Sir Edward also suggested that he should meet Professor Keilin, Sir Lawrence Bragg and Professor Roughton to talk about his ideas; it was made clear that he himself would have to work out his own programme. Dame Honor's view was that Crick would be able to gain the necessary experience at Strangeways, and it was arranged that he should start work there under her supervision in September 1947; he then embarked on post-graduate studies in the biological field to complement his research interests and at the same time began part-time research on aspects of the mechanism of cell division.[23]

The Strangeways Institute had been founded by F. S. P. Strangeways in 1909 to investigate the cause of arthritis. When it became clear that apart from the classification of the various forms of arthritis, no progress was being made, the trustees switched to a broader policy with more fundamental research. It was here that the technique of tissue culture which Ross Harrison had invented in 1905 was applied to cartilage cells from arthritic joints. Here R. G. Canti took the first film of cell division, Arthur Hughes used phase-contrast cinematography to demonstrate the movement of mitochondria, C. H. Waddington carried out the first induction in mammalian embryos, and Chambers and Honor Fell used micromanipulation to study the role of the nucleus in the cell. Today the institute is doing exciting work on membrane structure, but when Crick arrived, the threadbare topic of the viscosity of protoplasm was a favored subject. A relic from the days when the physical ultrastructure of protoplasm was thought to hold the key to vital functions, studies of the viscosity of protoplasm have consistently failed to yield hard facts except in the field of plant sensitivity, where it is thought to play a part in perception. The great names of the past on the structure of protoplasm—Lewis Heilbrunn, William Seifriz, and Herbert Freundlich—have been forgotten, and along with them the "brush-heap" (Seifriz) and "framework" (Frey-Wyssling) theories of structure.

Arrival in Cambridge

When Crick arrived on the scene Dr. Fell suggested that he work with Arthur Hughes, who was studying the movement of magnetic particles ingested by chick fibroplasts in tissue culture. Crick jumped at this opportunity, since it offered clear-cut experiments of a physical nature in the uncertain green-thumb world of tissue culture. With such work he reckoned he could come to grips with the strange world of biology without the distraction of having to learn more physics. Crick had escaped from the viscosity of water only to trap himself in the viscosity of protoplasm: just as he spent two years on water, he now spent two years on protoplasm. To this day his friends tease him about his "twist," "drag," and "prod" games with ingested particles, experiments which led to seventy-two pages in the journal *Experimental Cell Research* on what he has always regarded as inconsequential work. Be that as it may, his analysis of previous work and theories of structure exhibits the same incisive mind so characteristic of his later work.[24]

Also, the years 1947-1949 were used by Crick to read widely, and on the basis of this self-education in biology and chemistry Crick formed many of his ideas on molecular biology. This was no case of casual browsing through the literature, but sustained, intensive, and critical study, during which he formulated what might be called the protein version of the sequence and collinearity hypotheses. At this stage he recognized the relation between gene and gene product, an enzyme, as of crucial importance, but was thinking entirely in terms of proteins, the study of whose structure he regarded as the field most appropriate for him.

This work at Strangeways was not presented as a thesis, for he had deliberately not embarked on a Ph.D. for fear he would be tied down too much. Now he had to decide whether to stay on to join Murdoch Michison and M. M. Swann who were working on mitosis, or go to the Cavendish Laboratory to work under Max Perutz whom he had met through his friend George Kreisel, the mathematician. What happened we learn from the records of the Medical Research Council.

In the early part of 1949 Crick wrote to Sir Edward asking if he would see him again to discuss his work at Cambridge. Sir Edward agreed and during their subsequent meeting, Crick revealed that his interests were diverging from the biological aspects of biophysics to the purely physical side and that what he really wanted was to join the Council's Unit at the Cavendish Laboratory under Dr. Perutz—who was also very anxious to have him—and to carry out research using methods of X-ray analysis for determination of the structure of protein and allied molecules. Crick's studentship was nearing the end of its second year and Sir Edward told him that while the Council could not assume responsibility for looking after him when his training was over, he had better apply for a further year's extension and await events.

These were not long delayed; a few weeks later Sir Lawrence Bragg, then Head of the Cavendish Laboratory, wrote to Sir Edward as follows:

> "I have just been having a talk with F. H. C. Crick, who holds one of your Research Studentships. He has been working in the Strangeways, and now wishes to get some research experience in Perutz's team. If you wish him to do this I shall be very glad to have him in the Laboratory. I think he would be a useful member of the team, and also that he would be getting the kind of experience he appears to desire. I am leaving it to Perutz and Crick to consult you about the transfer; this is merely to say that as far as I am concerned I should welcome the arrangement."

Sir Lawrence's letter was closely followed by an application on Crick's behalf from Dr. Perutz, who wrote:

> ". . . I have known him ever since he decided to enter the field of biophysics and know that he has always been keenly interested in the problem of protein structure, and would have liked to join our Unit from the start, but was advised to gain some experience with living materials before making a final decision about his future line of research. After a thorough study of the subject he has now decided that X-ray analysis of protein structure really is the field that attracts him most.
>
> "I should be very glad to have Crick. I had many conversations with him and he has always struck me as an exceptionally intelligent person, with a lively interest, a remarkably clear analytical mind and a capacity for quickly grasping the essence of any problem. At yesterday's meeting of the Biophysics Committee Dr. Hughes reported on the results Crick had obtained in his research at the Strangeways Laboratory. This research is now completed and Crick is writing it up for publication. He thinks that he will be ready to join us in June or July.
>
> "I am not quite sure what form of appointment the Council would consider appropriate. Having worked for the Admiralty during the war, Crick is much older than a man in his position would normally be—he is 32—and has family obligations. If he were younger, a research studentship would undoubtedly be the right thing for him, but as it is he would be much happier to have a definite appointment. On the other hand, if the Council does not consider him ripe for one, I know that he would be glad to continue with his present studentship for another year."

In the event, Crick was offered an unestablished 3 year appointment on the scientific staff of the Council's Unit and started work there in June 1949, relinquishing his research studentship at the same time. He has not in fact registered for a Ph.D. degree while working at Strangeways and the reasons are explained in a letter from Dr. Perutz requesting the Council's permission for him to do so after joining the Unit, as follows:

> "I should like to consult your advice regarding the position of Mr. F. H. C. Crick. Crick has been with me for over a year now, and I am rather anxious that he should register as a research student and take a Combridge Ph.D. since this would considerably improve his future prospects. You may remember that he held an MRC studentship while he worked at the Strangeways Laboratory before he joined me. At that time, however, he did not make use of the possibility to become a research student, first because he did not think his research problem was a suitable subject for a Ph.D. thesis, and secondly, because difficulties arose through the Strangeways Laboratory not being a University Department.
>
> "I wonder if Crick could have the Council's permission to become a registered research student while holding his present position, and if so, whether the Council would be willing to assist him with the payment of fees."

Permission was given for Crick's registration.[25]

When Crick, a bold thirty-three-year-old graduate student of Gonville and Caius College, knowing no X-ray crystallography, having no

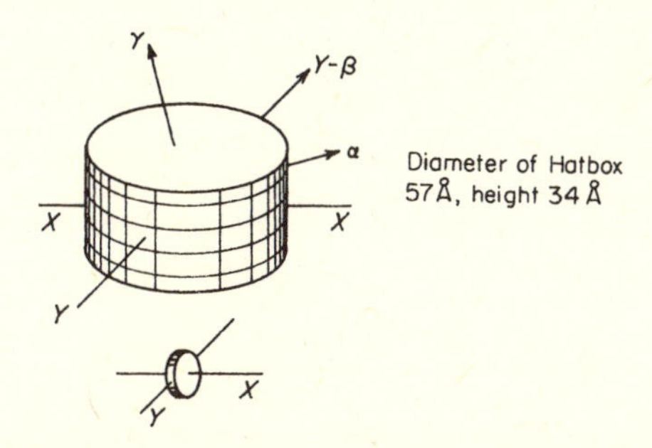

Figure 1. Diagrammatic model of haemoglobin molecule, showing its orientation with respect to the crystal axes. *Y* is the diad axis. The small disk underneath represents a haem group drawn on the same scale and its correct orientation with respect to the crystal axes. The four lines on the cylinder surface indicate the positions of the concentrations of scattering matter deduced from the Fourier projections. The directions of the principal refractive indices are indicated by arrows. [This is merely a simplified diagramatic picture, giving as it were the fuzzy outline of the molecule. (From J. Boyes-Watson, E. Davidson, and M. F. Perutz, "An X-ray Study of Horse Methaemoglobin, I," *Proceedings of the Royal Society A,* 191 [1947], 123.)]

research degree, entered the little MRC unit of the Cavendish Laboratory, he taught himself the subject and read through the early papers on the structure of hemoglobin. After a year he gave a seminar on his conclusions which he titled, on the advice of Perutz's co-worker John Kendrew, "What Mad Pursuit." Perutz, Kendrew, and Bragg listened while this newcomer exposed the inadequacies of their techniques and attacked their picture of the hemoglobin molecule. Those were the days of the "hat-box" model (see Figure 1) in which the polypeptide chains were stacked in four layers all in parallel array like a neat pile of logs. This was consistent with the over-all electron density as calculated by Perutz, which in a vertical section through the hat box showed regularly spaced peaks of high density where the chains had been sliced through, and in transverse section, parallel ribs of high density where the chains ran across the box.

Crick believed that the technique of counting vectors so far employed was too superficial, and since a three-dimensional structural analysis was prohibitively lengthy, he tried reducing this to a two-dimensional analysis in the direction of the rods. This exercise, which he completed some time after his seminar talk, revealed a tenfold discrepancy between the model and the diffraction data. Crick concluded that only half the protein in the molecule could be arranged in the manner of the hat-box model. He believed there were more kinks, shorter straight runs, and that even when broadly parallel the chains meandered. By the time his thesis was completed, most of these conclusions had been found less serious than he had thought at the time, but they did serve to jolt his colleagues out of a rut in which they had been working. His criticism and the subsequent analysis by Bragg marked the end of the belief in regular geometric structures for the globular proteins, a belief which had hitherto been widespread among protein crystallographers. Had crystallographers thought otherwise in those early days, they would never have attacked the proteins when they did. We are not to judge them for their oversimplified picture, for as Crick remarked: "It is one of the occupational hazards of the sort of crystallography in which you do not get results within a reasonable time, that those who work in it tend to deceive themselves after a bit; they get hold of an idea or an interpretation and unless there is someone there to knock it out of them, they go on along those lines, and I think that was the state of the subject when I went to the Cavendish."[26]

As we know from J. D. Watson's book *The Double Helix*, Crick's twenty-minute report on the techniques of structure analysis of proteins did not endear him to Sir Lawrence Bragg, although Kendrew and Perutz took it "remarkably well." The subsequent argument which Crick had with Sir Lawrence over priority for the "wave idea" exacerbated their relationship further. Nor could Crick settle down to one problem and see it through. Under Perutz's supervision he was supposed to find a protein smaller than hemoglobin and work out its structure. He considered secretin (an intestinal hormone which stimulates secretion of the pancreatic juice), which is about one-tenth the size of hemoglobin, but David Keilin, director of another Cambridge laboratory—the Molteno Institute—advised Perutz that this would be too difficult for a beginner to assay; Crick would have had to prepare it himself. So he turned to trypsin inhibitor. With this protein he was more successful and got as far as finding out the number of molecules in the unit cell. "What I should have done now, I realize pretty clearly," Crick said later, "is to have gone and looked at another species or done different salt conditions, but it was so discouraging, the number of molecules in the unit cell being so large, that I turned from it."[27] This was not just the impatience of a theoretician with the slow grind of experimental work, but the exasperation that many protein X-ray crystallographers feel from time to time.

As Crick turned from one trying topic to another, his thesis, "Polypeptides and Proteins: X-Ray Studies," became more and more of a "ragbag," and he considers it a miracle that he was able to make any thesis at all out of the material. On the other hand, there was in the pocket at the back of the thesis enough published material to make another one, "including of course the structure of DNA!"[28] In Great Britain it is the custom to submit published papers as accessory material; the reasons for excluding the work on DNA from the thesis itself were that "it would have meant recasting the whole plan of the dissertation and this would not have been easy at such a late stage. Secondly, the work was carried out in very intimate collaboration with Dr. Watson, and it would have been difficult to disentangle our respective contributions to it."[29]

It is surely also true to say that the section of his thesis dealing with the Fourier transform of a helix and the packing of helices in synthetic and natural polypeptides would have sufficed for a Ph.D. The Fourier transform of a helix was worked out in conjunction with the crystallographers V. Vand and William Cochran and predicted the characteristics of the diffraction patterns produced by continuous and discontinuous helices.[30] This theory enabled one to calculate rapidly the X-ray pattern predicted by a given model, a process which had hitherto consumed much time and which had normally been confined to the pattern along the equator (zero layer line) (see Figure 2). Bragg was mollified by this, the first evidence that Crick could see a good job through to a successful conclusion. By applying this theory to the early diffraction patterns obtained by W. T. Astbury, Crick was able for the first time to interpret them in terms of Pauling's α helices. Pauling had worked out the α helix for polypeptides while ill in London in 1948 but had refrained from publishing it because it did not account for the strongest feature of the diffraction pattern of keratin. Two years later he ventured a short note, because he had seen the patterns of synthetic polypeptides produced by the group at Courtaulds Ltd. in London and had found that they lacked the anomalous spot of α keratin.[31] His detailed papers followed in 1951.[32] In 1952 Crick was able to show that if the helices have a second twist of a higher order imposed upon them the keratin spot can be accounted for.[33] The same solution was arrived at independently by Pauling.[34]

The importance of the coiled-coil idea should not be underestimated, although it is still not unequivocally proven. The anomalous spot on the α keratin diagram had for years been the major stumbling block to any models so far advanced, including Pauling's α helix. Crick's coiled coil, which thus cut the Gordian knot for the students of proteins, is a good example of that physical intuition which enables Crick to brush aside the complexities of the situation and the plethora of possible alternatives and to concentrate on finding a simple physical model. The problem had been to account for the discrepancy between the meridional spot of 5.15 Å (angstroms) on the X-ray diagram of α keratin, and the 5.4 Å repeat of the Pauling-Corey helix. Knowing how difficult it would be for such a helix to pack down more tightly than the 5.4 Å pitch, Crick suggested that the helix is tilted in such a

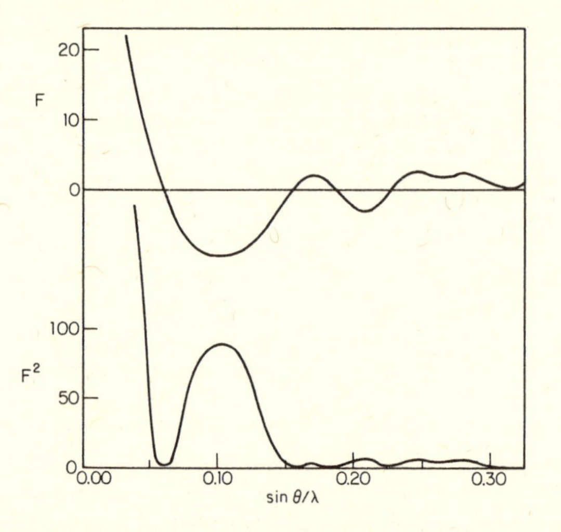

Figure 2a. The calculated X-ray form factor F and its square F^2 for equatorial reflections of nucleic acid. [Pauling's representation of the distribution of electron density moving out from the center of the X-ray diagram of DNA along the equator. (From L. Pauling and R. B. Corey, "A Proposed Structure for the Nucleic Acids," *Proceedings of the National Academy of Sciences,* 39 [1953], 94.)]

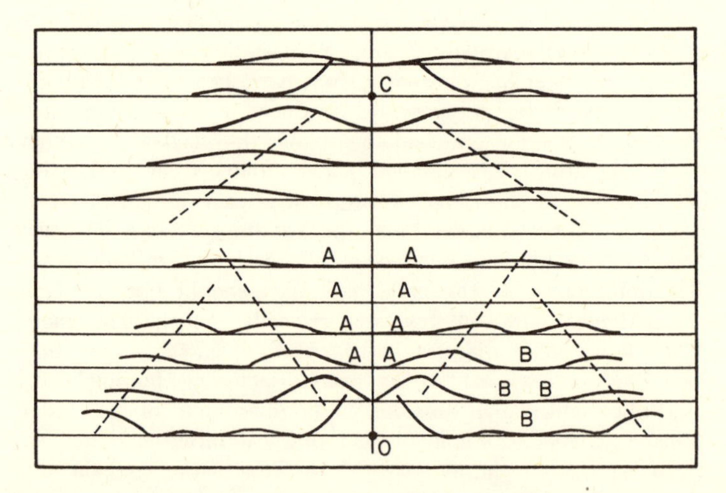

Figure 2b. Diffraction pattern of system of helices corresponding to structure of deoxypentose nucleic acid. The squares of Bessel functions are plotted about O on the equator and on the first, second, third, and fifth layer lines for half of the nucleotide mass at 20 Å diameter and remainder distributed along a radius, the mass at a given radius being proportional to the radius. About C on the tenth layer line similar functions are plotted for an outer diameter of 12 Å. [The more detailed picture obtained by Wilkins, Stokes, and Wilson for DNA using the helical transform theory. (From M. H. F. Wilkins, A. R. Stokes, and H. R. Wilson, "Molecular Structure of Deoxypentose Nucleic Acids," *Nature,* 171 [1953], 738.)]

way as to split the single meridional 5.4 Å reflection into two near-meridional spots at 5.1 Å. Simply by bending the helix into a coiled coil with a pitch angle of 18° he was able to generate the required spots, for 5.4 Å × cos 18° = 5.1 Å.

When Crick met Pauling in the summer of 1952, the structure of α keratin was mentioned, and they agreed that the solution may involve the coiling of α helices around one another. The details were not discussed because their analyses were still unsatisfactory, but by the autumn both men felt more confident. The result was that the editorial office of *Nature* received almost simultaneously a paper from Crick and one from Pauling and Corey on the solution of the same problem. Crick's paper, dated October 22, 1952, appeared in the November 22 issue; Pauling and Corey's, dated October 14, appeared in the January 10, 1953, issue. Unfortunately, the papers were not treated in strict order of receipt; it just happened that Pauling's paper had three diagrams which necessitated the making of blocks, Crick's paper had none and was supported by a letter from Bragg. A comparison of the dates of other papers in the January 10 issue of *Nature* shows that Pauling's paper was not delayed, but that Crick's paper was pushed ahead.

Leaving the unfruitful topic of priority to one side, a reading of both papers does bring out some of the differences between the authors' approaches and style of writing. Crick's paper is easier to read and introduces the "knobs and holes" concept for interleaving the side chains. He gives an approximate calculation of the force necessary to deform the helix into the coiled coil. Pauling, on the other hand, does a comparison of the observed and calculated X-ray intensities along the equator and postulates the packing of the helices into three- and seven-stranded ropes, which he illustrates. Whereas Pauling favored a super-helix pitch of 66 Å, Crick suggested 198 Å. Support for the latter alternative may be found in the recent discovery of two near-meridional spots on the α keratin pattern of mohair by Spei, Heidemann, and Zahn.[35]

The Structure of DNA

This background of helical expertise and detailed knowledge of the crystallography of hemoglobin served Crick well when he collaborated with Watson on DNA. The first time that Crick studied the literature on DNA was when he gave a talk about it at Strangeways Laboratory in the 1940's. All he can now remember of that event was that he noted the way in which radiation reduced viscosity. Clearly he had no grounds for regarding DNA as being as important as proteins or more so, and since single crystals of DNA cannot be produced, no crystallographer in his right mind would tackle it save for very compelling reasons. Indeed he often teased his friend Maurice Wilkins, of King's College, London, about the difficulties Wilkins was having with DNA, and once in the Embankment Gardens over tea he quipped, "What you ought to do is get yourself a good protein."[36] About this time Crick was present at a small protein meeting in Cambridge on July 6, 1950, at which Wilkins gave an account of his work.[37] Wilkins showed his early X-ray pictures and pointed out the correlation between the inclination of the "cross-ways" pattern and the slope of a helical or zig-zag molecule. Although he is sure he was present for the whole of the meeting, Crick cannot recall anything about Wilkins' lecture. Wilkins' former collaborator, William Seeds, does not find this surprising since he can distinctly remember that on more than one occasion Crick was seen to be in active conversation at the back of the hall.[38]

Although Crick was not aware of the overriding importance of DNA in 1950, he had immersed himself in biology to the point of developing interests outside the sphere of structural molecular biology to which the Cavendish group belonged. Thus it came about that when Watson came to Cambridge in the fall of 1951 Crick found him to be the first person who thought the same way as he did about biology. True, the X-ray crystallographers were working on biologically important molecules, and so were the biochemists; the geneticists, too, were hard at work on the fine-structure analysis of the genetic map. But Crick wanted to find the link between gene structure and protein structure. The specificity of proteins must be based on the unique character of their amino acid sequences and these must be related to the sequential arrangement of the hereditary material. In Watson he found a biologist steeped in genetics, anxious to discover how the gene works at the molecular level.

In a letter to Max Delbrück, Watson wrote of Crick: "He is no doubt the brightest person I have ever worked with and the nearest approach to Pauling I've ever seen—in fact he looks considerably like Pauling. He never stops talking or thinking, and since I spend much of my time in his house (he has a very charming French wife[39] who is an excellent cook) I find myself in a state of suspended animation. Francis has attracted around him most of the interesting young scientists and so at tea at his house I'm liable to meet many of the Cambridge characters."[40]

The circumstances which led Watson and Crick to work on DNA are too well known by now to need detailing, but the course of the scientific argument deserves repeating.[41] As Watson has recounted, the first of their two attempts at the structure was vitiated at the start by

Watson's faulty report of the data presented by Rosalind Franklin (who was working with Wilkins at King's College) in November 1951 when he mistook eight molecules of water per lattice point,[42] for eight per unit cell, thus making the sodium salt of DNA appear to have twenty-four times less water than Franklin had said. Crick admits that had he known more chemistry he would have realized that this figure is impossible, since the sodium ions must have water shells around them. With these shells, positively and negatively charged atoms or groups can be at a distance from one another in the crystal; without them, charged groups must lie close together. Since the only source of these is the phosphate group, clearly the metal ions must lie against the sugar-phosphate backbones of the DNA chains. Now from Astbury's density measurements it was known that in the molecules, more than one chain is present—probably three; what better means to hold these chains together than electrostatic attraction between the charged phosphate groups and the metal ions? In her seminar talk Franklin had used such a scheme to hold *neighboring* molecular chains together in the crystallites of the *A* form of DNA, and she had explained the loss of crystallinity incident on the addition of water which formed the *B* structure of DNA as due to the destruction of the sodium-phosphate linkages by interpenetration with water molecules. This scheme rested on her conviction that the sugar-phosphate backbones were on the outside of the molecule.[43] Watson and Crick reversed her argument, putting the backbones on the inside of the molecule and using the phosphate-sodium attractions as the force accounting for the known rigidity of the DNA molecule (see Figure 3). Whether in his report Watson confused inter- with intra-molecular attraction is unclear. What is clear is that Watson and Crick were convinced that electrostatic attraction between the metal ions and the phosphate groups was greater than any hydrogen bond formation, since only the bases could be active in this way.

Thus in an unpublished manuscript of late 1951, Crick wrote: "There are no atoms which can donate hydrogen bonds except in the basic rings and the water. Thus hydrogen bonding is unlikely to play the dominant part in the structure that it does in the polypeptide α helix. Electrostatic forces are so big relative to Van der Waals Forces that we may be confident that the Na^+ and the PO_2^- will mainly decide the arrangement."[44] Furthermore, there appeared to be evidence of variations in the structure of the bases such that precisely directed hydrogen-bond formation would be impossible. All in all, the case for a hydrogen-bonded structure like Pauling's α helix appeared weak.

The fiasco over the resulting three-stranded model is well known. Rosalind Franklin quickly spotted the error over the water content—there were only eight water molecules per twenty-four bases—and rightly

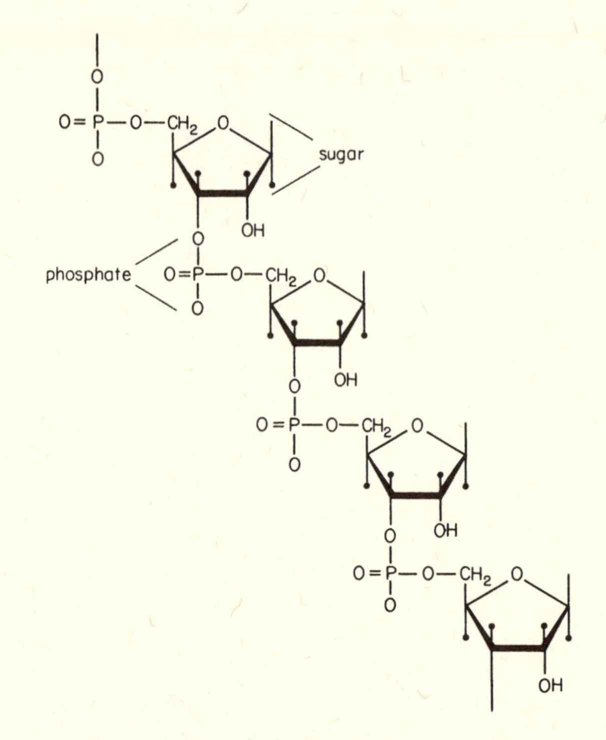

Figure 3a. Detail of Sugar-Phosphate backbone.

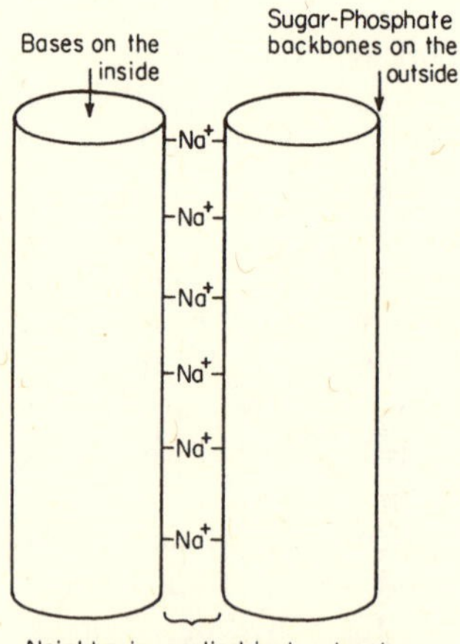

Figure 3b. Franklin's scheme in 1951 (not drawn to scale).

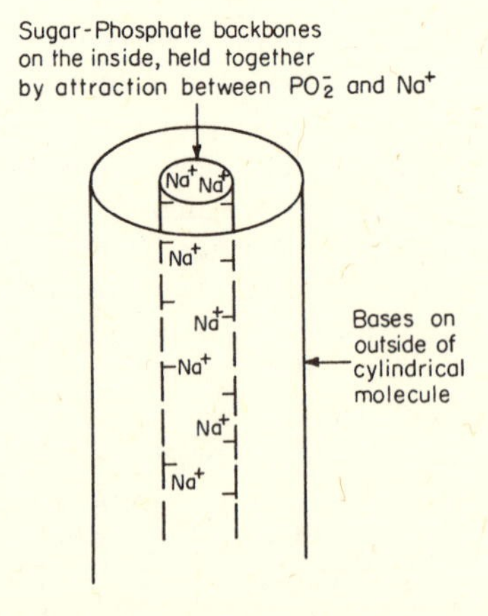

Figure 3c. Watson and Crick's scheme in 1951 (not drawn to scale).

insisted that the sugar-phosphate backbones must be on the outside, not the inside. Crick returned to his protein studies for his thesis and Watson turned to the other form of nucleic acid—RNA—in the tobacco mosaic virus. The paper, "The Structure of Sodium Thymonucleate: A Possible Approach," which Crick had written, along with the trial nets showing the topology of the phosphate-sodium spacings went back into his DNA file and was never published. But DNA continued to nag. Watson got very interested in the possible role of divalent ions in holding the chains in DNA together and had bacteriophage DNA from Copenhagen tested in the flame photometer for the presence of small quantities of magnesium.[45] Perhaps there was just enough magnesium in the sodium salts examined by Wilkins and Franklin to act as bridges between the sugar-phosphate backbones in the molecule.

This approach to finding the attractive forces in DNA did not appeal to Crick. Instead, he considered the electrostatic attraction between dipoles produced by a net shift of charge over the bases (now termed stacking forces). Through Professor Kemmer he was introduced to a young Cambridge mathematician, John Griffith, the nephew of Frederick Griffith, who discovered bacterial transformation *in vivo*. Griffith agreed to calculate what attractive forces there would be between *like* bases. That Crick had in mind attraction along the direction of the chains and not *across* from one to another is evident from his recollection of the event in which he said: "Since the bases are flat, perhaps that is so that they can stack on top of one another and attract. Why not work out if adenine attracts adenine, and so on?"[46] The idea was, of course, that the bases belonging to different chains of the same molecule are interleaved. (It is interesting to note here that in 1946 K. G. Stern, working at Yale, advanced a model with hydrogen bonding between amino and keto groups of interleaved bases.[47])

Later, Crick spotted Griffith in the tea queue and asked "Have you done the calculations?" Griffith said, "Yes, and I find that adenine attracts thymine, and guanine attracts cytosine." Crick replied, "Well, that is all right . . . *A* goes to make *B* and *B* goes to make *A;* you just have complementary replication." At this stage Crick had realized, and Griffith too, that preferential attraction between adenine and thymine and between guanine and cytosine would give rise to complementary replication. Neither had yet read Pauling and Delbrück's 1940 paper stating a preference for this type of replication rather than like-with-like schemes, but Griffith had already, unknown to Crick, worked out a scheme of hydrogen-bonded complementary base pairs (pyrimidine with pyrimidine and purine with purine) for nucleic acids, which he never published. That he did not mention this to Crick is understandable, for he was a young mathematics graduate taking an undergraduate course in biochemistry, introverted and unassuming; whereas Crick was older, extroverted, and in Griffith's eyes an established scientist. As Crick's questions all related to stacking pairing schemes, not planar schemes, there was no purpose in Griffith's bringing up his hydrogen-bonded scheme[48] (see Figure 4). Griffith had become interested in biochemical genetics from taking the Part II biochemistry course in the Cambridge Tripos in 1950-1951. It was probably around the spring of 1951 that he began to puzzle over gene replication. At first he had considered amino-acid matching for protein replication, but on learning about the possible role of nucleic acids in the process, he considered base-pairing schemes. From what source he learned of the hereditary role of DNA he cannot now recall, though he thinks that he probably read a review article.

Shortly after Crick's meeting with John Griffith in the tea queue, John Kendrew introduced Watson and Crick to the noted nucleic acid biochemist Erwin Chargaff, of the College of Physicians and Surgeons, Columbia University. Crick recalls the occasion as follows: "We were saying to him as protein boys, 'Well, what has all this work on nucleic acids led to? It has not told us anything we want to know.' Chargaff, slightly on the defensive, replied: 'Well, of course there are the 1:1 ratios.' So I said: 'What is that?' And he said: 'Well, it is all published!'

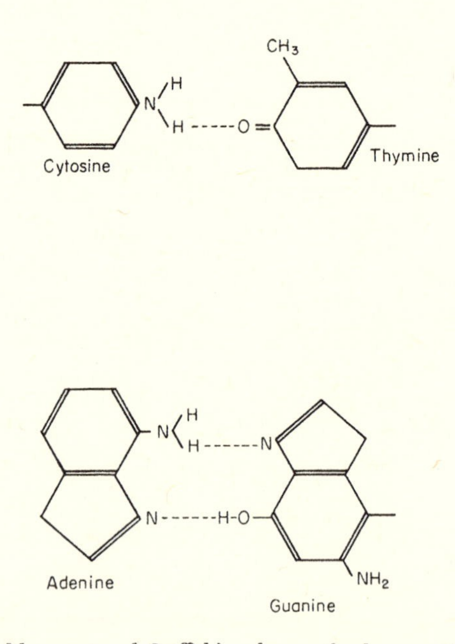

Figure 4. Probable nature of Griffith's schemes for base-pairing in 1951.

Of course I had never read the literature, so I would not know. Then he told me and the effect was electric. That is why I remember it. I suddenly thought: 'Why, my God, if you have complementary pairing, you are bound to get a 1:1 ratio.' [See Table 2.] By this stage I had forgotten exactly what Griffith had told me. I did not remember the names of the bases. Then I went to see Griffith and asked him which his bases were and wrote them down. Then I had forgotten what Chargaff had told me so I had to go back and look at the literature. And to my astonishment the pairs that Griffith said were the pairs that Chargaff said."[49] Hence at this stage, June 1952, Crick realized that base pairing (although the wrong sort) could be the cause of the Chargaff rules.

This simple inference which had escaped everyone except G. R. Wyatt (who first thought of it in March 1953)[50] is rather like the explanation of Mendelian ratios in terms of germinal segregation. Of the so-called three rediscoverers and the several other plant breeders who published such ratios in 1900, there are good grounds for believing that only one established the connection: Carl Correns. What seems an obvious step with the advantage of hindsight is rarely so without it. Whether Crick would have realized that Chargaff's ratios meant base pairing had he not known the outcome of Griffith's calculations cannot be decided, but Griffith's work made sure he did.

The strange aspect of this story is the fact that Griffith was able to arrive at the correct pairs of bases using calculations which he knew and which were generally known to be too approximate to give reliable answers except for the simplest of compounds. It is clear that he examined unlike pairs because of the likelihood that replication is complementary rather than identical. On the other hand, he would hardly have omitted telling Crick about Chargaff's data had he known it at

Table 2. Composition of two microbial desoxyribonucleic acids.*

Constituent	Yeast		Avian tubercle bacilli
	Prep. 1	Prep. 2	
Adenine	0.24	0.30	0.12
Guanine	0.14	0.18	0.28
Cytosine	0.13	0.15	0.26
Thymine	0.25	0.29	0.11
Recovery	0.76	0.92	0.77

* These figures show that the ratio of *A* to *T* and of *G* to *C* are close to 1. (From E. Chargaff, "Chemical Specificity of Nucleic Acids and Mechanism of Their Enzymatic Degradation," *Experientia*, 6 [1950], 205.)

this stage. The correlation would have been so exciting as to cause the quiet and calm young Griffith to proclaim it. The correct pairing must have been a fluke. At the time, Griffith told Watson and Crick that π orbital interaction might account for the adenine-thymine and guanine-cytosine preferences. In fact, specific pairing is brought about not by electrostatic attraction in the stacking direction—though such forces help to hold the molecule together—but by hydrogen bonding perpendicular to the fiber axis. *The Double Helix*, according to Crick, "gives the impression that we knew all about Chargaff's rules before we talked to John Griffith. Now that may be formally so, but it does not correspond with what was actually going on in our minds."[51]

This brings us to an even stranger episode in the story. Watson and Crick were now fully aware of Chargaff's rules and—on the basis of Griffith's calculations—of the possibility that a structural relationship between pairs of bases might explain them. Why did they not at once set to work to build a model incorporating base pairing? Here it is easy to telescope events and to imagine that they had in mind hydrogen-bonded base pairing in the plane of the bases, whereas in fact not once in 1952 was any attempt made to figure out such schemes, for they were thinking in terms of stacking pairing of interleaved bases. Hydrogen bonding was, they thought, too unreliable as a mechanism of replication. Here Crick made the mistake of assuming that both tautomeric forms of the bases exist in the same DNA molecule, that the proton could shift from one position to another, thus altering the sites for hydrogen bond formation. Now if these sites varied in the same base, how could one get constancy of matching in replication? Crick was not alone in holding this view, for L. Hunter had described the tautomeric distinction in certain cases as "meaningless"[52] and P. G. Owston got evidence of the easy migration of the proton in ice crystals.[53] Coming nearer home, June Broomhead, in her Cambridge Ph.D. thesis on the structure of purines, had found a curious electron density distribution near the amino group, reminiscent of Hunter's proton migration effect.[54] To cap it all, textbooks figured here one tautomeric form of the bases, and there another.[55]

Stacking pairing schemes also called for caution: both Crick and Griffith knew how unreliable are conclusions based on quantum mechanical arguments. Further, the Chargaff rules might yet prove to be a red herring; they might be the product of some feature of the code or of the metabolism of the bases. Crick hated using more than the minimum number of assumptions, for by so doing, it would be easy to exclude the correct approach. They had yet to find some direct evidence for base pairing.

At the end of July 1952 Crick tried to get such evidence by determining the ultraviolet absorption of mixed and pure bases. If the mixtures entered into pairing relationships, they would absorb less ultraviolet radiation than would the sum of their constituents irradiated separately. These experiments failed to show any difference, because Crick lacked small enough cells to achieve the required concentrations without bringing on opacity, and as he was in the last year of his thesis, he dropped this unpromising attempt. But he continued to feel the need for more evidence.

The trouble with X-ray crystallography is that there are only two ways to attack a structural problem. The conventional method is to use single crystals which will give sufficient data for an unambiguous structure determination providing that the phase problem can be solved. Alternatively, one can attempt a solution on the basis of slender data derived from powder or fiber diagrams, making the minimum number of assumptions and developing the argument in a systematic manner—a procedure resembling a logical argument in that if the premises are wrong the whole argument will be wrong. The Cavendish group had earlier failed to discover the α helix because they believed that the amide bond is not planar and they only allowed for integral screw axes in their models. The resulting fiasco in 1950 left its mark on all in the unit.[56] Bragg regarded their paper on the possible polypeptide helices as the worst they had ever written. Crick was determined not to slip up like this over DNA. The unpublished 1951 structure had been bad enough.

A further attempt on DNA was not made until 1953 and then only because Pauling had produced a structure shortly to be published.[57] As for the relative positions of Watson and Crick in this second attempt, it is clear that to sharply divide the roles into Crick the crystallographer and Watson the geneticist is not justified by the facts. The suggestion that replication of DNA is complementary, not identical, arose out of Griffith's calculation which Crick connected with Chargaff's ratios. The majority of geneticists were by no means sold on complementary replication, and their hero—H. J. Muller of Indiana University, where Watson studied as a graduate—had favored a like-with-like scheme. Either scheme suggested two chains, as did analogy with chromosomes, which sufficed to convince Watson; but Crick feared there might be a third chain—as indeed the density at that time suggested—for transmitting information to the cytoplasm.

Evidence *so far collected* suggests that this successful attempt in 1953 to determine the structure of DNA took from Friday, February 6, when Watson took the Pauling DNA manuscript with him to King's College, London, until Saturday, February 28, when Crick retired to bed exhausted after nearly a week of model building. At King's, Watson learned from Wilkins that the density data did not after all rule out two-chain models, and that the sugar-phosphate chains must, as Franklin had stated in Watson's presence in 1951,[58] be on the outside. At the end of the first week back in Cambridge, Watson had come round to this view. Crick recalls an earlier incident when Watson was complaining of the difficulties he was having in the attempt to intertwine the backbones on the inside of the structure. Crick's reaction was to ask: "Then why don't you put them on the outside?" Watson replied: "That would be too easy," which prompted the retort: "Then why don't you do it!"[59] Watson himself wrote: "Finally over coffee I admitted that my reluctance to place the bases inside partially arose from the suspicion that it would be possible to build an almost infinite number of models of this type. Then we would have the impossible task of deciding whether one was right. But the real stumbling block was the bases. As long as they were outside, we did not have to consider them. If they were pushed inside, the frightful problem existed of how to pack together two or more chains with irregular sequences of bases. Here, Francis had to admit that he saw not the slightest ray of light."[60]

At the same time, Watson's building of sugar-phosphate backbones had been vitiated by the narrow separation of the sugars. "Watson," said Crick, "was always trying to build it with the sugars too close. He went away one day and said: 'You try it.' and while he was playing tennis I didn't try his scheme; I was convinced it was wrong. Instead, I built one with a 36° rotation."[61] In the following week, while Crick continued to work on his thesis, Watson either played tennis or tried to figure out a scheme for packing the bases. On Thursday, February 19, he arrived at a like-with-like pairing scheme which Jerry Donohue (an American crystallographer who shared Watson and Crick's office) and Crick discredited on the following morning,[62] Donohue on the grounds that he had used the wrong tautomeric forms, Crick because the axis of symmetry of the molecule was in the wrong direction and the structure did not explain Chargaff's rules.

"The crucial point," said Crick, "was when we asked Donohue about the tautomeric forms, because when he put us right we could put the hydrogen bonds together. At that stage, and I remember this very clearly, Jerry and Jim were by the blackboard and I was by my desk, and we suddenly thought, 'Well perhaps we could explain 1:1 ratios by pairing the bases.' It seemed too good to be true. So at this point [Friday, February 20] all three of us were in possession of the idea we should put the bases together and do the hydrogen bonding."[63] At last the misjudgment over tautomeric shifts had been disposed of, so Crick as well as Watson could be enthusiastic about hydrogen-bonded base pairs; and they had been given an authoritative decision as to which tautomers are present in DNA.

Using the cardboard replicas of the DNA bases which he had cut out the previous afternoon, Watson arrived at the correct scheme on Saturday, February 21, but not by consciously trying to apply Chargaff's rules. He said: "Though I initially went back to my like-with-like prejudices, I saw all too well that they led nowhere. When Jerry came in I looked up, saw that it was not Francis, and began shifting the bases in and out of various other pairing possibilities. Suddenly I became aware that an adenine-thymine pair held together by two hydrogen bonds was identical in shape to a guanine-cytosine pair."[64] Both Donohue and Crick quickly gave their blessing to Watson's base pairs, and Crick pointed out that since the axis of symmetry of the unit cell was perpendicular to the fiber axis, the sugar-phosphate backbones must run in opposite directions, as indeed the direction of the links between base and backbone indicated in Watson's pairs. This meant that both purines and pyrimidines can be accommodated on the same chain without a change in the orientation of the glycosidic links. At this stage, the chief features of the *in vivo* form of DNA were discovered. Two weeks had elapsed since Watson's visit to London; a further week was required to build the model incorporating these features and to measure its parameters accurately.

The model building now was done not on several residues and base pairs, but on a single residue "and the appropriate base." Crick continued, "I worked out a theorem which now escapes me which expressed the geometrical restraint of the other base on this one. I knew that as long as I obeyed this geometrical rule, some projection of something on something else, one could do it . . . The model building was done in a period which I remember rather vividly. It started about Wednesday and finished on a Saturday evening [February 28], by which time I was so tired, I just went straight home and to bed."[65] Those three weeks of model building and discussion were marked by a suppressed excitement which became intense. They could talk and think of nothing else. Not even Hedy Lamarr's romps in *Ecstasy* could take Watson's mind off the bases.[66]

Watson's story of the discovery of the double helix differs from Crick's, for the obvious reason that what was going on in Watson's mind at the time differed from what was going on in Crick's mind. Crick had made a major error in thinking that the bases would suffer tautomeric changes sufficiently often to make specific hydrogen bonding impossible. It says much for Watson's strength of character that he did not allow Crick's opinion to deter him in the end from examining hydrogen-bonded schemes. So it was Watson who benefited from J. M. Gulland's remarks about hydrogen bonding[67] and Broomhead's patterns of hydrogen-bonded molecules in purine crystals. So Watson thought more about hydrogen bonding than did Crick. On the other hand, Watson did not appear to have understood the arguments which Crick developed from the knowledge that DNA crystallites belong to the space group C2, hence he continued to try to get a like-with-like pairing scheme after Crick had told him it would not square with the crystallographic evidence. In the collaboration they compensated for each other's blind spots. But their collaboration involved far more than this. Alexander Humboldt expressed the essence of such work well when he said: "Collaboration operates through a process in which the successful intellectual achievements of one person arouse the intellectual passions and enthusiasms of others, and through the fact that what was at first expressed only by one individual becomes a common intellectual possession instead of fading away into isolation."[68] At first, Watson did the work of model building and Crick interrupted his thesis writing from time to time only to advise him, but after base pairing had been discovered, Crick threw himself into model building on a full-time basis.

One other feature of the story deserves underlining. Because of the nature of the problem, biological arguments seem to have played a very minor role. It is true that Watson ruled out three-chain models for biological reasons, but not before Wilkins told him that the physical data on DNA did not exclude two-chain models. Second, Watson did not study base-pairing schemes until forced to do so by the physico-chemical evidence that the sugar-phosphate chains must be on the outside.

The next four weeks were occupied with writing draft after draft of the now famous *Nature* paper,[69] showing visitors the model, and awaiting the parallel contributions from the King's group. Then there was a second *Nature* paper written chiefly by Crick, and a paper for *Discovery* by Crick. In June the Cavendish had an open day for members of the University, for which Watson and Crick built their model from the table almost to the ceiling of their office. Watson wrote and presented a paper for a symposium held at the Cold Spring Harbor Laboratory in Long Island, New York, and he wrote the major part of a paper for the *Proceedings of the Royal Society*. On the order of names, they decided that since Watson had been closely associated with most of the work, his name should appear first on the April 25 *Nature* paper. The second *Nature* paper, written chiefly by Crick, was decided by the toss of a coin which Watson won. The order for the Cold Spring Harbor paper was decided by the fact that Watson was the one invited to give the paper, and for the Royal Society paper, custom dictated the alphabetic rule.[70]

The Genetic Code

In the summer of 1953 the cosmologist George Gamow sent Crick a copy of his short note to *Nature* which marked the beginning of work on the genetic code using the Watson-Crick model. Watson and Crick could say in 1953 about the genetical implications of the DNA structure that it offered a mechanism of replication. They wrote (in their second 1953 paper to *Nature*): "If the actual order of the bases on one of the pair of chains were given, one could write down the exact order of the bases on the other one, because of the specific pairing. Thus one chain is, as it were, the complement of the other, and it is this feature which suggests how the deoxyribonucleic acid molecule might duplicate itself."[71] They suggested that tautomeric shifts during the replication process might account for mutations, and with regard to protein synthesis they emphasized that the "open" nature of their structure meant that the bases, though "inside," are sufficiently accessible to allow DNA to exert its "highly specific influence on the cell."[72] Although at this stage they had the sequence hypothesis clearly in mind, they were very careful not to suggest how the information in the base sequence might be expressed. (This will be discussed later.)

Gamow forced the issue over the genetic code by presenting his scheme of geometric specificity of the "holes" in the DNA molecule. When Crick received from Gamow the text of the *Nature* paper[73] and the expanded version intended for the *Proceedings of the National Academy of Sciences*[74] which contained the list of Gamow's twenty standard amino acids, he sat down with Watson in the Eagle Hotel, where they regularly had lunch, and wrote out the correct list. Gamow's choice of twenty indicates that his knowledge of protein chemistry was not quite adequate for the task: While he excluded common amino acids like asparagine and glutamine, he included the rare amino acid hydroxyproline (see Table 3). He was also unaware that protein synthesis takes place almost exclusively in the cytoplasm (there was some evidence at the time suggesting that it also occurs in the nucleus, but it is doubtful that he knew about this). Nevertheless, although he was wrong Gamow proved a great stimulus. His DNA scheme could be applied equally to RNA in the cytoplasm, and it did contain the idea of degeneracy and triplet codons, as did the 1952 scheme of Alexander Dounce.[75]

Crick took up the subject of the code in the autumn of 1953. The way he handled it illustrates his deductive approach—reasoning from the minimum of assumptions and seeing where the argument would lead. Deductive though he is, he always seeks some hard evidence

Table 3. The twenty amino acids in protein synthesis.

Gamow's list	Watson and Crick's list
Glycine	Glycine
Alanine	Alanine
Valine	Valine
Leucine	Leucine
Isoleucine	Isoleucine
Proline	Proline
Phenylalanine	Phenylalanine
Tyrosine	Tyrosine
Serine	Serine
Threonine	Threonine
	Asparagine
	Glutamine
Aspartic Acid	Aspartic Acid
Glutamic Acid	Glutamic Acid
Arginine	Arginine
Lysine	Lysine
Histidine	Histidine
Tryptophane	Tryptophane
	Cysteine
Methionine	Methionine
Cystine	
Cysteic Acid	
Hydroxyproline	

before he will publish his conclusions. "I have always had great difficulty," he said, "in publishing theoretical ideas in a vacuum of evidence." One has to be cautious about publishing ideas, "partly because if you are a professional theoretician as I am, you cannot afford to publish too many wrong ones."[76] So later when Gamow, whom he had first met at Woods Hole in the summer of 1954, asked him to join his RNA Tie Club,[77] he was delighted to accept. Members of the club could circulate papers for discussion without publishing them: it was a sort of preprint club. Hence the adaptor hypothesis, which Crick put forward at the Gordon Conference in June 1956, he had first described in a paper for the RNA Tie Club. The "comma-less code" was also presented first to the club and published only because "three or four people wanted to quote it."[78] At the time Crick was very guarded about its importance—which turned out to be an appropriate attitude, for it was quite wrong.

Initially, Crick, like most other molecular biologists, did not perceive the weakness in Gamow's scheme. In 1954 interest in the "holes in RNA" was intense. Watson, Delbrück, and Rich were studying the holes in a hypothetical model of RNA in Pasadena, and Gamow, who visited them, wrote to Crick: "They have a model of RNA, big and nice looking, but they do not believe in it very much themselves (except Alex Rich, who conceived it). It has trapezoid holes formed by two bases, and two different 'sugar edges.' And there are twenty different holes. But I have found . . . that the combination rules do not work at all."[79] In fact, Crick had tested Gamow's scheme two months earlier while he was at the Brooklyn Polytechnic Institute by applying it to Frederick Sanger's data on insulin, and he found it demanded a greater correlation of neighboring amino acids than insulin possessed. By the summer of 1954, he was convinced that holes in the helix schemes were all wrong. In his RNA Tie Club paper "On Degenerate Templates and the Adaptor Hypothesis," Crick wrote: "I don't think that anybody looking at DNA or RNA would think of them as templates for amino acids were it not for other indirect evidence. What the DNA structure *does* show (and probably RNA will do the same) is a specific pattern of *hydrogen bonds*, and very little else. It seems to me, therefore, that we should widen our thinking to embrace this obvious fact."[80]

Crick went on to postulate adaptor molecules composed either of amino-sugars or of di-, tri-, or tetranucleotides to which specific enzymes could attach the right amino acid; the adaptor thus charged would then pair specifically with the codons in RNA.[81] At the time when Crick's adaptor hypothesis was still unproven, Paul Zamecnik's group at the Massachusetts General Hospital in Boston was studying the incorporation of labeled RNA precursors in the "cell-free system." They had been puzzled to find some radioactive counts in the RNA fraction of the control experiment to which radioactive amino acids had been fed, as well as in the system supplied with radioactive nucleotides. It was just possible that some RNA species might be present in the cell-free system to which the radioactive amino acids became attached. They quickly established this point and Mahlon Hoagland, who had come to work with Zamecnik on protein synthesis, identified the fraction as the nonribosomal RNA "which up to that time, had not been of interest to anyone and had been discarded as 'soluble RNA' . . . possibly a breakdown product of ribosomes!" Zamecnik's group went on to show that this substance is an intermediate in protein synthesis and, according to M. B. Hoagland, "published all this work in early 1957, *with no knowledge of Francis' hypothesis*! Jim Watson first told me of the adaptor hypothesis about Christmas 1956 and in January 1957 Francis visited us, enormously excited, and we quickly arranged that I spend a year with him in

Cambridge [1957-1958], which I did!"[82] This story was recalled somewhat differently by Zamecnik in 1969 when he said:

There were already in the air at this time [1955-1956] a few suggestions that something might be missing between the activation and the sequentialization steps . . . We had also found that a purified tryptophane activating enzyme . . . was inactive in promoting tryptophane incorporation into protein. A rumour had also reached us that at a recent meeting Francis Crick had expressed the opinion that there should be an oligonucleotide intermediate as a language translation piece between activated amino acid and RNA template.

It was at this point that we prepared a ^{14}C aminoacyl RNA and Mahlon Hoagland plugged it into the cell-free system to determine whether it could substitute for free ^{14}C amino acid and serve as an intermediate.

In retrospect, it may be asked why we allowed over a year to elapse before reporting on the existence of RNA, since from the first positive experiment in November 1955 we regarded this molecule as a possible new and dazzling transfer intermediate. We were, however, anxious to remove all doubts that (a) the amino acid was covalently bound, (b) that the aminoacyl nucleotide was a direct line intermediate in the pathway from free amino acid to peptide, and not a side path for storage of activated amino acids, and (c) that the entire complex polynucleotide—and not a much smaller triplet or oligomer which adventitiously aggregated with this larger type of RNA—was the active intermediate.[83]

Whatever the precise facts of the matter may have been, Hoagland in 1959 expressed the important point: "Here was one of those rare and exciting moments when theory and experiment snapped into soul-satisfying harmony."[84]

Because these soluble RNA molecules were larger than trinucleotides, Crick regarded them as "a half-way step in this process of breaking the RNA down to trinucleotides and joining on the amino acids."[85] He feared that if the adaptor molecules were as large as soluble RNA, there would not be enough space for them to function properly or they would take too long to get into the right positions. Crick and Hoagland were both embarrassed by the unexpectedly large size of soluble RNA species, and in their correspondence before Hoagland's arrival in Cambridge they discussed possible ways of getting around the problem.[86] Now it is recognized that the adaptor molecules are all large molecules.

Messenger RNA

Where Crick's thinking—and that of the majority of molecular biologists of the day—was completely at sea was over the function of the ribosomes, or "microsomal particles" as they used to be called. Crick describes this as the one big howler in molecular biology, for unlike the holes-in-the-helix idea, it lasted a long time—nearly seven years.

From the early work of Jean Brachet and Tobjorn Caspersson it was known that protein synthesis is associated with cytoplasmic RNA. So it was concluded that for every protein, or perhaps for every polypeptide chain, one RNA molecule is assembled on the chromosomal DNA and passes out into the cytoplasm to form a stable, metabolically active particle (the ribosome) on which one species of protein is synthesized. A major rethink had to be undertaken before this idea was disposed of and a new RNA species—messenger RNA—introduced, the ribosomes being recognized as nonspecific sites for protein synthesis. Here the old ribosomal "template RNA" theory would not bend to the results of experiments on phage infection and ribosomal longevity.

Matters came to a head in the autumn of 1959, and for Crick and Brenner the turning point came in the spring of 1960 when two informal meetings were held in Cambridge to discuss the "Pa-Ja-Mo" experiment of Arthur Pardee, François Jacob, and Jacques Monod. In 1958 these three, working at the Pasteur Institute, had succeeded in conjugating bacteria to produce heterozygotes for the *lac* operon (the genetic region associated with utilization of lactose sugar) and they found, much to their surprise, that when a recipient mutant bacterium unable to synthesize the enzyme β-galactosidase received the "wild-type" gene (the typical form in a natural environment), synthesis of the protein started up within three to four minutes. It did not build up gradually, as one would expect if ribosomes acting as β-galactosidase templates had to be synthesized *de novo*. Perhaps, they suggested, the DNA introduced was acting as the template directly; or some DNA-like RNA had been transferred with the DNA; or there was some other short-lived RNA like the soluble RNAs but big enough to encode the protein, to act as messengers between DNA and the sites of protein synthesis (see Figure 5).[87]

When Jacob spoke at a conference in Copenhagen in the autumn of 1959 his audience, which included Crick, proved unsympathetic to the idea of an unstable RNA species—a "tape" or "X" as Jacob recalls they then termed it[88]—because the Pa-Ja-Mo experiment alone was not thought adequate to exclude other explanations. But by 1960 the situation had changed: experiments of Monica Riley supervised by Pardee at Berkeley showed that the destruction of the same gene for β-galactosidase synthesis by ^{32}P decay quickly depressed the rate of synthesis of this enzyme, and these experiments together with the earlier one ruled out ribosomes and any long-lived nucleic acid intermediate, and defined what the authors described as an "interesting dilemma."[89] Either the gene introduced acts directly, or there must be some functionally unstable intermediate involved in the information

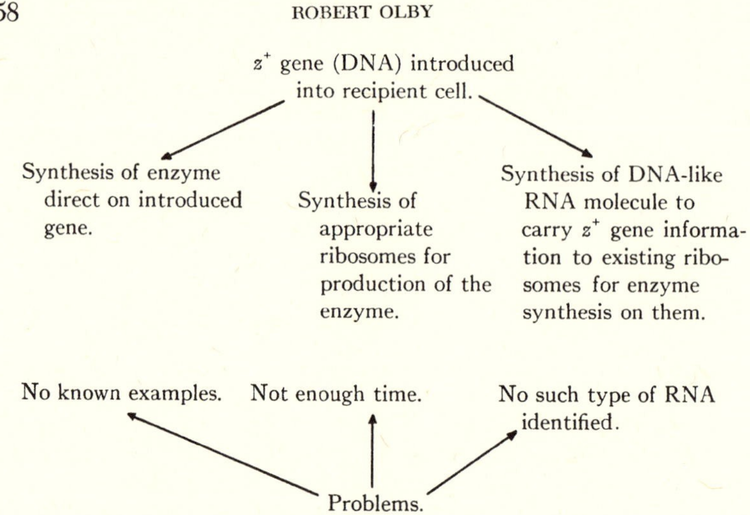

Figure 5. Possible explanations of the "Pa-Ja-Mo" experiment.

transfer from DNA to the site of protein synthesis. Unfortunately, there appeared to be no hard evidence for the presence of such a species of RNA, and the natural tendency was to try to find other explanations of the mechanism involved. Thus when Pardee had suggested "an intermediate carrier of information (perhaps RNA)" in 1958 he doubted how such a mechanism could be squared with the demonstration of constant rate of protein synthesis after mating: "a number of possible but not too plausible explanations exist based on the idea that the cell is saturated with templates in a few minutes, or that the number of templates reaches equilibrium in a few minutes, or that the template might be unstable and rapidly reach a steady state concentration as a balance of formation and inactivation [now accepted as the correct explanation]."[90]

When the Pa-Ja-Mo experiment was first performed in 1958 the sudden switching on of protein synthesis caused Pardee, Jacob, and Monod to alter their hypothesis as to the *control mechanism* of the synthesis. A year elapsed before they became seriously concerned about the relevance of the Pa-Ja-Mo experiment for *protein synthesis itself*. The irony of this situation was that as early as 1954 Pardee had carried out experiments designed to determine whether protein synthesis continues in the absence of nucleic acid synthesis, and he had come to the then surprising conclusion that "continuous formation of RNA is essential for protein formation and that the RNA found in the bacterial cell [that is, ribosomal RNA] is largely an inert metabolic product."[91]

While for Jacob and Monod the year 1959 was crucial in the development of the messenger concept for Crick the decisive step "was taken on Good Friday (March 31, 1960) in [Sidney] Brenner's rooms in Gibbs House, King's College, Cambridge, when a number of us, including Brenner and myself, were talking to Jacob. We were asking him about this [Pa-Ja-Mo] experiment, and he was explaining that it had been improved, done in ways such that the argument was now much crisper; and at some point Brenner and I both suddenly realized the following thing: Volkin and Astrachan [working at the Oak Ridge National Laboratory] had previously shown that there was an unstable RNA in phage infection, which was DNA-like. At first they thought it was a DNA precursor, although later they got away from that idea.[92] We suddenly realized that the Volkin-Astrachan RNA was *the message.* All the experiments were done later by Brenner and his co-workers with the heavy isotope. They were planned that evening at a party in my house. Half the people were enjoying the party and half the people were in little corners trying to devise all these experiments which were later to show that messenger RNA really exists. At the same time Jim Watson and François Gros, by a quite independent line, arrived at the idea of messenger in ordinary cells . . . The only thing one is thankful for is that it wasn't all done by someone, as it were, outside the magic circle, because we would all have looked so silly. As it was, nobody realized just how silly we were."[93]

This manner of presenting the story may appear to give too much credit to the theoreticians in Cambridge, for it was Pardee who first supplied the evidence that the ribosome is not protein specific, and it was Jacob and Monod at the Pasteur Institute who were worried about the ribosome model in the light of their experiments on the switching on and off of genes. But not one of them appears to have given the unstable intermediary RNA species *exclusive preference,* and the only chemical evidence of the existence of such a class of substances (shown in Volkin and Astrachan's experiments) escaped them. But to avoid giving the wrong impression let it be emphasized that the initial concern with the ribosomal template theory and the destructive evidence came not from Cambridge but from Paris and from Harvard.

The Coding Problem

At the same time that the mechanism of information transfer from DNA to the site of protein synthesis was causing such intensive interest, the nature of the genetic code itself was a "hot" subject. In 1958 two Russian biochemists, A. N. Belozersky and A. S. Spirin, brought to a close what Crick has called the "optimistic phase" of the coding problem (that is, how the bases in DNA determine the sequence of amino acids in proteins). Crick and Watson had already noted that the number of bases in the RNA of the tobacco mosaic virus exceeded the number of amino acids in the protein units by such an extent (50:1) that there was no clear evidence from this source as to the number of bases which code for one amino acid.[94] All that seemed likely was that most of the RNA in the virus coded for proteins not represented in the virus coat, the amino acids of which constitute only a part of the total synthesized. The Russians showed that very substantial differences in the base composition of bacterial DNA was not associated with similar variations in the base composition of the bacterial RNA, though there did exist a correlation between the DNA and RNA constitution of a series of species with increasing guanine and cytosine content.[95] The messenger concept, by making ribosomal RNA nonspecific, alleviated the embarrassment to some extent. Nonetheless, cause for continuing concern was the fact that the average amino acid constitution of a group of organisms may vary little while the base composition of the DNA varies greatly. Fortunately, with the introduction of Crick's wobble hypothesis this difficulty can in theory be overcome.

Just how was the code to be broken? The direct approach of making synthetic RNA polymers, introducing them into a cell-free system and identifying the product, was not entertained at this time (1959-1960) for a variety of reasons, chiefly because the messenger concept had not been introduced. The attempt to arrive at the code by invoking stereochemical arguments about the specificity of fit between the twenty amino acids and defined sequences of bases (Gamow's idea) had been disposed of by Crick. His paper on this subject carried the following quotation on the title page: "Is there anyone so utterly lost as he that seeks a way where there is no way?"[96] Coding ideas treated in a formal mathematical manner had led to the biologically convenient and neat solution embodied in the "comma-less code," about which Crick had strong reservations.[97] With this scheme the existence of more codors (sixty-four in a triplet code of the four bases) than there are amino acids to be coded (twenty) was overcome together with the problem of how to punctuate the message by establishing twenty "sense" triplets, the remaining forty-four combinations making "nonsense."

The remaining line of attack—the attempt to demonstrate collinearity (that is, a direct relationship between the linear sequence of bases in the genetic map and the linear sequence of amino acids in the protein of the gene produce)—continued to be a very active field from the time of Seymour Benzer's classic experiments at Purdue University on the r_{II} region of phage T4.[98] The "handle" used by Brenner to open up this line of attack was mutagenesis. Exploration of the mode of action of certain mutagens, rather than attempts to demonstrate collinearity, led the Cambridge group on to discover that the genetic message is read from a fixed point in one direction only and that some mutagens act by addition or deletion of bases. At this time there were two ideas concerning the mechanism by which a mutant back-mutates or "reverts" to the normal "wild type." If the change which caused the mutation is simply reversed, then it is assumed that the mutation and the "back mutation" take place at the *same site* on the genetic map. If, on the other hand, the original mutation has merely been suppressed by a second mutation at a *different site* on the map, then this reversion is attributed to the action of a "suppressor." Working at Harvard, Ernest Freese discovered that the mutagens he used could be divided into two groups, no mutations induced by compounds in one group being capable of reversion by those in the other. This led Freese to support the idea of back-mutation by base substitution rather than by suppression, one class producing purine-purine and pyrimidine-pyrimidine substitutions, the other purine-pyrimidine or pyrimidine-purine substitutions.[99]

On examining earlier work carried out on the r_{II} region when Benzer had been with them in Cambridge in 1958,[100] Brenner, Leslie Barnett, and Crick found that the sites of the mutations caused by these two classes of mutagens never overlapped. This made them doubt Freese's explanation and consider another mechanism for one class of the mutagens, namely the addition and deletion of bases. The suggestion that one class of Freese's mutagens caused substitutions while the other class caused deletions and additions had been put to Freese by Bruce Alberts and his colleague Jacques Fresco, whose paper describing their model for such a mechanism appeared in 1960.[101] That January, Crick received a copy of the paper from Fresco and replied, saying that they were not very happy about the extension of the argument from synthetic RNA to DNA. "Nor do we find your discussion of mutations very illuminating. What you say is plausible enough, but it doesn't solve any of our problems, such as 'how does proflavine act as a specific mutagen?' or 'What causes hot-spots?' "[102] Crick was not satisfied with Fresco's evidence for looping-out in RNA, and was therefore loth to apply it to DNA. Today, the existence of looping-out in cellular RNAs is still unproven.

Evidently the idea of deletions and additions was foreign to their way of thinking at that time, partly because of the attractive base-change idea put forward by Freese, but also because of the presence of great differences in the mutability of different regions of the genetic map of the r_{II} region of phage T4. There are "hot spots," so-called because many of the mutations are located in them, and also regions, the deletion of which causes no impairment of function. And Fresco's scheme included no suggestions as to how the message was punctuated.

The trouble was that no one was considering mutation from the point of view of a message which is *read from a fixed point*. Although Crick placed little reliance on the comma-less code, this theory, which did without initiator and terminator codons, had sufficient impact to restrict the ways in which investigators thought about mutagenesis. Finally, there was the overriding belief that a large proportion of the DNA is "nonsense," which meant that the odds were against a deletion at one site on the genetic map restoring the function of a gene which has undergone an addition at another site. And for those who assumed that the message is read from a fixed point there was the conviction that any deletion or addition, by causing a frame shift, would yield so much confusion as to prevent any protein being formed.

By the autumn of 1960, Brenner was driven by the data on acridines to explore a deletion-addition mechanism of mutation, and L. S. Lerman, working at the Medical Research Council Laboratory, Cambridge, in 1959 gave Brenner and Crick confidence in such a mechanism by suggesting that proflavine is occasionally intercalated between the bases of DNA on one chain only, thus answering Crick's question about how the mutagen works. Lerman thus extended the general argument of Fresco and Bruce Alberts who had shown by model building that mistakes in replication leading to mismatching of base pairs do not destroy a double RNA helix if they are looped out (see Figure 6).[103] Another paper which Fresco sent to Crick in the autumn of 1960 for submission to *Nature*[104] drew attention to the possible coding roles for looped-out segments in RNA. In Cambridge in the winter of 1960-1961 Crick and Leslie Orgel used the concept of "looping out" to devise a coding system in which messenger RNA folds back on itself to form a double helix. During translation both ascending and descending segments are then "read off" simultaneously. Paired bases could act as the code and looped-out bases as the punctuation marks or vice versa, but in either case the base composition of the *coding portion* of the molecule would be approximately the same from one organism to another although the composition of the *whole molecule* differed. Further, they noted that if base substitutions yield reversible mutations, as Fresco had suggested, then in an RNA molecule folded back on itself during translation, one would expect "a rather regular pattern of 'suppressor' mutations within a gene" (see Figure 7). "I got so excited about this," said Crick, "that I thought 'I'll do the experimental work myself,' and I started to work on the r_{II} gene in phage T4 to try and pick up internal suppressors of mutants."[105]

Starting in February 1961 he looked at quite a number of mutants but at first had no success. When he did discover the sites of some of the suppressors they did not yield the pattern predicted by the "loopy code." Early in May Crick decided to see if there were many different

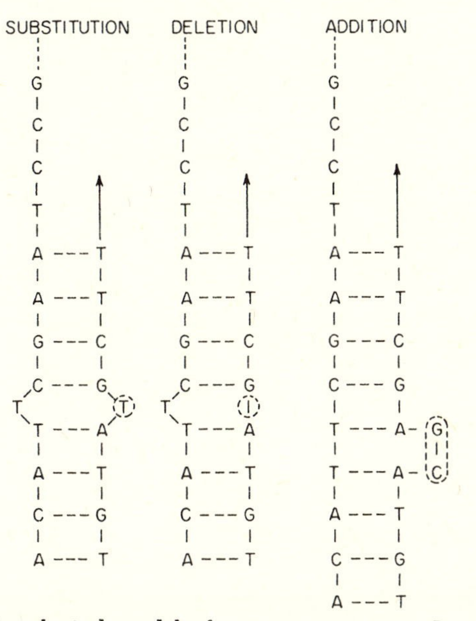

Figure 6. Hypothetical models for point mutations. In each case, the template chain is the one on the left. Note that while the templates are identical in each case, the sequence in the growing chains varies, depending on the kind of "mistake" made during replication. The "mistake" in the growing chain is indicated by a dotted circle. [The concept of "looping-out" used to suggest three types of mutation by Fresco and Alberts. (From J. Fresco and B. M. Alberts, "The Accommodation of Non-Complementary Bases in Helical Polyribonucleotides and Deoxyribonucleic Acids," *Proceedings of the National Academy of Sciences,* 46 [1960], 319.)]

suppressors. He had by this time realized that if suppressors are additions or deletions, such mutations at a number of sites could restore the wild-type function to the same mutant. The next step was to call the original mutation "plus" (a quite arbitrary notation), its suppressors "minus," to isolate the double mutants + −, and to show that they are wild or pseudo-wild (normal function restored incompletely). By the end of June, Crick was able to state that the results "are compatible with our suggestion that the action of acridines [the mutagens] is to add or delete bases. An attractive additional hypothesis is that the code is read in short groups, starting from one end of the gene. The exact starting point is supposed to determine which group is read. The deletion of a base would then alter the active reading from this point onwards. The double mutants produced by the reversion of acridine mutants would then, on this hypothesis, be altered not just in two, separated amino acids, but in a short stretch of amino acids in sequence. If this were so it would be very important for decoding."[106]

That September Crick went to the International Biochemical Congress in Moscow, where he heard Marshall Nirenberg describe how he and Heinrich Matthaei, using the cell-free system, had discovered the synthesis of a specific polypeptide in the presence of a specific polynucleotide. Back in Cambridge Crick returned to the r_{II} mutants with renewed vigor. The double mutants were successfully isolated. Next Brenner suggested putting three pluses together—since in a triplet code the addition or deletion of three bases should restore the correct reading of the code (see Figure 8)—and Crick, realizing that it mattered whether you shifted to the right or to the left, predicted which combinations should and which should not work. What followed was reported in *Nature* at the end of the year.[107]

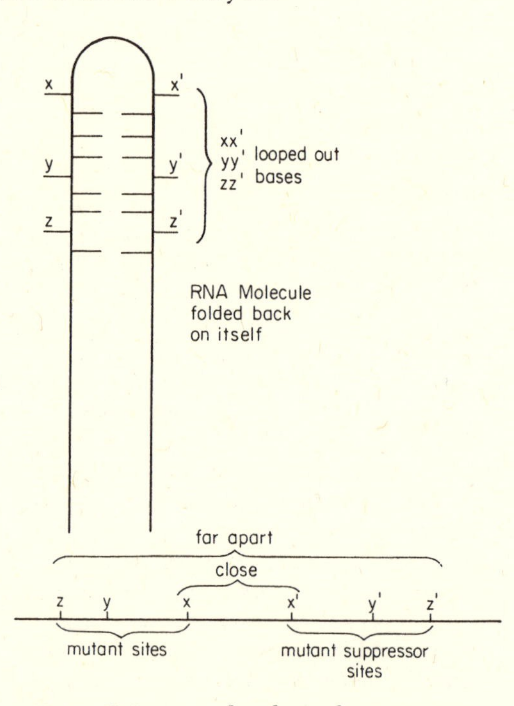

Figure 7. Loopy code scheme for suppressors.

This work, widely acclaimed as a classic piece of research, had its origin in an attempt to analyze the fine structure of the gene and to explore theories of mutagenesis. Crick, who from the beginning had played a part on the theoretical side, took up experimentation in order to test the completely erroneous hypothesis of loopy codes; he and his colleagues finished by demonstrating that the code is composed of

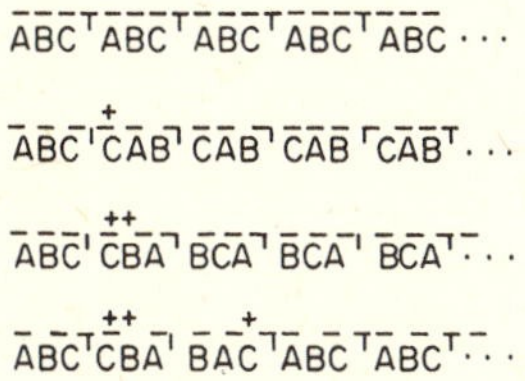

ABC ABC ABC ABC ABC · · ·

ABC CAB CAB CAB CAB · · ·

ABC CBA BCA BCA BCA · · ·

ABC CBA BAC ABC ABC · · ·

Figure 8. Diagram to show how three + mutations (assumed to be additions) restore the phased reading of the code.

triplet or multiple triplet codons, and that the message is read starting at a fixed point and traveling always in the same direction. This is a particularly striking example of how tortuous scientific research can be.

As with Watson, so with Brenner, Crick discussed ideas and plans for experiments day after day. They never collaborated in the sense of doing experiments together at the same time, and their particular research interests have been complementary, only at a very general level forming parts of the same research program. One hour's discussion of their problems per day is the rule. On such occasions they "say anything without having to justify it up to the hilt"; in this way problems get thrashed out, Crick insisting on formulating them very clearly in an analytic sense before he goes into them.[108] They interrupt each other, to continue either in dialogue or "duologue," the ideas tumbling helter-skelter from Crick to be met by a relentless questioning from Brenner. The doubts and uncertainties are explored, and they may end up with a feeling that a fresh start will have to be made. The idea was no good after all—throw it out. On this point, says Brenner, "Crick is as ruthless with himself as he is with other people."[109] In science one needs more than the ability to theorize; one must be able to relinquish theories, even neat, attractive ones like the comma-less code.

The T4 phage experiments apart, Crick's post-1953 work gives the impression that he was usually standing on the sidelines urging on the workers, interpreting their results for them and telling them what experiments to undertake next. As in the early days in Cambridge, so in the late fifties, experimentalists were not entirely happy about the intrusion of the quick mind of Crick. Some objected to his confident manner of making predictions and his insistence in letting them know when he had been proved right. Thus, speaking of Vernon Ingram's discovery that a single amino acid change in hemoglobin can prove lethal, he remarked: "It may surprise the reader . . . for my part, Ingram's result is just what I expected."[110] And again, he advanced statements such as the "wobble hypothesis" with such an air of conviction that he offended those cautious or iconoclastic members of the scientific community to whom generalizations are anathema. But, as Crick points out, since the professional theoretician must avoid publishing a hypothesis until he is pretty sure of it, when he has reached that state, why should he not show it?

As far as the central dogma and the sequence hypothesis are concerned, he very rightly pointed out: "It is an instructive exercise to attempt to build a useful theory without using them. One generally ends in the wilderness."[111] And on the same occasion he said: "It is remarkable that one can formulate principles such as the Sequence Hypothesis and the Central Dogma, which explain many striking facts and yet for which proof is completely lacking."[112] The sequence hypothesis states that "the specificity of a piece of nucleic acid is expressed solely by the sequence of its bases, and that this sequence is a (simple) code for the amino acid sequence of a particular protein"; the central dogma states that "once 'information' has passed into protein *it cannot get out again.*"[113] It is not possible for an alteration in protein sequence to cause a subsequent alteration in nucleic acid sequence.

These two principles—which in many popular accounts of molecular biology are referred to together as the central dogma—constitute a statement at the molecular level of the "hardness" of the hereditary material and its chemical characterization as nucleic acid and *only* nucleic acid. These statements by Crick have served to define the assumptions which were becoming implicit in a wide program of research, from mutagenesis to protein synthesis. They distinguish the Watson-Crick era from that which preceded it, when nucleoproteins rather than nucleic acids were identified with the hereditary material and when ideas about the specificity of sequences were nearly all confined to those of the amino acids in proteins.

Crick's own evaluation of the role of theory is well expressed in the following passage from his 1963 paper "The Recent Excitement in the Coding Problem":

> It does not seem to be appreciated that theoretical work is often of two rather distinct types. There is first the deduction from experiment: the weighing of the data and the reasoned assessment of, say, the evidence that a particular codon represents a particular amino acid. This I would call interpretation, and it needs to be done critically.
>
> Second we have theory proper. This may take several forms; for example, Wall's demonstration that a *partially* overlapping code is not yet eliminated; or Woese's attempt to deduce the whole structure of the code from only part of it. These theories may not be correct but they are both sensible and useful, in that they enable us to tighten up our logic and make us scrutinize the experimental evidence to some purpose. Moreover, even if Woese's code is wrong, his careful exploration of its consequences may enable us to see something about the general character of the genetic code. But, most important of all, these ideas are not merely useful, they are novel. If their authors had

not suggested them, they might not have occurred to many people working on the problem.

The bad theoretical paper takes an obvious technique, applies it to shaky data, and reaches a solution that even on general grounds can be seen to be unlikely. What is gained by this it is difficult to discover.

In the long run we do not want to *guess* the genetic code, we want to *know* what it is. It is, after all, one of the fundamental problems of biology. The time is rapidly approaching when the serious problem will be not whether, say, UUC is *likely* to stand for serine, but what evidence can we accept that establishes this beyond reasonable doubt. What, in short, constitutes proof of a codon? Whether theory can help by suggesting the general structure there is little doubt that its discovery would greatly help the experimental work. Failing that, the main use of theory may be to suggest novel forms of evidence and to sharpen critical judgement. In the final analysis it is the quality of the experimental work that will be decisive.[114]

It is clear that Crick's chief contribution to biology has been his physical sense and his ability to see through to the essence of the problem. Many ideas which were already floating around were picked up by Crick, who recognized the really important ones and made these precise, then presented them as fundamental generalizations. He had the advantage over many of his contemporaries in being self-educated in biology and biochemistry and therefore not overburdened by the trappings of the educational machine of the 1930's and 1940's. In his Cherwell Simon Lecture of 1967 he emphasized the importance to him of the ideas of symmetry and simplicity which he got from physics: these ideas, combined with a rigorous examination of the evidence, made it possible for him to see quickly the fallacies in theories of the code based on the amino acid specificity of holes in the RNA helix. Later he exposed the errors in the work of S. R. Pelc and M. G. E. Welton, purporting to show that amino acids fit their transfer RNA codons specifically. When Pelc showed Crick his models, Crick found that all the polynucleotide sequences had been built backwards.[115]

Recently Crick has turned his attention to diffusion models as mechanisms for differentiation. In typical fashion he has picked out the significant fact, already noted by Lewis Wolperts, that embryonic fields appear to involve only short distances of up to one hundred cells, and he is examining a scheme involving facilitated diffusion between the members of a two-dimensional sheet of cells, of the kind found in the insect cuticle, which his postdoctoral student, Peter Lawrence, is studying in insect cuticle. Crick's recent paper in *Nature* on the subject could only have been written by him: the style of the arguments, the frequent recourse to quantitative data, and the confident predictions ("it is my belief that mechanisms based on diffusion are not only plausible but rather probable") are unmistakable.[116]

Conclusion

It is now seventeen years since the discovery of the Watson-Crick model for DNA. Crick is now codirector (with Brenner) of the Cell Biology Division of the Medical Research Council Laboratory with thirty-three research graduates under him, some doctoral and some postdoctoral. In the old days he had very little correspondence, no administrative duties, and he deliberately avoided teaching. At first this was because he wanted to get his Ph.D. thesis completed as soon as possible. Later when the opportunity for graduate teaching came again, he no longer felt a need for the money. At University College, London, he had done his spell of demonstrating (laboratory supervision and instruction); at Cambridge he was happy to be free from it. In those days his journey to the lab at the Cavendish was only a short walk; there he worked long hours, including Saturday mornings and often Saturday afternoons. After lunches at the Eagle there was always a walk along the "Backs," when many an animated discussion took place. Now he drives to work at the new lab in Hills Road, three miles from the center of the town. He has a sizable correspondence and much of his time goes in deciding the research program of the division and in finding the right people to work on it (most of the research staff are appointed for three- to five-year terms).

Crick himself is trying to escape from the domination of his life by science. He gets through much more work in less time, yet feels uncomfortable about the delegation of the work to others. (One has to get used to the idea of enjoying in a vicarious manner the progress of the work in the hands of a younger man.) Of course his style has always been toward the theoretical rather than the practical side, but it would be a mistake to regard model building, for instance, as not deserving the label "experimental," for with models one manipulates the environment of the molecule and observes the result. The model is used as an analogue computer. It is often assumed that bench work is "work" and thinking is not. Here again, Crick would argue that thinking, discussing, and communicating are all work, and they all take time. Nor is Crick satisfied with the formal correspondence of a mathematical model with biological reality; for him a physical and chemical reality must also be achieved. The "black box" situation in biology, though of interest to Crick as the starting point of an investigation, has always been for him a preliminary to discovering the machinery within. No model for pattern formation during differentiation will satisfy him until an actual "morphogen" has been isolated and shown to possess the required properties.

Many theoreticians are less concerned about the experimental evidence than Crick. At the same time Crick's demands for evidence did prevent him from publishing his adaptor hypothesis when he first thought of it in 1955, which was, on reflection, a pity.

At a time when molecular biology is being attacked as too narrow and when the trumpet call is for hierarchical laws, Crick remains undisturbed. He doubts that if these critics knew more chemistry they would maintain their opposition. The place of DNA in biology may be challenged, but such threats are due more to misunderstanding and misrepresentation of the claims of the subject than to the program of research as such. When Michael Polanyi states that vitalism, as he defines it, "can be ignored only by a truculently bigotted mechanistic outlook,"[117] it is clear that he is concerned with rather different matters than the molecular biologist, who would not claim that a piece of Shakespeare's writing is reducible to physics and chemistry though he would claim that the thought processes which generate them will be explicable in terms of those sciences. Meanwhile, evidence for the structure of DNA, the sequence hypothesis, and the central dogma is accumulating. Some facts—reiterated DNA sequences, for instance[118]—do not appear at first sight to square with the theory; but the weight of such evidence is unclear until more is known about it.

Finally, we may try to predict how this body of theory will look at the end of the century. Will it have lost some of its charisma, its public appeal? Clearly the latter quality stems partly from its medical importance, just as does the Curies' discovery of radium. Will the DNA discovery be looked upon as a transforming principle? Undoubtedly, the connection it has with the work that preceded it will become clearer with the passage of time, but what Gardner Quarton called the "psychological factors associated with a transforming principle"[119] will remain; to anyone who goes back over the pre-1950 literature the contrast in outlook is very striking, for it was in the 1950's that DNA rose to a special position of importance due to the work of Erwin Chargaff, the phage biologists, and the Watson-Crick structure for DNA.

That the reproduction of the gene could be based on so simple an idea as complementary hydrogen bonding was almost too good to be true, and this caught the imagination of scientists, funding institutions, the Nobel Prize Committee, and, finally, the general public. The discovery will also retain a certain aura, due to Watson's book *The Double Helix.*

From the technically oriented physics at the Admiralty, Crick has moved to X-ray crystallography and on to genetics, protein synthesis, and now to differentiation. "He is," to quote Brenner, "the only one in the MRC Lab at Cambridge who can encompass all the work there even at a very deep level . . . He is a very great force there." His career is proof of the fact "one learns the field best by helping to create it."[120]

This year Crick stated confidently that "nature usually has such difficulty evolving elaborate biochemical mechanisms (for example, those used in protein synthesis) that the underlying processes are often rather simple."[121] The physicist in Crick is still hard at work, and molecular biology is by no means an exhausted subject.

POSTSCRIPT

I am grateful to Dr. Crick for reading and commenting on several drafts of this essay and for supplying the following comments on the version printed in *Dædalus*.

Comment on p. 941. "The special mines to blow up a sweep had the standard amount of explosive. The detector was made insensitive by putting a resistance across the very sensitive relay they contained. These mines were too insensitive to blow up a ship because the magnetic field of a ship [as opposed to that of a minesweeper or "sweep"] would have been too weak to activate the mechanism."

Comment on p. 948. "The lecture 'What Mad Pursuit' did not merely claim that the current ideas about protein structure were wrong, but that all the *methods* then being used were inadequate, except for the isomorphous replacement method. Incidentally I think I was the first person who ever tried to estimate how heavy an atom was needed for a protein. Amusingly enough it was also the first structure factor calculation I had ever done."

Comment on p. 952. The knobs and holes of α keratin. "This I realized later was a naïve idea. What I should emphasize now if I were doing it would be the new symmetry produced, that is, the fact that relative backbone configurations repeat every seven residues."

Comment on p. 957. "It never occurred to me until I read your account that John Griffith looked at me with deference. Curiously enough I always felt deference to him, because of his very considerable manipulative mathematical ability. How shyness can mislead!"

Comment on p. 967. On soluble RNA molecules being too big for the role of adaptor molecules. "I had to read my *SEB* article again recently and was taken aback by what I had written there. Of course we now know (or think we know) that the 'size' is not so big because tRNA molecules are long and thin. How can I have worried about diffusion rate I do not know, since the increase in diffusion constant could only be a matter of times two or times three, and we know far too little to worry about that."

Comment on p. 977. "I realized some years ago that the definition of the sequence hypothesis which I gave in the *SEB* paper is back-to-front. It should be stated as saying that all (= almost all) amino acid sequences in protein are coded for by a nucleic acid sequence. As stated in the *SEB* paper it implies that all nucleic acid sequences must be translated into protein. This is certainly not true (consider the genes for tRNA and rRNA, and the control and spacer sequences now known in RNA viruses) and I do not think we believed it then. I think it was simply careless drafting.

Is this why you worry about repetitive sequences?" (p. 980)

There are a few differences between this version of the essay and the text as published in the Fall 1970 issue of *Dædalus*. The passages in question have been altered on the advice of Dr. Crick and Mrs. Arnold Dickens.

I would like to comment further on two remarks made in the body of the essay: "These statements [defining the Central Dogma] distinguish the Watson-Crick era from that which preceded it, when nucleoproteins rather than nucleic acids were identified with the hereditary material" (p. 977) and "to anyone who goes back over the pre-1950 literature the contrast in outlook is very striking" (p. 980).

With respect to the first remark it is very important to see that debates over the chemical identity of the gene did not consist of the straight issue between DNA or protein, but of the less decisive one of DNA or *nucleoprotein*. The idea that protein and nucleic acid are both required for gene replication and possibly for gene mutation was attractive at a time when conjugated molecules were receiving publicity. It did not matter that the most effective region of the ultraviolet spectrum in producing mutation was that absorbed by the purine and pyrimidine rings of DNA. Thus George Beadle was not convinced in 1946 that the chemical substance which mutates is nucleic acid. "It is possible," he remarked, "that the nucleic acid is extragenic and that it transfers the mutation-producing energy [of ultraviolet irradiation] to the gene."[122] Nine years later Rollin Hotchkiss, looking back on the subject, recalled that the idea of an energy transfer from nucleic acid to protein, "and the recognition that added substances and environmental factors could considerably modify the number of mutations detected after ultraviolet irradiation, tended to reject any general acceptance of this phenomenon as an indication of the chemical nature of genes,"[123] and he went on to point out that the discovery of photoreactivation emphasized the indirect and complex nature of the process of ultraviolet mutagenesis.

Jack Schultz, in his essay "The Evidence of the Nucleoprotein Nature of the Gene," likened the functioning of the conjugated protein of the gene to that of the nucleoprotein enzymes of respiration described by Otto Warburg in 1938. Here, he said, "the active group is the conjugated protein; neither component without the other is effective . . . the specifities of the gene reside in the nucleoprotein, and the continuous structure of the chromosome is a protein fibre."[124] Erwin Chargaff, who had come to the discovery of base ratios by seeing the equivalence of 6-amino to 6-keto, recalled that he recognized at the time the existence of some pattern of structure of a type not to be found in the proteins. As a former student of an important class of conjugated molecules—the lipoproteins—it is natural that he should have later inclined to view such a structural basis for the ratios as involving the protein moiety. In 1956 he stated that the "most satisfactory construction that could impose this uniformity [6-amino to 6-keto] is one in which two polynucleotide chains are bonded to a polypeptide chain, so that each peptide carbonyl is linked by a hydrogen bond to the 6-amino group of adenine or cytosine and each peptide amino group to the 6-keto group of guanine or uracil (or thymine)."[125] After remarking on the hypothetical character of this structural arrangement he said: "It will, however, be clear that I should prefer to consider the protein moiety of a nucleoprotein as not merely an ornament or protective supplement."[126] These quotations have not been given to expose the "errors" of the past, but to recall the climate of opinion in which DNA research was evaluated before the modern era of the Central Dogma.

The contrast in outlook before and after the 1950's provides an indication of the transforming role played by the work of Erwin Chargaff the phage biologists and the Watson-Crick model for DNA. Especially associated with the latter is the assumption that hereditary information resides solely in the primary sequence of nucleotides in DNA and that the flow of such information is unidirectional, from DNA to protein. It is instructive to inquire what view of the transmission of hereditary information was expressed by those who established the one gene–one enzyme hypothesis in the 1940's. We find that George Beadle was thinking in terms of a gene controlling the *conformation* of an enzyme. Seeing the direct relation between gene and antigen specificities he said, "it seems reasonable to suppose that the gene's primary and possibly sole function is in directing the final configurations of protein molecules . . . such a view does not mean that genes directly 'make' proteins . . In the synthesis of a single protein molecule probably at least several hundred different genes contribute. But the final molecule corresponds to only one of them and this is the gene we visualize as being in primary control."[127] David Bonner expressed himself in a similar vein at the Cold Spring Harbor meeting in 1945 when he said:

> There is quite general agreement at present that genes contain nucleoprotein as an essential component of their structure. One should expect, therefore, that genes, like other proteins, have specific configurations, the configuration of a single gene being characteristic of itself alone. These various considerations suggest the view that the gene controls biochemical reactions by imposing, directly or indirectly, a specific configuration on the enzymes essential for the specific reactions.[128]

The first reference to amino acid sequences in the writings of Beadle that I have noted comes in 1950.[129] While we have to go to Alexander Dounce's paper of 1952[130] for the detailing of information transfer from a polynucleotide code, it is possible to find the conception of a linear sequence code in terms of the protein concept of the gene much earlier. In 1936 Dorothy Wrinch developed a scheme of this type. She recognized that a mutation could result from so slight a change as that of an amino acid side chain—corresponding to an axial length of only 3.4 A—and she saw the relevance of the model to the position effect. She added: "It is but one aspect of the suggestion that the genetic constitution of a chromosome is to reside not only in the nature of the residues of which it is composed, not only on the proportions of these different residues, but essentially and fundamentally in their *linear arrangement in a sequence* of sequences."[131]

Francis Crick arrived at the amino acid sequence hypothesis of the gene in the late 1940's. When he learned from Watson the great importance of DNA for heredity, he was quickly converted to the nucleotide sequence hypothesis. It took a good two decades to establish the new theory and bury the old. As late as 1955 an eminent biochemist could write that "sequence cannot be the sole agent of biological information. Even if the arrangement of an entire polynucleotide could be written, a third dimension would be lacking—the operative three-dimensional shape of the molecular aggregate."[132]

REFERENCES

1. The source material for this account consists of published papers, unpublished correspondence and papers, and a number of oral interviews and informal conversations. Extracts from taped, oral interviews are referred to as interviews and are dated, whereas informal conversations by telephone and in person are simply described as personal communications and are undated. Those readers who desire more explanation of the subject matter of molecular biology may like to consult J. C. Kendrew's lucid book *The Thread of Life* (London: Bell, 1966). They will also find in Gunther Stent's essay on DNA in this volume a very clear exposition of the subject of molecular genetics.

It is a pleasure to express my thanks to all those who have helped me in collecting the material for this essay, and especially to Dr. and Mrs. Francis Crick, Dr. Sidney Brenner, Mrs. Arnold Dickens of Northampton, and Mrs. Anne Sanderson of the Medical Research Council, London. I am also grateful to Dr. J. D. Watson for reading this paper. This work has been carried out with support from the Royal Society and the American Philosophical Society.

2. Letter from Michael Hart, May 14, 1969.

3. Obituary notices in *Journal of the Northampton Natural History Society,* 12 (1903-1904), 134-144, and *Quarterly Journal of the Geological Society,* 60 (1904), lxxx.

4. C. R. Darwin, "On the Dispersal of Freshwater Bivalves," *Nature,* 25 (1882), 529-530.

5. W. Crick and F. Soddy, *Abolish Private Money, or Drown in Debt, Two Amended Addresses to Our Bosses* (London: Joseph Sault, 1939). Soddy (1877-1956), Nobel Prize winner and originator of the term "isotope," attributed the social ills of scientific progress to money. See "The Accursed Element," *Youth* (Cambridge, Eng., February 1922), p. 156.

6. Obituary notice in *Northampton Independent,* January 17, 1958, p. 9.

7. Obituary notice in *Northampton Independent,* February 6, 1948, p. 8.

8. Letter from Francis Crick, June 5, 1969.

9. Mrs. Arnold Dickens, personal communication, April 7, 1970.

10. University College, London, finals papers, 1937.

11. Report prepared by Professor Andrade for transfer of Crick to Naval Intelligence, London, March 28, 1946, University College, London, Archives.

12. E. N. da C. Andrade, "A Theory of the Viscosity of Liquids," *Philosophical Magazine,* 17 (1934), 497-511, 698-732.

13. See note 11.

14. Interview with Sir Harry Massey, May 8, 1969.

15. Linus Pauling taught elementary chemical analysis for one year between his sophomore and senior years at Corvallis because of the financial circumstances of the family. (Interview with Pauling, October 1968.)

16. L. Pauling, "Modern Structural Chemistry," *Chemical and Engineering News,* 24 (1946), 1789 (Willard Gibbs Medal presentation).

17. E. Schrödinger, *What Is Life? The Physical Aspect of the Living Cell* (Cambridge, Eng.: University Press, 1944), p. 2.

18. *Ibid.*, p. 81. For a discussion of Schrödinger's contribution see D. Fleming, "Emigré Physicists and the Biological Revolution,"in D. Fleming and B. Bailyn, eds., *The Intellectual Migration: Europe and America, 1930-1960* (Cambridge, Mass: Harvard University Press, 1969), 152-189. For an analysis of Schrödinger's book *What Is Life?* and the status of contemporary knowledge see R. C. Olby, "Schrödinger's Problem: What is Life?" *Journal of the History of Biology* (forthcoming).

19. Letter from Crick, June 5, 1969.

20. Cyril Darlington, personal communication.

21. Letter from Crick, June 5, 1969.

22. Crick, personal communication.

23. Letter from Mrs. Anne Sanderson, Information Section, Medical Research Council, London, March 26, 1970.

24. F. H. C. Crick (with A. A. W. Hughes), "The Physical Properties of Cytoplasm, A Study by Means of the Magnetic Particle Method, Part I: Experimental; F. H. C. Crick (alone), Part II: Theoretical Treatment," *Experimental Cell Research,* 1 (1950), 37-64, 505-542.

25. Letter from Mrs. Sanderson, March 26, 1970.

26. Interview with Crick, March 8, 1968.

27. *Ibid.*

28. *Ibid.*

29. F. H. C. Crick, *Polypeptides and Proteins: X-Ray Studies* (Ph.D. thesis, Cambridge University, 1953), p. v.

30. W. Cochran and F. H. C. Crick, "Evidence for the Pauling-Corey α Helix in Synthetic Polypeptides," *Nature,* 169 (1952), 234, and W. Cochran, F. H. C. Crick, and V. Vand, "The Structure of Synthetic Polypeptides, I: The Transformation of Atoms on a Helix," *Acta Crystallographic,* 5 (1952), 581-586.

31. L. Pauling and R. B. Corey, "Two Hydrogen-Bonded Spiral Configurations of the Polypeptide Chain," *Journal of the American Chemical Society,* 72 (1950), 5349.

32. L. Pauling, R. B. Corey, and H. R. Branson, "The Structure of Proteins: Two Hydrogen-Bonded Helical Configurations of the Polypeptide Chain," *Proceedings of the National Academy of Sciences,* 37 (1951), 205-211, 235-285.

33. F. H. C. Crick, "Is α-Keratin a Coiled Coil?" *Nature,* 170 (1952), 882-883.

34. L. Pauling and R. B. Corey, "Compound Helical Configurations Polypeptide Chains: Structure of Proteins of the α-Keratin Type," *Nature,* 171 (1953), 59-61.

35. M. Spei, G. Heidermann, and H. Zahn, "X-Ray Evidence of the 198 Å Period in α-Keratin," *Naturwissenschaften,* 55 (1968), 346.

36. Dr. Geoffrey Brown, personal communication, supported by Dr. William Seeds.

37. Date obtained from Wilkins' letter to Dr. Roy Markham, June 15, 1950.

38. Dr. Seeds, personal communication.

39. Mrs. Crick is English by birth but was brought up in the French manner by her French mother.

40. Letter from J. D. Watson to M. Delbrück, December 9, 1951.

41. It might be useful here to recall the components of deoxyribonucleic acid: four nitrogenous bases consisting of two purines (adenine and guanine) and two pyrimidines (thymine and cytosine), a five-carbon sugar, and a phosphate group.

42. Strictly speaking, Franklin was not quite correct in using this term for one nucleotide in DNA.

43. R. E. Franklin, MS of Colloquium, King's College, London (November 1951). For an account of her work see A. Klug, "Rosalind Franklin and the Discovery of the Structure of DNA," *Nature,* 219 (1968), 808-810, 843-844, corrigenda, 879, 1192, correspondence, 880.

44. F. H. C. Crick, "The Structure of Sodium Thymonucleate: A Possible Approach" (Unpublished MS, 1951), p. 4. Professor Donohue comments that the sugar hydroxyls can donate H bonds, and have since been shown to do so in several crystalline nucleotides.

45. Watson, personal communication, also mentioned in his letter to Delbrück, May 20, 1952.

46. Interview with Crick, March 8, 1968.

47. K. G. Stern, "Nucleoproteins and Gene Structure," *Yale Journal of Biology and Medicine,* 19 (1947), 937; see also *Scientia,* 82 (1947), 74-78.

48. Interview with John Griffith, May 3, 1968.

49. Interview with Crick, March 8, 1968.

50. G. R. Wyatt and S. S. Cohen, "The Bases of the Nucleic Acids of Some Bacterial and Animal Viruses: The Occurrence of 5-Hydroxymethylcytosine," *Biochemical Journal,* 55 (1953), 780: "One is tempted to speculate that regular structural association of nucleotides of adenine with those of thymine and of guanine with those of cytosine (or its derivatives) in the DNA molecule requires that they be equal in number."

51. Interview with Crick, March 8, 1968. Nevertheless, Watson is sure that he knew of Chargaff's ratios before meeting him in Cambridge. The journal containing them was in the zoology library where he went regularly to read.

52. L. Hunter, "Mesohydric Tautomerism," *Journal of the Chemical Society* (1945), pp. 806-809.

53. P. G. Owston, "Diffuse Scattering of X-rays by Ice," *Acta Crystallographica,* 2 (1949), 222-228. Donohue comments that "Hunter's view is worthless. The proton shift in ice does not lead to different tautomers, so the ice structure is not germane." W. Cochran localized all the protons in adenine ("The Structure of Pyrimidines and Purines. V. The Electron Distribution in Adenine Hydrochloride," *Acta Crystallographer,* 4 [1951], 81-92). Miss Broomhead, whose paper follows Cochran's does not appear to be influenced by his work.

54. J. M. Broomhead, "An X-Ray Investigation of Certain Sulphonates and Purines" (Ph.D. thesis, Cambridge University, 1950). She did note that the order of distance involved was about four or five times that considered by Hunter and Owston. Hence the hybrid tautomer explanation was unlikely for purines.

55. For example, of the four bases in DNA only one (adenine) is correctly shown in J. N. Davidson, *The Biochemistry of the Nucleic Acids,* 1st ed. (London, 1950), pp. 7-8.

56. In 1949, before Watson came to Europe and shortly after Crick had joined the Cavendish MRC unit, Bragg, Kendrew, and Perutz made a systematic exploration of possible helical conformations of polypeptide chains. Unfortunately all were wrong since they had decided against the planar configuration of the amide bond. See Sir Lawrence Bragg, J. C. Kendrew, and M. F. Perutz, "Polypeptide Chain Configurations in Crystalline Proteins," *Proceedings of the Royal Society A,* 203 (1950), 321-357.

57. Pauling had sent to Cambridge copies of his manuscript detailing a structure for DNA which was very like Watson and Crick's first structure of 1951. Pauling's model appeared in March 1953: L. Pauling and R. B. Corey, "A Proposed Structure for the Nucleic Acids," *Proceedings of the National Academy of Sciences,* 39 (1953), 84-97.

58. Franklin, MS of Colloquium (November 1951).

59. Interview with Crick, July 10, 1968.

60. J. D. Watson, *The Double Helix* (New York: Atheneum, 1968), pp. 177-178.

61. Interview with Crick, March 8, 1968.

62. The dating here is based on the letter which Watson wrote to Delbrück referring to his "very pretty model" for DNA. The letter is dated February 20, and according to *The Double Helix,* p. 190, was sent just before Donohue "protested that the idea [of his like-with-like scheme] would not work."

63. Interview with Crick, March 8, 1968.

64. Watson, *The Double Helix,* p. 194.

65. Interview with Crick, March 8, 1968.

66. Hedy Lamarr's film *Ecstasy* was shown at the Rex Cinema on February 26-28, 1953. According to *The Double Helix,* Watson had not discovered the correct base-pairing scheme when he saw this film.

67. J. M. Gulland and D. O. Jordan, "The Macromolecular Behaviour of Nucleic Acids," *Symposia of the Society for Experimental Biology,* 1 (1947), 58, and D. O. Jordan, "Physicochemical Properties of the Nucleic Acids," *Progress in Biophysics,* 2 (1951), 51-89.

68. A. von Humboldt, "On the Spirit and the Organizational Framework of Intellectual Institutions in Berlin," trans. Edward Shils, *Minerva,* 8 (1970), 243.

69. J. D. Watson and F. H. C. Crick, "Molecular Structure of Nucleic Acids," *Nature,* 171 (1953), 737-738; Watson and Crick, "Genetical Implications of the Structure of Deoxyribonucleic Acid," *Nature,* 171 (1953), 964-967; Watson and Crick, "The Structure of DNA," *Cold Spring Harbor Symposium on Quantitative Biology,* 18 (1953), 123-131; Crick, "Structure and Function of DNA," *Discovery,* 15 (1954), 12-17; Crick and Watson, "The Complementary Structure of Deoxyribonucleic Acid," *Proceedings of the Royal Society A,* 223 (1954), 80-96.

70. Letter from Crick, June 5, 1969.

71. J. D. Watson and F. H. C. Crick, "Genetical Implications of the Structure of Deoxyribonucleic Acid," *Nature,* 171 (1953), 966.

72. J. D. Watson and F. H. C. Crick, "The Structure of DNA," *Cold Spring Harbor Symposium on Quantitative Biology,* 18 (1953), 127.

73. G. Gamow, "Possible Relation between Deoxyribonucleic Acid and Protein Structure," *Nature,* 173 (1954), 318.

74. The editorial staff of the *Proceedings* refused to publish this paper under the joint authorship of Gamow and the fictitious Tompkins, so Gamow published it elsewhere: "Possible Mathematical Relation between Deoxyribonucleic Acid and Protein," *Biologiske Meddelelser Kongelige Danske Videnskabernes Selskabs,* 22 (1954), 1-13.

75. A. Dounce, "Duplicating Mechanism for Peptide Chain and Nucleic Acid Synthesis," *Enzymologia,* 15 (1952), 251-258.

76. Interview with Crick, March 8, 1968.

77. This club had been formed by Gamow to serve as a forum of discussion on the problems relating to the genetic code and protein synthesis.

78. Interview with Crick, March 8, 1968.

79. Letter from G. Gamow to Crick, March 8, 1954.

80. F. H. C. Crick, "On Degenerate Templates and the Adaptor Hypothesis," A Note for the RNA Tie Club, undated, p. 7.

81. These suggestions as to chemical identity were made two years after the RNA Tie Club paper. Three bases (a "codon") code one amino acid.

82. Letters from M. B. Hoagland, March 2, 1970, and April 1, 1970.

83. P. C. Zamecnik, "An Historical Account of Protein Synthesis," *Cold Spring Harbor Symposia on Quantitative Biology,* 39 (1969), 6.

84. M. B. Hoagland, "Nucleic Acids and Proteins," *Scientific American,* 201 (1959), 59.

85. F. H. C. Crick, "On Protein Synthesis," *Symposia of the Society for Experimental Biology,* 12 (1958), 156.

86. Letter from Hoagland, April 1, 1970.

87. A. Pardee, "Experiments on the Transfer of Information from DNA to Enzymes," *Experimental Cell Research,* Supplement 6 (1958), 150.

88. Interview with François Jacob, September 17, 1969.

89. M. Riley, A. Pardee, F. Jacob, and J. Monod, "On the Expression of a Structural Gene," *Journal of Molecular Biology,* 2 (1960), 225.

90. Pardee, "Experiments on the Transfer of Information from DNA to Enzymes," p. 150.

91. A. Pardee, "Nucleic Acid Precursors and Protein Synthesis," *Proceedings of the National Academy of Sciences,* 40 (1954), 263.

92. E. Volkin and L. Astrachan, "Phosphorus Incorporation in *Escherichia Coli* Ribonucleic Acid After Infection With Bacteriophage T2," *Virology,* 2 (1956), 149-161, and Astrachan and Volkin, "Properties of Ribonucleic Acid Turnover in T2 Infected *Escherichia Coli,*" *Biochimica et Biophysica Acta,* 29 (1958), 536: "These data suggested that bacteriophage infection induces the bacteria to synthesize a new kind of RNA."

93. Interview with Crick, March 8, 1968.

94. F. H. C. Crick and J. D. Watson, "Virus Structure: General Principles," *Ciba Foundation Symposium on the Nature of Viruses* (London: Churchill, 1957), p. 9. (The symposium was held March 26-28, 1956.)

95. A. N. Belozersky and A. S. Spirin, "A Correlation Between the Composition of Deoxyribonucleic and Ribonucleic Acids," *Nature,* 182 (1958), 112.

96. F. H. C. Crick, "On Degenerate Templates and the Adaptor Hypothesis," title page.

97. F. H. C. Crick, J. S. Griffith, and L. E. Orgel, "Codes Without Commas," *Proceedings of the National Academy of Sciences,* 43 (1957), 420: "The arguments and assumptions which we have had to employ to deduce this code are too precarious for us to feel much confidence in it on purely theoretical grounds."
    See also Crick, "On Protein Synthesis," p. 160; "I must confess that I find it impossible to form any considered judgment of this idea. It may be complete nonsense, or it may be the heart of the matter. Only time will show."

98. S. Benzer, "Fine Structure of a Genetic Region in Bacteriophage," *Proceedings of the National Academy of Sciences,* 41 (1955), 344-354.

99. E. Freese, "On the Molecular Explanation of Spontaneous and Induced Mutations," *Brookhaven Symposia in Biology,* 12 (1959), 63-75.

100. S. Brenner, S. Benzer, and L. Barnett, "Distribution of Proflavin-Induced Mutations in the Genetic Fine Structure," *Nature,* 182 (1858), 983-985.

101. Jacques Fresco, personal communication, and J. R. Fresco and B. M. Alberts, "The Accommodation of Non-Complementary Bases in Helical Polyribonucleotides and Deoxyribonucleic Acids," *Proceedings of the National Academy of Sciences,* 46 (1960), 319-320.

102. Letter from Crick to Fresco, January 7, 1960.

103. Fresco and Alberts, "The Accommodation of Non-Complementary Bases in Helical Polyribonucleotides and Deoxyribonucleic Acids," pp. 311-321.

104. J. R. Fresco, B. M. Alberts, and P. Doty, "Some Molecular Details of the Secondary Structure of Ribonucleic Acid," *Nature*, 188 (1960), 1-10.

105. Interview with Crick, March 8, 1968.

106. Crick, in the discussion following the paper by George Streisinger and others, in "Deoxyribonucleic Acid: Structure, Synthesis and Function," *Proceedings of the 11th Réunion of the Société Chimie Physique*, June 1961 (Oxford: Pergamon Press, 1962), p. 188.

107. F. H. C. Crick, L. Barnett, S. Brenner, and R. J. Watts-Tobin, "General Nature of the Genetic Code for Proteins," *Nature*, 192 (1961), 1227-1232.

108. Interview with Sidney Brenner, March 19, 1970.

109. *Ibid.*

110. "On Protein Synthesis," p. 143.

111. *Ibid.*, p. 152.

112. *Ibid.*, p. 161.

113. *Ibid.*, p. 153.

114. F. H. C. Crick, "The Recent Excitement in the Coding Problem," *Progress in Nucleic Acid Research* (1963), 213-214.

115. S. R. Pelc and M. G. E. Welton, "Stereochemical Relationship Between Coding Triplets and Amino-Acids," *Nature*, 209 (1966), 868-870, 870-872, and F. H. C. Crick, "An Error in Model Building," *Nature*, 213 (1967), 798.

116. F. H. C. Crick, "Diffusion in Embryogenesis," *Nature*, 225 (1970), 422.

117. M. Polanyi, *Personal Knowledge Towards a Post-Critical Philosophy* (New York: Harper Torchbook, 1964), p. 390.

118. R. J. Britten and E. H. Davidson, "Gene Regulation for Higher Cells: A Theory," *Science*, 165 (1969), 349-357.

119. G. Quarton, remark at Bellagio conference on Transforming Conceptions of Modern Science.

120. Interview with Sidney Brenner, March 19, 1970.

121. Crick, "Diffusion in Embryogenesis," p. 422.

122. G. W. Beadle, "Genes and the Chemistry of the Organism," *American Scientist*, 34 (1946), 52.

123. R. D. Hotchkiss, "The Biological Role of the Desoxypentose Nucleic Acids," in E. Chargaff and J. N. Davidson, eds., *The Nucleic Acids: Chemistry and Biology* (New York: Academic Press, 1955), p. 440.

124. J. Schultz, "The Evidence of the Nucleoprotein Nature of the Gene," *Cold Spring Harbor Symposium on Quantitative Biology*, 9 (1941), 62.

125. E. Chargaff, "The Base Composition of Desoxyribonucleic Acid and Pentose Nucleic Acid in Various Species," in W. D. McElroy and B. Glass, eds., *The Chemical Basis of Heredity* (Baltimore: Johns Hopkins Press, 1957), p. 525 (symposium held in 1956).

126. *Ibid.*, p. 526.

127. G. W. Beadle, "Genetics and Metabolism in *Neurospera*," *Physiological Reviews*, 25 (1945), 660.

128. D. Bonner, "Biochemical Mutations in *Neurospera*," *Cold Spring Harbor Symposium on Quantitative Biology*, 11 (1946), 21.

129. G. W. Beadle, "Chemical Genetics," in L. C. Dunn, ed., *Genetics in the 20th Century: Essays on the Progress of Genetics During Its First 50 Years* (New York: Macmillan, 1951), p. 225.

130. A. L. Dounce, "Duplicating Mechanism for Peptide Chain and Nucleic Acid Synthesis," *Enzymologia*, 15 (1952), 251-258.

131. D. M. Wrinch, "On the Molecular Structure of the Chromosomes," *Protoplasma*, 25 (1936), 558 (my italics).

132. Chargaff, "The Base Composition of Desoxyribonucleic Acid," p. 526.

LINUS PAULING

Fifty Years of Progress in Structural Chemistry and Molecular Biology

THE INVITATION to prepare an essay for this volume has stimulated me to think about the changes that have taken place in physics, chemistry, and biology, especially structural chemistry and molecular biology, during my lifetime.

I became interested in chemistry in 1914, when I was thirteen years old, on the day that a fellow high-school student of my own age (Lloyd A. Jeffress, now professor of psychology in the University of Texas) showed me some chemical experiments in the laboratory that he had set up in the corner of his bedroom. I decided then to be a chemist, and to study chemical engineering, which was, I thought, the profession that chemists followed.

By 1919 I had achieved a moderately good grasp of classical chemistry, but only a small acquaintance with the developments that were taking place in atomic physics at that time. During the academic year 1919-1920 I was a full-time teacher of quantitative chemical analysis in the Oregon State Agricultural College, where I had been a student of chemical engineering during the preceding two years. I read some of the chemical journals, and studied the early papers of Gilbert Newton Lewis and Irving Langmuir on the electronic structure of molecules. It was then, some fifty years ago, that I developed a strong desire to understand the physical and chemical properties of substances in relation to the structure of the atoms and molecules of which they are composed. This desire has largely determined the course of my work for fifty years.

As I try to remember the state of my development at that time, I am led to believe that this desire was the result of pure intellectual curiosity, and did not have any theological or philosophical basis. I was skeptical of dogmatic religion, and had passed the period when it was a cause of worry; and my understanding of the experiential world was so fragmentary as to be unsatisfactory as the basis for the development of a philosophical system. I was simply entranced by chemical phenomena, by the reactions in which substances disappear and other substances, often with strikingly different properties, appear; and I hoped to learn more and more about this aspect of the world. It has turned out, in fact, that I have worked on this problem year after year, throughout my life; but I have worked also on other problems, some closely related, such as the structure of atomic nuclei and the molecular basis of disease, and others less closely, such as the pollution of the earth with radioactive fallout and carbon 14 from the testing of nuclear weapons, the waste of life and the earth's resources through war and militarism, and the maldistribution of wealth.

During the period between 1780 and 1880 a great body of empirical information about the physical and chemical properties of substances was amassed, and it was coordinated in an effective way by the classical chemical structure theory that was developed during the same period. The correct ratios of atoms of different elements in compounds had been determined, and chemical structural formulas, written with use of lines to represent the bonds between atoms in a molecule, had come into use. This theory provided a sort of explanation of properties of substances, especially chemical properties, and was of great value in the work of chemists, especially in guiding them in the preparation of new substances with special properties.

During the following period of several decades, beginning about 1880, many aspects of chemistry were clarified by the application of physical principles, especially the principles of thermodynamics and statistical mechanics. The principles of thermodynamics were applied to chemical equilibrium by J. Willard Gibbs, J. H. van't Hoff, and many other investigators, and by 1922, when I became a graduate student in the California Institute of Technology, I learned that some aspects of the subject had been very well worked out. Experimental values of the standard molar enthalpy, entropy, and free energy had by that time been measured and tabulated for many substances. During the following year I came more thoroughly under the influence of Richard Chace Tolman, who was applying statistical mechanics and quantum theory to some chemical problems. The third law of thermodynamics had been discovered some years earlier but was still not well understood. This law can be formulated in the statement that the entropy of every pure crystalline substance can be taken as zero at temperatures close to the absolute zero. The German chemist Walther Nernst, who had discovered the third law of thermodynamics in 1906, had found that the heat capacity of crystalline substances becomes very small at low temperatures. A quantum theory of the heat capacity of crystals had been formulated by Einstein and refined by Peter Debye. It was recognized that the quantum theory provided the basis for the third law, but because of the imperfections of the old quantum theory the law remained somewhat unclear. As late as 1924 a leading physical chemist felt justified in suggesting that a crystalline substance with a large unit cell should be assigned greater entropy at the absolute zero than one with a small unit cell. Tolman and I published a paper in 1925, still based on the old quantum theory, in which we showed that a reasonable discussion of the quantum states of a crystal with use of the methods of statistical mechanics leads to the conclusion that the entropy should be zero for all perfect crystals, even those with large unit cells, but greater than zero for crystals with some sort of disorder and for supercooled liquids, also characterized by molecular disorder. In 1930 I published a paper on the residual entropy of molecular hydrogen and other crystalline substances that might be described as having molecules with considerable freedom of rotational motion in the crystal, even at very low temperatures. During a period of a few years, then, in the 1920's, there had been developed what seemed to me to be an essentially complete theory of the thermodynamic properties of substances. This development had resulted in part from the theoretical study of the problem, based upon statistical mechanics and quantum mechanics, and in part from the low-temperature experimental studies of a number of investigators, especially Gilbert Newton Lewis and members of his school, of whom William F. Giauque was outstanding.

One of the great chemical problems, however—the nature of the chemical bond—remained a puzzling one throughout this period. When I was teaching chemistry in 1919 I made use of a popular method of describing the chemical bond: atoms were assigned a certain number of hooks and eyes which could be hooked into one another to represent the bonds between the atoms. It is my memory that I felt reasonably well satisfied with this explanation of chemical bonding—an explanation that seems to me now, with my much greater experience, to be completely unsatisfactory.

As I recall my early life, I recognize that I was often satisfied with a very incomplete sort of understanding of the various aspects of nature. For example, during my first year as a graduate student I was asked by Tolman, in a seminar that he was conducting, why most substances are diamagnetic. My answer was that diamagnetism is a general property of matter. I had not heard about the theory of diamagnetism formulated by Paul Langevin in 1905, which involves a precessional motion of electrons around the atomic nuclei induced by the magnetic field as it builds up. It had not occurred to me that I might be interested in an explanation of diamagnetism.

I may give another example. I participated in a seminar, conducted by Tolman, in which the subject of discussion was the material in the third edition of Arnold Sommerfeld's book *Atombau und Spektrallinien*, published in 1922. In this book Sommerfeld discussed his theory of the fine structure of atomic energy levels, based upon the idea that the electrons move in elliptical orbits. The eccentricity of an orbit was said to be determined by the value of the azimuthal quantum number, which specified the angular momentum of the electron. In atoms containing more than one electron there is a large difference in energy levels for orbits with the same total quantum number and different azimuthal quantum numbers, because of the penetration of the inner electron shells by these electrons. Sommerfeld explained the fine structure of the energy levels as resulting from a relativistic change in mass of the electrons, different for electron orbits with different eccentricity. In this calculation he ignored the different amount of penetration of the inner shells; in fact, he assigned to an electron two different values of the azimuthal quantum number, which he called the inner quantum number and the outer quantum number. It is my memory that I was not troubled by this discrepancy at all: I tried to understand the theory as presented by Sommerfeld, but I did not try to resolve the puzzle, if indeed I was even aware that there was one. I know now that other people were troubled by this matter. Later on, in 1926-1927, I spent a year in Sommerfeld's Institute, in Munich, and I got to know him well. It is clear that he himself knew that there was a puzzle here, and it is probable that he made his students aware of it, in particular Werner Heisenberg, who was of course responsible in part for its resolution. But in his book Sommerfeld ignored the paradox presented by his use simultaneously of two numerical values for what seemed to be the same quantity—he wrote that the reader should concentrate on the equations, not on their interpretation in terms of a model.

I believe that I now understand why I did not have an interest in some of these questions during my first year or two in Pasadena. When I came to the California Institute of Technology in 1922 I became rapidly aware of the great amount of existing knowledge in mathematics, physics, and chemistry, and I strove to become familiar with as much of it as possible. There were so many gaps in my understanding that the two problems mentioned above did not stand out among the hundreds that I failed to understand; often I did not know whether to attribute this failure to myself or to the existing state of development of science. Soon, however, I began to strive to get the answers to certain questions that interested me greatly, those relating to molecular structure and the chemical bond, as I shall describe later. One episode impressed itself on my memory so strongly that I conclude that it had a significant impact on my development. In the spring of 1923 Tolman asked me a question during a seminar. My answer was "I don't know; I haven't taken a course in that subject." At the end of the seminar Dr. Richard M. Bozorth, who had received his Ph.D. degree the year before, took me to one side and said, "Linus, you shouldn't have answered Professor Tolman the way you did; you are a graduate student now, and you are supposed to know everything."

In addition to the gaps in my knowledge that made it difficult for me to distinguish the real problems, to which no one knew the answers, from the apparent ones, to which the answers were known, although not to me, there was another reason for my lack of interest in puzzling questions such as those mentioned above: the old quantum theory was recognized as only partially satisfactory; there were hundreds of problems for which it gave only partial solutions, and it was acknowledged that the complete and correct solutions had to await the development of a better quantum theory.

I began experimental work on the determination of the structure of crystals by the X-ray diffraction method in the fall of 1922, under the direction of Roscoe G. Dickinson. The first structure completed by us was that of the mineral molybdenite, molybdenum disulfide. The structure turned out to be interesting in several respects. Each molybdenum atom is surrounded by six sulfur atoms lying at the corners of a trigonal prism, a sort of coordination that had not been previously observed. Moreover, the smallest interatomic distances were found not to agree completely with a set of atomic radii that had been formulated by W. L. Bragg in 1920. Bragg had assigned to the molybdenum atom a radius equal to half the contact distance of the atoms in metallic molybdenum, and to the sulfur atom a radius equal to half the sulfur-sulfur distance in pyrite, iron disulfide. The molybdenum-sulfur contact distance in molybdenite was found to be equal to the sum of the molybdenum radius and sulfur radius, as given by Bragg, but the sulfur-sulfur contact distance between adjacent layers of sulfur atoms was far larger than twice the sulfur radius. In an effort to understand this lack of agreement I immediately began a study of the known crystal structures and an analysis of the observed interatomic distances. It soon became clear that the effective radius of an atom would be less in the direction in which it forms a covalent bond (a shared-electron-pair bond, as described by G. N. Lewis in 1916) than in the directions in which it has unshared pairs of electrons. The understanding of atomic sizes developed rapidly during the next few years. In 1923 J. A. Wasastjerna evaluated the radii for alkali and halide ions by use of the assumption that the electric polarizability of the ions, as given by the index of refraction of crystals containing them, is proportional to the cube of the radius for isoelectronic ions. In 1926 V. M. Goldschmidt made use of a large number of experimental values of interatomic distances and of Wasastjerna's ionic radii by assigning two sets of radii, ionic and atomic (covalent, metallic) radii, to a large number of elements.

In March 1926 I went to Europe to work for a year and a half as a Fellow of the John Simon Guggenheim Memorial Foundation. In my application I had proposed to apply quantum mechanics, discovered only a few months earlier, to the problem of the structure of molecules and the nature of the chemical bond. I began my work in Sommerfeld's Institute for Theoretical Physics in Munich. The success that Heisenberg and Schrödinger had had in applying quantum mechanics to simple atoms made me hopeful that similar success could be achieved for more complex atoms, containing many electrons, and also for molecules and crystals. Shortly after reaching Munich I read a paper by Gregor Wentzel in the *Zeitschrift für Physik* on a quantum mechanical calculation of the values of the screening constants for electrons in complex atoms that had been introduced by Sommerfeld to explain the fine structure of X-ray levels. Wentzel reported poor agreement between the calculated and experimental values, but I found that his calculation was incomplete and that when it was carried out correctly it led to values of the screening constants in good agreement with the experimental values. It was evident that a quantum mechanical screening-constant method of discussing the electronic structure of complex atoms could be developed, and in 1927 I published in the *Proceedings of the Royal Society of London* a paper with the title "The Theoretical Prediction of the Physical Properties of Many-Electron Atoms and Ions; Mole Refraction, Diamagnetic Susceptibility, and Extension in Space." The screening-constant method was found to lead to values of ionic radii, with a reasonably sound theoretical basis, which were in very close agreement with the Wasastjerna-Goldschmidt empirical values and also to lead to reasonably satisfactory theoretical values of various other atomic and ionic properties. During the next few years more refined and accurate theories of this sort were formulated, especially by Hartree, Fock, and Slater.

Max Born and other investigators had developed a basic theory of the structure of crystals composed of cations and anions in the period around 1920. With the reliable information about ionic radii and interionic forces that resulted from the application of quantum mechanics to the problem, it had become possible to discuss in a straightforward way the properties of crystals involving bonds with a large amount of ionic character. I found, as Goldschmidt and others had already pointed out, that many properties of ionic substances, such as the choice of structure, were determined by the relative sizes of cations and anions. W. L. Bragg in 1926 pointed out that many oxides and silicates have structures in which the oxygen atoms are arranged approximately in closest packing, with the metal atoms in tetrahedral or octahedral interstices. In 1928 I formulated a set of principles determining the structure of complex ionic crystals, including the silicate minerals. The most important of these principles is based on the idea that the valence of a cation, equal to its electric charge, is divided equally among the anions that are coordinated about it, and that the sum of the strengths of the ionic bonds (the ratio of the electric charge of the cation to its coordination number) reaching an anion is approximately equal to the negative charge of the anion. This rule is equivalent to saying that in a stable structure the nature of the ligancy is such that the lines of electric force between cations and ions have the minimum length.

Thus by 1928 it had become possible for the first time to say that the properties of substances with bonds of large ionic character were well understood as resulting from their structure.

The development of the theory of the covalent bond and of bonds of intermediate type was not quite so rapid. In 1916 Lewis had proposed that the chemical bond is a pair of electrons held jointly between two nuclei, and during the next three or four years Irving Langmuir developed this idea and applied it in many ways. It was difficult, however, to see how the electronic structures of molecules described by Lewis and Langmuir were related to the electronic structure of atoms as described by the old quantum theory on the basis of the analysis of line spectra. Many people, including Heisenberg and Pauli, attempted to apply the old quantum theory to the simplest molecule, the hydrogen molecule ion, but without success. Then in 1927 O. Burrau, in Copenhagen, published his quantum mechanical treatment of the hydrogen molecule ion. He obtained theoretical values for the bond energy, bond length, and vibrational frequency in excellent agreement with the spectroscopic values. E. U. Condon immediately published a treatment of the hydrogen molecule in which he assigned to it two electrons in the normal state as calculated by Burrau, with a semi-empirical evaluation of the mutual energy of the two electrons. W. Heitler and F. London carried out their treatment of the electron-pair bond in the hydrogen molecule at the same time. In the Heitler-London treatment the wave function represents first one assignment, and then the other, of the two electrons, with opposed spins, to the two nuclei. A satisfactory basic theory of the covalent bond had accordingly been developed by 1927.

During the next six years most of the puzzling problems about the chemical bond and the structure of molecules were resolved by the application of quantum mechanical principles. One simple aspect of chemical bonding, the equivalence of the four bonds of the tetrahedral carbon atom, had seemed to contradict the evidence from spectroscopy and quantum theory that the outer electron orbitals of the carbon atom are of two kinds, the $2s$ orbital and the three $2p$ orbitals. In 1928 I pointed out that a resonance structure, equivalent to the assignment of the four binding electrons to hybrid sp^3 orbitals, can be assigned to the carbon atom. The four sp^3 orbitals are equivalent to one another, and each one is directed to one of the corners of a regular tetrahedron. This idea accordingly brought the existing knowledge about the electronic structure of atoms as discovered by the physicists into agreement with the firmly based chemical theory of the tetrahedral carbon atom.

The idea of hybridization of bond orbitals was also discussed by John C. Slater in 1930, and again more thoroughly by me in 1931. Also, the recognition of a general principle of quantum mechanics—that wave functions corresponding to structures with the same character (especially the same number of unpaired electrons) can be combined linearly to a more satisfactory function—was found to lead directly to the answers to a number of other questions. For example, it was immediately seen that the transition from extreme covalent bonding to extreme ionic bonding can occur continuously if the extreme structures correspond to the same number of unpaired electrons. During the preceding decade there had been vigorous controversy as to whether a molecule such as hydrogen chloride should be assigned a structure involving the ions H^+ and Cl^- held together by electrostatic attractions or the atoms H and Cl held together by a shared-electron-pair bond. This question was answered: the molecule has an intermediate structure, which might be described as a resonance combination (a hybrid) of the two extreme structures. Also, it was seen that experimental values of the magnetic moment of molecules and ions provide information about the nature of the atomic orbitals involved in bond formation. The concept of the electronegativity of atoms, correlated with the amount of ionic character of their bonds, was then developed and the conditions for formation of unusual bonds, involving one or three shared electrons, were also formulated.

The theory of quantum mechanical resonance of molecules among several valence-bond structures constituted a major addition to the classical structure theory of organic chemistry. This theory was developed in the period from 1931 on by a number of investigators, including Slater, E. Hückel, G. W. Wheland, and me.

In the application of the classical structure theory of organic chemistry the effort was made to assign to the substance a molecular structure in which atoms are attached to one another by single bonds, double bonds, or triple bonds. The theory was thoroughly satisfactory for many substances, but it was unsatisfactory for benzene and other aromatic molecules and for molecules to which structures with alternating single and double bonds would conventionally be assigned (conjugated systems). To the benzene molecule, for example, the classical structure theory would assign a structure with alternating single and double bonds, but the substance does not show the easy addition of hydrogen characteristic of other molecules containing carbon-carbon double bonds, and, moreover, there had been early evidence that the six carbon-carbon bonds in the benzene ring are equivalent. In 1889 J. Thiele had suggested a theory of partial valences, and in the period around 1924 a number of investigators (F. Arndt, T. M. Lowry, H. J. Lucas, Robert Robinson, and especially C. K. Ingold and E. H. Ingold) suggested that a molecule in its normal state may have an intermediate or mesomeric structure resembling two or more valence-bond structures of the classical sort. The general principles of quantum mechanics provided a straightforward and acceptable way of describing these molecules. Slater had shown how to write a wave function for a molecule corresponding to a classical valence-bond structure, refining and simplifying earlier attacks on this problem by Heitler and London. It was clear from the minimum-energy principle of quantum mechanics that for some molecules, such as the benzene molecule, the normal state would not be represented by the wave function for any single valence-bond structure, but could be represented by a linear combination of these wave functions, with the individual contributions determined by the nature and stability of the structures. Such a molecule may be described as resonating among the several valence-bond structures. The properties of the molecule would then be expected to be a sort of average of the properties for the individual structures. Moreover, the minimum-energy principle requires that the actual normal state of the molecule be more stable than any of the hypothetical states corresponding to the individual valence-bond structures. This idea of resonance stabilization had not been discovered empirically before the development of quantum mechanics, although a great number of other aspects of chemical structure theory had been. The semi-empirical exploitation of the concept of resonance and other new concepts, such as stabilization through partial ionic character of bonds, was accomplished during the next few years. In addition, extensive quantum mechanical calculations about molecular structure, especially of aromatic and conjugated substances, were carried out in the 1930's, along lines initiated by Hückel.

An important factor in the rapid acceptance of the extended chemical structure theory was the application to many molecules of two approximate quantum mechanical methods during the 1930's (and, in refined form, in later years). These are the valence-bond method (called the Heitler-London-Slater-Pauling or HLSP method) and the molecular-orbital method (called the Hund-Mulliken-Hückel or HMH method).

These methods of calculation in their simple form lead to conclusions about molecular structure that are for the most part in good agreement with one another. Slater and I soon pointed out that these approximate theories, which give somewhat different results when applied in their simplest forms, become identical when they are refined by the introduction of ionic structures in the HLSP treatment and configuration-mixing in the HMH treatment. The correlation of calculated values of resonance energies and other molecular properties, such as bond length, with the empirical information was excellent.

A very important contribution was being made in these years to the understanding of the structure of simple molecules by the development of molecular spectroscopy, especially by F. Hund and Robert S. Mulliken. Also, the information about bond lengths and bond angles obtained by the electron-diffraction study of gas molecules was especially useful in providing evidence about the nature of the bonds in the molecules and the extent of agreement with the general theories formulated on quantum mechanical principles.

I had been hoping in the late 1920's that a method of determining interatomic distances and bond angles could be discovered that would be more generally and easily applicable than the X-ray study of crystals as it had been developed at that time. In 1930 I visited Herman Mark in Ludwigshafen and saw the apparatus that he and R. Wierl had constructed to determine the structure of gas molecules by the diffraction of electrons. This technique is somewhat similar to that of determining the structure of crystals by X-ray diffraction but makes use of a beam of electrons instead of X-rays, and a jet of gas in the evacuated apparatus instead of the crystal. Mark encouraged me to build a similar apparatus, and it was built by my student L. O. Brockway during the next couple of years. This technique permitted a large amount of experimental information about molecular structure to be gathered very quickly; during the following twenty-five years structures of the molecules of 225 substances were determined in our laboratory, and many more were determined, with the same technique, in other laboratories. In this as in my other work, I was fortunate in having the collaboration of many very able students and postdoctoral coworkers (as is described in my article "Fifty Years of Physical Chemistry in the California Institute of Technology," the prefatory chapter in the *Annual Review of Physical Chemistry* for 1965).

I remember clearly how much different my own thinking about molecular structure and the chemical bond was in 1935 from what it had been ten years earlier. In 1925 I had accepted the idea that the covalent bond consists of a pair of electrons shared jointly by two atoms. I had no sound theoretical basis, however, for any detailed consideration of chemical bonding. In that year I wrote a paper describing some hypothetical electronic structures for many molecules, involving electrons in elliptical orbits circling two nuclei, and I attempted to assign single valence-bond structures to benzene, naphthalene, the carbonate ion, and other molecules and ions. Many of my suggested structures were wrong: for example, I described structures as reasonable in which the carbon atom was surrounded by six shared electron pairs or by only three. Moreover, although I was fairly familiar with the existing knowledge about molecular structure, including interatomic distances and dissociation energies of molecules, and with the old quantum theory and atomic physics, I had no way of distinguishing between the good ideas and the poor ideas about the electronic structure of molecules. By 1935, however, I felt that I had an essentially complete understanding of the nature of the chemical bond. This understanding had been developed in large part through the direct application of quantum mechanical principles to the problem of the electronic structure of molecules, and also in large part through the formulation of new empirical principles, based upon the observed properties of substances (especially thermodynamic properties and bond lengths, bond angles, and other details of molecular configuration), and usually suggested by quantum mechanical considerations.

I think that the most important ideas were those of the hybrid character of bond orbitals (relative to the central-field orbitals of atomic theory) and the theory of resonance of molecules among two or more valence-bond structures, both of which have already been mentioned.

The theory of resonance led to a surprising development: it was subjected to vigorous criticism on ideological or philosophical grounds. The critics were following the example of the Soviet agriculturalist Lysenko, who for many years had great personal success through his advocacy of the rejection of modern genetics on ideological grounds. The criticism of the resonance theory began in 1949 with the publication of a paper by V. F. Tatevskiĭ and M. I. Shakhparanov entitled "About a Machistic Theory in Chemistry and Its Propagandists."[1] The criticism seems to have been based largely upon the fact that the contributing resonance structures do not have real independent existence. The criticism was repeated in 1950 in the "Report of the Commission of the Institute of Organic Chemistry of the Academy of Sciences, U.S.S.R., for the Investigation of the Present State of the Theory of Chemical Structure."[2] Essentially the same criticism of the theory was made by the German chemist W. H. Hückel and by L. H. Long, translator of the English edition, in the book *Structural Chemistry of Inorganic Compounds*. The criticism ended with the sentence "It must further never be forgotten that the theory ultimately depends upon the use of limiting structures which, by admission, have no existence in reality."[3]

This criticism rests upon a misunderstanding about the nature of the theory of resonance and of chemical structure theory as a whole. I have pointed out that the theory of resonance is no more artificial than the classical structure theory of organic chemistry, and that the contributing valence-bond structures in the theory of resonance are not more ideal (imaginary) than the structural elements of classical theory, such as the double bond.[4] For example, in the description of the molecule of cyclohexene, to which a single classical valence-bond structure is assigned by all chemists, the carbon-carbon single bond, the carbon-carbon double bond, and the other structural elements that are used are idealizations, having no independent existence in reality. There is no rigorous way of showing by experiment that two of the carbon atoms in the cyclohexene molecule are connected by a double bond, unless one uses a formal and vague definition of the double bond. We may say that the cyclohexene molecule is a system that can be shown by experiment to be resolvable into six carbon nuclei, ten hydrogen nuclei, and forty-six electrons, and it can be shown to have certain other structural properties, such as values 133 pm (picometers), 154 pm, and so on, for the average distances between nuclei in the molecule in its normal state; but it is not resolvable by any experimental technique into one carbon-carbon double bond, five carbon-carbon single bonds, and ten carbon-hydrogen single bonds; these bonds are theoretical constructs, idealizations, which have aided chemists during the past one hundred years in developing the convenient and extremely valuable classical structure theory of organic chemistry. The theory of resonance constitutes an extension of this theory. It is based upon the use of the same idealizations—the bonds between atoms—as used in classical structure theory, with the important extension that in describing the benzene molecule two arrangements of these bonds are used, rather than only one.

I think that the critics of the theory of resonance had not taken the simple action of making the foregoing comparison of the two theories, the theory of resonance and classical chemical structure theory, and did not recognize how closely related they are in their philosophical and logical basis. I do not understand either Ernst Mach's phenomenalistic positivism or Karl Marx's dialectical materialism well enough to know what their relation is to the theory of resonance in chemistry, and I doubt that anyone else understands these philosophies and chemical structure theories well enough. My skepticism on this point is supported by the fact that in his book *The Marxist Philosophy and the Sciences* J. B. S. Haldane referred to my work on the theory of resonance as a beautiful example of dialectical thinking in science;[5] I was astounded to learn later that other authorities thought quite the opposite.

While I was attending the Lomonosov celebration in 1961 I gave a public lecture in Moscow on the subject of the theory of resonance,[6] at the invitation of the Institute of Organic Chemistry of the Academy of Sciences, U.S.S.R. At that time I learned that a Commission of the Institute had in 1954 published a revised report on the state of the theory of chemical structure, in which most of the objections to the theory of resonance were withdrawn. There are still many chemists in the Soviet Union, however, who have an incomplete understanding of the present state of structure theory and are not able to apply it with confidence.

In 1935 there remained one branch of structural chemistry that was poorly understood—the structural chemistry of metals and alloys (including also the structural chemistry of other electron-deficient substances, such as the boranes). A free-conduction theory of metals had been advanced by H. A. Lorentz in 1916. The modern electronic theory of metals was initiated then by Wolfgang Pauli in 1927 in his discussion of the small temperature-independent paramagnetism shown by many metals, and it was further developed by Sommerfeld, Fermi, and many other investigators. This theory permits many properties of metals, especially the electric and magnetic properties, to be discussed in a moderately satisfactory way. It is, however, less satisfactory with respect to the choice of structure by metals and intermetallic compounds and to the relation of stability to composition of alloys.

In 1938 and the following years I attempted to develop a somewhat different treatment of metals and alloys, resembling the structure theories of inorganic and organic chemistry. Goldschmidt in 1926 had mentioned the close relation of the metal-metal interatomic distances in metals and the effective covalent radii of the metals in their compounds, and J. D. Bernal in 1928 had emphasized the structural similarity of metals and intermetallic compounds to covalent compounds. My own work led me to the conclusion that metals and intermetallic compounds contain covalent bonds that resonate among many interatomic positions, with the substances assuming structures such as to give atoms the maximum ligancy permitted by their relative sizes. The number of bonds is determined by the metallic valence, which for transition elements is larger than the stable oxidation numbers. The metallic valence is for some elements restricted by the number of available bond orbitals. A curious structural feature seems to be determinative for metallic character—the presence of an extra orbital for most of the metal atoms (about 72 per cent). The significance of this extra orbital, the metallic orbital, is that it permits the resonance of the valence bonds to be largely uninhibited, by permitting a large fraction of the atoms to have one more or one less electron than the average. This largely uninhibited resonance gives rise to the observed extra stability of metals and intermetallic compounds, and it accounts for the electronic conductivity of the substances.

This chemical theory of metals and intermetallic compounds, which seems to me to be reasonably satisfying, has not attracted much attention from the physicists and has not been brought into good correlation with the conventional energy-band (free-electron) theory. In particular, there has not been developed, so far as I am aware, any energy-band theory of metals and intermetallic compounds in which the significance of the metallic orbital is made evident. The structure theory of metals and alloys, as well as that of other electron-deficient substances, must accordingly be described as still far less complete than the structure theories of other branches of chemistry.

The rapid development in the study of the nature of the chemical bond and the structure of molecules and crystals that occurred during the decade following 1925 was a direct result of the discovery of quantum mechanics in that year, which made an impact on scientific thinking in several ways. First, it became possible to answer certain questions; for instance, the nature of the one-electron bond in the hydrogen molecule-ion could be determined by the straightforward solution of the quantum mechanical equations for the system. Also some new ideas, such as the concept of quantum mechanical resonance (which had been introduced by Heisenberg in 1926 in his treatment of the helium atom), suggested the possibility of formulating new semi-empirical structural principles compatible with quantum mechanics and supported by agreement with the facts obtained by experiment. Third, scientists developed a new attitude toward those aspects of the world that had not been explained: before 1925 it was often said that the failure to find a satisfactory theoretical explanation of some observed facts could be attributed to the defects of the old quantum theory; after the middle of 1925 this excuse was not acceptable, and a search was made for the reason for the failure. This search led, for example, to the discovery of the spin of the electron by G. E. Uhlenbeck and S. Goudsmit in 1925 in the course of their effort to resolve the mystery of the two sets of azimuthal quantum numbers, mentioned above; it also led to the discovery of the spin of the proton and the slow equilibrium between two forms of molecular hydrogen (ortho-H_2 and para-H_2) in 1927 by D. M. Dennison in the course of his effort to explain the observed heat capacity of hydrogen gas at low temperatures. This third type of impact of quantum mechanics on scientific thought was, I think, as valuable as the other two, perhaps even more valuable in its relation to molecular biology.

My serious interest in what is now called molecular biology began about 1935. I had started to learn something about biology in 1929, when Thomas Hunt Morgan came to the California Institute of Technology, bringing with him a number of younger members of the new Biology Division. This Division was strong in genetics, and by 1931 I had become interested enough in genetics to present a seminar describing a theory of the crossing-over of chromosomes that I had developed. Our work on the structure of organic molecules, as well as inorganic, had been progressing rapidly, and by 1935 I had begun to speculate more generally about the properties of the large molecules found in living organisms. In 1936 I published a paper on a structural theory of the oxygen equilibrium of hemoglobin, and also began the study of the magnetic properties of hemoglobin, in collaboration with Charles D. Coryell. I had initiated the work on magnetic properties of hemoglobin in order to find by experiment whether or not the two unpaired electrons of the oxygen molecule remain unpaired when oxygen combines with hemoglobin. The result of our investigation was a surprise: we found that the oxygen molecule no longer has its unpaired electron spins in oxyhemoglobin, but we also found that each of the four iron atoms of the hemoglobin molecule has four odd electrons in hemoglobin itself, and none in oxyhemoglobin or carbonmonoxyhemoglobin. These studies of the magnetic properties of hemoglobin and its compounds led to a great increase in understanding of the structure of the hemoglobin molecule in the neighborhood of the heme groups.

There were two immediate personal consequences of this hemoglobin work: I became interested in the general problem of the structure of proteins and also in the problem of the structure of antibodies and the nature of serological reactions.

Coryell and I consulted for a year with Alfred Mirsky, who was in Pasadena on leave from the Rockefeller Institute for Medical Research. Mirsky had had extensive experience with proteins and was interested in the problem of the difference between a native protein and a denatured protein. After we had discussed the various studies that had been made of proteins and protein denaturation, Mirsky and I, in 1936, published a paper in which we presented a general theory of protein structure, to the effect that the polypeptide chain is coiled in a specific configuration in the native molecule, stabilized largely by hydrogen bonds between one part and another of the chain, and that denaturation consists in the loss of a well-defined configuration and the assumption of more random configurations by the polypeptide chains.

At about that time many people became interested in the problem of the three-dimensional structure of proteins. Early X-ray studies of silk fibroin and beta-keratin (stretched hair) suggested that the polypeptide chains are in a stretched-out configuration, and the work of

W. T. Astbury in England on alpha-keratin was interpreted as indicating that the polypeptide chains are coiled in some way. There was, however, even some uncertainty about the presence of polypeptide chains. The idea that hexagonal rings might be present in proteins had been proposed by H. S. Frank in 1933 and was strongly advocated by D. M. Wrinch and Irving Langmuir (the cyclol theory). Carl Niemann and I published a paper in 1939 in which we marshaled the evidence of various sorts (X-ray diffraction, thermochemical bond energies) for the polypeptide-chain structure and against the cyclol theory. Our X-ray work with crystals and electron-diffraction work with gas molecules had provided a large amount of information about bond lengths in simple molecules, and I decided to see to what extent this information could be made compatible with the X-ray diffraction diagrams of the fibrous proteins. Although no direct experimental information was available about the molecular dimensions of simple peptides or closely related substances, I thought that the general structure theory should permit predictions to be made not only of the values of bond lengths and bond angles, but also about the formation of hydrogen bonds and the planarity of the amide group. I spent the summer of 1937 in an unsuccessful effort to find a way of coiling a polypeptide chain in three dimensions compatible with the apparent identity distance of alpha-keratin, 510 pm, as reported by Astbury. This effort was unsuccessful, which led me to conclude that I was making some unjustified assumption about the structural properties of the molecules. Eleven years later I discovered that this conclusion was wrong; the structural properties that I had assumed were right, but the apparent identity distance 510 pm reported by Astbury turned out not to be a true identity distance, but a misinterpretation of the X-ray pattern.

In October 1937 Dr. Robert B. Corey, who had just come to Pasadena from the Rockefeller Institute for Medical Research and who also was interested in the structure of proteins, talked with me about the possibility of making a successful attack on this structural problem. At that time no correct crystal structure determination had been carried out for any amino acid or any simple peptide or any other substance closely related to proteins. We decided that it was worthwhile to attack the problem of the structure of the amino acids and related substances by the X-ray investigation of their crystals, with the hope that the structural parameters could be determined with greater accuracy and that any unknown structural feature that might be present in proteins could be identified in the simpler substances. During the next year Corey determined the structure of diketopiperazine, a simple cyclic dipeptide, and he and others in our laboratory then determined the structures of a good number of amino acids and other peptides. By 1948 it had become clear that there was nothing surprising about the dimensions of these molecules, and that the dimensions that I had assumed eleven years earlier for the polypeptide chain were to be accepted as correct.

In the spring of 1948 I attacked again the problem of coiling the polypeptide chain and found that two structurally satisfactory helical configurations could be formulated. These configurations, the alpha helix and the gamma helix, have a nonintegral number of residues per turn of the helix, and successive turns are connected by hydrogen bonds. I was assisted in the study of these structures by Professor Corey and Dr. Herman R. Branson. In 1950 Professor Corey and I published a brief description of the helical structures, and a more detailed description was published by us and Dr. Branson a few months later.

We were not alone in the effort to discover a satisfactory structure of the fibrous proteins. Dr. M. L. Huggins had been active in this field, as had also W. L. Bragg, John Kendrew, and M. Perutz, but they had been led to formulate only incorrect structures. Our success resulted, I think, from our great emphasis on the preservation of the known structural features, including planarity of the amide group, and our rejection of the idea, taken over from crystallography, that an acceptable helical structure should have an integral number of residues per turn. The alpha helix, which has been found to be one of the most important ways of folding the polypeptide chain in proteins, has about 3.61 residues per turn.

Our delay of two years in publishing a description of these helical structures was the result of the difficulty that we had in correlating the alpha helix with the X-ray diffraction pattern of alpha-keratin (hair, horn, fingernail, muscle). The pitch (axial distance per turn) of the alpha helix is 540 pm, which is not in satisfactory agreement with the value 510 pm that seemed to be given for the identity distance by the X-ray diffraction patterns of hair and related substances. Later it was recognized that the axes of the alpha helixes in hair deviate somewhat from the fiber direction. The repeat distance for synthetic polypeptides was found to be 540 pm, in good agreement with the value for the alpha helix.

We also made a careful study of two forms of silk-fibroin and were able to show that these fibrous proteins have a pleated-sheet structure, agreeing closely with the configuration predicted from the structural features of the polypeptide chain.

Recent work on the determination of the structure of globular proteins, such as myoglobin, studied by Kendrew and his collaborators, has shown that the polypeptide chain in these proteins contains segments with the alpha-helix structure. Many globular proteins are now being subjected to detailed structural analysis by the X-ray diffraction method, and there is no doubt that a completely satisfactory structural basis of the activity of many enzymes and other physiologically active proteins will be obtained during the coming years. Thus over a period of thirty years I have witnessed the change from almost complete ignorance to a significant amount of knowledge and understanding of the structure of proteins.

My ideas about the structural basis of biological specificity were developed during the years 1936 to 1939. I had, of course, speculated in a rather vague way about the possible mechanisms of transfer of hereditary characters from one generation to the following generation of a species of living organisms, and had thought about some possible explanations of the observed self-sterility in the sea urchin, which Morgan had been studying. In May 1936, after I had given a seminar on hemoglobin at the Rockefeller Institute, Karl Landsteiner asked me to come to his laboratory to talk with him about immunology. He told me about serological phenomena, including his observations on the properties of synthetic antigens (azoproteins with haptenic groups such as benzenearsonic acid[7]) and asked how I would explain these phenomena in terms of molecular structure. I had no explanation; but a few days later I read Landsteiner's book *The Specificity of Serological Reactions* and some other publications in this field, and began a serious effort to think about this problem. In the fall of 1937, while I was serving as George Fisher Baker Lecturer in Chemistry in Cornell University, Landsteiner visited there for several days, which were devoted to giving me a thorough survey of the field of serology and Landsteiner's opinions about the reliability of some apparently contradictory experimental observations by different people.

At that time I found that Landsteiner and I had a much different approach to science: Landsteiner would ask, "What do these experimental observations force us to believe about the nature of the world?" and I would ask, "What is the most simple, general, and intellectually satisfying picture of the world that encompasses these observations and is not incompatible with them?" I think that my attitude can be described as essentially that of the theoretical physicist.

By 1939 I had reached some conclusions about the structure of antibodies and the nature of their interaction with antigens. With Landsteiner's encouragement I published a paper on this subject in 1940 and, with the collaboration of Dan H. Campbell, David Pressman, and a number of students and young postdoctoral fellows, began a series of experiments to check on the ideas.

The existing information in 1939 indicated clearly that the forces operating between the combining regions of antibodies and the haptenic groups of antigens are weak intermolecular forces, including electronic van der Waals attraction, the formation of hydrogen bonds, and the attraction of oppositely charged groups, such as the carboxylate ion and the ammonium ion. In order for these weak forces to constitute an effective bond between antibody and antigen it is necessary that there be a close complementariness in structure between the combining group of the antibody and the haptenic group of the antigen. The idea that antibody and homologous antigen have complementary structures had been suggested in the period 1930 to 1932 by F. Breinl and F. Haurowitz, Jerome Alexander, and Stuart Mudd. There was some intimation of it in the early work of Ehrlich and Bordet. Landsteiner's observations seemed to me to provide strong support for the idea, and in our laboratory during the years 1940 to 1948 we gathered additional experimental information, by the study of cross reactions for hundreds of chemically similar haptens, that supported the idea of complementariness so strongly as to require its acceptance. We were able to show that the fit between the combining groups of antibody and antigen is such as to bring the atomic surfaces to within a fraction of an atomic diameter of one another over an area of ten or twenty atoms, and, moreover, that complementary hydrogen-bond-forming groups are present, and groups with opposite electrical charges, in the relative positions to give a significant contribution to the binding energy.

For example, we found by our quantitative studies of the binding powers of various specific antibodies with many different haptens (verifying Landsteiner's qualitative observations) that antibodies can distinguish between a molecule containing a hydrogen atom (with van der Waals radius 115 pm) and a similar molecule containing a bromine atom (radius 195 pm) or a methyl group (rather knobby, average radius about 200 pm) in place of the hydrogen atom, but could not distinguish between the bromine atom and the methyl group. We concluded from this observation and many other similar ones that the closeness of fit of the combining region of the antibody to the haptenic group is better than to within about 50 pm, half the radius of an atom.

In my 1940 paper I had supported the idea, proposed by J. R. Marrack and by M. Heidelberger, that precipitating and agglutinating antibodies are multivalent, and I had further proposed, as the simplest way of explaining their properties of forming precipitates and agglutinates, that their multivalence is restricted to two. Our later experimental results verified the value two for the number of combining groups per molecule for precipitating and agglutinating antibodies.

I also supported the idea that the combining regions of antibodies consist of polypeptide chains with such amino-acid sequence as to permit the assumption of any one of a large number of spatial configurations, with nearly the same energy, in the absence of the haptenic group of the immunizing antigen; further, that the specific combining power resulted from the folding of the chain of the antibody in the presence of the antigen into a configuration complementary to that of the combining group of the antigen, and that this configuration had greater stability than other configurations by virtue of the interaction with the antigen.

It is now known, of course, that antibodies with different specificities have different sequences of amino-acid residues in the chains that produce the combining regions, so that my assumption that antibodies have the same sequence of amino-acid residues is known to be wrong. The mechanism of stimulation of cells to form clones (groups of genetically identical cells) that manufacture antibody molecules homologous to an injected antigen is, so far as I know, not yet known; I feel confident that this stimulation results from the combination of the antigen with a globulin molecule with a complementary structure in its combining group.

In July 1940 Max Delbrück and I published a paper in *Science* with the title "The Nature of the Intermolecular Forces Operative in Biological Processes." During the preceding two years the German physicist P. Jordan had published several papers in which he advanced the idea that there exists a quantum mechanical stabilizing interaction, the resonance phenomenon, that operates preferentially between identical or nearly identical molecules or parts of molecules and is able to influence the process of biological molecular synthesis in such a way that replicas of molecules present in the cell are formed. He used the idea in suggesting explanations of the reproduction of genes, the growth of bacteriophage, the formation of antibodies, and other biological phenomena showing specificity. The novelty in his work lay in his suggestion that quantum mechanical resonance would lead to attraction between molecules containing identical groups great enough to cause autocatalytic reproduction of the molecules. Delbrück and I analyzed the argument and showed that it was not possible for resonance stabilization in living organisms to operate in such a way as to explain the autocatalytic reproduction of molecules and other phenomena of biological specificity. We said that biological specificity must instead result from the existence of molecules with complementary structure, and that autocatalysis, as in the duplication of genes, is also to be explained in terms of two mutually complementary molecules.

During the next few years I amplified the arguments about molecular complementariness and biological specificity in relation to crystallization, the activity of enzymes (complementariness of the active region of the enzyme to the transition state configuration of the molecules involved in the reaction that is catalyzed), and the structure of genes. In my Royal Institution lecture on February 27, 1948, I repeated some of my earlier statements, in the following words:

I believe that the same mechanism, dependent on a detailed complementariness in molecular structure, is responsible for all biological specificity. I think that enzymes are molecules that are complementary in structure to the activated complexes of the reactions that they catalyse, that is, to the molecular configuration that is intermediate between the reacting substances and the products of reaction for these catalysed processes. The attraction of the enzyme molecule for the activated complex would thus lead to a decrease in their energy, and hence to a decrease in the energy of activation of the reaction and to an increase in the rate of the reaction. Although convincing evidence is not yet at hand, I believe that it will be found that the highly specific powers of self-duplication shown by genes and viruses are due to the same intermolecular forces, dependent upon atomic contact, and the same processes of replica formation through complementariness in structure as are operative in the formation of antibodies under the influence of an antigen. I believe that it is molecular size and shape, on the atomic scale, that are of primary importance in these phenomena, rather than the ordinary chemical properties of the substances, involving their power of entering into reactions in which ordinary chemical bonds are broken and formed.

Three months later, on May 28, 1948, in the Sir Jesse Boot Foundation Lecture "Molecular Architecture and the Processes of Life" that I gave in Nottingham, I amplified the argument as follows:

This concept thus gives us an automatic method of producing a substance with a specific biological property, that of combining with the molecules of the antigen. The mechanism of obtaining this property is one of moulding a plastic material, the coiling chain, into a die or mould, the surface of the antigen molecule. I believe that the same process of moulding of plastic materials into a configuration complementary to that of another molecule, which serves as a template, is responsible for all biological specificity. I believe that the genes serve as the templates on which are moulded the enzymes that are responsible for the chemical characters of the organisms, and that they also serve as templates for the production of replicas of themselves. The detailed mechanism by means of which a gene or a virus molecule produces replicas of itself is not yet known. In general the use of a gene or virus as a template would lead to the formation of a molecule not with identical structure but with complementary structure. It might happen, of course, that a molecule could be at the same time identical with and complementary to the template on which it is moulded. However, this case seems to me to be too unlikely to be valid in general, except in the following way. If the structure that serves as a template (the gene or virus molecule) consists of, say, two parts, which are themselves complementary in structure, then each of these parts can serve as the mould for the reproduction of a replica of the other part, and the complex of two complementary parts thus can serve as the mould for the production of duplicates of themselves.

At about that time I became interested in the problem of the structure of deoxyribonucleic acid (DNA). Several investigations, such as the work of O. T. Avery, M. McCarty, and C. M. McLeod on the transformation of pneumococcus types, had indicated strongly that DNA is the bearer of genetic information, and Alexander Todd and his co-workers had just succeeded in discovering the positions on the deoxyribose residue to which the phosphate group and the purine and pyrimidine groups are attached. I began an effort to determine the structure of DNA by analysis of X-ray diffraction photographs of DNA fibers. I made use of rather poor photographs that had been published by W. T. Astbury and F. O. Bell in 1938 and of some equally poor photographs made in our laboratory. M. H. F. Wilkins and his co-workers in Kings College, London, in 1951 published mention of much better photographs that they had prepared, of more highly crystalline material, but my efforts to obtain prints of these photographs (by writing to Wilkins) were not successful.*

Our rather poor X-ray photographs, which in fact represented the superposition of two patterns, corresponding to two forms of DNA fibers, led me to the conclusion that the fibers contained helical complexes of three DNA molecules, twisted about one another, instead of the double helix that I expected on the basis of the arguments about complementariness given in my earlier papers. Although I knew about the work of Erwin Chargaff, who had shown that pairs of purines and pyrimidines are present in equal numbers in DNA, the idea of hydrogen-bonded complementary structures involving pairing of adenine with thymine and of guanine with cytosine did not occur to me. I thought that the three-chain structure that Corey and I described in February 1953 probably represented an artifact, produced in the process of fiber formation and not showing the DNA molecules in their physiologically active state. A few months later J. D. Watson and F. H. Crick reported the double-helix structure for DNA that they had discovered by the use of

---

* Editor's Note. At the conference interest was expressed on this point; Linus Pauling was not present but Robert Olby suggested that when Pauling had requested prints of X-ray diffraction photographs of DNA fibers, Wilkins replied that he had not reached the stage when he wished to show them. Olby continued to say, however, that Pauling had planned to attend a meeting in England in 1952 and was arranging to visit Wilkins' laboratory. At the last minute, Pauling was denied a passport by the U. S. government. Dr. Olby felt sure that if Pauling had made the trip, Wilkins would have shown him something. He also said that most people believe that Wilkins, given sufficient time, would have arrived at the double helix structure for DNA.

Olby would like to add (June 1970) that he has recently seen the X-ray diffraction photographs taken after Pauling's suggested structure for DNA had been published. These photographs were taken by Alexander Rich, a postdoctoral fellow working at Caltech with Pauling. Though taken with a more primitive camera than those available later, they show considerable detail, at least in the light of later knowledge. If a camera designed for fiber work had been available, plus an appropriate X-ray source—or if Pauling had had access to Wilkins' photographs—he would then no doubt have had the data in his hands on which to base a correct model for the structure of DNA.

Wilkins' X-ray photographs, and the period of tremendous activity in DNA molecular biology was begun.

In their work Watson and Crick made use of the same principles of molecular structure that I had used in the discovery of the alpha helix for proteins. For some reason that is not very clear they assumed that two hydrogen bonds, different in character, are formed between cytosine and guanine, as well as between adenine and thymine. Corey and I then pointed out that there is strong evidence for three hydrogen bonds between cytosine and guanine, and this small refinement of the double helix was immediately accepted.

The discovery of the DNA structure by Watson and Crick has had a tremendous impact on molecular biology. The simplicity of the structural complementariness of the two pyrimidines and their corresponding purines was a surprise to me—a pleasant one, of course, because of the great illumination that it threw on the problem of the mechanism of heredity and also because for years I had emphasized the significance of hydrogen-bond formation and of complementariness in relation to gene duplication and other phenomena involving biological specificity. I was surprised at the simplicity of the DNA structure, because the studies of complementariness of antigens and antibodies that we had carried out had shown that for the combining regions of these molecules the complementariness is not of a simple sort, but involves cooperation of various weak forces of different kinds, including hydrogen bonds. DNA is, of course, far simpler than proteins: it is constructed of only four different structural units, whereas proteins are constructed of about twenty units, the amino-acid residues.

Here again, in this field of biological specificity, as in the field of the nature of the chemical bond and the explanation of the physical and chemical properties of substances on the basis of their molecular structure, I feel now a great satisfaction in the extent to which, over a period of about thirty years, we have passed through the transition from the condition of almost complete lack of understanding, with the possibility that mysticism or vitalism might be operating in living organisms, to the condition of having a thoroughly satisfying understanding of many of the properties of living organisms in terms of the structure of the molecules of which they are composed. There are, of course, many aspects of life that are not yet accounted for in a detailed way on a molecular basis; but our experience during the last thirty years leads me to believe that with each passing decade the field of our understanding will encompass more and more of the phenomena that are not yet understood on a molecular basis.

For about twenty years I have been deeply interested in understanding the structure and function of the brain, and I have worked on a molecular theory of general anesthesia and also on the problem of the molecular basis of disease, especially mental disease. Many other people are also working in these fields at the present time, and I am confident that great progress will be made, which will not only contribute to our intellectual satisfaction but will also make possible a significant decrease in the amount of human suffering through the prevention and amelioration of mental disease as well as physical disease.

My active effort in the field of the molecular basis of disease began in 1945. On the evening of February 6 I was having dinner in the Century Club in New York with the half-dozen other members of a committee on medical research that later contributed a section to the Bush Report, "Science, the Endless Frontier." One of the members of the group, Dr. William B. Castle, described some work that he was doing on the disease sickle-cell anemia. When he mentioned that the red cells of patients with the disease are deformed (sickled) in the venous circulation but resume their original shape in the arterial circulation, the idea occurred to me that sickle-cell anemia was a molecular disease, involving an abnormality of the hemoglobin molecule determined by a mutated gene. I thought at once that the abnormal hemoglobin molecules that I postulated to be present in the red cells of these patients would have two mutually complementary regions on their surfaces, such as to cause them to aggregate into long columns, which would be attracted to one another by van der Waals forces, causing the formation of a needle-like crystal which, as it grew longer and longer, would cause the red cell to be deformed and would thus lead to the manifestations of the disease.

I asked one of my students, Dr. Harvey Itano, who had already received his M.D. degree and was working for a Ph.D. in chemistry, to get some blood from a patient and to investigate it, to determine whether or not the hemoglobin was different from that in ordinary blood. The problem was at that time a technically difficult one, but by 1949 Dr. Itano, with the assistance of Dr. S. J. Singer and Dr. I. C. Wells, had solved it. The hemoglobin turned out to be different in its electrophoretic properties from normal adult human hemoglobin, and in the course of time the abnormality was found to involve a point mutation in the gene responsible for the synthesis of the beta chains of hemoglobin, which leads to the substitution of one of the 146 amino-acid residues in this chain by a residue of a different sort. Our paper reporting this discovery had the title "Sickle-cell Anemia, a Molecular Disease." This was, I think, the first time that the expression "molecular disease" had been used. Dr. Itano and his co-workers soon discovered other abnormal human hemoglobins, and over one hundred are now known. Abnormal proteins of other sorts manufactured by certain human beings, as well as by other animals, have also been discovered.

I think that there were several reasons for my having immediately reached the conclusion that sickle-cell anemia is a molecular disease, a disease of the hemoglobin molecule. First, I had for many years worked on the structure and properties of hemoglobin, and I knew that hemoglobin and water are the principal constituents of erythrocytes. I also knew, from the work of Brown and Reichart on the crystallographic form of crystals of hemoglobin from animals of different species and from the work of Landsteiner on the serological properties of the hemoglobins, that there are a great number of kinds of hemoglobin molecules, presumably one for each animal species, and I was sure that these different kinds of hemoglobin molecules are manufactured under genetic control. The properties of antibodies and the mechanism of serological precipitation had led me to look for aggregation of molecules with complementary surface regions as the cause of precipitation, and the sort of deformation observed for sickled red cells—an elongated or crescent shape—suggested as a mechanism the formation of a needle-like crystal of hemoglobin. The possibility that oxygenation of the hemoglobin would destroy the complementariness, by introducing steric hindrance to the apposition of the mutually complementary regions, resulted from knowledge of similar phenomena in antibody-antigen interactions.

During recent years I have become interested in a novel aspect of molecular medicine: the possibility that substances that are normally present in the human body and that in some cases are required for life may be used in the treatment of disease. Two well-known examples are the use of insulin for the control of diabetes and the restriction of intake of phenylalanine for the control of phenylketonuria. I have become especially interested in the possibility that some of the vitamins may be used to improve the health and wellbeing of human beings by a large change in their concentration in body fluids. Vitamins are physiologically active substances, necessary for life. If the intake of a vitamin becomes low a person becomes ill with a disease described as an avitaminosis. A certain intake, called the minimum daily requirement or recommended daily intake, suffices to prevent overt manifestation of the avitaminosis. For ascorbic acid, for example, this amount is about 50 milligrams per day for an adult. But ascorbic acid is essentially nontoxic. Ingestion of an amount one hundred or even one thousand times 50 milligrams is not harmful. There is the possibility that improvement in health would result from the regular ingestion of perhaps one hundred times the so-called daily requirement, and that for some persons, of special genotype, the improvement in physical or mental health resulting from this sort of megavitamin therapy would be great.

My ideas about mental disease have been presented in a paper to which I gave the title "Orthomolecular Psychiatry" (*Science,* 160

[1968], 265-271). In that paper I say that orthomolecular psychiatric therapy is the treatment of mental disease by the provision of the optimum molecular environment for the mind, especially the optimum concentrations of substances normally present in the human body, such as the vitamins and the essential amino acids. The argument is summarized in the concluding paragraph of the paper:

> The functioning of the brain is affected by the molecular concentrations of many substances that are normally present in the brain. The optimum concentrations of these substances for a person may differ greatly from the concentrations provided by his normal diet and genetic machinery. Biochemical and genetic arguments support the idea that orthomolecular therapy . . . may be the preferred treatment for many mentally ill patients. Mental symptoms of avitaminosis sometimes are observed long before any physical symptoms appear. It is likely that the brain is more sensitive to changes in concentration of vital substances than are other organs and tissues. Moreover, there is the possibility that for some persons the cerebrospinal concentration of a vital substance may be grossly low at the same time that the concentration in the blood and lymph is essentially normal. A physiological abnormality such as decreased permeability of the blood-brain barrier for the vital substance or increased rate of metabolism of the substance in the brain may lead to a cerebral deficiency and to a mental disease. Diseases of this sort may be called localized cerebral deficiency diseases. It is suggested that the genes responsible for abnormalities (deficiencies) in the concentration of vital substances in the brain may be responsible for increased penetrance of the postulated gene for schizophrenia, and that the so-called gene for schizophrenia may itself be a gene that leads to a localized cerebral deficiency in one or more vital substances.

Molecular medicine, as described above, may turn out to be of great practical value. It may be considered to be a branch of molecular biology, and investigations in this field may lead to greatly increased understanding of the nature of living organisms, in terms of the molecules of which they are composed, and especially of the nature of memory and consciousness.

I conclude this discussion of my association with the development of structural chemistry and molecular biology during the last fifty years by saying that the great curiosity that I had a half-century ago about the nature of the physical and biological world has been in part satisfied in a manner that has given me much happiness. In the course of the developments that have provided answers to many interesting questions about both the inorganic world and the organic world, however, new questions have made themselves evident, and we can now transfer our curiosity to them. The existing theories of the structure of atomic nuclei are not very satisfying to me; I hope that it will be possible to obtain greater insight in this field. Also, the rapidly developing subject of the fundamental particles and their relationship to one another is certain to excite our interest for many years to come. The problem of the functioning of the brain and the nature of thinking, briefly mentioned above, is one that will without doubt stimulate exciting discoveries in the future.

In the course of our work on the amino-acid sequences in hemoglobin molecules I collaborated with Emile Zuckerkandl in 1962 in the discussion of molecular disease, evolution, and genic heterogeneity. We found that the amino-acid sequences in hemoglobin molecules manufactured by animals of different species provide strong evidence about the evolution of species, supporting the evolutionary history as formulated by paleontologists on the basis of macroscopic characters. We were led to make the statement that "Once more Biology will show what it can do without any 'élan vital' or entelechy." This experience and my other experiences during the last fifty years, involving the ever-increasing understanding of the world on the basis of rational principles, have led me to reject all dogma and revelation, all authoritarianism. It is possible that the greatest contribution of the new world view that has resulted from the progress of science will be the replacement of dogma, revelation, and authoritarianism by rationality—even greater than the contribution to medicine or that to technology.

REFERENCES

1. V. F. Tatevskiĭ and M. I. Shakhparanov, "About a Machistic Theory in Chemistry and Its Propagandists," trans. Irving S. Bengelsdorf, *Journal of Chemical Education,* 29 (1952), 13.

2. D. N. Kursanov, M. G. Gonikberg, B. M. Dubinin, M. I. Kabachnik, E. D. Kaverzneva, E. N. Prilezhaeva, N. D. Sokolov, and R. Kh. Freidlina, "The Present State of the Chemical Structural Theory," trans. Irving S. Bengelsdorf, *Journal of Chemical Education,* 29 (1952), 2; see also I. M. Hunsberger, "Theoretical Chemistry in Russia," *Journal of Chemical Education,* 31 (1954), 504.

3. W. H. Hückel, *Structural Chemistry of Inorganic Compounds,* trans. L. H. Long (New York: Elsevier Publishing Company, Inc., 1950), I, 437.

4. L. Pauling, "The Nature of the Theory of Resonance," in Sir Alexander Todd, ed., *Perspectives in Organic Chemistry* (New York: Interscience, 1956), pp. 1-8.

5. J. B. S. Haldane, *The Marxist Philosophy and the Sciences* (New York: Random House, 1939), p. 101.

6. L. Pauling, "The Theory of Resonance in Chemistry," *Journal of the Mendeleev All-Union Chemical Society,* 7 (1962), 462.

7. Haptens are the active groups of antigens. When free, they cannot cause cells to produce antibodies although, like antigens, they can neutralize antibodies once formed.

SAUL BENISON

The History of Polio Research in the United States: Appraisal and Lessons

There is a fond belief that not too long ago, somewhere between 1890 and 1940, there was a golden age when medical scientists were free to think and work in universities, medical schools, and hospitals, unburdened by the requirements and plans arranged by an overbearing bureaucracy either of the federal government or of private foundations and voluntary health agencies. The belief is illusory. These agencies have long been central to the development of medical science in the United States. Their importance does not stem solely from their distribution of research funds, but also from the fact that their relationships to medical scientists have always been more than a fiduciary one. The history of polio research in the United States reveals that the development of the transforming conceptions of that research, as well as of other significant biological research, was in large measure fostered by a scientific environment created by the plans, programs, and activities of private foundations and voluntary health agencies.[1]

I

Although polio as a disease entity was recognized in Europe early in the nineteenth century, it did not appear in epidemic form in the United States until 1894.[2] In the decade that followed, seventeen epidemics of varying degrees of severity occurred in every region of the country. In spite of the increasing frequency of these epidemics, American physicians and public health officials paid scant attention to the disease. During this period it was not even a reportable disease.[3] If reporting had been made mandatory, it is doubtful whether physicians could have complied with such regulations. Many physicians simply could not make a diagnosis of polio and all too often confused it with other diseases that had similar symptoms. In part these difficulties were the result of poor medical education and training. In part they existed because there was little hard information about the disease itself. Although European investigators during the early years of the twentieth century uncovered many important features of the pathology and epidemiology of polio, their findings, by and large, remained unavailable to most American physicians. There were several reasons for this state of affairs. First, many of these early discoveries were initially published in either French or German, and hence largely useless to the majority of monolingual American physicians. Second, and equally important, because epidemic polio was then still a relatively new phenomenon, these findings were not yet integrated or discussed extensively in the medical textbooks which American physicians in the main depended on for information. Third, there was no experimental polio research in the United States to make use of or foster interest in these findings.[4]

The first systematic effort to study polio experimentally in the United States was begun during the fall of 1907 in Simon Flexner's laboratory at the Rockefeller Institute for Medical Research following an unusually severe epidemic in the northeastern United States.[5] Both Flexner and the Rockefeller Institute merit further examination because they are inextricably linked with the subsequent progress and development of polio research.

When Simon Flexner died in 1946, at the age eighty-three, he was full of honors.[6] He was widely recognized as one of the nation's leading scientific administrators, and revered as a medical discoverer who had made important contributions to an understanding of the pathology and bacteriology of more than a score of diseases. Those who knew him when he was growing up in Louisville would have been surprised. Never did a scientific career begin more inauspiciously.

As a boy Simon Flexner was considered dull. His formal education ended at the age of fourteen, when, with his parents' consent, he was apprenticed to a plumber. At the end of a week the plumber returned him to his father with the blunt evaluation that he was too dumb to be a plumber. Thereafter he held a series of short-term jobs and finally was apprenticed to a local druggist. The new apprenticeship proved a turning point. His master introduced him to the mysteries of chemistry and encouraged him to continue his education at a local school of pharmacy. At the age of nineteen he completed these studies and took a job with his older brother Jacob, who also maintained a drugstore. Here he compounded medicines and on occasion helped his brother make microscopic analyses of blood and urine for local physicians. Although the work was profitable, Flexner found the life of a drug clerk boring and, at the age of twenty-four, decided to become a physician. In 1887 he entered the University of Louisville Medical School. Two years later, after a program of study that included a course in anatomy, occasional lectures in medicine, surgery, and physiology, and two lectures in bacteriology, he received the M.D. degree. The value of his medical education was perhaps best summed up by Flexner himself in his unpublished autobiography: "I did not learn to practice medicine, indeed, I cannot say that I was particularly helped by the school. What it did for me was give me the M.D. degree."[7]

Realizing the worthlessness of his degree for medical practice and stimulated by a reading of John Tyndall's *Floating Matter of Air,* Flexner decided to become a pathologist. In 1890, at the suggestion of his younger brother Abraham, a graduate of Johns Hopkins University, he entered Johns Hopkins Medical School to study pathology with William H. Welch.[8] The choice was a good one. In 1890 pathology was still a young discipline in the United States, and William Welch was in fact its most distinguished academic representative. Educated in Germany, Welch was trained in the older school of cellular pathology established by Rudolph Virchow as well as in the then rapidly emerging discipline of experimental pathology, which was guided by the tenet that the disease process, as well as normal function, was subject to the experimental approach. Equally important, Welch had acquired from his association with Robert Koch an appreciation of regarding bacteria as the direct incitants of disease.[9] In coming to Johns Hopkins, Flexner fell heir to Welch's conviction that bacteriology and experimental pathology held out the promise of solving age-old problems of disease.

Flexner's career as a pathologist developed very rapidly. In a measure the pace of that development was the result of good fortune. When Flexner arrived in Baltimore, he was ignorant. He knew little medicine and, save for some understanding of Pasteur's germ theory of disease and some elementary knowledge of microscopy, knew little more about pathology or bacteriology. His good fortune was not in what he knew, but in the time of his arrival. He went to Johns Hopkins Medical School while it was still in the process of organization and so could obtain the full benefit of the revolutionary innovations in medical education then being instituted by William Welch, Franklin Mall, and William Osler.

Flexner received little formal training in pathology and bacteriology at Johns Hopkins. Instead, from the moment he came to Baltimore, he was given the responsibility of coping with the myriad problems in histology, parasitology, and bacteriology that daily came to the pathological laboratories from physicians in the Johns Hopkins Hospital. Then, in informal association with William Councilman, Franklin Mall, John Abel, and William Thayer, who generously shared their ideas as well as their special technical abilities with him, Flexner began to learn by doing. Within two years he independently made an important contribution to an understanding of the etiology of diphtheria, a problem then in sharp debate in the American medical community. Subsequently he added significantly to the growing experimental literature on toxalbumin intoxication, pneumonia, amebic abscess, pancreatitis, dysentery, and plague.[10] Eight years after arriving at Johns Hopkins, Flexner was advanced to professor of pathological anatomy. Two years later he was appointed to a chair in pathology at the University of Pennsylvania Medical School.[11] In 1903 he was chosen as the first director of the Rockefeller Institute, a post he held until his retirement in 1935.[12]

It is noteworthy that the idea for the Rockefeller Institute did not originate with a physician or a scientist, but rather with a layman—in this instance Frederick Gates, the remarkable Baptist minister who served as business and philanthropic adviser to John D. Rockefeller. Gates's interest in medical research was no idle curiosity brought on by his work as Rockefeller's almoner. It was a long-term interest. In part it was evoked by the experience of his early ministry which brought him daily to the sickbeds of his parishioners. Equally, it was nourished by his skepticism of the medicine he saw practiced beside these same beds. In 1897, following a reading of Sir William Osler's *Principles and Practice of Medicine,* Gates became convinced that if medicine in the United States was to progress, it would be necessary to establish a research institute where qualified men could devote themselves with singlemindedness to the experimental investigation of disease. He carefully nurtured the idea until it took root with Rockefeller. In 1901, after a false start when both Gates and Rockefeller for a time thought to make the new institute a part of the University of Chicago, it was chartered as an independent entity in New York City. The government of the newly organized Rockefeller Institute was put in the hands of a director and two special boards—one, a lay board of trustees, whose function was to guide its economic fortunes, the other, a board of scientific advisers, who were to take responsibility for overseeing its research programs. Initially, the directorship was offered to the distinguished animal pathologist Theobald Smith, who refused it. Subsequently, on the advice of Dr. Welch, it was offered to Flexner, who accepted.[13]

If Gates created the Rockefeller Institute, it was Flexner who developed the environment which permitted it to flourish. The nature of his contribution is in part reflected by the staff he recruited. Flexner believed that medicine derived from such basic sciences as pathology, physiology, chemistry, and bacteriology. His original appointments to the Institute, therefore, were not of physicians interested in pursuing problems in clinical medicine, but rather of investigators skilled in the basic sciences who sought to cast light on medical problems through experimental research. Although every original member of the Rockefeller Institute possessed the M.D. degree, all were also capable of pursuing research in one or more of the basic sciences. Flexner's first work at the Institute reflected this principle as well. It is not only important in itself, it is one of the keys necessary to an understanding of the early development of polio research.[14]

Some months after the Rockefeller Institute opened its doors, a serious epidemic of cerebrospinal meningitis broke out in New York City. In response Flexner began an investigation of the disease using the rhesus monkey as his experimental animal. The choice of the monkey was important, because it was dictated by Flexner's conviction as a pathologist that such animals would permit him to conduct an experimental investigation of the disease under conditions which were similar to or resembled those which occurred in human beings. This was not Flexner's first encounter with cerebrospinal meningitis. A decade earlier he had studied the disease in Maryland, though without success. This time he not only cultivated the meningococcus in vitro (a task he had failed at before), he successfully transmitted the disease to monkeys as well. Subsequently he produced an antiserum in horses, which averted or modified the experimental disease when it was introduced into the body of the monkey by means of lumbar puncture. Later, when the antiserum was used intrathecally in humans, it proved equally successful. The mortality of those who contracted cerebrospinal meningitis was dramatically reduced from a high of 80 per cent to a new low of 15 per cent.[15]

Flexner's triumph had a transcendent effect. Its practical nature confirmed Gates's vision of the utility of medical research and solidified his support for the Institute, a venture that had no precedent in the United States. Its impact on Flexner was equally profound. Psychologically, it convinced him that there were no boundaries to what bacteriological and pathological techniques could achieve when applied to the experimental investigation of disease. It was the victory over cerebrospinal meningitis that encouraged Flexner to answer the call of the New York Neurological Society in 1907 to undertake an investigation of polio.[16] Indeed, the architecture of Flexner's subsequent polio research, beginning with his selection of the rhesus monkey as his experimental animal, was based on his research experience with cerebrospinal meningitis.[17]

The first important breakthrough in polio research did not occur at the Rockefeller Institute but at the Wilhelminen 'Spital in Vienna. In the fall of 1908 Dr. Karl Landsteiner, a pathologist at that hospital, using an emulsion made from the spinal cord of a polio victim who had come to autopsy, succeeded in transferring polio to monkeys. It was the first time such a transfer had been accomplished. Landsteiner's success, however, proved only a partial triumph, for, later, when he tried to pass the disease from monkey to monkey, he failed.[18] Late in September of 1909, Flexner, using all of Landsteiner's techniques save the mode of inoculation, transferred polio from human victims to monkeys, and then from monkey to monkey as well. In so doing, he demonstrated what many physicians had long suspected but had never been able to prove, namely, that polio was an infectious disease.[19] In the weeks and months that immediately followed, Flexner also showed that, although the monkey was naturally refractory to polio, the experimental infection in such animals was far more severe and lethal than that which occurred naturally in man. He also discovered that no matter how he inoculated his monkeys, whether intracerebrally, interperitoneally, or into the sheath of the sciatic nerve, the disease invariably manifested itself in the spinal cord or the medulla of its victim. Finally he demonstrated that polio was not caused by a bacterium, protozoan, or toxin, but rather by a virus. Little, however, was known about viruses at the time, except that they were ultramicroscopic and passed through the filters then generally used in laboratories. All of these discoveries were accomplished in a space of five months.[20]

From 1910 until 1913 Flexner and his associates at the Rockefeller Institute gathered much new information about polio in monkeys and man. At the time it appeared that Flexner had but to ask a question in order to find new features of the activity and effects of poliovirus. One of his findings during this period merits special discussion because it is representative of the way he worked and thought, and because in later years it also became the focus of an important controversy among polio investigators.

A fundamental problem that engaged early polio researchers was how poliovirus was transmitted and how it entered and left the human body. In 1905 Dr. Ivar Wickman, by very careful clinical and epidemiological study of a polio epidemic which ravaged Sweden, came to the conclusion that the disease was spread by direct contact between infected patients and new victims, as well as by indirect contact through intermediate carriers who showed no morbid symptoms of the disease. Although Wickman agreed with other physicians that paralysis was the one sign that conclusively established the diagnosis of polio, his clinical examinations of polio victims persuaded him that the disease could also appear in a number of other forms, including an abortive form. In the latter state, Wickman maintained that a patient could be infected with polio, outwardly present a series of vague symptoms common to other diseases, and never become paralyzed at all. For Wickman all polio epidemics were made up of both paralyzed and abortive cases, and he argued that unless physicians took into account this varied symptomatology of polio infection, they could never truly understand the way the disease spread.[21]

Not everyone accepted Wickman's conclusions. In the United States some public health authorities, after examining and analyzing the character of American polio epidemics, advised the medical community that the disease was not contagious.[22] Flexner, however, was not of this persuasion; he too believed in contagion. The sources for his belief were diverse. In part they stemmed from his faith in Wickman's studies, and in part, as we shall see, they were rooted in his research experience with cerebrospinal meningitis.

In January 1910, Flexner devised a series of experiments designed to demonstrate how poliovirus entered and left the human body. His previous research had shown that it was possible to bring down an experimental monkey in a wide variety of ways, ranging from intracerebral to subcutaneous inoculation of poliovirus. It was clear, however, that this was not the way the infection spontaneously occurred in man. The following month Flexner reported that he had discovered poliovirus in the nasal and pharyngeal mucous membrane of a polio infected monkey and that he had subsequently been able to bring down other monkeys with paralysis using a filtrate of such material.[23] This finding, among other things, suggested to Flexner that in the course of polio infection the membranes of the nose and throat become more severely infested with virus and, given the close lymphatic connection existing between the nose and brain, this connection might well serve as the pathway through which the virus found entrance into the brain. He further reasoned that it was highly probable that this connection also served as a portal of exit for the virus from the human body.[24]

In all of this Flexner was guided by his previous research experience with cerebrospinal meningitis. "It is difficult," he wrote at the time, "if not impossible to establish in human beings the fact that the diplococcus passes from the meninges by a reverse lymph current into the nasopharynx, and yet such a migration is not only highly probable, but would most readily and satisfactorily explain the persistent intracellularis infection of these mucous membranes, which is regularly present in epidemic cerebrospinal meningitis. The case is quite different in monkeys infected with diplococcus intracellularis by injection of cultures into the lumbar spinal canal, in which the migration into the nasopharynx of the diplococcus contained in leucocytes, and free also, has been followed with the microscope. It may therefore be regarded as established that this mucous membrane serves both as the site of escape from and of entrance into the meninges of the diplococcus intracellularis in man. The question arises: Does this membrane serve a similar double function in respect to the virus of poliomyelitis?"[25]

When Flexner subsequently was able to demonstrate that by traumatizing the nasal mucosa of a monkey and then swabbing it with a filtrate containing poliovirus he could bring the animal down with polio, it appeared that the answer to his hypothetical question was affirmative.

To all intents and purposes he had, by experimental means, established the probable mode of infection.[26]

In 1913 Flexner capped his early work on polio with an announcement that he and Hideyo Noguchi, by utilizing fragments of brain tissue from human polio victims in a culture medium, had succeeded in cultivating a peculiar globoid-shaped minute organism that could reproduce the symptoms and lesions of polio when inoculated into experimental monkeys. Although Flexner and Noguchi were unwilling to state whether the microorganism they had cultivated belonged to bacteria or protozoa, they allowed that the culture methods they had used were those that applied more particularly to bacteria. They also warned their colleagues that it was extraordinarily difficult to cultivate these "new" microorganisms.[27]

Flexner's and Noguchi's announcement was acclaimed by medical scientists throughout the world. William W. Keen, the doyen of American surgery, was so moved by this and other discoveries made by his former colleague at the University of Pennsylvania that he nominated Flexner for a Nobel Prize.[28] The achievements were impressive. In a space of four years, from 1909 to 1913, Flexner had not only delineated important features of the pathology of polio, he also established, among other things, that the disease was infectious and passed by human contact. Although two of his findings—that the portal of entry and exit of poliovirus was through the nasal and pharyngeal mucosa and that the causative agent was a "globoid microorganism"—conflicted with the results of other experimental and epidemiological investigations, most polio investigators had high hopes that these conflicts would soon be resolved. For the next quarter of a century Flexner's concepts guided polio research in the United States.

II

Yet development of that research was halting and slow. In 1954 Dr. Thomas Rivers, the late dean of American virologists, in a speech before the American Philosophical Society, explained the slow progress of polio research after 1913:

> At the roots of progress in poliomyelitis research, as in all research, lie proper concepts and adequate techniques. These have been acquired more slowly in relation to poliomyelitis than is the case regarding other viruses. Because of this, many errors have been made and workers have taken part in many wild goose chases. Those who have made the errors, unless they refuse to admit them and to give up false concepts, should not be held in low repute, for not infrequently their mistakes, if recognized, make easier the paths of other investigators. There are times, however, when workers of great scientific repute continue to misconstrue the meaning of their data or will not admit inadequacies in the techniques employed by them. When this happens, progress may be materially

impeded and much effort must be expended in tearing down the false edifice before a true one can be built. "Thus no one has the right to encumber science with premature assertions, for an erroneous affirmation which has taken a day to construct requires sometimes twenty years to overthrow."[29]

Although Rivers in his speech delicately refrained from naming the authorities he had in mind, it is perfectly clear that he meant Flexner. Do Rivers' charges have merit or can they be dismissed as a misreading of the early history of polio research?

First, there can be little doubt that for approximately a period of three decades a number of Flexner's concepts guided polio research in the United States. Second, it is equally true that two of these concepts, namely, that the portal of entry and exit of poliovirus in man was through the nasal and pharyngeal mucosa, and that the causative agent of the disease was a "globoid microorganism" which could be cultivated in artificial culture media, were later proved to be mistaken. It does not follow from these facts, however, that Flexner's authority inhibited the pace of polio research.

One index of Flexner's impact on the course and development of polio research is to be found in his relations with Peter Olitsky of the Rockefeller Institute. During the early 1920's Flexner persuaded Olitsky, one of his protégés, to turn his talents to the study of virus diseases.[30] It was a momentous move, for in the end it was the investigations conducted by Olitsky and his associates which destroyed Flexner's most cherished concepts of polio. In 1930 Olitsky, in association with C. P. Rhoads and P. H. Long, published a paper which demolished Flexner's and Noguchi's thesis of a "globoid microorganism" as the causative agent of polio infection.[31] In view of Rivers' charges, it is significant that Olitsky, a man who was timid and gentle by nature and who, by the testimony of contemporaries at the Institute, was highly susceptible to even the smallest pressure by Flexner, had no difficulty in publishing his paper in the *Journal of Experimental Medicine,* then edited by Flexner and Peyton Rous.[32] In 1939 Albert Sabin, another member of Olitsky's laboratory, after several years of research on various aspects of the pathology of polio, published the first of a series of papers which showed that there was an absence of demonstrable virus in the nasal mucosa and olfactory bulbs in human cases of polio—extraordinarily strong evidence that Flexner's construction of the olfactory pathway as the portal of entry of poliovirus in man was also mistaken.[33] It must be remembered that until the publication of Sabin's paper, and that of Dr. Charles Swan in Australia the year before,[34] the experimental evidence on the problem of the portal of entry of poliovirus in man was equivocal. Although it is true that Carl Kling and Constantine Levaditi in 1929 and again in 1931[35] published papers which suggested that the portal of entry of poliovirus was through the alimentary tract, later experiments conducted by Kling and Levaditi, as well as by Flexner, to prove this point remained inconclusive.[36] As late as 1936 Flexner, in very careful feeding experiments of both rhesus and cynomolgous monkeys, discovered that monkeys which resisted polio infection after being fed polio virus invariably were successfully brought down with polio after a nasal installation of the virus.[37]

Flexner's construction of the mode of polio infection was based on seemingly substantive experimental evidence. Its vitality rested on that fact and not on his status as a polio authority. Actually the problems to which Olitsky's laboratory addressed itself are clear indication that Flexner's authority and ideas were freely questioned. By any canon, authoritarian domination precludes examination and questioning of ideas, especially by subordinates. The lack of progress in early polio research cannot be placed on Flexner's shoulders; its causes were far more fundamental.

Flexner's knowledge of polio was a product of his training as a pathologist. It derived from a study of the results of polio infection in experimental monkeys and man. Put another way, while Flexner knew a great deal about the effects of poliovirus, he did not know what poliovirus was. In this he was no different from his contemporaries. For approximately the first three decades of the twentieth century, most pathologists and bacteriologists did not recognize that there were fundamental differences between bacteria and viruses and looked upon viruses as simply being a very small species of bacteria.[38] The conception was hardly outlandish. Many of the problems that investigators encountered during their research of virus diseases appeared to be similar to those they had previously found in their investigations of known bacterial and protozoan diseases. For example, why should the difficulty in cultivating poliovirus in artificial media necessarily have suggested that there were fundamental differences between viruses and bacteria? The root of the difficulty might just as easily have been explained by inadequate laboratory techniques. Historical experience certainly pointed in the latter direction. Following Anton Weichselbaum's isolation of the meningococcus in 1884, it had taken bacteriologists almost two decades to learn how to cultivate that fastidious diplococcus in the laboratory.[39] Again, had not Hideyo Noguchi encountered similar difficulties in his attempts to cultivate treponema pallidum in artificial media?[40] Or again, if investigators initially had great difficulty in transmitting polio to experimental animals, was not this experience similar to the problems they had encountered in their early efforts to transmit syphilis and meningitis experimentally? The plain fact is that the limitations of the evidence from which Flexner developed his concepts of polio were not understood. They were not understood because during this period virology was still a nascent science.

One measure of the state of its development is to be found in the manner in which virologists were trained. During the first three decades of the twentieth century formal instruction in virology was still part and parcel of general instruction in pathology and bacteriology. Not until 1922 was the first independent course in virology in the United States offered at Johns Hopkins School of Public Health and Hygiene. It is significant that Dr. Charles Simon, who organized the course, had no textbook on which he or his students could rely.[41] The bulk of the literature of virology at the time was still a scattered journal literature, or else encompassed in brief sections of such textbooks of bacteriology and preventive medicine as Hans Zinsser's *Infection and Immunity*, and Edward Rosenau's *Public Hygiene and Preventive Medicine*. In the end Simon and his students had to depend on the lecture material prepared by the working virologists whom he had invited to instruct his classes. It was not until 1928 that a textbook exclusively devoted to virus diseases was published in the United States. The text, titled *Filterable Viruses* and edited by Thomas Rivers of the Rockefeller Institute, was notable on several counts.[42] First, like Charles Simon's course it was a cooperative venture that included, among others, such pioneer investigators of virus diseases as Harold Amoss, Peter Olitsky, Ernest Goodpasture, Rudolf Glaser, Louis Kunkel, and Jacques Bronfenbrenner. Second, it was not limited to a discussion of animal viruses and included material on plant, insect, and bacterial viruses as well. Third, in an introductory chapter Rivers offered a new synthesizing concept which for the first time sharply differentiated viruses from bacteria—namely, that viruses, unlike bacteria, were obligate parasites that could live and multiply only in the presence of a living susceptible cell.[43]

As with many other subsequent important developments in virology, Rivers' insight into the nature of viruses was in large measure the product of research programs initiated and fostered by the Rockefeller Institute and the Rockefeller Foundation in the two decades following the First World War. It is ironic that Flexner, who Rivers believed inhibited the pace of early polio research, in his capacity as director of the Rockefeller Institute and as a member of the Board of Trustees of the Rockefeller Foundation played a key role in the development of these research programs. For example, in 1923 Rivers, who was originally trained as a pediatrician at Johns Hopkins, was brought by Flexner to the Rockefeller Institute Hospital and given a mandate to develop a virus disease research program. Rivers' research not only resulted in a new concept of viruses, it subsequently advanced understanding of the nature of chickenpox, psittacosis, Rift Valley Fever, louping-ill, and vaccinia virus as well.[44] Flexner's support of virus research at the Institute was not restricted to the hospital. As mentioned earlier, during the 1920's Peter Olitsky with Flexner's aid began studies of viral and rickettsial diseases which in a period of a little more than a decade was to make his laboratory at the Institute preeminent in the investigation of polio and other neurotropic viral infections.[45] Again in 1929, following Hideyo Noguchi's death from yellow fever, the Rockefeller Foundation with Flexner's encouragement established a special laboratory at the Institute for the investigation of yellow fever. Eight years later the research programs initiated and conducted by this laboratory culminated in Max Theiler's perfection of a live virus vaccine against yellow fever. These examples can be multiplied.[46]

Although the above investigations marked an increasing momentum in virus research, it must be borne in mind that until the decade of the 1930's viruses were still essentially known by the pathology they created. The biochemical and biophysical techniques necessary for an understanding of the structure and composition of viruses were either still very crude or in the offing. Not until 1935 did Dr. Wendell Stanley, a chemist in L. O. Kunkel's laboratory at the Rockefeller Institute, succeed in crystallizing tobacco mosaic virus.[47] In so doing, he demonstrated that this ultramicroscopic virus was in essence a large protein molecule, and, as other organic molecules, subject to biochemical analysis. It was not until the late 1930's that Dr. Theodore Svedberg at Uppsala and Dr. Ernest E. Pickels at the yellow fever laboratories of the Rockefeller Foundation perfected the ultracentrifuge,[48] and Dr. Arne Tiselius at Uppsala described the method for the separation of proteins from body fluids by means of electrophoresis, instruments and techniques, which because they provided means for accurately determining particle sizes, analyzing complex fluids, obtaining molecular homogeneous material, and separating and analyzing proteins in a solution, gave impetus to more sophisticated examination of the structure and composition of viruses.[49] Again, it was not until 1938 that Bodo von Borries and Ernest Ruska in Germany, using an electron microscope developed at Siemen's Allgemeine Gesellschaft, succeeded in making an electron micrograph of a virus.[50] Save for the work in electron microscopy, all of these developments were the direct outcome of research programs initiated and fostered either by the Rockefeller Institute or the Rockefeller Foundation. Actually such instruments as the ultracentrifuge, Tiselius' apparatus, and the electron microscope so necessary for the visualization and biochemical analysis of viruses did not become part of the everyday armamentarium of virus laboratories until the early 1940's. Given these circumstances it is difficult to see how polio investigation could have proceeded at a much faster pace than it did during the first three decades of the twentieth century.

There is one aspect of the research programs sponsored and supported by the Rockefeller Institute and the Rockefeller Foundation that bears further comment. By 1940, the Rockefeller Institute had become one of the most important centers for virus research in the United States. Its investigators dominated the field. In the process of casting new light on the nature of virus disease they had also helped train a new cadre of virologists who in subsequent years were to make even more fundamental contributions to the growth of virology. One of the key figures in this development was Thomas Rivers. In a deep sense the evolution of Rivers' career as a virologist parallels the development of Flexner's career as a pathologist. Rivers came to virus research when virology was still a young discipline. Although he was not a young man (he was thirty-five when he began), Rivers' career, as Flexner's before him, developed very rapidly. In the relatively brief period of fourteen years he too made notable and fundamental contributions to his discipline. At the age of fifty his career as an active investigator was over.[51] These fourteen years marked an important transition in virus research from a preoccupation with the effects of virus infection to a fundamental concern with the nature of the structure and composition of viruses. Although Rivers, a pathologist by training, never quite mastered biochemistry and physics, he nevertheless understood and appreciated their importance for the future development of virus research. Later in his career as a scientific administrator he used this knowledge in structuring and organizing the polio research programs sponsored by the National Foundation for Infantile Paralysis.

III

On January 3, 1938, President Franklin D. Roosevelt, with the aid of a distinguished group of lawyers, philanthropists, and businessmen, organized the National Foundation for Infantile Paralysis. Its stated purpose was: "to lead, direct, and unify the fight against every aspect of the killing and crippling infection of poliomyelitis."[52]

The struggle against polio was not new to Franklin Roosevelt in 1938. Stricken by polio shortly after the First World War, he knew at first hand the terrible physical and psychological cost of the disease. In 1926, with the aid of his law partner, Basil O'Connor, he helped establish a foundation for the care and treatment of polio patients at Warm Springs, Georgia.[53] Never did a man have a more reluctant helper. James Roosevelt, the son of the late President, recalling the origins of Warm Springs Foundation, recently remarked:

> My father and Mr. O'Connor often disagreed. One of their first disagreements occurred when my father indicated that he planned to buy the old Meriwether Inn at Warm Springs in Georgia. He wished to convert it into a foundation for the care and treatment of polio patients. Mr. O'Connor doubted

the wisdom of that move. He thought the purchase of the inn a poor investment. He was also skeptical of the motives of the distinguished banker and philanthropist, George Foster Peabody, who owned the property and first interested my father in Warm Springs.

Mr. O'Connor's views were of some moment because he also served as my father's attorney. The negotiations for the property proceeded slowly. No item was too small to escape Basil's attention; his skepticism had made him cautious. For example, when he heard rumors that there had been thefts at the inn, he held up negotiations until he got an itemized accounting of the silverware and bedsheets . . . Even though Mr. O'Connor was initially skeptical of my father's venture at Warm Springs, Georgia, he devoted all his efforts to making it a success. He organized the Warm Springs Foundation and then proceeded to breathe life into it.[54]

At first O'Connor raised funds for the Warm Springs Foundation through his personal contacts in the business and financial community. Following Roosevelt's election to the presidency in 1932, the yearly birthday balls held in the President's honor became the main vehicle for acquiring funds in the struggle against polio. In the period between 1934 and 1938, the President's Birthday Balls raised a total of $3,362,000. Of this sum, $1,665,000 was expended for the care and rehabilitation of polio victims in various local communities, $1,456,000 was granted in support of activities at Warm Springs, Georgia, while $241,000 was spent in support of polio research.[55]

The research programs supported by the President's Birthday Balls, although laudable, had limited and, in two cases, disastrous results. In part, the responsibility for this failure rests with Dr. Paul de Kruif, who as secretary to a special scientific advisory board of the President's Birthday Ball Commission organized and administered the research programs. The cause for the failure is rooted in the evolution of de Kruif's career.

In 1919 de Kruif, who had been trained as a bacteriologist by Friedrich Novy at the University of Michigan, joined the laboratories of the Rockefeller Institute. His work at the Institute by all accounts was superior and promising. However, in 1923 Simon Flexner fired him when he discovered that de Kruif was the author of an anonymous attack on the Institute published in *Century* magazine.[56] Subsequently, de Kruif became a science writer and though he achieved great success in this new career, he never forgave Flexner. He took his revenge in a variety of ways. Following his separation from the Institute, de Kruif served as scientific consultant to Sinclair Lewis, then in process of preparing his novel on medical research, *Arrowsmith.* Lewis' uncompromising delineation of the character of medical research and the personnel at the McGurk Institute portrayed in the novel was in essence de Kruif's vision and judgment of his contemporaries at the Rockefeller Institute.[57]

When de Kruif later came to distribute the research funds of the President's Birthday Balls, he overlooked the laboratories of the Rockefeller Institute which had pioneered in polio and other virus research. Instead, the bulk of the funds went to Dr. William H. Park of the New York City Health Department and to Dr. John Kolmer of Temple University to develop vaccines against polio. The research that de Kruif chose to support is also revealing. De Kruif viewed medical research as a kind of organized empiricism where heroic workers in the mold of Jenner, Pasteur, Koch, and Martin Arrowsmith by brilliant insight found cures to baffling diseases. It was not really necessary to know anything about poliovirus to solve the problem of polio. After all, what did Jenner know about smallpox and vaccine virus when he developed vaccination against that disease? It worked, that's what mattered. Unfortunately, the polio vaccines developed by Park and Kolmer later proved to be premature and killed a number of children.[58]

In 1937, following the debacle of the Park and Kolmer vaccines, President Roosevelt became convinced that polio could only be conquered through a broad and sustained program of scientific education and research. The organization of the National Foundation for Infantile Paralysis was in essence the first step toward the realization of that goal. It was also something more. At a time when deadly assaults had already been launched against the human spirit and life itself in Europe, the new Foundation in addition stood as an affirmation of the value of conserving human life and dignity. Ordinary people everywhere recognized this quality and quietly and emphatically made its cause their own. At the end of 1967 it was estimated that in a period of thirty years the American public, through the March of Dimes, had voluntarily contributed approximately $775 million in support of the National Foundation and its activities.[59]

Apart from the care and rehabilitation of polio victims, one of the most important achievements of the National Foundation has been in the development of medical research. Since its organization the Foundation has spent $96,400,000, or 12.5 per cent of all funds contributed to it, in support of such research. While these figures in part reveal the order of the Foundation's commitment to research, the significance and import of that support can only be appreciated when one understands the nature of the scientific problems that the Foundation faced at its inception.[60]

In 1938 knowledge about polio was still essentially scant. At that time, for example, no one actually knew how polio virus was transmitted, how it entered the human body, how it spread to the nervous system, how many types of poliovirus existed, how far it was disseminated in nature, or whether it would be possible to immunize humans against its effects.

The development of research programs to grapple with these and other fundamental problems became the first order of business of the new Foundation.[61] The responsibility for the formulation of these programs did not rest with the organizers of the Foundation, but rather with a specially appointed, voluntary Scientific Research Committee, composed of virologists and other medical scientists.[62] In the spring of 1939 this committee, then under the direction of Thomas Rivers of the Rockefeller Institute, prepared an eleven-point research program which urged the Foundation to devote its support to studies of such problems as:

1. Pathology of poliomyelitis in human beings
2. Portal of entry and exit of virus
3. Purification and concentration of the virus
4. What is to be called poliomyelitis
5. Mode of transmission of virus from man to man
6. Transmission of virus along the nerves
7. Further attempts to establish poliomyelitis in small laboratory animals
8. Settlement of the question of chemical blockade
9. Chemotherapy of poliomyelitis
10. Relation of constitution to susceptibility
11. Production of a good vaccine[63]

The importance of the eleven-point program cannot be overemphasized. By arranging problems in order of importance, it in effect established a long-term guide for polio research. In so doing, it also blueprinted what at the time was not known about polio. Equally important it established the principle that the Foundation give priority of support to such research as concerned itself with exploring fundamental problems of the nature of polio. As if to emphasize this policy guideline, the production of a vaccine against polio was made the last order of business. In implementing this program the National Foundation not only gave extraordinary impetus to the development of virus research, it created a revolution in the process of philanthropic giving as well.

The National Foundation was not the first voluntary health agency in the United States; it was, however, the first to open the field of philanthropy to the so-called "common man." Before the organization of the Foundation, philanthropy was generally an attribute and activity of the wealthy.[64] By calling upon men and women everywhere to help rectify the inequalities between men created by disease, the Foundation based itself on the whole population and not just a wealthy minority. As a result it also gave the public a direct interest in basic research, a stake which, as we shall see, was to have an extraordinary impact on the development of polio investigation in the United States.

One of the most important obstacles that the Foundation faced during its early years was finding enough competent investigators to deal with the research problems suggested by the Scientific Research Committee. In 1938 there were less than a score of laboratories in the United States engaged in virus research. Of these, only a small fraction were interested in working on problems of polio. The earliest research grants made by the Foundation quite naturally went to this handful of investigators.[65] The Scientific Research Committee, however, realized that such a policy, if long continued, would be self-limiting. In 1940, in an effort to cope with this problem, the committee arranged a substantive long-term grant of five years to Johns Hopkins University School of Public Health and Hygiene. Its purpose was to establish a virus research center which would not only engage in studying the fundamental nature of polio and other viruses, but would also train young virologists.[66] In quick succession similar long-term grants were also made to the University of Michigan School of Public Health and Yale University Medical School.[67]

These grants, in effect, helped revolutionize the art of supporting medical research in the United States. Prior to this time, most grants made for medical research by private philanthropies and the federal government were modest in size and rarely made for periods of more than one year. By making substantive long-term grants, the Foundation insured a continuity to polio as well as other virus research by giving investigators the precious uninterrupted time they needed to develop their work. By providing for the training of young virologists they further insured research continuity by helping to organize a cadre of young scientific investigators who could carry on in the laboratory when senior workers either retired or died.

This was not the only innovation in philanthropic giving developed by the Foundation. Shortly after World War II, Foundation officials noticed that many university and medical school administrators were reluctant for members of their faculties to accept research grants. Upon inquiry, many administrators complained that while it was true that such grants added to the renown of the university, in the long run they also acted as financial liabilities because of the indirect costs which accrued to the university in administering them.

Recognizing this problem as a distinct threat to the development of the Foundation's future research programs, Dr. Harry Weaver, then director of the Division of Medical Research, devised an accounting system whereby such indirect administrative costs could be calculated. Beginning in 1948, whenever the Foundation awarded a grant for medical research, it also made a special supplementary grant to the investigator's university or medical school to cover the indirect costs of administering such a grant.[68] The wisdom of making supplementary grants to insure the financial stability of institutions devoted to medical research was immediately recognized by administrators everywhere. Today, other philanthropic foundations, as well as the federal government, follow this system in supporting research.

If the grants made by the National Foundation created the opportunity for doing fundamental research on problems of polio, the result of that research, in the last analysis, was the product of the wit, ingenuity, and insights of the investigators doing the research.

Thomas Rivers, in speaking of the relationship between the Foundation and the investigators it supported, once said, "It has never been the policy of the Foundation to tell an investigator how to do an experiment. The Foundation could turn down a man's proposals or they could accept a man's proposals, but once they accepted they let him alone, which is as it should be."[69] From the beginning, the Foundation adhered to that rule. Its basic wisdom is seen in the fact that by 1946, in spite of a world war, investigators supported by the Foundation made important and substantive contributions, among them: basic studies in the epidemiology of polio, including such work as the refinement of techniques to isolate poliovirus from sewage and fecal matter (Francis, Melnick, Paul, and Gard);[70] the solution of the problem of the role of the fly in the transmission of polio (Sabin and Ward);[71] the discovery of MEF[1], a new type 2 poliovirus reference strain (Olitsky, Morgan, and Schlesinger);[72] a basic understanding of the neuropathology of polio (Bodian and Howe);[73] and singular work on the nutrition of laboratory animals used in polio research (Elvejhem and Clark).[74]

These achievements, among others, were important because they helped furnish a developmental base for breakthroughs in polio research from 1946 to 1953, especially in the area of immunization. The problem of immunization was particularly difficult in 1945 for a variety of reasons. First, many things about the nature of poliovirus were then not yet fully understood. Second, the tragic results of the Park and Kolmer polio vaccines were still fresh in the minds of polio investigators, and many simply shied away from the responsibility of dealing with such problems. Third, and perhaps most important, there was an information crisis.

In 1945 many of the new advances in virus research that had occurred during the Second World War had still not been synthesized or integrated into textbook literature. At the time virologists in the main depended on two textbooks prepared before World War II: C. E. Van Rooyen and A. J. Rhodes, *Virus Diseases of Man*, and the Harvard School of Public Health Symposium, *Virus and Rickettsial Diseases*.[75] Recognizing that some of the reluctance of polio investigators to deal with problems of immunization existed because of a lack of information, the Foundation sponsored the production of new textbooks both in virology and bacteriology[76] and began to hold special conferences where polio investigators could exchange their latest research experiences and discuss mutual problems.[77] The move was warranted, not only because of general developments in immunology, but more especially by the growth of new concepts of vaccination against virus diseases.

During the 1920's and early 1930's, most virologists and bacteriologists believed that it was not possible to vaccinate against a virus disease with an inactivated vaccine. Experience with smallpox and yellow fever had certainly pointed in this direction. In 1936, an important breakthrough occurred when Dr. Peter Olitsky and his associates at the Rockefeller Institute demonstrated that it was possible to immunize mice against western equine encephalitis with an inactivated vaccine.[78] Olitsky's research opened the door. During the early 1940's, Dr. Thomas Francis and other investigators working on problems of influenza extended Olitsky's findings by developing an effective inactivated vaccine against influenza.[79] At an immunization conference sponsored by the Foundation in the fall of 1946, Dr. Isabel Morgan (who had previously trained with Olitsky and was then serving in the laboratories of Dr. Howard Howe and Dr. David Bodian at Johns Hopkins Medical School), demonstrated that it was also possible to immunize a monkey against polio with a formalinized inactivated vaccine.[80] Morgan's work was a milestone in polio research. It not only established that previous theories holding it was impossible to immunize animals against polio with an inactivated vaccine were erroneous, it also had the virtue of reopening the question of whether it was feasible to immunize humans in a like manner. Although a number of virologists, after Morgan's presentation, were anxious to start work on such a vaccine, the Foundation instead directed its support to efforts aimed at discovering more basic knowledge about poliovirus, especially the number of types of virus that existed.

When Park and Kolmer prepared their polio vaccines in 1935, it was generally assumed that there was only one type of poliovirus. It must be stressed, however, that not all investigators adhered to such a point of view. In 1931, Dr. Frank Burnet in Australia published his suspicion that there was probably more than one type of poliovirus.[81] During the 1930's Dr. John Paul and Dr. James Trask of Yale also published their belief that more than one type of poliovirus existed.[82] By 1940, a number of polio investigators were certain that there was in fact more than one type of poliovirus. In 1948 and 1949, Dr. John Kessel of the University of Southern California and Dr. David Bodian of Johns Hopkins announced (almost simultaneously) that they had discovered three types of poliovirus.[83] At the time, however, no virologist actually knew if there were more than three types of poliovirus among the myriad strains that investigators had isolated for nearly fifty years. In 1949, the Foundation granted more than $1,250,000 to Dr. Louis Gebhard, Dr. John Kessel, Dr. Jonas Salk, and Dr. Herbert Wenner, to type all extant strains of poliovirus in the United States. By 1951, these laboratories conclusively established that there were only three types of poliovirus.[84]

Scientific research rarely progresses in a linear fashion. At approximately the same time the poliovirus typing program was getting under way in 1949, Dr. John Enders, Dr. Thomas Weller, and Dr. Frederick Robbins at Harvard University reported that they had successfully cultivated poliovirus in tissue cultures made up of nonnervous tissue.[85] The importance of this achievement cannot be overemphasized. Prior to the research of Enders and his associates, most virologists believed that poliovirus could only be cultivated in cultures composed of nervous tissue. That belief was a formidable barrier to the creation of a polio vaccine because virologists knew that if viruses which were grown on nervous tissue were used in a vaccine they would surely produce an allergic encephalomyelitis. The belief, moreover, was firmly grounded on experiments which had been carefully carried out fourteen years before by Dr. Sabin and Dr. Olitsky at the Rockefeller Institute.[86] In one stroke Enders and his associates showed the possibility of a practical solution for the production of large quantities of poliovirus suitable for the making of a vaccine. Equally important they revolutionized laboratory procedures in isolating, titering, and typing poliovirus. In 1954, in recognition of this extraordinarily fruitful research, the Karolinska Institute in Sweden awarded a Nobel Prize to Enders and his two young assistants.

Although by 1951 the research of Isabel Morgan and John Enders, as well as the results of the poliovirus typing program, pointed to the possibility of making an inactivated polio vaccine, an important barrier remained to be hurdled. Virologists at that time still did not know how poliovirus traveled from the portal of entry to the nervous system of man. The necessity of having this knowledge for the creation of an inactivated polio vaccine was crucial. Earlier Dr. Morgan's experiments had indicated that immunizing monkeys with an inactivated vaccine required a high titer of antibody, a factor which made use of such a vaccine impractical for human vaccination purposes.[87] Most virologists were resigned to the fact that if poliovirus reached the nervous system via the nerves, they would need a very high titer of antibody to give protection against the virus. However, if the virus reached the nervous system via the bloodstream, all were agreed that the titer of antibody could be much smaller. The answer to this problem emerged from research in passive immunization.

During World War II, Dr. Edwin Cohn and his associates at Harvard Medical School, while working on problems of blood substitutes, developed a unique process for fractionating human blood.[88] One of the fractions that Cohn obtained, gamma globulin, was very quickly discovered to have unusual ability in halting infections. In 1943, shortly after Cohn's discovery, Dr. David Kramer, at the laboratories of the Michigan State Department of Health, was able to demonstrate that gamma globulin gave marked protection to cotton rats and monkeys against an intracerebral inoculation of poliovirus. Although Kramer's animal experiments were promising, because of the exigencies of war, he never received an opportunity to hold a field trial to test whether gamma globulin had a prophylactic effect against poliomyelitis in humans as well.[89]

There the matter stood. In the immediate postwar years, save for Dr. Kramer and Dr. Joseph Stokes, Jr., of Philadelphia, few investigators devoted themselves to extending these promising beginnings in passive immunization.[90] In 1950, however, Dr. William Hammon of the University of Pittsburgh Medical School took up the cudgels for a field trial to test the effectiveness of gamma globulin. Although many virologists had doubts of the value of passive immunization against polio, the Foundation, after conducting an extensive inquiry, finally agreed to support Dr. Hammon in making such a trial.[91]

In 1951 and again in 1952, Dr. Hammon, with the assistance of Dr. Lewis Coriell and Dr. Joseph Stokes, Jr., conducted a series of field trials with gamma globulin in Provo, Utah, Houston, Texas, and Sioux City, Iowa. They soon discovered that a small dose of gamma globulin given intramuscularly into the buttock would give children temporary protection against paralytic polio.[92] The significance of Hammon's findings for the problem of titer of antibody was immediately and independently perceived by Dr. Dorothy Horstmann of Yale and Dr. David Bodian of Johns Hopkins. These investigators assumed that if Hammon's findings were valid it meant that poliovirus traveled from the portal of entry to the central nervous system in man by means of the bloodstream, and they began to search for evidence of a viremia. Their subsequent discovery of poliovirus in the bloodstream early in the disease (before paralysis set in) showed beyond cavil that even a small amount of antibodies could give protection against paralytic polio.[93]

Once this obstacle to the possiblility of making an effective inactivated vaccine against polio was overcome, the task of perfecting such a vaccine was almost immediately undertaken by Dr. Jonas Salk of the University of Pittsburgh Medical School.

In 1942, soon after finishing his medical internship at the Mount Sinai Hospital in New York, Salk received a special fellowship underwritten by the National Foundation to undertake training in virus research with Dr. Thomas Francis, Jr., at the University of Michigan School of Public Health. The training that Salk received in Francis' laboratory working with influenza and other viruses proved invaluable and eventually led to appointment as an assistant professor of virology at the University of Pittsburgh Medical School. In 1948, shortly after moving to Pittsburgh, Salk became involved in polio research through participation in the poliovirus typing program. His successful work in this project, and his later perfection of an inactivated vaccine against influenza, began to attract the favorable attention of his contemporaries in virology. In 1951, when Salk indicated to Foundation officials that he wished to undertake research on problems of polio immunization, the Scientific Research Committee readily agreed to support him in these efforts.[94]

The technical problems that Salk faced in making a polio vaccine, such as inactivating poliovirus with formalin so as to retain its antigenicity, or cultivating poliovirus in roller tube cultures of monkey kidney tissue, were formidable and trying. However, in the spring of 1952, just a year after beginning his work, he was able to demonstrate that he could successfully immunize monkeys against polio by inoculating them intramuscularly with tissue culture fluids containing poliovirus inactivated with formalin and emulsified with mineral oils.[95] When limited tests with similar preparations later that year showed like success in immunizing children at the Polk State School and the D. T. Watson Home in Pennsylvania,[96] the Foundation called a series of conferences with leading virologists and public health authorities to discuss the advisability of conducting large-scale field trials to test the safety and effectiveness of Salk vaccine.[97]

The conferences underscored both the validity and necessity of conducting such trials. As a result, in the spring of 1953 the National Foundation organized a special vaccine advisory committee composed of eminent virologists, members of the United States Public Health Service, and public health authorities,[98] for the express purpose of devising a program for the commercial development and testing of Salk vaccine. It was this committee which subsequently established the requirements and specifications for both the commercial production and safety testing of Salk vaccine.[99] Only the problem of organizing the field trials remained. It was not an easy problem since there were differences of opinion on the design of the trials and who should carry them out.[100]

In January 1954, after several months of debate the Foundation appointed Dr. Thomas Francis of the University of Michigan to conduct the field trials. To insure objective results, it subsequently also made a special grant to the University of Michigan to establish an independent center for the express purpose of evaluating Francis' results.[101]

During the spring of 1954, Francis, with the aid of other virologists, tested Salk vaccine in more than a million and a half children. These trials marked the largest controlled medical experiment ever conducted in human history. A year later, Francis was able to report that the vaccine was both safe and effective. Public health authorities have since termed Francis' evaluation of the Salk vaccine one of the finest achievements in the history of American public health.[102]

Two weeks after Dr. Francis' announcement that Salk vaccine was safe and effective, a number of children in California who had been inoculated with the vaccine succumbed to paralytic polio.[103] Was Francis' evaluation of Salk vaccine mistaken? Was the vaccine dangerous? Who was responsible for the tragedy? At a congressional hearing held soon after the tragedy, a number of distinguished investigators, including Albert Sabin, John Enders, and Wendell Stanley, cast doubt on the efficacy of the techniques used by Dr. Salk to inactivate poliovirus and urged that use of Salk vaccine be stopped pending further investigation. Equally distinguished researchers including Thomas Rivers, Thomas Francis, Jr., Frank Horsfall, as well as James Shannon, the director of the National Institutes of Health, reiterated their conviction that the vaccine was safe and that the program of vaccination be continued.[104] Special investigations of the commercial production of Salk vaccine launched by the National Institutes of Health subsequently discovered that all commercial producers had difficulty in inactivating poliovirus, and ordered a modification in the techniques then being used in the preparation of the vaccine.[105]

There can be no doubt that Dr. Francis' evaluation of the safety and effectiveness of Salk vaccine used in the field trials was accurate. The vaccine used in the trials was guided by the production requirements and specifications originally laid down by Dr. Salk and Dr. Joseph Smadel and Dr. William Workman of the National Institutes of Health.[106] Further, all commercially produced vaccine for the trials was subjected to a rigid tripartite testing on the part of the vaccine producer, the Division of Biological Control of the National Institutes of Health, and Dr. Salk's laboratory. No batch of vaccine was deemed acceptable unless thirteen previous consecutive batches had been passed by each of the three testing laboratories.[107] Under these conditions there were no incidents during the vaccine field trials. Following the licensing of Salk vaccine the duty of overseeing the production and safety testing of the vaccine became the responsibility of the producer and of the Division of Biological Control of NIH. It is here that the breakdown occurred. The huge public demand for vaccine was easily predictable. Yet during the months following the trials NIH made no move to increase the staff of the Division of Biological Control to cope with the responsibility it knew it would have of safety testing the vaccine. As a result, there was no adequate check of commercial production of the vaccine following the completion of the trials. It is apparent that the production protocols of vaccine submitted by commercial producers were largely accepted at face value. Although it is true that virologists subsequently ordered modifications in the procedures used in inactivating poliovirus for use in Salk vaccine, it is also clear that if commercial producers had used more care in producing the vaccine, and if the government had inspected and tested the commercial product, the incident could have been avoided.[108]

In 1957, Dr. David Bodian, who served on the ad hoc committee formed by NIH to investigate the commercial manufacture of Salk vaccine, told an audience at the Fourth International Poliomyelitis Congress that in spite of differences of opinion among scientists on the efficacy of the techniques used in the inactivation of poliovirus with formaldehyde, there was no doubt "that a completely satisfactory inactivation could be obtained by means of the present method of inactivation at 37°C with 1:4000 formaldehyde."[109] For Bodian, assuring the safety of biologic products prepared from potentially virulent materials required an intimate association of research and developmental production and, above all, a rigorous control over the commercial production process. "The outstanding fact learned from the experiences of the past two years," he reminded his listeners, "is that the future of immunization against poliomyelitis no longer depends on the proof of immunologic theory but rather on the certainty of control of the safety and potency of immunizing products . . . It is now clear that the variability inherent in the preparation of biologics generally has special significance for a vaccine with a potential hazard from its viral components. Only after careful study of the details of procedure, equipment, and interpretation of specifications in each laboratory in the United States has it been learned that experience, as well as theory and controlled experiments, is necessary to establish the key features of the production process which must be duplicated for each batch and the desired characteristics of safety which must be tested for in the final product."[110]

Although the successful development of Salk vaccine in 1955 held out a prospect for the early conquest of polio, at no time did that prospect alter the Foundation's basic policy to support fundamental polio research, including the development of other vaccines. Actually, as early as 1952, the Foundation also began to support a number of research programs designed to perfect a live poliovirus vaccine.[111] When Dr. Albert Sabin, one of the directors of such a program, reported in the fall of 1953 that he had succeeded in transforming all three types of poliovirus to avirulent variants by making repeated passage of large amounts of poliovirus in tissue cultures, the response of the Foundation in support of that work was prompt and generous.[112] It should be added that support for Sabin's work was not limited to the award of research grants. Thus in 1955, when Sabin had difficulty in choosing stable nonpathogenic virus strains for his vaccine, the Foundation arranged a special conference so that he might have a chance to discuss these particular problems with other virologists and geneticists.[113] As a result of this particular conference, Sabin later successfully adapted Dr. Renato Dulbecco's plaquing techniques for the selection of attenuated virus strains suitable for his vaccine.[114] In 1958, soon after Sabin had perfected his vaccine for purposes of testing, the vaccine advisory committee of the Foundation gave him permission to send his vaccine to the Soviet Union and other foreign countries for large-scale trials.[115] Although the Foundation played no part in the subsequent testing of Sabin vaccine in the United States, it nevertheless continued vigorous support of Sabin's polio research. It was not until the federal government licensed the commercial production of Sabin vaccine in 1961 that Foundation support of Dr. Sabin's research ceased.[116]

IV

The foregoing account of the contributions of the Rockefeller Foundation and the National Foundation to the development of the transforming conceptions of polio research is not offered as an inclusive history, but rather as a point of departure for an interpretation of such a history. The account as it now stands is incomplete, for it has focused on the contributions of but two foundations and then almost exclusively on their successes. Much has remained unsaid. Nothing, for example, has been said of the substantive impetus to polio research provided by other foundations, such as the Harvard Infantile Paralysis Commission, the Milbank Fund, and the Commonwealth Fund. More important, little or nothing has been said of failure. It is naïve to believe that all of the research projects supported by the Rockefeller Foundation and the National Foundation always ended successfully. Many projects, in spite of the individual brilliance of the investigators who conducted the research, failed. Nor should it be supposed that all of the policies adopted by both foundations were always wise and beneficent. Some policies caused sharp conflicts. The conflict that erupted between the National Foundation and some senior virologists, when the Foundation appointed a special vaccine advisory committee to guide the development and production of Salk vaccine, or when that special committee later approved the use of Mahoney type 1 polio virus as a component of Salk vaccine, was so fierce that the passions it aroused have to this date not yet completely subsided. There were other differences as well. The existence of these differences underscores the fact that foundation policy in support of polio research was like the process of polio research itself, a compound of imagination, error, failure, and success.

To whom does one then assign the credit for the extraordinary success in the conquest of polio? The answer at first glance seems simple enough: the credit belongs to those scientists who by their ingenuity in the laboratory solved the myriad and outstanding problems presented by polio virus. What then of the Rockefeller Foundation and the National Foundation; what credit do they deserve? Some scientists claim that they deserve very little credit, since their contributions were in the main limited to supplying the wherewithal to conduct research. In defense of this position they argue that not all of the scientists who made important contributions to the solution of the polio problem were supported by either foundation. On the surface this latter claim appears to have merit. In Europe, many of the early investigators, like Paul Römer, Karl Kling, and Wilhelm Wernstedt, who made important contributions to an understanding of the pathogenesis of polio, either worked independently or, like Karl Landsteiner and Constantin Levaditi, did their research under the auspices of the Pasteur Institute. In Australia, Dr. Frank Burnet, who early suggested that there were different types of polio virus, was supported in his research by the Walter and Eliza Hall Institute. It is equally true that in the United States a number of important contributions to an understanding of the epidemiology of polio were made by Public Health Service physicians like Wade Frost, Claude Lavinder, and Allen Freeman. These examples, however, do not modify the fact that from the second decade of the twentieth century most of the important contributions to polio research in the United States were made by investigators who were supported and trained by private foundations, led by the Rockefeller Foundation and the Rockefeller Institute.[117] After the third decade of the twentieth century, save for Dr. Charles Armstrong of the United States Public Health Service who successfully passed type 2 polio virus to cotton rats and guinea pigs in 1940, it is difficult to find any important advances in polio research in the United States which were not sponsored or supported by the National Foundation.

The import of the National Foundation's support of polio research is perhaps best understood by examining the development of Salk vaccine. Although the vaccine was the result of Salk's scientific and technical skill, it was no less the end product of scores of research projects, initiated and supported by the Foundation. The rapid perfection and production of the vaccine itself could not have been accomplished without the cooperation of other virus laboratories, many of whose personnel (like Salk himself) were the product of scientific training programs begun by the Foundation. While it is true that scientists have long cooperated with one another in their research, the cooperation that led to the development of Salk vaccine was different in that it emerged from an environment which was specially created by the Foundation. In essence it grew out of the discussions of mutual problems by polio investigators at

Foundation sponsored symposia and meetings, as well as from a unique program organized by the research division of the Foundation whereby grantees of the Foundation exchanged research findings well before publication.

A special word must be said here of the medical scientists and physicians of the research division of the Foundation. Although many of these men and women often did not possess the individual brilliance and scientific skills of the investigators whom the Foundation supported, they nevertheless made an important contribution to maintaining both the continuity and tempo of polio research. They achieved this in a variety of ways, through their perceptiveness in organizing meetings to discuss outstanding research problems, by facilitating communications among various polio laboratories throughout the country, and, above all, by their cogent analysis of the validity of the myriad research projects which the scientific community almost weekly presented to the Foundation.

The success of the field trials conducted by Dr. Francis to test the safety and effectiveness of Salk vaccine was no less a product of this cooperative environment. It is doubtful that Dr. Francis could have conducted the trials without the selfless cooperation of colleagues in other virus laboratories, or indeed without the cooperation of the general public. One need hardly add, that the public cooperation Dr. Francis received during the trials was the result of more than fifteen years of Foundation organized and sponsored polio education programs.

This last fact contains the kernel of the National Foundation's unique contribution to the conquest of polio, namely, that through the totality of its various programs the Foundation succeeded in establishing the social usefulness and appropriateness of supporting and conducting polio research. That social definition proved to be an extraordinary goad and support to polio investigators. Its strength was enormous. For example, throughout the cold war period of the 1950's, American polio investigators met openly with their eastern European counterparts at Foundation sponsored international polio congresses and freely exchanged their latest research findings. One result of these meetings was that Albert Sabin's laboratory in Cincinnati provided the seed lots of the live virus polio vaccines which the Russians subsequently used in their mass polio immunization programs. Dr. Sabin was never accused of disloyalty. He was in fact applauded by men and women throughout the United States for a humanitarian act that all immediately recognized as being right and just. The gift of the vaccine was important—the acceptance of the giving of such a gift was no less important. That acceptance also stands as measure of the National Foundation's contribution to the conquest of polio.

REFERENCES

1. For a recent discussion of a similar theme see W. Weaver, ed., *U.S. Philanthropic Foundations: Their History, Structure, Management and Record* (New York: Harper and Row, 1967), esp. pp. 223-251, 260-275.

2. Much of the early history of polio may be found in P. H. Römer, *Epidemic Infantile Paralysis,* trans. H. R. Prentice (New York: William Wood and Company, 1913). For early developments in the United States see C. S. Caverly, "Preliminary Report of an Epidemic of Paralytic Disease, Occurring in Vermont in the Summer of 1894," *Yale Medical Journal,* 1 (1894), 4.

3. L. E. Holt and F. H. Bartlett, "The Epidemiology of Acute Poliomyelitis, A Study of Thirty-Four Epidemics," *American Journal of Medical Science,* 135 (1908), 74. Poliomyelitis was first made a reportable disease by the state of Massachusetts in 1909. Although twenty-four states made polio reportable in 1910, only three states, Massachusetts, Minnesota, and Pennsylvania, made such reporting mandatory by name. In other states it was reportable under the broad rubric of infectious and contagious diseases. In part, the problem of reporting poliomyelitis was related to the general neglect at that time in collecting morbidity statistics. See further, J. Collins, "The Epidemiology of Poliomyelitis: A Plea That It May Be Considered a Reportable Quarantinable Disease," *Journal of the American Medical Association,* 54 (1910), 1925.

4. For a detailed discussion of this point see S. Benison, "The Enigma of Poliomyelitis: 1910," in L. W. Levy and H. Hyman, eds., *Freedom and Reform: Essays in Honor of Henry Steele Commager* (New York: Harper and Row, 1967), pp. 235-241.

5. S. Flexner and I. Strauss, "The Pathology and Pathological Anatomy of Epidemic Poliomyelitis," in *Epidemic Poliomyelitis: Report on the New York Epidemic of 1907* (Nervous and Mental Disease Monograph Series no. 6) (New York: The Journal of Nervous and Mental Disease Publishing Company, 1910), pp. 57-104. The Rockefeller Institute for Medical Research is now Rockefeller University.

6. The best short published account of Dr. Flexner's life is P. Rous, "Simon Flexner," *Obituary Notices of Fellows of the Royal Society,* 6 (1949), 409.

7. S. Flexner, "Autobiographical Notes" (manuscript), chap. 1, p. 13. These notes have never been published and are currently preserved in the library of the American Philosophical Society in Philadelphia.

8. *Ibid.,* p. 14; and A. Flexner, *An Autobiography* (New York: Simon and Schuster, 1960), pp. 58-60.

9. The best detailed biography of Welch is S. Flexner and J. T. Flexner, *William H. Welch and the Heroic Age of American Medicine* (New York: Viking Press, 1941); a brief but useful biography is D. Fleming, *William H. Welch and the Rise of Modern Medicine* (Boston: Little, Brown, 1954), see esp. pp. 32-56.

10. Flexner, "Autobiographical Notes," chaps. 2-5.

11. G. W. Corner, *Two Centuries of Medicine* (Philadelphia: J. B. Lippincott, 1965), pp. 205-207.

12. G. W. Corner, *A History of the Rockefeller Institute: Origins and Growth, 1901-1953* (New York: Rockefeller Institute Press, 1964), pp. 50-53.

13. *Ibid.*, pp. 19-26; Frederick T. Gates's recollections on the origin of the Rockefeller Institute is reproduced in Appendix I, pp. 575-584.

14. *Ibid.*, pp. 56-59; Flexner, "Autobiographical Notes," chap. 8.

15. Flexner, "Autobiographical Notes," chap. 8, esp. pp. 6-8.

16. *Epidemic Poliomyelitis: Report on the New York Epidemic of 1907* (Nervous and Mental Disease Monograph Series no. 6) (New York: The Journal of Nervous and Mental Disease Publishing Company, 1910), pp. 3-9.

17. This theme is reiterated time and again in Flexner's early papers on poliomyelitis. A typical example is to be found in S. Flexner, "Experimental Poliomyelitis," *New York State Journal of Medicine*, 10 (1910), 330.

18. K. Landsteiner, "Untitled Abstract" in *Wiener Klinische Wochenschrift*, no. 52 (1908); K. Landsteiner and E. Popper, "Übertragung der Poliomyelitis Acuta auf Affen," *Zeitschrift für Immunitätsforschung und Experimentelle Therapie*, 2 (1910), 377.

19. S. Flexner and P. A. Lewis, "The Transmission of Acute Poliomyelitis to Monkeys," *Journal of the American Medical Association*, 53 (1909), 1639. It should be noted that although Landsteiner inoculated his monkeys interperitoneally, Flexner used an intracerebral mode of inoculation.

20. S. Flexner and P. A. Lewis, "The Transmission of Acute Poliomyelitis to Monkeys, a Further Note," *Journal of the American Medical Association*, 53 (1909); "The Nature of the Virus of Epidemic Poliomyelitis," *Journal of the American Medical Association*, 53 (1909), 2095.

21. I. Wickman, *Acute Poliomyelitis*, trans. W. J. M. A. Maloney (Nervous and Mental Disease Monograph Series no. 16) (New York: The Journal of Nervous and Mental Disease Publishing Company, 1913), pp. 38-83.

22. Dr. H. W. Hill epidemiologist of the Minnesota Board of Health at that time was one of those who steadfastly opposed the notion that poliomyelitis was a contagious disease. See especially H. W. Hill, "The Contagiousness of Poliomyelitis," *Journal of the Minnesota State Medical Association and Northwestern Lancet*, 30 (1910), 111. Some of the early epidemiological reports from Massachusetts also suggested that the disease was only mildly contagious; H. C. Emerson, "An Epidemic of Infantile Paralysis in Western Massachusetts in 1908," *Monthly Bulletin of the Massachusetts State Board of Health* (July 1909), p. 25.

23. S. Flexner and P. A. Lewis, "Experimental Epidemic Poliomyelitis in Monkeys," *Journal of Experimental Medicine*, 12 (1910), 227.

24. S. Flexner and P. A. Lewis, "Epidemic Poliomyelitis in Monkeys: A Mode of Spontaneous Infection," *Journal of the American Medical Association*, 54 (1910), 535.

25. S. Flexner, "The Contribution of Experimental to Human Poliomyelitis," *Journal of the American Medical Association*, 55 (1910), 1105.

26. Flexner and Lewis, "Epidemic Poliomyelitis in Monkeys," pp. 536-537.

27. S. Flexner and H. Noguchi, "Experiments on the Cultivation of the Virus of Poliomyelitis," *Journal of the American Medical Association*, 60 (1913), 362.

28. W. W. Keen to S. Flexner, March 23, 1913, Simon Flexner papers, library of the American Philosophical Society.

29. T. M. Rivers, "The Story of Research on Poliomyelitis," *Proceedings of the American Philosophical Society*, 98 (1954), 254.

30. S. Benison, *Peter K. Olitsky: An Oral History Memoir*, 1962 (manuscript in author's possession).

31. P. H. Long, P. K. Olitsky, and C. P. Rhoads, "Survival and Multiplication of the Virus of Poliomyelitis in Vitro," *Journal of Experimental Medicine*, 52 (1930), 361.

32. For an appraisal of Peter K. Olitsky see S. Benison, *Tom Rivers: Reflections on a Life in Medicine and Science* (Cambridge, Mass.: MIT Press, 1967), pp. 111-114.

33. A. B. Sabin, "The Olfactory Bulbs in Human Poliomyelitis," *American Journal of Diseases of Children*, 16 (1940), 1313; "Etiology of Poliomyelitis," *Journal of the American Medical Association*, 117 (1941), 267.

34. C. Swan, "The Anatomical Distribution and Character of the Lesions of Poliomyelitis," *Australian Journal of Experimental Biological Medical Sciences*, 17 (1939), 345.

35. C. Kling, C. Levaditi, and P. Lépine, "La pénétration du Virus Poliomyélitique, à travers la muqueuse du tube digesitif chez le singe, et sa conservation daus l'eau," *Bulletin Académie de Médecine, Paris*, 102 (1929), 158; C. Levaditi, C. Kling, and P. Lépine, "Nouvelles Recherches expérimental sur la transmission de la poliomyélite par la voie digestive. Action du chlore sur le virus poliomyélitique," *Bulletin Académie de Médecine, Paris*, 105 (1931), 190.

36. C. Kling, C. Levaditi, and G. Hornus, "Comparaison entres les divers modes de contamination du singe par le virus poliomyélitique (voies digestive et nasopharyngée)," *Bulletin Académie de Médecine, Paris*, 111 (1934), 709.

37. S. Flexner, "Respiratory Versus Gastro-Intestinal Infection in Poliomyelitis," *Journal of Experimental Medicine*, 63 (1936), 209.

38. This conception has given rise to a problem, which for lack of a better term can be called the Noguchi problem. Recent analysis of Noguchi's work has been sharply critical, citing his research failures in polio, yellow fever, and trachoma. All of these failures have one element in common, all are virus diseases. Noguchi's failures all occurred before bacteria and viruses were differentiated. In essence, Noguchi is blamed for not having made the icthyian leap from bacteriology to virology. The criticism is nothing more than a species of Monday morning quarterbacking. Two examples of such criticism are: P. F. Clark, "Hideyo Noguchi, 1876-1929," *Bulletin of the History of Medicine*, 33 (1959), 1; and Benison, *Tom Rivers*, pp. 93-98. For a recent discussion of the historical development of virology see H. A. Lechevalier and M. Solotorovsky, *Three Centuries of Microbiology* (New York: McGraw-Hill, 1965), pp. 280-332.

39. H. Albrecht and A. Ghon, "Über die Aetiologie und Pathologische Anatomie der Meningitis Cerebrospinalis Epidemica," *Wiener Klinische Wochenschrift*, 14

(1901), 984; W. T. Councilman, F. B. Mallory, and J. H. Wright, "Epidemic Cerebrospinal Meningitis and Its Relation to Other Forms of Meningitis," *Report of the Massachusetts State Board of Health* (1898); S. Flexner, "Contributions to the Biology of Diplococcus Intercellularis," *Journal of Experimental Medicine*, 9 (1907), 105.

40. H. Noguchi, "Cultivation of Pathogenic Treponema Pallidum," *Journal of the American Medical Association*, 57 (1911), 102; "A Method for the Pure Cultivation of Pathogenic Treponema Pallidum," *Journal of Experimental Medicine*, 14 (1911), 99.

41. Benison, *Tom Rivers*, pp. 117-118. A prospectus for Dr. Simon's course can be found in Peter Olitsky's papers now preserved in the library of the American Philosophical Society.

42. T. M. Rivers, ed., *Filterable Viruses* (Baltimore: Williams and Wilkins, 1928).

43. Rivers first expressed this point of view in an article a year before. T. M. Rivers, "Filterable Viruses, a Critical Review," *Journal of Bacteriology*, 14 (1927), 228.

44. Benison, *Tom Rivers*, pp. 67-179; Corner, *History of the Rockefeller Institute*, pp. 264-267.

45. Corner, *History of the Rockefeller Institute*, pp. 384-390.

46. M. Theiler, "The Virus" in G. K. Strode, ed., *Yellow Fever* (New York: McGraw-Hill, 1951), pp. 37-136; Benison, *Tom Rivers*, pp. 413-416. It should be noted that Theiler's yellow fever research began before he joined the Rockefeller Foundation laboratories. In 1929, while still at Harvard, he made a significant breakthrough in that research by successfully transmitting the disease experimentally to mice.

47. W. M. Stanley, "Isolation of a Crystalline Protein Possessing the Properties of Tobacco Mosaic Virus," *Science*, 81 (1935), 644.

48. R. Fosdick, *The Story of the Rockefeller Foundation* (New York: Harper and Brothers, 1952), p. 151; Benison, *Tom Rivers*, p. 217; Corner, *History of the Rockefeller Institute*, pp. 185, 361-363.

49. *Ibid.*, pp. 357-360.

50. B. von Borries, E. Ruska, and H. Ruska, "Bacterium und virus in übermikroscopischer Aufnahme," *Klinische Wochenschrift*, 17 (1938), 94; see also Benison, *Tom Rivers*, pp. 220-221.

51. Benison, *Tom Rivers*, p. 205.

52. Certificate of Incorporation of the National Foundation for Infantile Paralysis Inc. Pursuant to Membership Corporation Law, Organizational Files, National Foundation Archives, New York.

53. T. Walker, *Roosevelt and the Warm Springs Story* (New York: A. A. Wyn Incorporated, 1953).

54. J. Roosevelt, "A Tribute to Basil O'Connor," in *Memorials to Basil O'Connor on His 75th Birthday* (New York: National Foundation, 1967).

55. R. Carter, *The Gentle Legions* (New York: Doubleday and Company, 1961); Benison, *Tom Rivers*, p. 179, n.1.

56. P. De Kruif, *The Sweeping Wind* (New York: Harcourt Brace and Company, 1962), pp. 16-51; Benison, *Tom Rivers*, pp. 180-183, esp. note on pp. 181-182; Corner, *History of the Rockefeller Institute*, pp. 160-161.

57. For those interested in examining further de Kruif's role in the writing of Sinclair Lewis' *Arrowsmith*, see M. Schorer, *Sinclair Lewis, an American Life* (New York: McGraw Hill, 1961); C. Rosenberg, "Martin Arrowsmith, the Scientist as Hero," *American Quarterly*, 15 (1963), 448.

58. Benison, *Tom Rivers*, pp. 183-190.

59. Private communication from Ruth Andrews, accounting staff of the National Foundation, June 1969.

60. *Ibid.*

61. Benison, *Tom Rivers*, pp. 229-230.

62. The Committee on Scientific Research was initially organized on July 6, 1938. It was subsequently reorganized several times with accompanying name changes reflecting new functions and responsibilities. On May 13, 1940, it became the Committee on Virus Research; on September 30, 1947, the Committee on Virus Research and Epidemiology; on April 8, 1959, the Committee on Research; and on October 15, 1959, the Committee on Research in the Basic Sciences. The original members of the committee were Paul de Kruif, Donald Armstrong, Charles Armstrong, George McCoy, Karl Meyer, and Thomas Rivers.

63. Minutes of the Committee on Scientific Research, April 18, 1939, National Foundation Archives.

64. R. H. Bremner, *American Philanthropy* (Chicago: University of Chicago Press, 1960); F. E. Andrews, *Philanthropic Foundations* (New York: Russell Sage Foundation, 1956); V. D. Bornet, *Welfare in America* (Norman, Okla.: University of Oklahoma Press, 1960); W. Weaver, ed., *U.S. Philanthropic Foundations*, pp. 19-89.

65. Benison, *Tom Rivers*, pp. 229-230.

66. *Ibid.*, pp. 240-249.

67. *Ibid.*, pp. 249-261, 261-271.

68. *Ibid.*, pp. 444-445.

69. *Ibid.*, p. 480.

70. J. R. Paul and J. D. Trask, "Occurrence and Recovery of Virus of Infantile Paralysis from Sewage," *American Journal of Public Health*, 32 (1942), 235; J. L. Melnick, "Ultracentrifuge as an Aid in Detection of Poliomyelitis Virus," *Journal of Experimental Medicine*, 77 (1943), 195; S. Gard, "Über mikroskopische Beobachtungen angereinigten Poliomyelitis virus präparaten: Ein Vergleich mit den physikalisch chemischen Versuchsergebnissen," *Archiv für die gesamte Virusforschung*, 3 (1943), 1; T. Francis, "Epidemiological Study of Poliomyelitis following Tonsillectomy," *Transactions of the Association of American Physicians*, 57 (1942), 277.

71. A. B. Sabin and R. Ward, "Flies as Carriers of Poliomyelitis Virus in Urban Epidemics," *Science*, 94 (1941), 590; R. Ward, J. L. Melnick, D. M. Horst-

mann, "Poliomyelitis Virus in Fly Contaminated Food, Collected at an Epidemic," *Science,* 101 (1945), 491.

72. R. W. Schlesinger, I. Morgan, and P. K. Olitsky, "Transmission to Rodents of Lansing Type Poliomyelitis Virus Originating in the Middle East," *Science,* 98 (1943), 452.

73. H. A. Howe and D. Bodian, *Neural Mechanisms in Poliomyelitis* (New York: Commonwealth Fund, 1943).

74. H. A. Waisman, A. F. Rasmussen, C. A. Elvekjem, and P. F. Clark, "Studies on the Nutritional Requirements of the Rhesus Monkey," *Journal of Nutrition,* 26 (1943), 205; A. F. Rasmussen, H. A. Waisman, C. A. Elvekjem, and P. F. Clark, "Inference of the Level of Thiamine Intake on the Susceptibility of Mice to Poliomyelitis Virus," *Journal of Infectious Diseases,* 74 (1944), 41.

75. C. E. Van Rooyen and A. J. Rhodes, *Virus Diseases of Man* (London: Oxford University Press, 1940); Harvard School of Public Health Symposium, *Virus and Rickettsial Diseases* (Cambridge, Mass.: Harvard University Press, 1940).

76. T. M. Rivers, ed., *Viral and Rickettsial Infections of Man* (Philadelphia: J. B. Lippincott, 1948); R. Dubos, *Bacterial and Mycotic Infections of Man* (Philadelphia: J. B. Lippincott, 1948). These volumes have now gone through four editions, the last appearing in 1965. For an account of the development of those volumes see Benison, *Tom Rivers,* pp. 397-403.

77. Benison, *Tom Rivers,* pp. 403-404.

78. H. R. Cox and P. K. Olitsky, "Active Immunization of Guinea Pigs with the Virus of Equine Encephalomyelitis," *Journal of Experimental Medicine,* 63 (1936), 745; I. M. Morgan and P. K. Olitsky, "Immune Response of Mice to Active Virus and Formalin-Inactivated Virus of Eastern Equine Encephalomyelitis," *Journal of Immunology,* 42 (1941), 445; Benison, *Tom Rivers,* pp. 212-213, esp. Dr. Olitsky's account of the evolution of this research, p. 212, n. 9.

79. Members of the Commission of Influenza, Army Epidemiological Board, "A Clinical Evaluation of Vaccination against Influenza," *Journal of the American Medical Association,* 124 (1944), 982; J. E. Salk, H. E. Pearson, P. H. Brown, and T. Francis, Jr., "Protective Effect of Vaccination against Induced Influenza B," *Proceedings of the Society of Experimental Biology and Medicine,* 55 (1944), 106.

80. I. M. Morgan, "The Role of Antibody in Immunity to Poliomyelitis," *Proceedings of the Round Table Conference on the Mechanics of Immunity in Poliomyelitis* (Baltimore: National Foundation, Typescript, 1946), pp. 16-20, National Foundation Archives.

81. F. M. Burnet and J. Macnamara, "Immunological Differences between Strains of Poliomyelitis Virus," *British Journal of Experimental Pathology,* 12 (1931), 57.

82. J. R. Paul and J. D. Trask, "Comparative Study of Recently Isolated Human Strains and Passage Strain of Poliomyelitis Virus," *Journal of Experimental Medicine,* 53 (1933), 513; J. D. Trask, J. R. Paul, A. R. Beebe, and W. J. German, "Viruses of Poliomyelitis: Immunologic Comparison of Six Strains," *Journal of Experimental Medicine,* 65 (1937), 687.

83. J. F. Kessel and C. F. Pait, "Resistance of Convalescent Macaca Mulatta to Challenge with Homologous and Heterologous Strains of Poliomyelitis Virus," *Proceedings of the Society of Experimental Biology and Medicine,* 68 (1948), 606; D. Bodian, I. M. Morgan, and H. A. Howe, "Differentiation of Types of Poliomyelitis Viruses: The Grouping of Fourteen Strains into Three Basic Immunologic Types," *American Journal of Hygiene,* 49 (1949), 234.

84. Benison, *Tom Rivers,* pp. 419-421.

85. J. F. Enders, T. H. Weller, and F. C. Robbins, "Cultivation of the Lansing Strain of Poliomyelitis Virus in Cultures of Various Human Embryonic Tissues," *Science,* 109 (1949), 85.

86. A. B. Sabin and P. K. Olitsky, "Cultivation of Poliomyelitis Virus in Vitro in Human Embryonic Nervous Tissue," *Proceedings of the Society of Experimental Biology and Medicine,* 34 (1936), 357. For an evaluation of this research see Benison, *Tom Rivers,* pp. 236-237.

87. Morgan, "The Role of Antibody."

88. E. J. Cohn, "The History of Plasma Fractionation," in C. E. Andrus and others, eds., *Advances in Military Medicine,* I (Boston, Mass.: Little, Brown, 1946), pp. 364-443.

89. Benison, *Tom Rivers,* pp. 470-472.

90. *Ibid.,* see esp. the comments of Dr. Joseph Stokes, Jr., on the problems of gaining support for passive immunization, pp. 472-474n.

91. *Minutes, Round Table Conference on Gamma Globulin,* February 3, 1950, National Foundation Archives; Benison, *Tom Rivers,* pp. 476-482.

92. W. McD. Hammon, L. L. Coriell, and J. Stokes, Jr., "Evaluation of Red Cross Gamma Globulin as a Prophylactic Agent for Poliomyelitis, I, Plan of Controlled Field Tests and Results of 1951 Pilot Study in Utah," *Journal of the American Medical Association,* 150 (1952), 739; "Evaluation of Red Cross Gamma Globulin as a Prophylactic Agent for Poliomyelitis, II, Conduct and Early Follow-Up of 1952 Texas and Iowa-Nebraska Studies," *Journal of the American Medical Association,* 150 (1952), 750.

93. D. Bodian, "A Reconsideration of the Pathogenesis of Poliomyelitis," *American Journal of Hygiene,* 55 (1952), 414; D. Horstmann, "Poliomyelitis Virus in Blood of Orally Infected Monkeys and Chimpanzees," *Proceedings of the Society of Experimental Biology and Medicine,* 79 (1952), 417.

94. The development of Dr. Salk's early career as an investigator can be traced in R. Carter, *Breakthrough: The Saga of Jonas Salk* (New York: Trident Press, 1966), pp. 34-53, 95-99; Benison, *Tom Rivers,* pp. 492-493.

95. The process of Dr. Salk's research on an inactivated vaccine in 1952-1953 can best be followed by a perusal of Grant Files, Jonas Salk, University of Pittsburgh, CRBS #105 (1952-1953), National Foundation Archives.

96. J. E. Salk, with the collaboration of B. L. Bennet, L. J. Lewis, E. H. Ward, and J. S. Younger, "Studies in Human Subjects on Active Immunization Against Poliomyelitis, I, A Preliminary Report of Experiments in Progress," *Journal of the American Medical Association,* 151 (1953), 1081. See also Carter, *Breakthrough,* pp. 137-140.

97. Carter, *Breakthrough,* pp. 141-169; Benison, *Tom Rivers,* pp. 495-501. See also Minutes of the Meeting of the Committee on Immunization, National Foundation, New York, January 23, 1953; Memorandum, Harry Weaver to Basil O'Connor, January 30, 1953 (Folder, Vaccine, Polio Salk: Development and Promotion, 1952); Minutes of a Meeting of the National Foundation, February 26, 1953, all in National Foundation Archives.

98. Members of the Vaccine Advisory Committee were David Price, Thomas Murdock, Ernest Stebbins, Thomas Turner, Norman Topping, Joseph Smadel, and Thomas Rivers. For the debate growing out of the creation of this committee see Benison, *Tom Rivers,* pp. 502-505.

99. Development of specifications for production and safety testing can be followed in Benison, *Tom Rivers,* pp. 512-534.

100. Carter, *Breakthrough,* pp. 176-202; Benison, *Tom Rivers,* pp. 506-512.

101. Benison, *Tom Rivers,* pp. 534-537.

102. T. Francis, Jr., and others, *Evaluation of the 1954 Field Trial of Poliomyelitis Vaccine* (Final Report), Michigan University. Poliomyelitis Evaluation Center (1957).

103. For two opposing views on this problem see John R. Wilson, *Margin for Safety* (New York: Doubleday and Company, 1963), and Benison, *Tom Rivers,* pp. 545-563; Carter, *Breakthrough,* pp. 301-354.

104. House of Representatives, Interstate and Foreign Commerce Committee, 84th Congress, 1st Session, *Hearings on Poliomyelitis Vaccine,* June 22-23, 1955.

105. Benison, *Tom Rivers,* pp. 560-561.

106. *Ibid.,* pp. 514-515.

107. *Ibid.,* pp. 530-534.

108. There are some commentators and scientists who do not adhere to these views. See especially J. R. Wilson, *Margin for Safety,* pp. 67-115, and Paul Meier, "Safety Testing of Poliomyelitis Vaccine," *Science,* 125 (1957), 1067, who stresses the culpability of the Foundation in the Cutter incident.

109. David Bodian, "Control of the Manufacture of Poliomyelitis Vaccine," *Papers and Discussions Presented at the Fourth International Poliomyelitis Congress* (Philadelphia: Lippincott, 1958), p. 79.

110. *Ibid.,* p. 77.

111. Grant Files, Herbert Wenner, University of Kansas, CRBS #142 (1952-1955), National Foundation Archives.

112. The development of Dr. Sabin's work on his live poliovirus vaccine is best followed in Grant Files, Albert Sabin, University of Cincinnati, CRBS #139 (1952-1962), National Foundation Archives; see also Benison, *Tom Rivers,* pp. 540-543. For a good review of Dr. Sabin's research in the early years of his work on the vaccine see A. B. Sabin, *Immunity in Poliomyelitis, with Special Reference to Vaccination,* WHO Monograph Series, no. 26, pp. 297-334 (1955).

113. Proceedings of the Round Table Conference on Genetic Aspects of Virus Host-Relationships with Particular Reference to Polioviruses, February 23-24, 1956, National Foundation Archives.

114. Benison, *Tom Rivers,* p. 566. See also A. B. Sabin, "Properties of Attenuated Polioviruses and Their Behavior in Human Beings," in *Cellular Biology, Nucleic Acids and Viruses,* Special Publication of the New York Academy of Sciences, 5 (1957), 113-140.

115. Benison, *Tom Rivers,* pp. 568-574.

116. Grant Files, Albert Sabin, University of Cincinnati, CRBS #139 (1952-1962), National Foundation Archives, shows very clearly the nature of Foundation support for Sabin in the years during the development of his vaccine.

117. The significance of foundation support for the development of polio research is perhaps best illustrated by the evolution of the investigative career of Dr. Lloyd Aycock of Harvard Medical School. Dr. Aycock, who made important contributions to an understanding of the epidemiology of polio, began his research career as a member of a special polio laboratory of the Vermont Board of Health. This laboratory was organized and supported through the beneficence of the Proctor family of Rutland, Vermont, and was directed by Dr. Harold Amoss of the Rockefeller Institute. In 1916, Dr. Aycock, for a brief period of time, left his post in Vermont and studied a polio outbreak in Westchester, New York, as a member of a special research team organized and supported by the Rockefeller Foundation. In the decade that followed, Dr. Aycock continued his polio research under the auspices of the Harvard Infantile Paralysis Commission. Subsequently his polio research was supported by the President's Birthday Ball Commission (the predecessor of the National Foundation for Infantile Paralysis) and finally by the Commonwealth Fund. These foundations did not teach Dr. Aycock how to think. They did, however, help train him in polio research and even more important gave that research continuity.

GERALD HOLTON

Mach, Einstein, and the Search for Reality

IN THE history of ideas of our century, there is a chapter that might be entitled "The Philosophical Pilgrimage of Albert Einstein," a pilgrimage from a philosophy of science in which sensationism and empiricism were at the center, to one in which the basis was a rational realism. This essay, a portion of a more extensive study,[1] is concerned with Einstein's gradual philosophical reorientation, particularly as it has become discernible during the work on his largely unpublished scientific correspondence.[2]

The earliest known letter by Einstein takes us right into the middle of the case. It is dated 19 March 1901 and addressed to Wilhelm Ostwald.[3] The immediate cause for Einstein's letter was his failure to receive an assistantship at the school where he had recently finished his formal studies, the Polytechnic Institute in Zürich; he now turned to Ostwald to ask for a position at his laboratory, partly in the hope of receiving "the opportunity for further education." Einstein included a copy of his first publication, "Folgerungen aus den Capillaritätserscheinungen" (*Annalen der Physik*, 4 [1901], 513), which he said had been inspired *(angeregt)* by Ostwald's work; indeed, Ostwald's *Allgemeine Chemie* is the first book mentioned in all of Einstein's published work.

Not having received an answer, Einstein wrote again to Ostwald on 3 April 1901. On 13 April 1901 his father, Hermann Einstein, sent Ostwald a moving appeal, evidently without his son's knowledge. Hermann Einstein reported that his son esteems

Ostwald "most highly among all scholars currently active in physics."[4]

The choice of Ostwald was significant. He was, of course, not only one of the foremost chemists, but also an active "philosopher-scientist" during the 1890's and 1900's, a time of turmoil in the physical sciences as well as in the philosophy of science. The opponents of kinetic, mechanical, or materialistic views of natural phenomena were vociferous. They objected to atomic theory and gained great strength from the victories of thermodynamics, a field in which no knowledge or assumption was needed concerning the detailed nature of material substances (for example, for an understanding of heat engines).

Ostwald was a major critic of the mechanical interpretation of physical phenomena, as were Helm, Stallo, and Mach. Their form of positivism—as against the sophisticated logical positivism developed later in Carnap and Ayer's work—provided an epistemology for the new phenomenologically based science of correlated observations, linking energetics and sensationism. In the second (1893) edition of his influential textbook on chemistry, Ostwald had given up the mechanical treatment of his first edition for Helm's "energetic" one. "Hypothetical" quantities such as atomic entities were to be omitted; instead, these authors claimed they were satisfied, as Merz wrote around 1904, with "measuring such quantities as are presented directly in observation, such as energy, mass, pressure, volume, temperature, heat, electrical potential, etc., without reducing them to imaginary mechanisms or kinetic quantities." They condemned such conceptions as the ether, with properties not accessible to direct observation, and they issued a call "to consider anew the ultimate principles of all physical reasoning, notably the scope and validity of the Newtonian laws of motion and of the conceptions of force and action, of absolute and relative motion."[5]

All these iconoclastic demands—except anti-atomism—must have been congenial to the young Einstein who, according to his colleague Joseph Sauter, was fond of calling himself "a heretic."[6] Thus, we may well suspect that Einstein felt sympathetic to Ostwald who denied in the *Allgemeine Chemie*[7] that "the assumption of that medium, the ether, is unavoidable. To me it does not seem to be so. . . . There is no need to inquire for a carrier of it when we find it anywhere. This enables us to look upon radiant energy as independently existing in space." It is a position quite consistent with that shown later in Einstein's papers of 1905 on photon theory and relativity theory.

In addition, it is worth noting that Einstein, in applying to Ostwald's laboratory, seemed to conceive of himself as an experimentalist. We know from many sources that in his student years in Zürich Einstein's earlier childhood interest in mathematics had slackened considerably. In the *Autobiographical Notes,*[8] Einstein reported: "I really could have gotten a sound mathematical education. However, I worked most of the time in the physical laboratory, fascinated by the direct contact with experience" (p. 15). To this, one of his few reliable biographers adds: "No one could stir him to visit the mathematical seminars. . . . He did not yet see the possibility of seizing that formative power resident in mathematics, which later became the guide of his work. . . . He wanted to proceed quite empirically, to suit his scientific feeling of the time. . . . As a natural scientist, he was a pure empiricist." (Anton Reiser, *Albert Einstein* [New York, 1930], pp. 51-52.)

Ostwald's main philosophical ally was the prolific and versatile Austrian physicist and philosopher Ernst Mach (1838-1916), whose main work Einstein had read avidly in his student years and with whom he was destined to have later the encounters that form a main concern of this paper. Mach's major book, *The Science of Mechanics,*[9] first published in 1883, is perhaps most widely known for its discussion of Newton's *Principia,* in particular for its devastating critique of what Mach called the "conceptual monstrosity of absolute space" (preface, 7th edition, 1912)—a conceptual monstrosity because it is "purely a thought-thing which cannot be pointed to in experience." Starting from his analysis of Newtonian presuppositions, Mach proceeded in his announced program of eliminating all metaphysical ideas from science. As Mach said quite bluntly in the preface to the first edition of *The Science of Mechanics:* "This work is not a text to drill theorems of mechanics. Rather, its intention is an enlightening one—or to put it still more plainly, an anti-metaphysical one."

It will be useful to review briefly the essential points of Mach's philosophy. Here we can benefit from a good, although virtually unknown, summary presented by his sympathetic follower, Moritz Schlick, in the essay "Ernst Mach, Der Philosoph."[10]

Mach was a physicist, physiologist, and also psychologist, and his philosophy . . . arose from the wish to find a principal point of view to which he could hew in any research, one which he would not have to change when going from the field of physics to that of physiology or psychology. Such a firm point of view he reached by going back to that which is given before all scientific research: namely, the world of sensations. . . . Since all our testimony concerning the so-called external world rely only on sensations, Mach held that we can and must take these sensations and complexes of sensations to be the sole contents [*Gegenstände*] of those testimonies, and, therefore, that there is no need to assume in addition an unknown reality hidden behind the sensations. With that, the existence *der Dinge an sich* is removed as an unjustified and unnecessary assumption. A body, a physical object, is nothing else than a complex, a more or less firm [we would say, invariant] pattern of sensations, i.e., of colors, sounds, sensations of heat and pressure, etc.

There exists in this world nothing whatever other than sensations and their connections. In place of the word "sensations," Mach liked to use rather the more neutral word "elements." . . . [As is particularly clear in Mach's book *Erkenntnis und Irrtum*,] scientific knowledge of the world consists, according to Mach, in nothing else than the simplest possible description of the connections between the elements, and it has as its only aim the intellectual mastery of those facts by means of the least possible effort of thought. This aim is reached by means of a more and more complete "accommodation of the thoughts to one another." This is the formulation by Mach of his famous "principle of the economy of thought."[11]

The influence of Mach's point of view, particularly in the German-speaking countries, was enormous—on physics, on physiology, on psychology, and on the fields of the history and the philosophy of science[12] (not to mention Mach's profound effect on the young Lenin, Hofmannsthal, Musil, among many others outside the sciences). Strangely neglected by recent scholarship—there is not even a major biography—Mach has in the last two or three years again become the subject of a number of promising studies. To be sure, Mach himself always liked to insist that he was beleaguered and neglected, and that he did not have, or wish to have, a philosophical system; yet his philosophical ideas and attitudes had become so widely a part of the intellectual equipment of the period from the 1880's on that Einstein was quite right in saying later that "even those who think of themselves as Mach's opponents hardly know how much of Mach's views they have, as it were, imbibed with their mother's milk."[13]

The problems of physics themselves at that time helped to reinforce the appeal of the new philosophical attitude urged by Mach. The great program of nineteenth-century physics, the reconciliation of the notions of ether, matter, and electricity by means of mechanistic pictures and hypotheses, had led to enormities—for example, Larmor's proposal that the electron is a permanent but movable state of twist or strain in the ether, forming discontinuous particles of electricity and possibly of all ponderable matter. To many of the younger physicists of the time, attacking the problems of physics with conceptions inherited from classical nineteenth-century physics did not seem to lead anywhere. And here Mach's iconoclasm and incisive critical courage, if not the details of his philosophy, made a strong impression on his readers.

Mach's Early Influence on Einstein

As the correspondence at the Einstein Archives at Princeton reveals, one of the young scientists deeply caught up in Mach's point of view was Michelange (Michele) Besso—Einstein's oldest and closest friend, fellow student, and colleague at the Patent Office in Bern, the only person to whom Einstein gave public credit for help *(manche wertvolle Anregung)* when he published his basic paper on relativity in 1905. It was Besso who introduced Einstein to Mach's work. In a letter of 8 April 1952 to Carl Seelig, Einstein wrote: "My attention was drawn to Ernst Mach's *Science of Mechanics* by my friend Besso while a student, around the year 1897. The book exerted a deep and persisting impression upon me . . ., owing to its physical orientation toward fundamental concepts and fundamental laws." As Einstein noted in his *Autobiographical Notes*[8] written in 1946, Ernst Mach's *The Science of Mechanics* "shook this dogmatic faith" in "mechanics as the final basis of all physical thinking. . . . This book exercised a profound influence upon me in this regard while I was a student. I see Mach's greatness in his incorruptible skepticism and independence; in my younger years, however, Mach's epistemological position also influenced me very greatly (p. 21)."

As the long correspondence between those old friends shows, Besso remained a loyal Machist to the end. Thus, writing to Einstein on 8 December 1947, he still said: "As far as the history of science is concerned, it appears to me that Mach stands at the center of the development of the last 50 or 70 years." Is it not true, Besso also asked, "that this introduction [to Mach] fell into a phase of development of the young physicist [Einstein] when the Machist style of thinking pointed decisively at observables—perhaps even, indirectly, to clocks and meter sticks?"

Turning now to Einstein's crucial first paper on relativity in

1905, we can discern in it influences of many, partly contradictory, points of view—not surprising in a work of such originality by a young contributor. Elsewhere I have examined the effect—or lack of effect—on that paper of three contemporary physicists: H. A. Lorentz,[14] Henri Poincaré,[15] and August Föppl.[1] Here we may ask in what sense and to what extent Einstein's initial relativity paper of 1905 was imbued with the style of thinking associated with Ernst Mach and his followers—apart from the characteristics of clarity and independence, the two traits in Mach which Einstein always praised most.

In brief, the answer is that the Machist component—a strong component, even if not the whole story—shows up prominently in two related respects: first, by Einstein's insistence from the beginning of his relativity paper that the fundamental problems of physics cannot be understood until an epistemological analysis is carried out, particularly so with respect to the meaning of the conceptions of space and time[16]; and second, by Einstein's identification of reality with what is given by sensations, the "events," rather than putting reality on a plane beyond or behind sense experience.

From the outset, the instrumentalist, and hence sensationist, views of measurement and of the concepts of space and time are strikingly evident. The key concept in the early part of the 1905 paper is introduced at the top of the third page in a straightforward way. Indeed, Leopold Infeld in his biography of Einstein called them "the simplest sentence[s] I have ever encountered in a scientific paper." Einstein wrote: "We have to take into account that all our judgments in which time plays a part are always judgments of *simultaneous events.* If for instance I say, 'that train arrived here at seven o'clock,' I mean something like this: 'The pointing of the small hand of my watch to seven and the arrival of the train are simultaneous events.' "[17]

The basic concept introduced here, one that overlaps almost entirely Mach's basic "elements," is Einstein's concept of *events (Ereignisse)*—a word that recurs in Einstein's paper about a dozen times immediately following this citation. Transposed into Minkowski's later formulation of relativity, Einstein's "events" are the intersections of particular "world lines," say that of the train and that of the clock. The time (t coordinate) of an event by itself has no operational meaning. As Einstein says: "The 'time' of an event is that which is given simultaneously with the event by a stationary clock located at the place of the event" (p. 894). We can say that just as the *time* of an event assumes meaning only when it connects with our consciousness through sense experience (that is, when it is subjected to measurement-in-principle by means of a clock present at the same place), so also is the *place*, or space coordinate, of an event meaningful only if it enters our sensory experience while being subjected to measurement-in-principle (that is, by means of meter sticks present on that occasion at the same time).[18]

This was the kind of operationalist message which, for most of his readers, overshadowed all other philosophical aspects in Einstein's paper. His work was enthusiastically embraced by the groups who saw themselves as philosophical heirs of Mach, the Vienna Circle of neopositivists and its predecessors and related followers,[19] providing a tremendous boost for the philosophy that had initially helped to nurture it. A typical response welcoming the relativity theory as "the victory over the metaphysics of absolutes in the conceptions of space and time . . . a mighty impulse for the development of the philosophical point of view of our time," was extended by J. Petzoldt in the inaugural session of the *Gesellschaft für Positivistische Philosophie* in Berlin, 11 November 1912.[20] Michele Besso, who had heard the message from Einstein before anyone else, had exclaimed: "In the setting of Minkowski's space-time framework, it was now first possible to carry through the thought which the great mathematician, Bernhard Riemann, had grasped: 'The space-time framework itself is formed by the events in it.' "[21]

To be sure, rereading Einstein's paper with the wisdom of hindsight, as we shall do presently, we can find in it also very different trends, warning of the possibility that "reality" in the end is not going to be left identical with "events." There are premonitions that sensory experiences, in Einstein's later work, will not be regarded as the chief building blocks of the "world," that the laws of physics themselves will be seen to be built into the event-world as the undergirding structure "governing" the pattern of events.

Such precursors appear even earlier, in one of Einstein's early letters in the Archives. Addressed to his friend Marcel Grossmann, it is dated 14 April 1901, when Einstein believed he had found a connection between Newtonian forces and the forces of attraction between molecules: "It is a wonderful feeling to recognize the unity of a complex of appearances which, to direct sense experience, seem to be separate things." Already there is a hint here of the high value that will be placed on intuited unity and the limited role seen for evident sense experience.

But all this was not yet ready to come into full view, even to the author. Taking the early papers as a whole, and in the context of the physics of the day, we find that Einstein's philosophical pilgrimage did start on the historic ground of positivism. Moreover, Einstein thought so himself, and confessed as much in letters to Ernst Mach.

The Einstein-Mach Letters

In the history of recent science, the relation between Einstein and Mach is an important topic that has begun to interest a number of scholars. Indeed, it is a drama of which we can sketch here four stages: Einstein's early acceptance of the main features of Mach's doctrine; the Einstein-Mach correspondence and meeting; the revelation in 1921 of Mach's unexpected and vigorous attack on Einstein's relativity theory; and Einstein's own further development of a philosophy of knowledge in which he rejected many, if not all, of his earlier Machist beliefs.

Happily, the correspondence is preserved at least in part. A few letters have been found, all from Einstein to Mach. Those of concern here are part of an exchange between 1909 and 1913, and they testify to Einstein's deeply felt attraction to Mach's viewpoint, just at a time when the mighty Mach himself—forty years senior to the young Einstein whose work was just becoming widely known—had for his part embraced the relativity theory publicly by writing in the second (1909) edition of *Conservation of Energy:* "I subscribe, then, to the principle of relativity, which is also firmly upheld in my *Mechanics* and *Wärmelehre*."[22] In the first letter, Einstein writes from Berne on 9 August 1909. Having thanked Mach for sending him the book on the law of conservation of energy, he adds: "I know, of course, your main publications very well, of which I most admire your book on Mechanics. You have had such a strong influence upon the epistemological conceptions of the younger generation of physicists that even your opponents today, such as Planck, undoubtedly would have been called Mach followers by physicists of the kind that was typical a few decades ago."

It will be important for our analysis to remember that Planck was Einstein's earliest patron in scientific circles. It was Planck who, in 1905, as editor of the *Annalen der Physik*, received Einstein's first relativity paper and thereupon held a review seminar on the paper in Berlin. Planck defended Einstein's work on relativity in public meetings from the beginning, and by 1913 had succeeded in persuading his German colleagues to invite Einstein to the Kaiser-Wilhelm-Gesellschaft in Berlin. With a polemical essay "Against the New Energetics" in 1896, he had made clear his position, and by 1909 Planck was one of the few opponents of Mach, and scientifically the most prominent one. He had just written a famous attack, *Die Einheit des physikalischen Weltbildes.* Far from accepting Mach's view that, as he put it, "Nothing is real except the perceptions, and all natural science is ultimately an economic adaptation of our ideas to our perceptions," Planck held to the entirely antithetical position that a basic aim of science is "the finding of a *fixed* world picture independent of the variation of time and people," or, more generally, "the complete liberation of the physical picture from the individuality of the separate intellects."[23] At least by implication in Einstein's remarks to Mach, he dissociated himself from allegiance to Planck's view. It may also not be irrelevant that just at that time Einstein, who since 1906 had been objecting to inconsistencies in Planck's quantum theory, was preparing his first major invited paper before a scientific congress, the eighty-first meeting of the Naturforscherversammlung, announced for September 1909 in Salzburg. Einstein's paper called for a revision of Maxwell's theory to accommodate the probabilistic character of the emission of photons—none of which Planck could accept—and concluded: "To accept Planck's theory means, in my view, to throw out the bases of our [1905] theory of radiation."

Mach's reply to Einstein's first letter is now lost, but it must have come quickly, because eight days later Einstein sends an acknowledgment:

Berne, 17 August 1909. Your friendly letter gave me enormous pleasure. . . . I am very glad that you are pleased with the relativity theory. . . . Thanking you again for your friendly letter, I remain, your student [indeed: *Ihr Sie verehrender Schüler*], A. Einstein.

Einstein's next letter was written as physics professor in Prague, where Mach before him had been for twenty-eight years. The post had been offered to Einstein on the basis of recommendations of a faction (Lampa, Pick) who regarded themselves as faithful disciples of Mach. The letter was sent out about New Year's 1911-12, perhaps just before or after Einstein's sole (and, according to P. Frank's account in *Einstein, His Life and Times,* not very successful) visit to Mach, and after the first progress toward the general relativity theory:

. . . I can't quite understand how Planck has so little understanding for your efforts. His stand to my [general relativity] theory is also one of refusal. But I can't take it amiss; so far, that one single epistemological argument is the only thing which I can bring forward in favor of my theory.

Here, Einstein is referring delicately to the Mach Principle, which he had been putting at the center of the developing theory.[24] Mach responded by sending Einstein a copy of one of his books, probably the *Analysis of Sensations.*

In the last of these letters to Mach (who was now seventy-five years old, and for some years had been paralyzed), Einstein writes from Zürich on 25 June 1913:

Recently you have probably received my new publication on Relativity and Gravitation which I have at last finished after unending labor and painful doubt. [This must have been the "Entwurf einer verallgemeinerten Relativitätstheorie und einer Theorie der Gravitation," written with Marcel Grossmann.[25]] Next year at the solar eclipse it will turn out whether the light rays are bent by the sun, in other words whether the basic and fundamental assumption of the equivalence of the acceleration of the reference frame and of the gravitational field really holds. If so, then your inspired investigations into the foundations of mechanics—despite Planck's unjust criticism—will receive a splendid confirmation. For it is a necessary consequence that inertia has its origin in a kind of mutual interaction of bodies, fully in the sense of your critique of Newton's bucket experiment.[26]

The Paths Diverge

The significant correspondence stops here, but Einstein's public and private avowals of his adherence to Mach's ideas continue for several years more. For example, there is his well-known, moving eulogy of Mach, published in 1916.[13] In August 1918, Einstein writes to Besso quite sternly about an apparent—and quite temporary—lapse in Besso's positivistic epistemology; it is an interesting letter, worth citing in full:

28 August 1918.
Dear Michele:

In your last letter I find, on rereading, something which makes me angry: That speculation has proved itself to be superior to empiricism. You are thinking here about the development of relativity theory. However, I find that this development teaches something else, that it is practically the opposite, namely that a theory which wishes to deserve trust must be built upon generalizable facts.

Old examples: Chief postulates of thermodynamics [based] on impossibility of perpetuum mobile. Mechanics [based] on grasped [*ertasteten*] law of inertia. Kinetic gas theory [based] on equivalence of heat and mechanical energy (also historically). Special Relativity on the constancy of light velocity and Maxwell's equation for the vacuum, which in turn rest on empirical foundations. Relativity with respect to uniform [?] translation is a *fact of experience.*

General Reality: *Equivalence of inertial and gravitational mass.* Never has a truly useful and deep-going theory really been found purely speculatively. The nearest case is Maxwell's hypothesis concerning displacement current; there the problem was to do justice to the fact of light propagation. . . . With cordial greetings, your Albert. [Emphasis in the original]

Careful reading of this letter shows us that already here there is evidence of divergence between the conception of "fact" as understood by Einstein and "fact" as understood by a true Machist. The impossibility of the *perpetuum mobile,* the first law of Newton, the constancy of light velocity, the validity of Maxwell's equations, the equivalence of inertial and gravitational mass—none of these would have been called "facts of experience" by Mach. Indeed, Mach might have insisted that—to use one of his favorite battle words—it is evidence of "dogmatism" not to regard all these conceptual constructs as continually in need of probing reexamination, thus, Mach had written:[27]

. . . for me, matter, time and space are still *problems,* to which, incidentally, the physicists (Lorentz, Einstein, Minkowski) are also slowly approaching.

Similar evidence of Einstein's gradual apostasy appears in a letter of 4 December 1919 to Paul Ehrenfest. Einstein writes:

I understand your difficulties with the development of relativity theory. They arise simply because you want to base the innovations of 1905 on epistemological grounds (nonexistence of the stagnant ether) instead of empirical grounds (equivalence of all inertial systems with respect to light).

Mach would have applauded Einstein's life-long suspicion of formal epistemological systems, but how strange would he have found this use of the word *empirical* to characterize the hypothesis of the equivalence of all inertial systems with respect to light! What we see forming slowly here is Einstein's view that the fundamental role played by experience in the construction of fundamental physical theory is, after all, not through the "atom" of experience, not through the individual sensation or the protocol sentence, but through some creative digest or synthesis of "*die gesammten Erfahrungstatsachen*," the *totality* of physical experience.[28] But all this was still hidden. Until Mach's death, and for several years after, Einstein considered and declared himself a disciple of Mach.

In the meantime, however, unknown to Einstein and everyone else, a time bomb had been ticking away. Set in 1913, it went off in 1921, five years after Mach's death, when Mach's *The Principles of Physical Optics* was published at last. Mach's preface was dated July 1913—perhaps a few days or, at most, a few weeks after Mach had received Einstein's last, enthusiastic letter and the article on general relativity theory. In a well-known passage in the preface (but one usually found in an inaccurate translation), Mach had written:

I am compelled, in what may be my last opportunity, to cancel my views [*Anschauungen*] of the relativity theory.

I gather from the publications which have reached me, and especially from my correspondence, that I am gradually becoming regarded as the forerunner of relativity. I am able even now to picture approximately what new expositions and interpretations many of the ideas expressed in my book on Mechanics will receive in the future from this point of view. It was to be expected that philosophers and physicists should carry on a crusade against me, for, as I have repeatedly observed, I was merely an unprejudiced rambler endowed with original ideas, in varied fields of knowledge. I must, however, as assuredly disclaim to be a forerunner of the relativists as I personally reject the atomistic doctrine of the present-day school, or church. The reason why, and the extent to which, I reject [*ablehne*] the present-day relativity theory, which I find to be growing more and more dogmatical, together with the particular reasons which have led me to such a view—considerations based on the physiology of the senses, epistemological doubts, and above all the insight resulting from my experiments—must remain to be treated in the sequel [a sequel which was never published].

Certainly, Einstein was deeply disappointed by this belated disclosure of Mach's sudden dismissal of the relativity theory. Some months later, during a lecture on 6 April 1922 in Paris, in a discussion with the anti-Machist philosopher Emile Meyerson, Einstein allowed in a widely reported remark that Mach was "*un bon méchanicien*," but a "*deplorable philosophe*."[29]

We can well understand that Mach's rejection was at heart very painful, the more so as it was somehow Einstein's tragic fate to have the contribution he most cared about rejected by the very men whose approval and understanding he would have most gladly had —a situation not unknown in the history of science. In addition to Mach, the list includes these four: *H. Poincaré,* who, to his death in 1912, only once deigned to mention Einstein's name in print, and then only to register an objection; *H. A. Lorentz,* who gave Einstein personally every possible encouragement—short of fully accepting the theory of relativity for himself; *Planck,* whose support of the special theory of relativity was unstinting, but who resisted Einstein's ideas on general relativity and the early quantum theory of radiation; and *A. A. Michelson,* who to the end of his days did not believe in relativity theory, and once said to Einstein that he was sorry that his own work may have helped to start this "monster."[30]

Soon Einstein's generosity again took the upper hand and resulted, from then to the end of his life, in many further personal testimonies to Mach's earlier influence.[31] A detailed analysis was provided in Einstein's letter of 8 January 1948 to Besso:

> As far as Mach is concerned, I wish to differentiate between Mach's influence in general and his influence on me. . . . Particularly in the *Mechanics* and the *Wärmelehre* he tried to show how conceptions arose out of experience. He took convincingly the position that these conceptions, even the most fundamental ones, obtained their warrant only out of empirical knowledge, that they are in no way logically necessary. . . .
>
> I see his weakness in this, that he more or less believed science to consist in a mere ordering of empirical material; that is to say, he did not recognize the freely constructive element in formation of concepts. In a way he thought that theories arise through *discoveries* and not through *inventions.* He even went so far that he regarded "sensations" not only as material which has to be investigated, but, as it were, as the building blocks of the real world; thereby, he believed, he could overcome the difference between psychology and physics. If he had drawn the full consequences, he would have had to reject not only atomism but also the idea of a physical reality.
>
> Now, as far as Mach's influence on my own development is concerned, it certainly was great. I remember very well that you drew my attention to his *Mechanics* and *Wärmelehre* during my first years of study, and that both books made a great impression on me. The extent to which they influenced my own work is, to say the truth, not clear to me. As far as I am conscious of it, the immediate influence of Hume on me was greater. . . . But, as I said, I am not able to analyze that which lies anchored in unconscious thought. It is interesting, by the way, that Mach rejected the special relativity theory passionately (he did not live to see the general relativity theory [in the developed form]). The theory

was, for him, inadmissibly speculative. He did not know that this speculative character belongs also to Newton's mechanics, and to every theory which thought is capable of. There exists only a gradual difference between theories, insofar as the chains of thought from fundamental concepts to empirically verifiable conclusions are of different lengths and complications.[32]

Antipositivistic Component of Einstein's Work

Ernst Mach's harsh words in his 1913 preface leave a tantalizing mystery. Ludwig Mach's destruction of his father's papers has so far made it impossible to find out more about the "experiments" (possibly on the constancy of the velocity of light) at which Ernst Mach hinted. Since 1921, many speculations have been offered to explain Mach's remarks.[33] They all leave something to be desired. Yet, I believe, it is not so difficult to reconstruct the main reasons why Mach ended up rejecting the relativity theory. To put it very simply, Mach had recognized more and more clearly, years before Einstein did so himself, that Einstein had indeed fallen away from the faith, had left behind him the confines of Machist empiriocriticism.

The list of evidences is long. Here only a few examples can be given, the first from the 1905 relativity paper itself: What had made it really work was that it contained and combined elements based on two entirely different philosophies of science—not merely the empiricist-operationist component, but the courageous initial postulation, in the second paragraph, of two thematic hypotheses (one on the constancy of light velocity and the other on the extension of the principle of relativity to all branches of physics), two postulates for which there was and can be no direct empirical confirmation.

For a long time, Einstein did not draw attention to this feature. In a lecture at King's College, London, in 1921, just before the posthumous publication of Mach's attack, Einstein still was protesting that the origin of relativity theory lay in the facts of direct experience:

> . . . I am anxious to draw attention to the fact that this theory is not speculative in origin; it owes its invention entirely to the desire to make physical theory fit observed fact as well as possible. We have here no revolutionary act, but the natural continuation of a line that can be traced through centuries. The abandonment of certain notions connected with space, time, and motion, hitherto treated as fundamentals, must not be regarded as arbitrary, but only as conditioned by observed facts.[34]

By June 1933, however, when Einstein returned to England to give the Herbert Spencer Lecture at Oxford entitled "On the Method of Theoretical Physics," the more complex epistemology that was in fact inherent in his work from the beginning had begun to be expressed. He opened this lecture with the significant sentence: "If you want to find out anything from the theoretical physicists about the methods they use, I advise you to stick closely to one principle: Don't listen to their words, fix your attention on their deeds." He went on to divide the tasks of experience and reason in a very different way from that advocated in his earlier visit to England:

We are concerned with the eternal antithesis between the two inseparable components of our knowledge, the empirical and the rational. . . . The structure of the system is the work of reason; the empirical contents and their mutual relations must find their representation in the conclusions of the theory. In the possibility of such a representation lie the sole value and justification of the whole system, and especially the concepts and fundamental principles which underlie it. Apart from that, these latter are free inventions of the human intellect, which cannot be justified either by the nature of that intellect or in any other fashion *a priori.*

In the summary of this section, he draws attention to the "purely fictitious character of the fundamentals of scientific theory." It is this penetrating insight which Mach must have smelled out much earlier and dismissed as "dogmatism."

Indeed, Einstein, in his 1933 Spencer Lecture—widely read, as were and still are so many of his essays—castigates the old view that "the fundamental concepts and postulates of physics were not in the logical sense inventions of the human mind but could be deduced from experience by 'abstraction'—that is to say, by logical means. A clear recognition of the erroneousness of this notion really only came with the general theory of relativity."

Einstein ends this discussion with the enunciation of his current credo, so far from that he had expressed earlier:

Nature is the realization of the simplest conceivable mathematical ideas. I am convinced that we can discover, by means of purely mathematical constructions, those concepts and those lawful connections between them which furnish the key to the understanding of natural phenomena. Experience may suggest the appropriate mathematical concepts, but they most certainly cannot be deduced from it. Experience remains, of course, the sole criterion of physical utility of a mathematical construction. But the creative principle resides in mathematics. In a certain sense, therefore, I hold it true that pure thought can grasp reality, as the ancients dreamed.[35]

Technically, Einstein was now at—or rather just past—the mid-stage of his pilgrimage. He had long ago abandoned his youthful allegiance to a primitive phenomenalism that Mach would have commended. In the first of the two passages just cited and others like it, he had gone on to a more refined form of phenomenalism which many of the logical positivists could still accept. He has, however, gone beyond it in the second passage, turning toward interests that we shall see later to have matured into clearly metaphysical conceptions.

Later, Einstein himself stressed the key role of what we have called thematic rather than phenomenic elements[36]—and thereby he fixed the early date at which, in retrospect, he found this need to arise in his earliest work. Thus he wrote in his *Autobiographical Notes* of 1946 that "shortly after 1900 . . . I despaired of the possibility of discovering the true laws by means of constructive efforts based on known facts. The longer and the more despairingly I tried, the more I came to the conviction that only the discovery of *a universal formal principle* could lead us to assured results."[37]

Another example of evidence of the undercurrent of disengagement from a Machist position is an early one: It comes from Einstein's article on relativity in the 1907 *Jahrbuch der Radioactivität und Elektronik* (Vol. 4, No. 4), where Einstein responds, after a year's silence, to W. Kaufmann's paper in the *Annalen der Physik* (Vol. 19, 1906). That paper had been the first publication in the *Annalen* to mention Einstein's work on the relativity theory, published there the previous year. Coming from the eminent experimental physicist Kaufmann, it had been most significant that this very first discussion was announced as a categorical, experimental disproof of Einstein's theory. Kaufmann had begun his attack with the devastating summary:

> I anticipate right here the general result of the measurements to be described in the following: *the measurement results are not compatible with the Lorentz-Einsteinian fundamental assumption.*[38]

Einstein could not have known that Kaufmann's equipment was inadequate. Indeed, it took ten years for this to be fully realized, through the work of Guye and Lavanchy in 1916. So in his discussion of 1907, Einstein had to acknowledge that there seemed to be small but significant differences between Kaufmann's results and Einstein's predictions. He agreed that Kaufmann's calculations seemed to be free of error, but "whether there is an unsuspected systematic error or whether the foundations of relativity theory do not correspond with the facts one will be able to decide with certainty only if a great variety of observational material is at hand."

Despite this prophetic remark, Einstein does not rest his case on it. On the contrary, he has a very different, and what for his time and situation must have been a very daring, point to make: He acknowledges that the theories of electron motion given earlier by Abraham and by Bucherer do give predictions considerably closer to the experimental results of Kaufmann. But Einstein refuses to let the "facts" decide the matter: "In my opinion both theories have a rather small probability, because their fundamental assumptions concerning the mass of moving electrons are not explainable in terms of theoretical systems which embrace a greater complex of phenomena."[39]

This is the characteristic position—the crucial difference between Einstein and those who make the correspondence with experimental fact the chief deciding factor for or against a theory: Even though the "experimental facts" at that time very clearly seemed to favor the theory of his opponents rather than his own, he finds the *ad hoc* character of their theories more significant and objectionable than an apparent disagreement between his theory and their "facts."[40]

So already in this 1907 article—which, incidentally, Einstein mentions in his postcard of 17 August 1909 to Ernst Mach, with a remark regretting that he has no more reprints for distribution—we have explicit evidence of a hardening of Einstein against the epistemological priority of experiment, not to speak of sensory experience. In the years that followed, Einstein more and more openly put the consistency of a simple and convincing theory or of a thematic conception higher in importance than the latest news from the laboratory—and again and again he turned out to be right.

Thus, only a few months after Einstein had written in his fourth letter to Mach that the solar eclipse experiment will decide "whether the basic and fundamental assumption of the equivalence of the acceleration of the reference frame and of the gravitational field really holds," Einstein writes to Besso in a very different vein (in March 1914), before the first, ill-fated eclipse expedition was scheduled to test the conclusions of the preliminary version of the general relativity theory: "Now I am fully satisfied, and I do not doubt any more the correctness of the whole system, may the observation of the eclipse succeed or not. The sense of the thing [*die*

Vernunft der Sache] is too evident." And later, commenting on the fact that there remains up to 10 per cent discrepancy between the measured deviation of light owing to the sun's field and the calculated effect based on the general relativity theory: "For the expert, this thing is not particularly important, because the main significance of the theory does not lie in the verification of little effects, but rather in the great simplification of the theoretical basis of physics as a whole."[41] Or again, in Einstein's "Notes on the Origin of the General Theory of Relativity,"[42] he reports that he "was in the highest degree amazed" by the existence of the equivalence between inertial and gravitational mass, but that he "had no serious doubts about its strict validity, even without knowing the results of the admirable experiment of Eötvös."

The same point is made again in a revealing account given by Einstein's student, Ilse Rosenthal-Schneider. In a manuscript "Reminiscences of Conversation with Einstein," dated 23 July 1957, she reports:

> Once when I was with Einstein in order to read with him a work that contained many objections against his theory . . . he suddenly interrupted the discussion of the book, reached for a telegram that was lying on the windowsill, and handed it to me with the words, "Here, this will perhaps interest you." It was Eddington's cable with the results of measurement of the eclipse expedition [1919]. When I was giving expression to my joy that the results coincided with his calculations, he said quite unmoved, "But I knew that the theory is correct"; and when I asked, what if there had been no confirmation of his prediction, he countered: "Then I would have been sorry for the dear Lord—the theory *is* correct."[43]

Minkowski's "World" and the World of Sensations

The third major point at which Mach, if not Einstein himself, must have seen that their paths were diverging is the development of relativity theory into the geometry of the four-dimensional space-time continuum, begun in 1907 by the mathematician H. Minkowski (who, incidentally, had had Einstein as a student in Zürich). Indeed, it was through Minkowski's semipopular lecture, "Space and Time," on 21 September 1908 at the eightieth meeting of the Naturforscherversammlung,[44] that a number of scientists first became intrigued with relativity theory. We have several indications that Mach, too, was both interested in and concerned about the introduction of four-dimensional geometry into physics (in Mach's correspondence around 1910, for example, with A. Föppl); according to F. Herneck,[45] Ernst Mach specially invited the young Viennese physicist Philipp Frank to visit him "in order to find out more about the relativity theory, above all about the use of four-dimensional geometry." As a result, Frank, who had recently finished his studies under Ludwig Boltzmann and had begun to publish noteworthy contributions to relativity, published the "presentation of Einstein's theory to which Mach gave his assent" under the title "Das Relativitätsprinzip und die Darstellung der physikalischen Erscheinungen im vierdimensionalen Raum."[46] It is an attempt, addressed to readers "who do not master modern mathematical methods," to show that Minkowski's work brings out the "empirical facts far more clearly by the use of four-dimensional world lines." The essay ends with the reassuring conclusion: "In this four-dimensional world the facts of experience can be presented more adequately than in three-dimensional space, where always only an arbitrary and one-sided projection is pictured."

Following Minkowski's own papers on the whole, Frank's treatment can make it nevertheless still appear that in most respects the time dimension is equivalent to the space dimensions. Thereby one could think that Minkowski's treatment based itself not only on a functional and operational interconnection of space and time, but also—fully in accord with Mach's own views—on the primacy of ordinary, "experienced" space and time in the relativistic description of phenomena.

Perhaps as a result of this presentation, Mach invoked the names of Lorentz, Einstein, and Minkowski in his reply of 1910 to Planck's first attack, citing them as physicists "who are moving closer to the problems of matter, space, and time." Already a year earlier, Mach seems to have been hospitable to Minkowski's presentation, although not without reservations. Mach wrote in the 1909 edition of *Conservation of Energy*[22]: "Space and time are here conceived not as independent entities, but as forms of the dependence of the phenomena on one another"; he also added a reference to Minkowski's lecture of 1908.[44] But a few lines earlier, Mach had written: "Spaces of many dimensions seem to me not so essential for physics. I would only uphold them if things of thought [*Gedankendinge*] like atoms are maintained to be indispensable, and if, then, also the freedom of working hypotheses is upheld."

It was correctly pointed out by C. B. Weinberg[47] that Mach may eventually have had two sources of suspicion against the Minkowskian form of relativity theory. As was noted above, Mach regarded the fundamental notions of mechanics as problems to be continually discussed with maximum openness within the frame of empiricism, rather than as questions that can be solved and settled —as the relativists, seemingly dogmatic and sure of themselves, were in his opinion more and more inclined to do. In addition, Mach held that the questions of physics were to be studied in a broader setting, encompassing biology and psychophysiology. Thus Mach wrote: "Physics is not the entire world; biology is there too, and belongs essentially to the world picture."[48]

But I see also a third reason for Mach's eventual antagonism against such conceptions as Minkowski's (unless one restricted their application to "mere things of thought like atoms and molecules, which by their very nature can never be made the objects of sensuous contemplations"[49]). If one takes Minkowski's essay seriously—for example, the abandonment of space and time separately, with identity granted only to "a kind of union of the two"—one must recognize that it entails the abandonment of the conceptions of experiential space and experiential time; and that is an attack on the very roots of sensations-physics, on the meaning of actual measurements. If identity, meaning, or "reality" lies in the four-dimensional space-time interval *ds*, one is dealing with a quantity which is hardly *denkökonomisch*, nor one that preserves the primacy of measurements in "real" space and time. Mach may well have seen the warning flag; and worse was soon to come, as we shall see at once.

In his exuberant lecture of 1908 (see note 44), Minkowski had announced that "three-dimensional geometry becomes a chapter in four-dimensional physics. . . . Space and time are to fade away into the shadows, and only *eine Welt an sich* will subsist." In this "world" the crucial innovation is the conception of the "*zeitartige Vektorelement*," *ds*, defined as $(1/c)\sqrt{c^2dt^2-dx^2-dy^2-dz^2}$ with imaginary components. To Mach, the word *Element* had a crucial and very different meaning. As we saw in Schlick's summary, elements were nothing less than the sensations and complexes of sensations of which the world consists and which completely define the world. Minkowski's rendition of relativity theory was now revealing the need to move the ground of basic, elemental truths from the plane of direct experience in ordinary space and time to a mathematicized, formalistic model of the world in a union of space and time that is not directly accessible to sensation—and, in this respect, is reminiscent of absolute space and time concepts that Mach had called "metaphysical monsters."[50]

Here, then, is an issue which, from the beginning, had separated Einstein and Mach even before they realized it. To the latter, the fundamental task of science was economic and descriptive; to the former, it was speculative-constructive and intuitive. Mach had once written: "If all the individual facts—all the individual phenomena, knowledge of which we desire—were immediately accessible to us, science would never have arisen."[51] To this, with the forthrightness caused perhaps by his recent discovery of Mach's opposition, Einstein countered during his lecture in Paris of 6 April 1922: "Mach's system studies the existing relations between data of experience: for Mach, science is the totality of these relations. That point of view is wrong, and in fact what Mach has done is to make a catalog, not a system."[52]

We are witnessing here an old conflict, one that has continued throughout the development of the sciences. Mach's phenomenalism brandished an undeniable and irresistible weapon for the critical reevaluation of classical physics, and in this it seems to hark back to an ancient position that looked upon sensuous appearances as the beginning and end of scientific achievement. One can read Galileo in this light, when he urges the primary need of *description* for the fall of bodies, leaving "the causes" to be found out later. So one can understand (or rather, misunderstand) Newton, with his too-well-remembered remark: "I feign no hypotheses."[53] Kirchhoff is in this tradition. Boltzmann wrote of him in 1888:

> The aim is not to produce bold hypotheses as to the essence of matter, or to explain the movement of a body from that of molecules, but to present equations which, free from hypotheses, are as far as possible true and quantitatively correct correspondents of the phenomenal world, careless of the essence of things and forces. In his book on mechanics, Kirchhoff will ban all metaphysical concepts, such as forces, the cause of a motion; he seeks only the equations which correspond so far as possible to observed motions.[54]

And so could, and did, Einstein himself understand the Machist component of his own early work.

Phenomenalistic positivism in science has always been victorious, but only up to a very definite limit. It is the necessary sword for destroying old error, but it makes an inadequate plowshare for cultivating a new harvest. I find it exceedingly significant that Einstein saw this during the transition phase of partial disengagement from the Machist philosophy. In the spring of 1917 Einstein wrote to Besso and mentioned a manuscript which Friedrich Adler had sent him. Einstein commented: "He rides Mach's poor horse to exhaustion." To this, Besso—the loyal Machist—responds on 5 May 1917: "As to Mach's little horse, we should not insult it; did it not make possible the infernal journey through the relativities? And who knows—in the case of the nasty quanta, it may also carry Don Quixote de la Einsta through it all!"

Einstein's answer of 13 May 1917 is revealing: "I do not inveigh against Mach's little horse; but you know what I think about it. It cannot give birth to anything living, it can only exterminate harmful vermin."

Toward a Rationalistic Realism

The rest of the pilgrimage is easy to reconstruct, as Einstein more and more openly and consciously turned Mach's doctrine upside down—minimizing rather than maximizing the role of actual details of experience, both at the beginning and at the end of scientific theory, and opting for a rationalism that almost inevitably would lead him to the conception of an objective, "real" world behind the phenomena to which our senses are exposed.

In the essay, "Maxwell's Influence on the Evolution of the Idea of Physical Reality" (1931), Einstein began with a sentence that could have been taken almost verbatim from Max Planck's attack on Mach in 1909, cited above: "The belief in an external world independent of the perceiving subject is the basis of all natural science." Again and again, in the period beginning with his work on the general relativity theory, Einstein insisted that between experience and reason, as well as between the world of sensory perception and the objective world, there are logically unbridgeable chasms. He characterized the efficacy of reason to grasp reality by the word *miraculous;* the very terminology in these statements would have been anathema to Mach.

We may well ask when and under what circumstances Einstein himself became aware of his change. Here again, we may turn for illumination to one of the hitherto unpublished letters, one written to his old friend, C. Lanczos, on 24 January 1938:

> Coming from sceptical empiricism of somewhat the kind of Mach's, I was made, by the problem of gravitation, into a believing rationalist, that is, one who seeks the only trustworthy source of truth in mathematical

simplicity. The logically simple does not, of course, have to be physically true; but the physically true is logically simple, that is, it has unity at the foundation.

Indeed, all the evidence points to the conclusion that Einstein's work on general relativity theory was crucial in his epistemological development. As he wrote later in "Physics and Reality" (1936): "the first aim of the general theory of relativity was the preliminary version which, while not meeting the requirements for constituting a closed system, could be connected in as simple a manner as possible with 'directly observed facts.'" But the aim, still apparent during the first years of correspondence with Mach, could not be achieved. In notes on the origin of the general relativity theory, Einstein reported:

> I soon saw that the inclusion of non-linear transformation, as the principle of equivalence demanded, was inevitably fatal to the simple physical interpretation of the coordinate—i.e., that it could no longer be required that coordinate differences [*ds*] should signify direct results of measurement with ideal scales or clocks. I was much bothered by this piece of knowledge . . . [just as Mach must have been].
>
> The solution of the above mentioned dilemma [from 1912 on] was therefore as follows: A physical significance attaches not to the differentials of the coordinates, but only to the Riemannian metric corresponding to them.[55]

And this is precisely a chief result of the 1913 essay of Einstein and Grossmann,[56] the same paper which Einstein sent to Mach and discussed in his fourth letter. This result was the final consequence of the Minkowskian four-space representation—the sacrifice of the primacy of direct sense perception in constructing a physically significant system. It was the choice that Einstein had to make—against fidelity to a catalogue of individual operational experiences and in favor of fidelity to the ancient hope for a unity at the base of physical theory.[57]

Enough has been written in other places to show the connections that existed between Einstein's scientific rationalism and his religious beliefs. Max Born summarized it in one sentence: "He believed in the power of reason to guess the laws according to which God has built the world."[58] Perhaps the best expression of this position by Einstein himself is to be found in his essay, "Über den gegenwärtigen Stand der Feld-Theorie," in the *Festschrift* of 1929 for Aurel Stodola:[59]

> Physical Theory has two ardent desires, to gather up as far as possible all

pertinent phenomena and their connections, and to help us not only to know *how* Nature is and *how* her transactions are carried through, but also to reach as far as possible the perhaps utopian and seemingly arrogant aim of knowing why Nature is *thus and not otherwise*. Here lies the highest satisfaction of a scientific person. . . . [On making deductions from a "fundamental hypothesis" such as that of the kinetic-molecular theory,] one experiences, so to speak, that God Himself could not have arranged those connections [between, for example, pressure, volume, and temperature] in any other way than that which factually exists, any more than it would be in His power to make the number 4 into a prime number. This is the promethean element of the scientific experience. . . . Here has always been for me the particular magic of scientific considerations; that is, as it were, the religious basis of scientific effort.

This fervor is indeed far from the kind of analysis which Einstein had made only a few years earlier. It is doubly far from the asceticism of his first philosophic mentor, Mach, who had written in his day book: "Colors, space, tones, etc. These are the only realities. Others do not exist."[60] It is, on the contrary, far closer to the rational realism of his first scientific mentor Planck, who had written: "The disjointed data of experience can never furnish a veritable science without the intelligent interference of a spirit actuated by faith. . . . We have a right to feel secure in surrendering to our belief in a philosophy of the world based upon a faith in the rational ordering of this world."[61] Indeed, we note the philosophical kinship of Einstein's position with seventeenth-century natural philosophers—for example, with Johannes Kepler who, in the preface of the *Mysterium Cosmographicum,* announced that he wanted to find out concerning the number, positions, and motions of the planets, "why they are as they are, and not otherwise," and who wrote to Herwart in April 1599 that, with regard to numbers and quantity, "our knowledge is of the same kind as God's, at least insofar as we can understand something of it in this mortal life."

Not unexpectedly, we find that during this period (around 1930) Einstein's nonscientific writings began to refer to religious questions much more frequently than before. There is a close relation between his epistemology, in which reality does not need to be validated by the individual's sensorium, and what he called "Cosmic religion,"[62] defined as follows: "The individual feels the vanity of human desires and aims, and the nobility and marvelous order which are revealed in nature and in the world of thought. He feels the individual destiny as an imprisonment and seeks to experience the totality of existence as a unity full of significance."

Needless to say, Einstein's friends from earlier days sometimes had to be informed of his change of outlook in a blunt way. For example, Einstein wrote to Moritz Schlick on 28 November 1930:

In general your presentation fails to correspond to my conceptual style insofar as I find your whole orientation so to speak too positivistic. . . . I tell you straight out: Physics is the attempt at the conceptual construction of a model of the *real world* and of its lawful structure. To be sure, it [physics] must present exactly the empirical relations between those sense experiences to which we are open; but only *in this way* is it chained to them. . . . In short, I suffer under the (unsharp) separation of Reality of Experience and Reality of Being. . . .

You will be astonished about the "metaphysicist" Einstein. But every four- and two-legged animal is de facto in this sense metaphysicist. [Emphasis in the original]

Similarly, P. Frank, Einstein's early associate and later his biographer, reports that the realization of Einstein's true state of thought reached Frank in a most embarrassing way, at the congress of German physicists in Prague in 1929, just as Frank was delivering "an address in which I attacked the metaphysical position of the German physicists and defended the positivistic ideas of Mach." The very next speaker disagreed and showed Frank that he had been mistaken still to associate Einstein's views with that of Mach and himself. "He added that Einstein was entirely in accord with Planck's view that physical laws describe a reality in space and time that is independent of ourselves. At that time," Frank comments, "this presentation of Einstein's views took me very much by surprise."[63]

In retrospect it is, of course, much easier to see the evidences that this change was being prepared. Einstein himself realized more and more clearly how closely he had moved to Planck, from whom he earlier dissociated himself in three of the four letters to Mach. At the celebration of Planck's sixtieth birthday, two years after Mach's death, Einstein made a moving speech in which, perhaps for the first time, he referred publicly to the Planck-Mach dispute and affirmed his belief that "there is no logical way to the discovery of these elementary laws. There is only the way of intuition" based on *Einfühlung* in experience.[64] The scientific dispute concerning the theory of radiation between Einstein and Planck, too, had been settled (in Einstein's favor) by a sequence of developments after 1911—for example, by Bohr's theory of radiation from gas atoms. As colleagues, Planck and Einstein saw each other regularly from 1913 on. Among evidences of the coincidence of these outlooks there is in the Einstein Archives a handwritten draft, written on or just before 17 April 1931 and intended as Einstein's introduction to Planck's hard-hitting article "Positivism and the Real External World."[65] In lauding Planck's article, Einstein concludes: "I presume I may add that both Planck's conception of the logical state of affairs as well as his subjective expectation concerning the later development of our science corresponds entirely with my own understanding."[66]

This essay gave a clear exposition of Planck's (and one may assume, Einstein's) views, both in physics and in philosophy more generally. Thus Planck wrote there:

> The essential point of the positivist theory is that there is no other source of knowledge except the straight and short way of perception through the senses. Positivism always holds strictly to that. Now, the two sentences: (1) *there is a real outer world which exists independently of our act of knowing* and (2) *the real outer world is not directly knowable* form together the cardinal hinge on which the whole structure of physical science turns. And yet there is a certain degree of contradiction between those two sentences. This fact discloses the presence of the irrational, or mystic, element which adheres to physical science as to every other branch of human knowledge. The effect of this is that a science is never in a position completely and exhaustively to solve the problem it has to face. We must accept that as a hard and fast, irrefutable fact, and this fact cannot be removed by a theory which restricts the scope of science at its very start. Therefore, we see the task of science arising before us as an incessant struggle toward a goal which will never be reached, because by its very nature it is unreachable. It is of a metaphysical character, and, as such, is always again and again beyond our achievement.[67]

From then on, Einstein's and Planck's writings on these matters are often almost indistinguishable from each other. Thus, in an essay in honor of Bertrand Russell,[68] Einstein warns that the "fateful 'fear of metaphysics' . . . has come to be a malady of contemporary empiricistic philosophizing." On the other hand, in the numerous letters between the two old friends, Einstein and Besso, each to the very end touchingly and patiently tries to explain his position, and perhaps to change the other's. Thus, on 28 February 1952, Besso once more presents a way of making Mach's views again acceptable to Einstein. The latter, in answering on 20 March 1952, once more responds that the facts cannot lead to a deductive theory and, at most, can set the stage "for intuiting a general principle" as the basis of a deductive theory. A little later, Besso is gently scolded (in

Einstein's letter of 13 July 1952): "It appears that you do not take the four-dimensionality of reality seriously, but that instead you take the present to be the only reality. What you call 'world' is in physical terminology 'spacelike sections' for which the relativity theory—already the special theory—denies objective reality."

In the end, Einstein came to embrace the view which many, and perhaps he himself, thought earlier he had eliminated from physics in his basic 1905 paper on relativity theory: that there exists an external, objective, physical reality which we may hope to grasp—not directly, empirically, or logically, or with fullest certainty, but at least by an intuitive leap, one that is only guided by experience of the totality of sensible "facts." Events take place in a "real world," of which the space-time world of sensory experience, and even the world of multidimensional continua, are useful conceptions, but no more than that. For a scientist to change his philosophical beliefs so fundamentally is rare, but not unprecedented. Mach himself underwent a dramatic transformation quite early (from Kantian idealism, at about age seventeen or eighteen, according to Mach's autobiographical notes). We have noted that Ostwald changed twice, once to anti-atomism and then back to atomism. And strangely, Planck himself confessed in his 1910 attack on Mach (note 23) that some twenty years earlier, near the beginning of his own career when Planck was in his late twenties (and Mach was in his late forties), he, too, had been counted "one of the decided followers of the Machist philosophy," as indeed is evident in Planck's early essay on the conservation of energy (1887).

In an unpublished fragment apparently intended as an additional critical reply to one of the essays in the collection *Albert Einstein, Philosopher-Scientist* (1949), Einstein returned once more—and quite scathingly—to deal with the opposition. The very words he used showed how complete was the change in his epistemology. Perhaps even without consciously remembering Planck's words in the attack on Mach of 1909 cited earlier—that a basic aim of science is "the complete liberation of the physical world picture from the individuality of the separate intellects"[23]—Einstein refers to a "basic axiom" in his own thinking:

> It is the postulation of a "real world" which so-to-speak liberates the "world" from the thinking and experiencing subject. The extreme positivists think that they can do without it; this seems to me to be an illusion, if they are not willing to renounce thought itself.

Einstein's final epistemological message was that the world of mere experience must be subjugated by and based in fundamental thought so general that it may be called cosmological in character. To be sure, modern philosophy did not gain thereby a major novel and finished corpus. Physicists the world over generally feel that today one must steer more or less a middle course in the area between, on the one hand, the Machist attachment to empirical data or heuristic proposals as the sole source of theory and, on the other, the aesthetic-mathematical attachment to persuasive internal harmony as the warrant of truth. Moreover, the old dichotomy between rationalism and empiricism is slowly being dissolved in new approaches.[69] Yet by encompassing in his own philosophical development both ends of this range, and by always stating forthrightly and with eloquence his redefined position, Einstein has helped us to define our own.

REFERENCES

1. Another recently published chapter for this study is the analysis of the probable sources of Einstein's first paper on relativity theory; see *American Scholar,* 37 (Winter 1967), 59-79. See also "Einstein, Michelson, and the 'Crucial' Experiment," *Isis,* 60 (Summer 1969), 132-197.

2. These documents are mostly on deposit at the Archives of the Estate of Albert Einstein at Princeton; where not otherwise indicated, citations made here are from those documents. In studying and helping to order for scholarly purposes the materials in the Archives, I have benefited from and am grateful for the help received from the Trustees of the Albert Einstein Estate, and particularly from Miss Helen Dukas.

   I thank the Executor of the Estate for permission to quote from the writings of Albert Einstein. I also wish to acknowledge the financial support provided by the Rockefeller Foundation for cataloguing the collection in the Archives at Princeton. The Institute for Advanced Study at Princeton and its director have been most hospitable throughout this continuing work. I am also grateful to M. Vero Besso for permission to quote from the letters of his father, Michelange Besso. All translations here are the author's, unless otherwise indicated.

   Early drafts of portions of this essay have been presented as invited papers at the *Tagung* of *Eranos* in Ascona (August, 1965), at the International Congress for the History of Science in Warsaw (August, 1965), and at the meeting, *Science et Synthèse,* at UNESCO in Paris (December, 1965.

3. This letter, as well as the next two letters mentioned in the text (those of 3 April and 13 April 1901), have been published by Hans-Günther Körber, *Forschungen und Forschritte,* 38 (1964).

4. The only other known attempt on Einstein's part to obtain an assistantship at that time was a request to Kammerlingh-Onnes (12 April 1901), to which, incidentally, he also seems to have received no response.

5. J. T. Merz, *A History of European Thought in the Nineteenth Century*, 2 (reprint, Dover Publishing Co., N. Y., 1965), 184, 199.

6. "Erinnerungen an Albert Einstein," issued by the Patent Office in Berne, about 1965 (n.d.. no pagination).

7. Wilhelm Ostwald, *Allgemeine Chemie*, 2 (2d ed., 1893), 1014.

8. P. A. Schilpp, ed., *Albert Einstein: Philosopher-Scientist* (Evanston, Ill., 1949).

9. *Die Mechanik in ihrer Entwicklung, historisch-kritisch dargestellt* (Leipzig, 1883).

10. In a special supplement on Ernst Mach in the journal *Neue Freie Presse* (Vienna), 12 June 1926.

11. Einstein himself, in a brief and telling analysis, published also in the *Neue Freie Presse* of Vienna on 12 June 1926 (the day of unveiling of a monument to Mach), wrote:

    Ernst Mach's strongest driving force was a philosophical one: the dignity of all scientific concepts and statements rests solely in isolated experiences [*Einzelerlebnisse*] to which the concepts refer. This fundamental proposition exerted mastery over him in all his research, and gave him the strength to examine the traditional fundamental concepts of physics (time, space, inertia) with an independence which at that time was unheard of.

12. Among many evidences of Mach's effectiveness, not the least are his five hundred or more publications (counting all editions—for example, seven editions of his *The Science of Mechanics* in German alone during his lifetime), as well as his large exchange of letters, books, and reprints (of which many important ones "carry the dedication of their authors," to cite the impressive catalogue of Mach's library by Theodor Ackermann, Munich, No. 634 [1959] and No. 636 [1960]). A glimpse of Mach's effect on those near him was furnished by William James, who in 1882 heard Mach give a "beautiful" lecture in Prague. Mach received James "with open arms. . . . Mach came to my hotel and I spent four hours walking and supping with him at his club, an unforgettable conversation. I don't think anyone ever gave me so strong an impression of pure intellectual genius. He apparently has read everything and thought about everything, and has an absolute simplicity of manner and winningness of smile when his face lights up, that are charming." From James's letter, in Gay Wilson Allen, *William James, a Biography* (New York, 1967), p. 249.

    The topicality of Mach's early speculations on what is now part of General Relativity Theory is attested by the large number of continuing contributions on the Mach Principle. Beyond that, Mach's influence today is still strong in scientific thinking, though few are as explicit and forthright as the distinguished physicist R. H. Dicke of Princeton University in his recent, technical book, *The Theoretical Significance of Experimental Relativity* ([London, 1964] pp. vii-viii): "I was curious to know how many other reasonable theories [in addition to General Relativity] would be supported by the same facts. . . . The reason for limiting the class of theories in this way is to be found in matters of philosophy, not in the observations. Foremost among these considerations was the philosophy of Bishop Berkeley and E. Mach. . . . The philosophy of Berkeley and Mach always lurked in the background and influenced all of my thoughts."

13. Albert Einstein, "Ernst Mach," *Physikalische Zeitschrift,* 17 (1916), 2.

14. Gerald Holton, "On the Origins of the Special Theory of Relativity," *American Journal of Physics,* 28 (October 1960), 627-636.

15. Gerald Holton, "Note on the Thematic Analysis of Science: The Case of Poincaré and Relativity," *Mélanges Koyré* (Paris, 1964).

16. For evidences that this insistence on prior epistemological analysis of conceptions of space and time are Machist rather than primarily derived from Hume and Kant (who had, however, also been influential), see Einstein's detailed rendition of Mach's critique of Newtonian space and time in Einstein, "Ernst Mach"; his discussion of Mach in the *Autobiographical Notes,* pp. 27-29; and the works cited in note 1.

17. Albert Einstein, "Zur Elektrodynamik bewegter Körper," *Annalen der Physik,* 17 (1905), 893.

18. See also P. Frank, in Schilpp, ed., *Albert Einstein,* pp. 272-273: "The definition of simultaneity in the special theory of relativity is based on Mach's requirement that every statement in physics has to state relations between observable quantities. . . . There is no doubt that . . . Mach's requirement, 'positivistic' requirement, was of great heuristic value to Einstein."

19. For example, see Philipp Frank, *Modern Science and Its Philosophy* (New York, 1955), pp. 61-89; V. Kraft, *The Vienna Circle* (New York, 1953); R. von Mises, *Ernst Mach und die Empiristische Wissenschaftsauffassung* (1938; printed as a fascicule of the series *Einheitswissenschaft*).

20. J. Petzoldt, "Gesellschaft für Positivistische Philosophie," reprinted in *Zeitschrift für Positivistische Philosophie,* 1 (1913), 4.

In that same speech, Petzoldt sounded a theme that became widely favored in the positivistic interpretation of the genesis of relativity theory—namely, that the relativity theory was developed in direct response to the puzzle posed by the results of the Michelson experiment:

> Clarity of thinking is inseparable from knowledge of a sufficient number of individual cases for each of the concepts used in investigation. Therefore, the chief requirement of positivistic philosophy: greatest respect for the facts. The newest phase of theoretical physics gives us an exemplary case. There, one does not hesitate, for the sake of a single

experiment, to undertake a complete reconstruction. The Michelson experiment is the cause and chief support of this reconstruction, namely, the electrodynamic theory of relativity. To do justice to this experiment one has no scruples to submit the foundations of theoretical physics as it has hitherto existed, namely, Newtonian mechanics, to a profound transformation.

In his interesting essay "Das Verhältnis der Machschen Gedankenwelt zur Relativitätstheorie," published as an appendix in the eighth German edition of Mach's *The Science of Mechanics* in the year 1921, Petzoldt faithfully attempts to identify and discuss several Machist aspects of Einstein's relativity theory:

(1) The theory "in the end is based on the recognition of the coincidence of sensations; and therefore it is fully in accord with Mach's worldview, which may be best characterized as a relativistic positivism" (p. 516).

(2) Mach's works "produced the atmosphere without which Einstein's Relativity Theory would not have been possible" (p. 494), and in particular Mach's analysis of the equivalence of rotating reference objects in Newton's bucket experiment prepared for the next step, Einstein's "equivalence of relatively moving coordinate systems" (p. 495).

(3) Mach's principle of economy is said to be marvelously exhibited in Einstein's succinct and simple statements of the two fundamental hypotheses. The postulate of the equivalence of inertial coordinate systems deals with "the simplest case thinkable, which now also serves as a fundamental pillar for the General Theory. And Einstein chose also with relatively greatest simplicity the other basic postulate [constancy of light velocity]. . . . These are the foundations. Everything else is logical consequence" (pp. 497-98).

21. Letter of Besso to Einstein, 16 February 1939. Among many testimonies to the effect of Einstein on positivistic philosophies of science, see P. W. Bridgman, "Einstein's Theory and the Operational Point of View," in Schilpp, ed., Albert Einstein.

22. Ernst Mach, *History and Root of the Principle of the Conservation of Energy* (Chicago, 1911), translation by P. Jourdain of second edition (1909), p. 95. For a brief analysis of Mach's various expressions of adherence as well as reservations with respect to the principle of relativity, see H. Dingler, *Die Grundlagen der Machschen Philosophie* (Leipzig, 1924), pp. 73-86.

F. Herneck (*Physikalische Blätter,* 17 [1961], 276) reports that P. Frank wrote him he had the impression during a discussion with Ernst Mach around 1910 that Mach "was fully in accord with Einstein's special relativity theory, and particularly with its philosophical basis."

23. Republished in M. Planck, *A Survey of Physical Theory* (New York, 1960), p. 24. We shall read later a reaffirmation of this position, in almost exactly the same words, but from another pen.

After Mach's rejoinder (*Scientia,* 7 [1910], 225), Planck wrote a second, much more angry essay, "Zur Machschen Theorie der physikalischen Erkenntnis," *Vierteljahrschrift für wissenschaftliche Philosophie,*

Vol. 34 (1910), p. 497. He ends as follows: "If the physicist wishes to further his science, he must be a Realist, not an Economist [in the sense of Mach's principle of economy]; that is, in the flux of appearances he must above all search for and unveil that which persists, is not transient, and is independent of human senses."

24. Later Einstein found that this procedure did not work; see *Ideas and Opinions* (New York, 1954), p. 286, and other publications. In a letter of 2 February 1954 to Felix Pirani, Einstein writes: "One shouldn't talk at all any longer of Mach's principle, in my opinion. It arose at a time when one thought that 'ponderable bodies' were the only physical reality and that in a theory all elements that are fully determined by them should be conscientiously avoided. I am quite aware of the fact that for a long time, I, too, was influenced by this fixed idea."

25. *Zeitschrift für Mathematik und Physik,* 62 (1913), 225-261.

26. For a further analysis and the full text of the four letters, see F. Herneck, *Forschungen und Fortschritte,* 37 (1963), 239-243, and *Wissenschaftliche Zeitschrift der Friedrich-Schiller-Universität Jena,* 15 (1966), 1-14; and H. Hönl, *Physikalische Blätter,* 16 (1960), 571. Many other evidences, direct and indirect, have been published to show Mach's influence on Einstein prior to Mach's death in 1916. For example, recently a document has been found which shows that in 1911 Mach had participated in formulating and signing a manifesto calling for the founding of a society for the positivistic philosophy. Among the signers, together with Mach, we find Joseph Petzoldt, David Hilbert, Felix Klein, George Helm, Sigmund Freud, and Einstein. (See F. Herneck, *Physikalische Blätter,* 17 [1961], 270.)

27. *Physikalische Zeitschrift,* 11 (1910), 605. To be sure, Mach was not at all times a rigid "Machist" himself.

28. See Albert Einstein, "Time, Space, and Gravitation" (1948), *Out of My Later Years* (New York, 1950). Einstein makes the distinction between constructive theories and "theories of principle." Einstein cites, as an example of the latter, the relativity theory, and the laws of thermodynamics. Such theories of principle, Einstein says, start with "empirically observed general properties of phenomena." See also "Autobiographical Notes," in Schilpp, ed., *Albert Einstein,* p. 53, and *The World as I See It* (New York, 1934), pp. 74-75.

29. *Bull. Soc. Franc. Phil.,* 22 (Paris, 1922), 111. In his 1913 preface rejecting relativity, Mach expressed himself perhaps more impetuously and irascibly than he may have meant. Some evidence for this possibility is in Mach's letters to J. Petzoldt. On 27 April 1914 Mach wrote: "I have received the copy of the positivistic *Zeitschrift* which contains your article on relativity; I liked it not only because you copiously acknowledge my humble contributions with respect to that theme, but also in general." And on 1 May 1914 Mach writes—rather more incoherently—to Petzoldt: "The enclosed letter of Einstein [a copy of the last of Einstein's four letters, cited above] proves the penetration of positivistic philosophy into physics; you can be glad about it. A year ago, philosophy was altogether sheer nonsense. The details prove it. The paradox of the clock would not have been noticed by Einstein a year ago."

I thank Mr. John Blackmore for drawing my attention to the Mach-Petzoldt letters and Dr. H. Müller for providing copies from the Petzoldt Archive in Berlin.

30. R. S. Shankland, *American Journal of Physics*, 31 (1963), 56.

31. A typical example is a letter of 18 September 1930 to Armin Weiner:

> . . . I did not have a particularly important exchange of letters with Mach. However, Mach did have a considerable influence upon my development through his writings. Whether or to what extent my life's work was influenced thereby is impossible for me to find out. Mach occupied himself in his last years with the relativity theory, and in a preface to a late edition of one of his works even spoke out in rather sharp refusal against the relativity theory. However, there can be no doubt that this was a consequence of a lessening ability to take up [new ideas] owing to his age, for the whole direction of thought of this theory conforms with Mach's, so that Mach quite rightly is considered as a forerunner of general relativity theory. . . .

I thank Colonel Bern Dibner for making a copy of the letter available to me from the Archives of the Burndy Library in Norwalk, Connecticut. Among other hitherto unpublished letters in which Einstein indicated his indebtedness to Mach, we may cite one to A. Lampa, 9 December 1935:

> . . . You speak about Mach as about a man who has gone into oblivion. I cannot believe that this corresponds to the facts since the philosophical orientation of the physicists today is rather close to that of Mach, a circumstance which rests not a little on the influence of Mach's writings.

Moreover, practically everyone else shared Einstein's explicitly expressed opinion of the debt of relativity theory to Mach; thus H. Reichenbach wrote in 1921 (*Logos*, Vol. 10, p. 331): "Einstein's theory signifies the accomplishment of Mach's program." Even Hugo Dingler agreed: "[Mach's] criticism of the Newtonian conceptions of time and space served as a starting point for the relativity theory. . . . Not only Einstein's work, but even more recent developments, such as Heisenberg's quantum mechanics, have been inspired by the Machian philosophy." *Encyclopedia of the Social Sciences*, 9 (New York, 1933), 653. And H. E. Hering wrote an essay whose title is typical of many others: "Mach als Vorläufer des physikalischen Relativitätsprinzips," *Kölner Universitätszeitung*, 1 (January 17, 1920), 3-4. I thank Dr. J. Blackmore for a copy of the article.

32. In the special supplement of *Neue Freie Presse* of Vienna on 12 June 1926, Einstein—then already disenchanted for some time with the Machist program—wrote immediately after the portion quoted in note 11:

> Philosophers and scientists have often criticized Mach, and correctly so, because he erased the logical independence of the concepts vis-à-vis the "sensations," [and] because he wanted to dissolve the Reality of Being, without whose postulation no physics is possible, in the Reality of Experience. . . .

There are additional resources, both published and unpublished, on the detailed aspects of the relation between Einstein and Mach, which, for lack of space, cannot be summarized here.

33. For example, by Einstein himself, by Joseph Petzoldt, and by Hugo Dingler (cf. H. Dingler, *Encyclopedia of the Social Sciences,* pp. 84-86). I assign relatively little weight to the possibility that the rift grew out of the difference between Einstein and Mach on atomism. Herneck provides the significant report that according to a letter from P. Frank, Mach was personally influenced by Dingler, whom Mach had praised in the 1912 edition of the *Mechanik* and who was from the beginning an opponent of relativity theory, becoming one of the most "embittered enemies" of Einstein (Herneck [see note 26, above], p. 14). The copies of letters from Dingler to Mach in the Ernst-Mach-Institute in Freiburg indicate Dingler's intentions; nevertheless, there remains a puzzle about Dingler's role which is worth investigating. It is significant that in his 1921 essay, Petzoldt (note 20) devotes much space to a defense of Einstein's work against Dingler's attacks. See also Joachim Thiele's detailed analysis of Mach's Preface, in *NTM, Schriftenreihe für Geschichte der Naturwissenschaften, Technik und Medizin,* 2 (Leipzig, 1965), 10-19.

34. "On the Theory of Relativity," *Mein Weltbild* (Amsterdam, 1934); republished in *Ideas and Opinions* (New York, 1954), p. 246. F. Herneck has given the texts of similar discussions on phonographic records by Einstein in 1921 and even in 1924; cf. *Forschungen und Fortschritte,* 40 (1966), 133-134.

35. Quotations from "On the Method of Theoretical Physics," *Mein Weltbild* (1934), as reprinted in translation in *Ideas and Opinions,* pp. 270-76, except for correction of mistranslation of one line. There are a number of later lectures and essays in which the same point is made. See, for example, the lecture, "Physics and Reality" (1936, reprinted in *Ideas and Opinions*), which states that Mach's theory of knowledge is insufficient on account of the relative closeness between experience and the concepts which it uses; Einstein advocates going beyond this "phenomenological physics" to achieve a theory whose basis may be further removed from direct experience, but which in return has more "unity in the foundations." Or see *Autobiographical Notes,* p. 27: "In the choice of theories in the future," he indicates that the basic concepts and axioms will continue to "distance themselves from what is directly observable."

    Even as Einstein's views developed to encompass the "*erlebbare, beobachtbare*" facts as well as the "*wild-spekulative*" nature of theory, so did those of many of the philosophers of science who also had earlier started from a more strict Machist position. This growing modification of the original position, partly owing to "the growing understanding of the general theory of relativity," has been chronicled by P. Frank, for example, in "Einstein, Mach, and Logical Positivism," in Schilpp, ed., *Albert Einstein.*

36. For a discussion of thematic and phenomenic elements in theory construction, see G. Holton, "The Thematic Imagination in Science," in G. Holton, ed., *Science and Culture* (Boston, 1965).

37. Schilpp, ed., *Albert Einstein,* p. 53. Emphases added. On pp. 9-11, Einstein

describes what may be a possible precursor of this attitude in his study of geometry in his childhood.

38. W. Kaufmann, "Concerning the Constitution of the Electron," *Annalen der Physik,* 19 (1906), 495. Emphasis in original.

39. *Jahrbuch der Radioaktivitaet und Elektronik,* 4 (1907), 28. Shortly after Kaufmann's article appeared, M. Planck (in *Physikalische Zeitschrift,* 7 [1906], 753-761) took it on himself publicly to defend Einstein's work in an analysis of Kaufmann's claim. He concluded that Kaufmann's data did not have sufficient precision for his claim. Incidentally, Planck tried to coin the term for the new theory that had not yet been named: *"Relativtheorie."*

40. It should be remembered that Poincaré, with a much longer investment in attempts to fashion a theory of relativity, was quite ready to give in to the experimental "evidence. See note 15.

41. Carl Seelig, *Albert Einstein* (Zürich, 1960), p. 195.

42. Albert Einstein, "Notes on the Origin of the General Theory of Relativity," *Mein Weltbild* (Amsterdam, 1934), reprinted in translation in *Ideas and Opinions* (New York, 1954), pp. 285-290.

43. *"Da könnt' mir halt der liebe Gott leid tun, die Theorie stimmt doch."* This semi-serious remark of a person who was anything but sacrilegious indeed illuminates the whole style of a significant group of new physicists. P. A. M. Dirac, in *Scientific American,* 208 (May 1963), 47-48, speaks about this, with special attention to the work of Schrödinger, a spirit close to that of his friend, Einstein, despite the ambivalence of the latter to the advances in quantum physics. We can do no better than quote in *extenso* from Dirac's account [pp. 46-47]:

> Schrödinger worked from a more mathematical point of view, trying to find a beautiful theory for describing atomic events, and was helped by deBroglie's ideas of waves associated with particles. He was able to extend deBroglie's ideas and to get a very beautiful equation, known as Schrödinger's wave equation, for describing atomic processes. Schrödinger got this equation by pure thought, looking for some beautiful generalization of deBroglie's ideas, and not by keeping close to the experimental development of the subject in the way Heisenberg did.
>
> I might tell you the story I heard from Schrödinger of how, when he first got the idea for this equation, he immediately applied it to the behavior of the electron in the hydrogen atom, and then he got results that did not agree with experiment. The disagreement arose because at that time it was not known that the electron has a spin. That, of course, was a great disappointment to Schrödinger, and it caused him to abandon the work for some months. Then he noticed that if he applied the theory in a more approximate way, not taking into account the refinements required by relativity, to this rough approximation his work was in agreement with observation. He published his first

paper with only this rough approximation, and in this way Schrödinger's wave equation was presented to the world. Afterward, of course, when people found out how to take into account correctly the spin of the electron the discrepancy between the results of applying Schrödinger's relativistic equation and the experiments was completely cleared up.

I think there is a moral to this story, namely, that it is more important to have beauty in one's equations than to have them fit experiment. If Schrödinger had been more confident of his work, he could have published it some months earlier, and he could have published a more accurate equation. That equation is now known as the Klein-Gordon equation, although it was really discovered by Schrödinger, and in fact was discovered by Schrödinger before he discovered his nonrelativistic treatment of the hydrogen atom. It seems that if one is working from the point of view of getting beauty in one's equations, and if one has really a sound insight, one is on a sure line of progress. If there is not complete agreement between the results of one's work and experiment, one should not allow oneself to be too discouraged, because the discrepancy may well be due to minor features that are not properly taken into account and that will get cleared up with further developments of the theory. That is how quantum mechanics was discovered . . . .

44. Published several times—for example, by Teubner, Leipzig, 1909.

45. F. Herneck, *Physikalische Blätter,* 15 (1959), 565. P. Frank's remark is reported by F. Herneck, in W. F. Merzkirch, ed., *Ernst Mach* (Freiburg, 1967), p. 50.

46. *Zeitschrift für Physikalische Chemie* (1910), pp. 466–495.

47. C. B. Weinberg, *Mach's Empirio-Pragmatism in Physical Science* (Thesis, Columbia University, 1937).

48. Ernst Mach, *Scientia,* 7 (1910), 225.

49. *Space and Geometry* (1906), p. 138. Mach's attempts to speculate on the use of n-dimensional spaces for representing the configuration of such "mere things of thought"—the derogatory phrase also applied to absolute space and absolute motion in Newton—are found in his first major book, *Conservation of Energy* (first edition, 1872).

50. Cf. J. Petzoldt, "Verbietet die Relativitätstheorie Raum und Zeit als etwas Wirkliches zu denken?," *Verhandlungen der Deutschen Physikalischen Gesellschaft,* Nos. 21-24 (1918), pp. 189-201. Here, and again in his 1921 essay (note 20), Petzoldt tries to protect Einstein from the charge—for example, that by Sommerfeld—that space and time no longer "are to be thought of as real."

51. Ernst Mach, *Conservation of Energy,* p. 54.

52. See note 29; also reported in "Einstein and the Philosophies of Kant and Mach," *Nature,* 112 (August 1923), 253.

53. That Einstein did not so misunderstand Newton can be illustrated, for example, in a comment reported by C. B. Weinberg: "Dr. Einstein further

maintained that Mach, as well as Newton, tacitly employs hypotheses—not recognizing their non-empirical foundations." (Weinberg, *op. cit.*, p. 55.) For an example of Mach's tacit presuppositions, see H. Dingler, *Encyclopedia of the Social Sciences*, pp. 69-71. Dingler also analyzed some of the nonempirical foundations of relativity theory in *Kritische Bemerkungen zu den Grundlagen der Relativitätstheorie* (Leipzig, 1921).

54. Cited by R. S. Cohen in his very useful essay in P. Schilpp (ed.), *The Philosophy of Rudolf Carnap* (LaSalle, Ill., 1963), p. 109. I am also grateful to Professor Cohen for a critique of parts of this paper in earlier form.

55. Albert Einstein, "Notes on the Origin of the General Relativity Theory," pp. 288-289.

56. *Zeitschrift für Mathematik und Physik*, 62 (1913), 230-231.

57. I am not touching in this essay on the effect of quantum mechanics on Einstein's epistemological development; the chief reason is that while from his "heuristic" announcement of the value of a quantum theory in 1905 Einstein remained consistently skeptical about the "reality" of the quantum theory of radiation, this opinion only added to the growing realism stemming from his work on general relativity theory. In the end, he reached the same position in quantum physics as in relativity; cf. his letter of 7 September 1944 to Max Born: "In our scientific expectations we have become antipodes. You believe in the dice-playing God, and I in perfect rules of law in a world of something objectively existing, which I try to catch in a wildly speculative way." (Reported by Max Born, *Universitas*, 8 [1965], 33.)

58. Max Born, "Physics and Relativity," *Physics in My Generation* (London, 1956), p. 205.

59. *Orell Füssli Verlag* (Zürich and Leipzig, 1929), pp. 126-132. I am grateful to Professor C. Lanczos and Professor John Wheeler for pointing out this reference to me.

60. H. Dingler, *Die Grundlagen der Machschen Philosophie*, p. 98.

61. *The Philosophy of Physics* (New York, 1936), pp. 122, 125.

62. In "Religion and Science," written for *The New York Times Magazine*, 9 November 1930; cf. *Mein Weltbild*, p. 39, and *Cosmic Religion* (New York, 1931), p. 48.

Possible reasons for Einstein's growing interest in these matters, partly related to the worsening political situation at the time, are discussed in P. Frank, *Einstein, His Life and Times* (New York, 1947). It is noteworthy that while Einstein was quite unconcerned with religious matters during the period of his early scientific publications, he gradually returned later to a position closer to that at a very early age, when he reported he had felt a "deep religiosity. . . . It is quite clear to me that the religious paradise of youth . . . was a first attempt to free myself from the chains of the 'merely personal.'" *Autobiographical Notes*, p. 5. For a discussion, see G. Holton, *The Graduate Journal*, 7 (Spring 1967), 417-420.

63. Frank, *Einstein, His Life and Times,* p. 215. Einstein's change of mind was, of course, not acceptable to a considerable circle of previously sympathetic scientists and philosophers. See, for example, P. W. Bridgman, "Einstein's Theory and the Operational Point of View," in Schilpp, ed., *Albert Einstein.*

    An interesting further confirmation of Einstein's changed epistemological position became available after this essay was first published. Werner Heisenberg, in *Physics and Beyond* (New York, 1971), pp. 62-69, writes about his conversation with Einstein concerning physics and philosophy. See, for example (p. 63), a portion of a conversation set in 1925-1926:

    "But you don't seriously believe," Einstein protested, "that none but observable magnitudes must go into physical theory?"

    "Isn't that precisely what you have done with relativity?" I asked in some surprise. "After all, you did stress the fact that it is impermissible to speak of absolute time, simply because absolute time cannot be observed; that only clock readings, be it in the moving reference system or the system at rest, are relevant to the determination of time."

64. Originally entitled "Motiv des Forschens" (in *Zu Max Planck,* Planck's 60. *Geburtstag* [Karlsruhe, Müller, 1918]), the talk had a fate not untypical of many of Einstein's essays. It was reprinted without date or source, under the title "Prinzipien der Forschung," in *Mein Weltbild,* and again, in perhaps unauthorized extension and translation by James Murphy, as a preface to M. Planck, *Where Is Science Going?* (London, 1933). In an earlier appreciation of Planck in 1913, Einstein had written only very briefly about Planck's epistemology, merely lauding Planck's essay of 1896 against energetics, and not mentioning Mach.

65. *International Forum,* 1 (1931).

66. Einstein sent his introduction to the editor of the journal on 17 April 1931, but it appears to have come too late for inclusion.

67. Max Planck, "Positivism and External Reality," *The International Forum,* 1 (Berlin 1931), 15-16. Emphasis in original.

68. In P. Schilpp, ed., *The Philosophy of Bertrand Russell* (Evanston, 1944), p. 289.

69. Toward the end, Einstein himself acknowledged a similar point, in his "Reply to Criticisms," in Schilpp, ed., *Albert Einstein,* pp. 679-680: " 'Einstein's position . . . contains features of rationalism and extreme empiricism . . .' This remark is entirely correct . . . A wavering between these extremes appears to me unavoidable."

GERALD HOLTON

The Roots of Complementarity

Como, 1927

EACH AGE is formed by certain characteristic conceptions, those that give it its own unmistakable modernity. The renovation of quantum physics in the mid-1920's brought into public view just such a conception, one that marked a turning point in the road from which our view of the intellectual landscape, in science and in other fields, will forever be qualitatively different from that of earlier periods. It was in September 1927 in Como, Italy, during the International Congress of Physics held in commemoration of the one-hundredth anniversary of Alessandro Volta's death, that Niels Bohr for the first time introduced in a public lecture his formulation of complementarity.[1] Bohr's audience contained most of the leading physicists of the world in this area of work, men such as Max Born, Louis de Broglie, A. H. Compton, Peter Debye, Enrico Fermi, James Franck, Werner Heisenberg, Max von Laue, H. A. Lorentz, Robert Millikan, John von Neumann, Wolfgang Pauli, Max Planck, Arnold Sommerfeld, and Otto Stern. It was a veritable summit meeting. Only Einstein was conspicuously absent.

In the introduction to his lecture, Bohr said he would make use "only of simple considerations, and without going into any details of technical, mathematical character." Indeed, the essay contained only few and simple equations. Rather, its avowed purpose was a methodological one that, at least in this initial announcement, did not yet confess its ambitious scope. Bohr stressed only that he wanted to describe "a certain general point of view . . . which I hope will be helpful in order to harmonize the apparently conflicting views taken by different scientists."

He was referring to a profound and persistent difference between the classical description and the quantum description of physical phenomena. To review it, we can give four brief examples of the dichotomy:

1. In classical physics, for example in the description of the motion of planets or billiard balls or other objects which are large enough to be

directly visible, the "state of the system" can (at least in principle) be observed, described, defined with arbitrarily small interference of the behavior of the object on the part of the observer, and with arbitrarily small uncertainty. In quantum description, on the other hand, the "state of the system" cannot be observed without significant influence upon the state, as for example when an attempt is made to ascertain the orbit of an electron in an atom, or to determine the direction of propagation of photons. The reason for this situation is simple: the atoms, either in the system to be observed or in the probe that is used in making the observation, are never arbitrarily fine in their response; the energy exchange on which their response depends is not any small quantity we please, but, according to the "quantum postulate" (Planck's fundamental law of quantum physics), can proceed only discontinuously, in discrete steps of finite size.

2. It follows that in cases where the classical description is adequate, a system can be considered closed although it is being observed, since the flow of energy into and out of the system during an observation (for example, of the reflection of light from moving balls) is negligible compared to the energy changes in the system during interaction of the parts of the system. On the other hand, in systems that require quantum description, one cannot neglect the interaction between the "system under observation," sometimes loosely called the "object," and the agency or devices used to make the observations (sometimes loosely called the "subject"). The best-known case of this sort is illustrated by Heisenberg's gamma-ray microscope, in which the progress of an electron is "watched" by scattering gamma rays from it, with the result that the electron itself is deflected from its original path.

3. In "classical" systems, those for which classical mechanics is adequate, we have both conventional causality chains and ordinary space-time coordination, and both can exist at the same time. In quantum systems, on the other hand, there are no conventional causality chains; if left to itself, a system such as an atom or its radioactive nucleus undergoes changes (such as emission of a photon from the atom or a particle from the nucleus) in an intrinsically probabilistic manner. However, if we subject the "object" to space-time observations, it no longer undergoes its own probabilistic causality sequence. Both these mutually exclusive descriptions of manifestations of the quantum system must be regarded as equally relevant or "true," although both cannot be exhibited at one and the same time.

4. Finally, we can refer to Bohr's own illustration in the 1927 essay of "the much discussed question of the nature of light . . . [I]ts propagation in space and time is adequately expressed by the electromagnetic theory. Especially the interference phenomena *in vacuo* and the optical proper-

ties of material media are completely governed by the wave theory superposition principle. Nevertheless, the conservation of energy and momentum during the interaction between radiation and matter, as evident in the photo-electric and Compton effect, finds its adequate expression just in the light quantum idea put forward by Einstein."[2] Unhappiness with the wave-particle paradox, with being forced to use in different contexts two such antithetical theories of light as the classical wave theory and the quantum (photon) theory was widely felt. Einstein expressed it in April 1924 by writing: "We now have two theories of light, both indispensable, but, it must be admitted, without any logical connection between them, despite twenty years of colossal effort by theoretical physicists."[3]

The puzzle raised by the gulf between the classical description and quantum description was: Could one hope that, as had happened so often before in physics, one of the two antithetical views would somehow be subsumed under or dissolved in the other (somewhat as Galileo and Newton had shown celestial physics to be no different from terrestrial physics)? Or would one have to settle for two so radically different modes of description of physical phenomena? Would the essential continuity that underlies classical description, where coordinates such as space, time, energy, and momentum can in principle be considered infinitely divisible, remain unyieldingly antithetical to the essential discontinuity and discreteness of atomic processes?

Considering the situation in 1927 in thematic terms, it was by that time clear that physics had inherited contrary themata from the "classical" period (before 1900) and from the quantum period (after 1900). A chief thema of the earlier period was continuity, although it existed side by side with the atomistic view of matter. A chief thema of the more recent period was discontinuity, although it existed side by side with the wave theory of electromagnetic propagation and of the more recent theories associated with de Broglie and Erwin Schrödinger.

In the older physics, also, classical causality was taken for granted, whereas in the new physics the concept of indeterminacy, statistical description, and probabilistic distribution as an inherent aspect of natural description were beginning to be accepted. In the older physics, the possibility of a sharp subject-object separation was not generally challenged; in the new physics it was seen that the subject-object coupling could be cut only in an arbitrary way. In Bohr's sense, a "phenomenon" is the description of that which is to be observed *and* of the apparatus used to obtain the observation.

Bohr's proposal of 1927 was essentially that we should attempt not to reconcile the dichotomies, but rather to realize the complementarity of representations of events in these two quite different languages. The separateness of the accounts is merely a token of the fact that, in the normal language available to us for communicating the results of our experiments, it is possible to express the wholeness of nature only through a complementary mode of descriptions.[4] The apparently paradoxical, contradictory accounts should not divert our attention from the essential wholeness. Bohr's favorite aphorism was Schiller's "Nur die Fülle führt zur Klarheit." Unlike the situation in earlier periods, clarity does not reside in simplification and reduction to a single, directly comprehensible model, but in the exhaustive overlay of different descriptions that incorporate apparently contradictory notions.

Summarizing his Como talk, Bohr in 1949 stressed that the need to express one's reports ultimately in normal (classical) language dooms any attempt to impose a clear separation between an atomic "object" and the experimental equipment.

> The new progress in atomic physics was commented upon from various sides at the International Physical congress held in September 1927, at Como in commemoration of Volta. In a lecture on that occasion, I advocated a point of view conveniently termed "complementarity," suited to embrace the characteristic features of individuality of quantum phenomena, and at the same time to clarify the peculiar aspects of the observational problem in this field of experience. For this purpose, it is decisive to recognize that, *however far the phenomena transcend the scope of classical physical explanation, the account of all evidence must be expressed in classical terms.* The argument is simply that by the word "experiment" we refer to a situation where we can tell others what we have learned and that, therefore, the account of the experimental arrangement and of the results of the observations must be expressed in unambiguous language with suitable application of the terminology of classical physics.
>
> This crucial point, which was to become a main theme of the discussions reported in the following, *implies the impossibility of any sharp separation between the behaviour of atomic objects and the interaction with the measuring instruments which serve to define the conditions under which the phenomena appear.* In fact, the individuality of the typical quantum effects finds its proper expression in the circumstance that any attempt of subdividing the phenomena will demand a change in the experimental arrangement, introducing new possibilities of interaction between objects and measuring instruments which in principle cannot be controlled. Consequently, evidence obtained under different experimental conditions cannot be comprehended within a single picture, but must be regarded as *complementary* in the sense that only the totality of the phenomena exhausts the possible information about the objects.[5]

What Bohr was pointing to in 1927 was the curious realization that in the atomic domain, the only way the observer (including his equipment) can be uninvolved is if he observes nothing at all. As soon as he sets up the observation tools on his workbench, the system he has chosen to put under observation and his measuring instruments for doing the job form one inseparable whole. Therefore the results depend heavily on the apparatus. In the well-known illustration involving a light beam, if the instrument of measurement contains a double pinhole through which the light passes, the result of observation will indicate that a wave phenomenon is involved; but if the "same" light beam is used when the measuring instrument contains a collection of recoiling scatterers, then the observation results will indicate that a stream of particles is involved. (Moreover, precisely the same two kinds of observations are obtained when, instead of the beam of light, one uses a beam of "particles" such as atoms or electrons or other subatomic particles.) One cannot construct an experiment which simultaneously exhibits the wave and particle aspects of atomic matter. A particular experiment will always show only one view or representation of objects at the atomic level.

The study of nature is a study of artifacts that appear during an engagement between the scientist and the world in which he finds himself. And these artifacts themselves are seen through the lens of theory. Thus, different experimental conditions give different views of "nature." To call light either a wave phenomenon or a particle phenomenon is impossible; in either case, too much is left out. To call light *both* a wave phenomenon and a particle phenomenon is to oversimplify matters. Our knowledge of light is contained in a number of statements that are seemingly contradictory, made on the basis of a variety of experiments under different conditions, and interpreted in the light of a complex of theories. When you ask, "What is light?" the answer is: the observer, his various pieces and types of equipment, his experiments, his theories and models of interpretation, *and* whatever it may be that fills an otherwise empty room when the lightbulb is allowed to keep on burning. All this, together, is light.

No objections seem to have been raised against Niels Bohr's paper at the Como meeting. On the other hand, at this first hearing the importance of the new point of view was not immediately appreciated. Apparently a typical comment overheard after Bohr's lecture was that it "will not induce any of us to change his own opinion about quantum mechanics."[6] A distinguished group of physicists, although a minority in the field, remained unconvinced by and indeed hostile to the complementarity point of view. Foremost among them was Einstein, who heard the first extensive exposition a month after the Como meeting, in October 1927, at the Solvay Congress in Brussels. Einstein had disliked even the earlier Göttingen-Copenhagen interpretations of atomic physics that were based on the themata of discontinuity and nonclassical causality. He had written to Paul Ehrenfest (August 28, 1926), "I stand before quantum mechanics with admiration and suspicion," and to Born (December 4, 1926) Einstein had said, "Quantum mechanics demands serious attention. But an inner voice tells me that this is not the true Jacob. The theory accomplishes a lot, but it does not bring us closer to the secrets of the Old One. In any case, I am convinced that He does not play dice."[7]

Almost a quarter of a century later Einstein was still in opposition, and added two objections to the complementarity principle: "to me it must seem a mistake to permit theoretical description to be directly dependent upon acts of empirical assertions, as it seems to be intended (for example) in Bohr's principle of complementarity, the sharp formulation of which, moreover, I have been unable to achieve despite much effort which I have expended on it."[8]

Bohr himself was aware from the beginning that the complementarity point of view was a program rather than a finished work; that is, it had to be extended and deepened by much subsequent work. It was to him "a most valuable incentive . . . to reexamine the various aspects of the situation as regards the description of atomic phenomena" and "a welcome stimulus to verify still further the role played by the measuring instruments."[9] However, as we shall see, over the years Bohr came to regard the complementarity principle as more and more important, extending far beyond the original context in which it had been announced. For his later, deep commitment to the conception, and for his awareness of the antiquity of some of its roots, we need cite here only an anecdotal piece of evidence. When Bohr was awarded the Danish Order of the Elephant in 1947, he had to supervise the design of a coat of arms for placement in the church of the Frederiksborg Castle at Hillerød. The device (see Figure) presents the idea of complementarity: above the central insignia, the legend says "Contraria sunt complementa," and at the center Bohr placed the symbol for Yin and Yang.

Lux versus Lumen

How did Bohr's complementarity point of view—so far from the older scientific tradition of strict separation between the observer and the observed—come to be developed? Finding the various roots of and the likely preparatory conditions for this transforming conception—those in physical theory and those in philosophical tradition—appears to me to be an interesting problem that is far from its unambiguous solution. However, there are already some useful results of the search, particularly insofar as they may have relevance for a better understanding of the mutual interaction of scientific and humanistic traditions.

The first direction to look is the development of the early ideas concerning the nature of light. That a modern thema was already inherent in the formulations that began in antiquity should not surprise us; we

Coat of arms chosen by Niels Bohr when he was awarded the Danish Order of the Elephant, 1947. From S. Rozental, ed., *Niels Bohr: His Life and Work as Seen by His Friends and Colleagues* (New York: John Wiley & Sons, Inc., 1967), facing p. 305.

know from other studies that despite all change and progress of science, the underlying, important themata are relatively few. In one guise or another they have been the mainstay of the imagination.

One of the favorite ancient ideas concerning the nature of light, originating in the Pythagorean school, postulated that rays are emitted by the eye to explore the world. Euclid spoke of the eye as if it were sending out visual rays whose ends probed the object, somewhat like the stick of a blind man tapping around himself. A somewhat more refined conception of this general sort is still found in Ptolemy in the second century A.D. in the *Almagest,* and so was transmitted to a later period.

There is in these emission theories of light clearly an intimate interaction through contact between the observer and the observed. This is also true for the emanation tradition in another, less materialistic form.[10] Here, objects are thought to impress themselves upon our sight owing to a contact force similar to touch—action at a distance being ruled out in classical physics—and this touch reaches our souls by the action of the *eidola* or images or shadows which the emitting bodies send out.

Plato held that as long as the eye is open it emits an inner light. For the eye to perceive, however, there must be outside the eye a "related other light," that of the sun or some other source that allows rays to come from the objects. Once more, a coupling between the outer and inner world is clearly attempted.

There were immense problems with emission theories. How, for example, can the eye pupil, only a few millimeters wide, admit the image that was emitted by a huge mountain? Nevertheless, the emanation theory was the take-off point for the optics developed in the seventeenth century. Here we find the modern idea that there is an infinite number of rays leaving from every point of an illuminated object in all directions. But the observer now stands off-stage, and he may or may not be the recipient of some of these ray bundles. The latter are no longer the *lux* of the ancients—*lux* being the word for light when it is regarded as a subjective phenomenon—but rather the *lumen,* a kind of stream of light "objects."

The modern period started effectively with Kepler, who in his writing on Witelo in 1604 and later in the *Dioptrics* of 1611 described how light is refracted by a sphere, for example in a spherical bottle filled with water; he applied his findings to the pupil of the eye. Here was the basic new idea in the optics of vision: the eyeball, and the lens in front of it, focus the ray bundles that come through the pupil, and at the focus the sensorium is stimulated in some way—*which is simply not discussed as part of optics.* In the *Dioptrics* Kepler showed for the first time how lenses really work. Significantly, most images that can be constructed diagrammatically by ray optics can, in fact, not be seen at all by an eye placed at the instrument. Gone are the eidola and species, the "recognition" of soul by soul in Neoplatonist discussion of optics—but gone also is the close coupling of the observer and the observed. The *lumen* had won over the *lux.*

We see how the science of optics became "modern": by an act of breaking the bonding that was self-evident for the ancients, by disengaging the conceptions of what goes on "out there, objectively speaking" on the one hand, and what the eye does with light on the other hand. At some point someone had to do what Kepler, in preparing for Newton, finally did, namely to get interested in bundles of light rays coming together on a screen outside an eye—or, what is for the physics of light significantly exactly the same thing, on the retina or screen in back of the eye—and to stop thinking about the sense impressions produced at such a focus at the same time. As Müller's influential *Lehrbuch der Physik* said in 1926, just one year before Bohr's formulation of the idea of complementarity, the first task of physical optics "is the sharp separation between the objective ray of light and the sensory impression of light. The subject of discussion of physical optics is the ray of light, whereas the inner processes between eye and brain"—says the *Lehrbuch,* dismissing the matter—"are in the domain of physiology, and perhaps also psychology."

We see here an attempt at precisely the same separation of primary and secondary qualities, between the numerical and affective aspects of nature, that, as it had turned out three centuries before, was the key with which Galileo and others at that time managed to go from the mechanics of antiquity to modern mechanics. We recall that it was Galileo who did for particles, such as falling stones, what Kepler did for light—namely, to remove the language of volition and teleology, and to fortify the notion of "impersonal," causal laws of motion. The Newtonian science of light has no primary place for the observer and his sense impression. In this manner, the important, basic properties of light could be discovered: the finite propagation speed, the existence of light rays outside the range to which the eye is sensitive, the analogy between light rays and other radiation such as X-rays, and so forth.

The decoupling between *lux* and *lumen,* between subject and object, observer and the observed, and with it the destruction of the earlier, holistic physics, was a painful and lengthy process. The reason why it was ultimately victorious is the reason why the same process in all other parts of science worked: once the separation was made, there ensued a dazzling enrichment of our intellectual and material world. By 1927, a reader of physics texts was bound to feel that the modern theory of light, from electromagnetic theory to the design of optical instruments, devoted its attention entirely to *lumen,* and was a field just as deanthropomorphized as all other parts of the developed physical sciences.

But the seed of a new view of light was present, carried in the early historic development which we have sketched, in the prescientific, common-sense notions that everyone begins with—and in the operational meaning of some of the main concepts of optics. Thus we turn to a second main line of ideas leading to the complementarity point of view.

Operational Meanings

One of the oldest and most elementary building blocks of optics is: light travels in any homogeneous medium in straight lines. But let us consider for a moment why we believe that this statement is true.

We can check it most directly in an experimental way by inserting a screen or scatterer, such as chalk dust, in different parts of the same beam. If we consider this closely, we notice that such a method destroys the light beam that we wanted to examine. The insertion of the apparatus interferes with the phenomenon.

This situation is typical on the atomic scale. There are no comparable problems when one wishes to check, say, Newton's first law of motion for ordinary physical objects, for example by watching or photographing a ball rolling on a flat table. We can verify that a material object in a force-free medium will travel in straight lines without drastically interrupting the object's path. The small effects of the apparatus can be removed by calculation. The fact that the observer and the "object" must share between them at least one indivisible quantum is here negligible, that is, can be made an arbitrarily small part of the phenomenon. From past observations we can therefore extrapolate with certainty the paths the object will take in the future. Space-time descriptions and classical causality apply without difficulty. Not so for beams of light and of other particles on the atomic scale. The more certainly we have ascertained their past, the less certainly we can follow their subsequent progress; the effects of the perturbing interaction with the apparatus cannot be taken out by calculation, but is intrinsically probabilistic. In fact, owing to the uncertainty principle, it is not even possible to define precisely the initial state of the system in the sense required by the classical view of causality.

If we do not wish to intercept the whole beam, we can try to discover whether a beam goes in a straight line by another method: by placing a number of slits at some distance from one another, but all along the same axis, then checking if light penetrates this whole set of collimators. But there are now two problems. First, how do we know whether the slits are indeed arranged in a straight line? We might check it with a straight edge—but we know the straight edge is straight because we can sight directly along it and see no curves or protrusions. Clearly, this process of sighting, or anything equally effective, relies on using a light beam to sight along the ruler. And that, of course, is circular reasoning, assuming, in setting up the instrument, what the experiment is designed to prove.

The paradox is not inescapable; there are other, although more cumbersome, methods for lining up the slits without assuming anything about light. But again, we run into trouble. The more closely we wish to define the line along which the beam is to travel, and consequently the narrower we make the slit, the more we find that the beam's energy is spread out into the "shadow," turning, as it were, a corner on going through the slit. This is the phenomenon of diffraction. It is exceedingly easy to demonstrate with the crudest equipment, even just by letting light from a candle pass through the narrow space between two fingers held closely to the eye.

We are dealing here with an instrumental coupling between observer (equipment) and the entity to be observed. As soon as we try to give an operational meaning to the phrase "light travels in any homogeneous medium in straight lines," we see what a poor statement it is.

As a result, a physicist is likely to prefer another statement, more general but which can be reduced in the limit to the one above. It is Fermat's Principle of Least Time, derived from a statement that dates from about 1650. Between any two points, light will go along that path in which the time spent in transit is less than the time that would be spent in any other path. This view explains why a light beam appears to go in a straight line in a homogeneous medium, and also how a beam is reflected or refracted at the interfaces of two media. But the statement harbors the curious idea that light is "exploring" to find the quickest path, as if light were scouting around in the apparatus. We get here a hint of instrumental coupling of the most intimate sort. The suspicion arises that the properties we assign to light are to some degree the properties of the boxes through which light has to find its way.

This becomes quite obvious and unmistakable when we turn to another well-known experiment. When light is sent through a double slit, an interference pattern characteristic of the geometry of the arrangement is obtained on a screen. If one of the two slits is blocked off, a rather different pattern of interference results. All this can be easily understood with elementary constructions from the classical theory of light. However, if a very weak beam of light is used with the double slit experiment, so that at any given time it is exceedingly unlikely that more than a single photon travels through the apparatus, a remarkable thing will be observed: even though one cannot help using classical language and thinking that a single photon will have to go through either one or the other of the two slits at a given time, it will be found that as long as both slits are kept open, the interference pattern accumulating in due course on a photographic plate placed at the screen has exactly the same characteristics as that for the earlier double slit experiment, when the beam was so strong that at any given moment some photons were passing through one of the slits and some photons were passing through the other. Equally remarkable, if one now closes one of the slits toward which the very weak beam of photons is being sent, the interference pattern accumulated over a period changes to the pattern characteristic of a strong light beam passing through a single slit. The fact that for a weak light beam the interference pattern depends on the number of slits available—even though there is no evident way in which the single photon can "know" if the other slit is open—is an indication that the experimental observations of light yield characteristics of the box and its slits as much as of light itself. In short, the experiments are made on the entity light + box. Here, then, in the operational examination of the laws of light propagation, is a second path leading to the complementarity idea.

From Correspondence to Complementarity

Yet another primary influence on Bohr was, of course, the achievement and failures of physics in his own work from about 1912 to about 1925. Bohr's model of the hydrogen atom of 1912-1913 is now usually remembered best for the magnificent accomplishment of predicting the frequencies of the emission spectrum. To do this, Bohr essentially tried to reconcile the two apparently antithetical notions about light, both of which had had their successes—the electromagnetic theory of Maxwell, according to which light propagates as a wavelike disturbance characterized by continuity, and, on the other hand, Einstein's theory that light energy is characterized by discreteness and discontinuity. As Einstein had put it in his 1905 paper presenting a "heuristic" point of view concerning the interaction of light and matter: "The energy in the light propagated in rays from a point is not smeared out continuously over larger and larger volumes, but rather consists of a finite number of energy quanta localized at space points, which move without breaking up, and which can be absorbed or emitted only as wholes."

By 1912 the indisputable evidence for Einstein's outrageous notion was not yet at hand, but some experiments on the photoelectric effect, including those with X-rays, began to make it plausible.[11] Indeed, it was not until Millikan's experiment, published in 1916, and A. H. Compton's experiment of 1922, that the quantum theory of light was seen everywhere to be unavoidable.

It is therefore, in retrospect, even more remarkable how courageous Niels Bohr's work of 1912-1913 was. Let us recall his model of the hydrogen atom in its initial form, even though it was soon made more accurate though more complex. Bohr's hydrogen atom had the nucleus at the center (where Ernest Rutherford, in whose Manchester laboratory Bohr was a guest, had just then discovered it to be) and the electron orbiting at some fixed distance around the nucleus. When the sample is heated or the atoms are otherwise excited by being given extra energy, the electron of the excited atom will not be in the normal, innermost orbit, or ground state, but will be traveling in a more distant orbit. At some point the electron will jump from the outer orbit to one of the allowed inner orbits, and in so doing will give up the energy difference between these orbits, or stationary states, in the form of a photon of energy $h\nu$. This corresponds to the emission of light at the observed frequency ν or the corresponding wave length $\lambda = c/\nu$ (where c is the speed of light). The various observed frequencies emitted from an excited sample of hydrogen atoms were therefore interpreted to be a stream of photons, each photon having the energy corresponding to the allowed transition between stationary states.

The success of the model in explaining all known spectrum lines of hydrogen, in predicting other series that were also found, and in giving a solid foothold on the explanation of chemical properties, could not hide the realization, fully apparent to Niels Bohr himself, that the model carried with it a number of grave problems. First of all, it used simultaneously two separate notions which were clearly conflicting: the classical notion of an identifiable electron moving in an identifiable orbit like a miniature planetary system, and the quantum notion that such an electron is in a stationary state rather than continually giving up energy while orbiting (as it should do on the basis of Maxwell's theory, amply tested for charges circulating in structures of large size). Bohr's postulate that the electron would not lose energy by radiation while in an orbit, but only on transition from one orbit to the other, was necessary to "save" the atom from gradually collapsing with the emission of a spectrum line of continuously changing frequency. Also, contrary to all previous ideas, the frequency of the emitted photon was not equal to the frequency of the model's orbiting electron, either in its initial or in its final stationary state.

Looking back later on the situation of about 1912, Merle A. Tuve noted that the Bohr atom was "quite irrational and absurd from the viewpoint of classical Newtonian mechanics and Maxwellian electrodynamics . . . Various mathematical formalisms were devised which simply 'described' atomic states and transitions, but the same arbitrary avoidance of detailed processes, for example, descriptions of the actual *process* of transition, were inherent in all these formulations."[12]

Niels Bohr himself took pains to stress these conflicts from the beginning. In fact, the explanation of the spectral lines, which were the most widely hailed achievement, more or less constituted an afterthought in his own work. His interest was precisely to examine the area of conflict between the conceptions of ordinary electrodynamics and classical mechanics on the one hand and quantum physics on the other. As Jammer pointed out, "Not only did Bohr fully recognize the profound chasm in the conceptual scheme of his theory, but he was convinced that progress in quantum theory could not be obtained unless the antithesis between quantum-theoretic and classical conceptions was brought to the forefront of theoretical analysis. He therefore attempted to trace the roots of this antithesis as deeply as he could. It was in this search for fundamentals that he introduced the revolutionary conception of 'stationary' states, 'indicating thereby that they form some kind of waiting places between which occurs the emission of the energy corresponding to the various spectral lines,' [as Bohr put it in an address of December 20, 1913, to the Physical Society in Copenhagen[13]]." At the end of his address, Bohr said, "I hope I have expressed myself sufficiently clearly so that you appreciate the extent to which these considerations conflict with the admirably coherent group of conceptions which have been rightly termed the classical theory of electrodynamics. On the other hand, by emphasizing this conflict, I have tried to convey to you the impression that it may also be possible in the course of time to discover a certain coherence in the new ideas."

This methodological strategy of *emphasizing conceptual conflict as a necessary preparation for its resolution* culminated, fourteen years later, in the announcement of the complementarity principle. In the meantime, Bohr formulated a proposal that turned out to be a moderately successful half-way house toward the reconciliation between classical and quantum mechanics, a conception which, from about 1918 on, became known as the correspondence principle.

In essence, Bohr still hoped for the resolution between opposites by attending to an area where they overlap, namely the extreme cases where quantum theory and classical mechanics yield to each other. For example, for very large orbits of the hydrogen atom's electron, the neighboring, allowed stationary states in Bohr's model come to be very close together. It is easily shown that a transition between such orbits, on the basis of quantum notions, yields a radiation of just the same frequency expected on classical grounds for a charged particle orbiting as part of a current in a circular antenna—and, moreover, the frequency of radiation would be equal to the frequency of revolution in the orbit. Thus for sufficiently large "atoms," and conversely for sufficiently small "circuits" scaled down from the normal size of ordinary electric experiments, a coincidence, or correspondence, of predictions is obtained from the two theories.

In this manner, classical physics becomes the limiting case of the more complex quantum physics: our more ordinary, large-scale experiments fail to show their inherently quantal character only because the transitions involved are between states characterized by high quantum numbers. In this situation the quantum of action relative to the energies involved in the system is effectively zero rather than having a finite value, and the discreteness of individual events is dissolved, owing to the large number of events, in an experienced continuum.

The correspondence principle came to be developed in the hands of Bohr and his collaborators into a sophisticated tool. The basic hope behind it was explained by Bohr in a letter to A. A. Michelson on February 7, 1924:

It may perhaps interest you to hear that it appears to be possible for a believer in the essential reality of the quantum theory to take a view which may harmonize with the essential reality of the wave-theory conception even more closely than the views I expressed during our conversation. In fact on the basis of the correspondence principle it seems possible to connect the discontinuous processes occurring in atoms with the continuous character of the radiation field in a somewhat more adequate way than hitherto perceived . . . I hope soon to send you a paper about these problems written in cooperation with Drs. Kramer and Slater.[14]

But shortly after the publication in 1924 of the paper by Bohr, Kramers, and Slater,[15] experiments were initiated by W. Bothe and H. Geiger, and by A. H. Compton and A. W. Simon—with unambiguously disconfirming results. The correspondence principle, it appeared now clearly, had been a useful patch over the fissure, but it was not a profound solution.

Even before that discovery, major problems known to be inherent in the Bohr atom included the following: the fact that the antithetical notions of the wave (implied in the frequency or wave length of light emitted) and of the particle (implied in the then current idea of the electron) were by no means resolved, but on the contrary persisted unchanged in the model of the atom; so did the conflict between the antithetical notions of classical causality on the one hand (as in the presumed motion of the electrons in their orbits) and of probabilistic features on the other (as for the transitions between allowed orbits); and even the notion of the "identity" of the atom had to be revised, for it was no longer even in principle observable and explorable as a separate entity without interfering with its state. Each different type of experiment produces its own change of state, so that different experiments produce different "identities."

Such questions remained at the center of discussion among the most concerned physicists. Schrödinger and de Broglie, for example, hoped to deal with the glaring contrast between the themata of continuity and discontinuity by providing a wave-mechanical explanation for phenomena that previously had been thought to demand a language of quantization. As Schrödinger wrote in his first paper on the subject,[16] "It is hardly necessary to point out how much more gratifying it would be to conceive a quantum transition as an energy change from one vibrational mode to another than to regard it as a jumping of electrons. The variation of vibrational modes may be treated as a process continuous in space and time and enduring as long as the emission process persists." Thus, space-time description and classical causality would be preserved.

The reception accorded to Schrödinger's beautiful papers was interesting. Heisenberg had obtained essentially the same results in a quite different way through his matrix mechanics; as Jammer notes, "it was an *algebraic* approach which, proceeding from the observed discreteness of spectral lines, emphasized the element of *discontinuity;* in spite of its renunciation of classical description in space and time it was ultimately the theory whose basic conception was the *corpuscle.* Schrödinger's, in contrast, was based on the familiar apparatus of differential equations, akin to the classical mechanics of fluids and suggestive of an easily visualizable representation; it was an *analytical* approach which, proceeding from a generalization of the classical laws of motion, stressed the element of *continuity.*"[17] "Those who in their yearning for continuity hated to renounce the classical maxim *natura non facit saltus* acclaimed Schrödinger as the herald of a new dawn. In fact, within a few brief months, Schrödinger's theory 'captivated the world of physics' because it seemed to promise 'a fulfillment of that long-baffled and insuppressible desire' [in the words of K. K. Darrow, *The Bell System Technical Journal,* 6 (1927)] . . . Planck reportedly declared 'I am reading it as a child reads a puzzle,' and Sommerfeld was exultant."[18] So, of course, was Einstein, who as early as 1920 had written to Born, "that one has to solve the quanta by giving up the continuum, I do not believe."

We are, of course, dealing here with the kind of intellectual commitment, or "insuppressible desire," that characterizes a true thematic attachment. Rarely has there been a more obvious fight between different themata vying for allegiance, or a conflict between the aesthetic criteria of scientific choice in the face of the same set of experimental data. And nothing is more revealing of the true and passionate motivation of scientists than their responses to each others' antithetical constructs. In a letter to Pauli, Heisenberg wrote: "The more I ponder about the physical part of Schrödinger's theory, the more disgusting *[desto abscheulicher]* it appears to me." Schrödinger, on his side, freely published his response to Heisenberg's theory: "I was discouraged *[abgeschreckt]* if not repelled *[abgestossen].*"[19]

Different aspects of thematic analysis and thematic conflict were the subject of previous articles.[20] In these studies I pointed out a number of other theme-antitheme couples, which may be symbolized by $(\theta, \bar{\theta})$. What Bohr had done in 1927, shortly after the Heisenberg-Schrödinger debates, was to develop a point of view which would allow him *to accept both members of the* $(\theta, \bar{\theta})$ *couple as valid pictures of nature,* accepting the continuity-discontinuity (or wave-particle) duality as an irreducible fact, instead of attempting to dissolve one member of the pair in the other as he had essentially tried to do in the development of the correspondence-principle point of view. Secondly, Bohr saw that the $(\theta, \bar{\theta})$ couple involving discrete atomism on the one hand and continuity on the other is related to other $(\theta, \bar{\theta})$ dichotomies that had obstinately refused to yield to bridging or mutual absorption (for example, the subject-object separation versus subject-object coupling; classical causality versus probabilistic causality). The consequence Bohr drew from these recognitions was of a kind rare in the history of thought: he introduced explicitly a new thema, or at least identified a thema that had not yet been consciously a part of contemporary physics. Specifically, Bohr asked that physicists accept both θ and $\bar{\theta}$—though both would not be found in the same plane of focus at any given time. Nor are θ and $\bar{\theta}$ to be transformed into some new entity. Rather, they both exist in the form Either θ/Or $\bar{\theta}$, the choice depending on the theoretical or experimental question which you may decide to ask. We see at once why all parties concerned, both those identified with θ and those identified with $\bar{\theta}$, would not easily accept a new thema which saw a basic truth in the existence of a paradox that the others were trying to remove.

Poul Martin Møller and William James

Another root of the complementarity conception can be discerned in Niels Bohr's work when we carefully read and reread his own statements of the complementarity point of view. For it is at first curious and then undeniably significant that from the very beginning in 1927, Niels Bohr cited experiences of daily life to make apparent the difficulty of distinguishing between object and subject, and, as Oskar Klein wrote in a retrospective essay, in order "to facilitate understanding of the new situation in physics, where his view appeared too radical or mysterious even to many physicists."[21] In this connection, according to Klein, Bohr chose a particularly simple and vivid example: the use one may make of a stick when trying to find one's way in a dark room. The man, the stick, and the room form one entity. The dividing line between subject and object is not fixed. For example, the dividing line is at the end of the stick when the stick is grasped firmly. But when it is loosely held, the stick appears to be an object being explored by the hand. It is a striking reminder of the situation described in the classical emanation theory of light in which we first noted the problem of coupling between observer and observed.

On studying Bohr's writings one realizes by and by that his uses of apparently "extraneous" examples or analogies of this sort are more than mere pedagogic devices. In his September 1927 talk, the final sentence was "I hope, however, that the idea of complementarity is suited to characterize the situation, which bears a deep-going analogy to the general difficulty in the formation of human ideas, inherent in the distinction between subject and object." Similar and increasingly more confident remarks continued to characterize Bohr's later discussions of complementarity. Thus in his essay on "Quantum Physics and Philosophy" (1958), the lead essay in the second collection of Bohr's essays under the title *Essays 1958-1962 on Atomic Physics and Human Knowledge,*[22] Bohr concluded, "It is significant that . . . in other fields of knowledge, we are confronted with situations reminding us of the situation in quantum physics. Thus, the integrity of living organisms, and the characteristics of conscious individuals, and human cultures, present features of wholeness, the account of which implies a typical complementarity mode of description . . . *We are not dealing with more or less vague analogies, but with clear examples of logical relations which, in different contexts, are met with in wider fields.*" It will be important for our analysis to try to discern clearly what Bohr means in such passages.

Some illumination is provided by a story which Niels Bohr loved to tell in order to illustrate and make more understandable the complementarity point of view. Léon Rosenfeld, a long-term associate of Niels Bohr, who has also been concerned with the origins of complementarity, told how seriously Bohr took his task of repeatedly telling the story. "Everyone of those who came into closer contact with Bohr at the Institute, as soon as he showed himself sufficiently proficient in the Danish language, was acquainted with the little book: it was part of his initiation."[23]

The "little book" which Bohr used was a work of the nineteenth-century poet and philosopher, Poul Martin Møller. In that light story, *The Adventures of a Danish Student,* Bohr found what he called a "vivid and suggestive account of the interplay between the various aspects of our position." A student is trying to explain why he cannot use the opportunity for finding a practical job, and reports the difficulties he is experiencing with his own thought process:

> My endless enquiries make it impossible for me to achieve anything. Furthermore, I get to think about my own thoughts of the situation in which I find myself. I even think that I think of it, and divide myself into an infinite retrogressive sequence of "I"s who consider each other. I do not know at which "I" to stop as the actual one, and in the moment I stop at one, there is indeed again an "I" which stops at it. I become confused and feel a dizziness, as if I were looking down into a bottomless abyss, and my ponderings result finally in a terrible headache.

Further, the student remarks:

> The mind cannot proceed without moving along a certain line; but before following this line, it must already have thought it. Therefore one has already thought every thought before one thinks it. Thus every thought, which seems the work of a minute, presupposes an eternity. This could almost drive me to madness. How could then any thought arise, since it must have existed before it is produced? . . . The insight into the impossibility of thinking contains itself an impossibility, the recognition of which again implies an inexplicable contradiction.[24]

Bohr used the situation in the story not as a distant, vague analogy; rather, it is one of those cases which, "in different contexts, are met with in wider fields." Moreover, the story seems appropriate for two other reasons. Bohr reports that conditions of analysis and synthesis of psychological experiences "have always been an important problem in philosophy. It is evident that words like thoughts and sentiments, referring to mutually exclusive experiences, have been used in a typical complementary manner since the very origin of language."[25] Also, the humane setting of the Danish story, and the fact that it renders a situation in words rather than scientific symbols, should not mislead us into thinking that it is thereby qualitatively different from the information supplied in scientific discourse. On the contrary: Bohr said, in defending the complementarity principle, "The aim of our argumentation is to emphasize that all experience, whether in science, philosophy, or art, which may be helpful to mankind, must be capable of being communicated by human means of expression, and it is on this basis that we shall approach the question of unity of knowledge."[26] We shall come back to this important statement presently.

Now one must confess that it is on first encounter curious, and at least for a professional physicist perhaps a little shocking, to find that the father of the complementarity principle, in these passages and others, should frequently have gone so far afield, by the standards of the scientific profession, in illustrating and extending what he took to be the full power of the complementarity point of view. In looking for the roots of the complementarity principle, we might grant more readily the three avenues shown so far, namely through the history of the concept of light, the operational definition of light behavior, and through Bohr's own work in physics. But in pursuing this new avenue, we seem to be leaving science entirely.

I imagine that many of Bohr's students and associates listened to his remarks with polite tolerance, perhaps agreeing that there might be a certain pedagogic benefit, but not a key to the "unity of knowledge." To the typical scientist, the student in Møller's story who becomes dizzy when he tries to think about his own thoughts, because precise "thought" and "thought *about* thought" are complementary with respect to each other and so mutually exclusive at the same time, would seem somehow to have a problem different from that of the experimenter who cannot simultaneously show both the wave characteristics and the particle characteristics of a light beam. Similarly the intrusion of the student as introspective observer upon his own thought processes seems to have after all only a thin connection with the intrusion of the macroscopic laboratory upon the submicroscopic quantum events being studied.

It was therefore surprising and revealing when it was found recently, almost by accident, that one of the roots of the modern complementarity point of view in Niels Bohr's own experience was probably just this wider, more humanistic context shown in the previous quotations. The discovery I speak of came about in a dramatic way. A few years ago, the American Physical Society and the American Philosophical Society engaged in a joint project to assemble the sources for the scholarly study of the history of quantum mechanics. This project, under the general directorship of Thomas S. Kuhn, spanned a number of years, and one of its functions was to obtain interviews with major figures on the origins of their contributions to quantum physics. An appointment for a number of interviews was granted by Niels Bohr, and the fifth interview was conducted on November 17, 1962, by Kuhn and Aage Petersen. In the course of the interview, Petersen, who was Niels Bohr's long-time assistant, raised the question of the relevance of the study of philosophy in Bohr's early thoughts. The following interchange occurred, according to the transcript:

AaP: How did you look upon the history of philosophy? What kind of contributions did you think people like Spinoza, Hume, and Kant had made?

NB: That is difficult to answer, but I felt that these various questions were treated in an irrelevant manner [in my studies].

AaP: Also Berkeley?

NB: No, I knew what views Berkeley had. I had seen a little in Høffding's writings, but it was not what one wanted.

TSK: Did you read the works of any of these philosophers?

NB: I read some, but that was an interest by [and here Bohr suddenly stopped and exclaimed]—oh, the whole thing is coming [back to me]! I was a close friend of Rubin [a fellow student, later psychologist], and, therefore, I read actually the work of William James. William James is really wonderful in the way he makes it clear—I think I read the book, or a paragraph, called . . . No, what is that called? It is called "The Stream of Thoughts," where he in a most clear manner shows that it is quite impossible to analyze things in terms of—I don't know what to call it, not atoms. I mean simply, if you have some things . . . they are so connected that if you try to separate them from each other, it just has nothing to do with the actual situation. I think that we shall really go into these things, and I know something about William James. That is coming first up now. And that was because I spoke to people about other things, and then Rubin advised me to read something of William James, and I thought he was most wonderful.

TSK: When was this that you read William James?

NB: That may be a little later, I don't know. I got so much to do, and it may be at the time I was working with surface tension [1905], or it may be just a little later. I don't know.

TSK: But it would be before Manchester [1912]?

NB: Oh yes, it was many years before.[27]

Niels Bohr clearly was interested in pursuing this further—"we shall really go into these things." But alas, the next day Bohr suddenly died.

There are enough leads to permit plausible speculations on this subject. K. T. Meyer-Abich reports in his interesting book, *Korrespondenz, Individualität und Komplementarität* (Wiesbaden, 1965) that among German scientists it was remembered that Bohr used to cite William James and only a few other western philosophers. Moreover, Niels Bohr himself, in an article in 1929[28] makes lengthy excursions into psychology in order to use analogies that, in Meyer-Abich's opinion, could well refer directly to William James's chapter on the "Stream of Thought" in James's book, *The Principles of Psychology* (1890). On the other hand, doubts have been raised about the timing. Rosenfeld[29] has expressed his strong belief that the work of William James was not known to Niels Bohr until about 1932. He recalls that in or about 1932, Bohr showed Rosenfeld a copy of James's *Principles of Psychology.* Rosenfeld believes that a few days earlier Bohr had had a conversation with Rubin, the psychologist and Bohr's former fellow student. Rubin may have sent the book to Bohr after their conversation. Bohr showed excited interest in the book, and especially pointed out to Rosenfeld the passages on the "stream of consciousness." During the next few days, Bohr shared the same excitement with several visitors, and Rosenfeld retained the definite impression that this was Bohr's first acquaintance with William James's work. In Rosenfeld's opinion, more relevant than speculation concerning an early influence of James was a remark made by Bohr: after discussing his "early philosophical meditations and his pioneering work of 1912-1913, he told me [Rosenfeld] in an unusually solemn tone of voice, 'and you must not forget that I was quite alone in working out these ideas, and had no help from anybody.' "[30]

In view of remarkable analogies or similarities between the ideas of James and of Bohr, to be shown below, one can choose either to believe, with Meyer-Abich and Jammer, that Bohr had read James early enough to be directly influenced, or to believe with Rosenfeld that Bohr had independently arrived at the analogous thoughts (perhaps brought to them by other forces such as those we have already cited, or additional ones such as contemplation of the concepts of multiform function and Riemann surfaces).[31] In some ways the second alternative is the more interesting though difficult one, for it hints that here may be a place to attack the haunting old question why and by what mechanisms the same themata attain prominence in different fields in nearly the same periods. Still, no matter which view one chooses to take at this time, reading William James's chapter on the "Stream of Thought" in the light of Bohr's remark in the interview of November 1962 comes as a surprise to a physicist familiar with Bohr's contributions to atomic physics.[32]

James first insists that thought can exist only in association with a specific "owner" of the thought. Thought and thinker, subject and object, are tightly coupled. The objectivization of thought itself is impossible. Hence one must not neglect the circumstances under which thought becomes the subject of contemplation. "Our mental reaction to every given thing is really a resultant of our experience in the whole world up to that date. From one year to another we see things in new lights . . . The young girls that brought an aura of infinity—at present hardly distinguishable existences; the pictures—so empty; and as for the books, what was there to find so mysteriously significant in Goethe?" One can here imagine the sympathetic response of Bohr, who wrote, "for objective description and harmonious comprehension it is necessary in almost every field of knowledge to pay attention to the circumstances under which evidence is obtained."

There is another sense in which consciousness cannot be concretized and atomized. James writes, "Consciousness does not appear to itself chopped up in bits; it flows. Let us call it the stream of thought, of consciousness, or of subjective life." Yet there does exist a discontinuous aspect: the "changes, from one moment to another, in the quality of the consciousness." If we use the vocabulary of quantum theory, James here proposes a sequence of individual changes between stationary states, with short periods of rest in these states—a metaphor that brings to mind Bohr's notion of 1912-1913 of the behavior of the electron in the hydrogen atom. To quote James, "Like a bird's life, [thought] seems to be made of an alternation of flights and perchings. The rhythm of language expresses this, where every thought is expressed in a sentence and every sentence closed by a period . . . Let us call the resting places the 'substantive parts,' and the places of flight the 'transitive parts,' of the stream of thought."

But here enters a difficulty; in fact, the same one that plagued the student in Møller's story. The difficulty is, in James's words, "introspectively, to see the transitive parts for what they really are. If they are but flights to conclusions, stopping them to look at them before a conclusion is reached is really annihilating them." However, if one waits until one's consciousness is again in a stationary state, then the moment is over. James says, "Let anyone try to cut a thought across in the middle and get a look at its section, and he will see how difficult the introspective observation of the transitive tract is . . . Or if our purpose is nimble enough and we do arrest it, it ceases forthwith to be itself . . . The attempt at introspective analysis in these cases is in fact like . . . trying to turn up the light quickly enough to see how the darkness looks." Letting thoughts flow, and making thoughts the subject of introspective analysis are, as it were, two mutually exclusive experimental situations.

It is from such a vantage point that one may attempt to interpret some of the novel features of Bohr's 1927 paper on complementarity to have been influenced either by a reading of James, or by thinking independently on parallel lines—and thereby understand better the final passage in Bohr's paper: "I hope, however, that the idea of complementarity is suited to characterize the situation, which bears a deep-going analogy to the general difficulty in the formation of human ideas, inherent in the distinction between subject and object."[33]

At this point, one might well ask where the term "complementarity" itself, which Bohr introduced into physics in 1927, may have come from. There are a number of fields from which the term may have been adapted, including geometry or topology. But both Meyer-Abich and Jammer point to a more provocative possibility, namely the chapter on "The Relations of Minds to Other Things," in William James's *Principles of Psychology* (1890), just one chapter prior to that on "The Stream of Thought." In the subsection "'Unconsciousness' in Hysterics," James relates cases of hysterical anaesthesia (loss of the natural perception of sight, hearing, touch, and so on), and notes that P. Janet and A. Binet "have shown that during the times of anesthesia, and coexisting with it, *sensibility to the anaesthetic parts is also there, in the form of a secondary consciousness* entirely cut off from the primary or normal one, but susceptible of being *tapped* and made to testify to its existence in various odd ways."[34]

The chief method for tapping was Janet's method of "distraction." If Janet put himself behind hysteric patients who were "plunged in conversation with a third party, and addressed them in a whisper telling them to raise their hand or perform other simple acts [including writing out answers to whispered questions] they would obey the order given, although their *talking* intelligence was quite unconscious of receiving it."[35] If interrogated in this way, hysterics responded perfectly normally when, for example, their sensibility to touch was examined on areas of skin that had been shown previously to be entirely anaesthetic when examined through their primary consciousness.

In addition, some hysterics could deal with certain sensations only in either one consciousness *or* the other, but not in both at the same time. Here James cites a famous experiment in a striking passage:

M. Janet has proved this beautifully in his subject Lucie. The following experiment will serve as the type of the rest: In her trance he covered her lap with cards, each bearing a number. He then told her that on waking she should *not see* any card whose number was a multiple of three. This is the ordinary so-called "post-hypnotic suggestion," now well known, and for which Lucie was a well-adapted subject. Accordingly, when she was awakened and asked about the papers on her lap, she counted and said she saw those only whose number was not a multiple of 3. To the 12, 18, 9, etc., she was blind. But the *hand,* when the sub-conscious self was interrogated by the usual method of engrossing the upper self in another conversation, wrote that the only cards in Lucie's lap were those numbered 12, 18, 9, etc., and on being asked to pick up all the cards which were there, picked up these and let the others lie. Similarly when the sight of certain things was suggested to the sub-conscious Lucie, the normal Lucie suddenly became partially or totally blind. "What is the matter? I can't see!" the normal personage suddenly cried out in the midst of her conversation, when M. Janet whispered to the secondary personage to make use of her eyes.[36]

James gives these and other examples to support a conclusion in which he defines the concept of complementarity in psychological research:

It must be admitted, therefore, that in *certain persons,* at least, *the total possible consciousness may be split into parts which coexist but mutually ignore each other,* and share the objects of knowledge between them. More remarkable still, they are *complementary.* Give an object to one of the consciousnesses, and by that fact you remove it from the other or others. Barring a certain common fund of information, like the command of language, etc., what the upper self knows the under self is ignorant of, and *vice versa.*[37]

The analogy with Bohr's concept of complementarity in physics is striking, quite apart from the question of the genetic connection between these two uses of the same word.

Christian Bohr and Harald Høffding

Bohr's affinity for ideas analogous to those of William James was preceded by a philosophical and personal preparation that goes back to his childhood. In his essay, "Glimpses of Niels Bohr as a Scientist and Thinker," Oskar Klein, one of Bohr's earliest collaborators, provides a revealing picture of the young man.

Niels Bohr himself and his brother Harald, a brilliant mathematician, liked to give examples of the innocently credulous—and at the same time resolute—way in which as a child he accepted what he saw and heard. They also spoke of geometrical intuition he developed so early . . . The first feature appeared for instance in believing literally what he learned from the lessons on religion at school. For a long time this made the sensitive boy unhappy on account of his parents' lack of faith. When later, as a young man, he began to doubt, he did so also with unusual resolution and thereby developed a deep philosophical bent similar to that which seems to have characterized the early Greek natural philosophers.[38]

Christian Bohr, Niels Bohr's father, was professor of physiology at the University of Copenhagen. His work involved him in one of the important philosophical debates of the last part of the nineteenth century, the differences between and relative merits of the "vitalistic" theories and the mechanistic conceptions of life processes. In several ways, Christian Bohr's interests shaped his son's ideas and preoccupations. We know that as a youth, Niels Bohr was allowed to work in the laboratory of his father, and to meet the scholars interested in philosophy with whom Christian Bohr kept close contact, such as Harald Høffding, professor of philosophy at the University in Copenhagen. Høffding often visited the Bohr household, and Niels Bohr attested to the profound influence he received from early childhood by being permitted to stay and listen during meetings of an informal club made up of his father, Høffding, the physicist Christian Christiansen, and the philologist Hans Thomsen. Høffding, in turn, described Christian Bohr as a scientist who recognized "strict application of physical and chemical methods of physiology" in the laboratory, but who, outside the laboratory, "was a keen worshipper of Goethe. When he spoke of practical situations or of views of life, he liked to do so in a dialectic manner."[39]

We may understand the implications of this description best through Oskar Klein, who remembers a characterization which Niels Bohr gave him: "He mentioned his father's idea that teleology, when we want to describe the behavior of living beings, may be a point of view on a par with that of causality. This idea was later to play an essential role in Bohr's attempt to throw light on the relation between the biologist's and the physicist's way of describing nature."[40]

Niels Bohr entered the university in 1903, and soon took Høffding's course in the history of philosophy and logic. He also belonged to a student's club in which the questions raised in Høffding's lectures on philosophy were discussed. (Another member was Rubin.) While Bohr, as indicated in his last interview, felt no great attraction to philosophical systems (such as those of "Spinoza, Hume, and Kant"), there is little doubt about the lasting impression Høffding made on Bohr—perhaps most of all because of Høffding's active interest in the applicability to philosophy of the work of what he called *philosophierende Naturforscher*, from Copernicus to Newton and from Maxwell to Mach. For example, the latter two are discussed at some length in Høffding's *Moderne Philosophen* which appeared in 1904 in Danish (1905 in German) as successor to his monumental *History of Modern Philosophy*.

There also appears to have been a personal sympathy between the older and the younger man. While still Høffding's student, Bohr pointed out some error in Høffding's exposition, and Høffding, in turn, allowed Bohr to help him correct proofs of the offending passage. A warm friendship developed eventually that was freely acknowledged on both sides, as indicated, for example, by Niels Bohr's acknowledgment of Harald Høffding's influence on him, on the occasion of Høffding's eighty-fifth birthday,[41] and conversely in letters of Høffding to Emile Meyerson in 1926 and 1928.[42] The first of these letters, incidentally, is dated December 13, 1926, shortly before Bohr's vacation trip to Norway in early 1927 during which, according to Heisenberg and others, Bohr's ideas on complementarity were developed in the form he announced later in 1927. Another letter was written half a year after the presentation of the complementarity principle at Como. In it, Høffding writes to Meyerson (March 13, 1928): "Bohr declares that he has found in my books ideas which have helped the scientists in the 'understanding' of their work, and thereby they have been of real help. This is great satisfaction for me, who feels so often the insufficience of my special preparation with respect to the natural sciences."[43]

Among all the philosophers and scientists discussed by Høffding, it is unlikely that any interested student of Høffding's will have failed to encounter some aspect of William James's work. An admirer, like James, of G. T. Fechner (the father of psychophysics), Høffding devoted his first book to psychology (Danish edition, 1882). At about the time Bohr took his philosophy course, Høffding used the occasion of the St. Louis meeting of 1904 to visit James in the United States. James, in turn, supplied an appreciative preface for the English translation (of 1905) of Høffding's *Problems of Philosophy*—a book which Høffding reported later to have originated in his university lectures in 1902.[44] And in the same year of Høffding's visit to James, Høffding expressed in his *Moderne Philosophen* his admiration for James's work, to whom the concluding chapter is devoted, with such comments as, "James belongs to the most outstanding contemporary thinkers . . . The most important of his writings is *The Principles of Psychology*."

Kierkegaard

In Høffding's own life, a crucial and early influence was the work of Kierkegaard, as he freely confessed.[45] Høffding reported that in a youthful crisis, in which he was near "despair," he had found solace and new strength through Kierkegaard's writings, and he mentions particularly Kierkegaard's work now known as *Stages on Life's Way*. Høffding became known as one of the prominent exponents and followers of Kierkegaard; indeed, the second major work Høffding published was the book, *Kierkegaard als Philosoph*.[46]

Whether Niels Bohr caught some of his own interest in Kierkegaard while a student of Høffding is not known, but the fact of this early interest is well documented. Thus it is remembered that in 1909 Niels sent his brother Harald as a birthday gift Kierkegaard's book, *Stages on Life's Way*, with a letter saying, "It is the only thing I have to send; but I do not believe that it would be very easy to find anything better. In any case I have had very much pleasure in reading it, I even think that it is one of the most delightful things I have ever read."[47] Then he added that he did not fully agree with all of Kierkegaard's views. One can well imagine that Niels Bohr could enjoy the aesthetic experience and the moral passion, without having to agree also with the antiscientific attitude of much of the work.

Bohr's remarks about Kierkegaard bring us to the last of the various possible avenues that prepared for the complementarity notion. While this is not the proper place for a searching examination of those elements in Kierkegaard's works for which analogous elements have been noted in Bohr's work,[48] it will be of interest to remind ourselves of one or two chief features that characterized the writing of both Kierkegaard and his chief interpreter in Denmark, Høffding.

Kierkegaard's existentialism was rooted in German Romanticism, upholding the individual and the momentary life situation in which he finds himself against the rationality and objective abstraction championed by eighteenth-century Enlightenment. The denial of the subjective, Kierkegaard argued, leads to self-contradictions, for even the most abstract proposition remains the creation of human beings. In a reaction to Hegel and to some aspects of Kant, Kierkegaard wrote about science in his journal: "Let it deal with plants and animals and stars, but to deal with the human spirit in that way is blasphemy, which only weakens ethical and religious passions." Truth cannot be found without incorporating the subjective, particularly in the essentially irrational, discontinuous stages of recognitions leading to the achievement of insight. As J. Passmore writes, "each major step on the way to truth is a free decision. Our progress, according to Kierkegaard, from the aesthetic to the scientific point of view, and then again from the scientific to the ethical and from the ethical to the religious, cannot be rationalized into an orderly, formally justifiable, step from premise to conclusions: It is in each case a leap to a quite new way of looking at things."[49]

What is perhaps of greatest interest to us is the accentuation of the role of discontinuity in Kierkegaard's work. Here we can do no better than cite at some length the section on Kierkegaard in Høffding's own chief work, *A History of Modern Philosophy:*

> [Kierkegaard's] leading idea was that the different possible conceptions of life are so sharply opposed to one another that we must make a choice between them, hence his catchword *either—or;* moreover, it must be a choice which each particular person must make for himself, hence his second catchword, *the individual.* He himself designated his thought "qualitative dialectic," by which he meant to bring out its opposition to the doctrine taught by Romantic speculation of continuous development by means of necessary inner transitions. Kierkegaard regarded this doctrine as pure fantasticalness—a fantasticalness, to be sure, to which he himself had felt attracted.[50]

What is essential for us to notice is that a main feature of Kierkegaard's "qualitative dialectic" is an acceptance of thesis and antithesis, *without* proceeding to another stage at which the tension is resolved in a synthesis. Thus he draws a line between thought and reality which must not be allowed to disappear. Høffding writes: "Even if thought should attain coherency it does not therefore follow that this coherency can be preserved in the practice of life . . . Such great differences and oppositions exist side by side that there is no thought which can embrace them all in a 'higher unity.' "[51] "Kierkegaard came more and more to regard the capability of embracing great contrasts and of enduring the suffering which this involves as the criterion of the sublimity and value of a conception of life."[52]

Kierkegaard's stress on discontinuity between incompatibles, on the "leap" rather than the gradual transition, on the inclusion of the individual, and on inherent dichotomy, was as "nonclassical" in philosophy as the elements of the Copenhagen doctrine—quantum jumps, probabilistic causality, observer-dependent description, and duality—were to be in physics.

Now it would be as absurd as it is unnecessary to try to demonstrate that Kierkegaard's conceptions were directly and in detail translated by Bohr from their theological and philosophical context to a physical context. Of course, they were not. All one should do is permit oneself the open-minded experience of reading Høffding and Kierkegaard through the eyes of a person who is primarily a physicist—struggling, as Bohr was, first with his 1912-1913 work on atomic models, and again in 1927, to "discover a certain coherence in the new ideas" while pondering the conflicting, paradoxical, unresolvable demands of classical physics and quantum physics which were the near-despair of most physicists of the time. It is in this frame of mind that one can best appreciate, for example, Høffding's discussion of Kierkegaard's indeterministic notion of the "leap":

In Kierkegaard's ethics the qualitative dialectic appears partly in his conception of choice, of the decision of the will, partly in his doctrine of stages. He emphatically denies that there is any analogy between spiritual and organic development. No gradual development takes place within the spiritual sphere, such as might explain the transition from deliberation to decision, or from one conception of life (or "stadium") to another. Continuity would be broken in every such transition. As regards the choice, psychology is only able to point out possibilities and approximations, motives and preparations. The choice itself comes with a jerk, with a leap, in which something quite new (a new quality) is posited. Only in the world of possibilities is there continuity; in the world of reality decision always comes through a breach of continuity.

But, it might be asked, cannot this jerk or this leap itself be made an object of psychological observation? Kierkegaard's answer is not clear. He explains that the leap takes place between two moments, between two states, one of which is the last state in the world of possibilities, the other the first state in the world of reality. It would almost seem to follow from this that the leap itself cannot be observed. But then it would also follow that it takes place unconsciously—and the possibility of the unconscious continuity underlying the conscious antithesis is not excluded.[53]

It is at this point that the writings of Høffding and Kierkegaard most evidently overlap with the teachings of William James. In fact, there are two specific periods where the overlapping conceptions of Kierkegaard, Høffding, and James can plausibly have been influential for Bohr in the sense of providing sympathetic preparation or support: one came in Bohr's work during the early period, from 1912 through the correspondence point of view (that is, in the analogy between Bohr's nonclassical transitions of the electron between stationary states on one hand, and Kierkegaard's "leaps" or James's transient flights and "transitive parts" on the other hand). The other came in the period from about 1926, when Bohr's complementarity point of view was being developed; and here we have already pointed to possible sources or antecedents for Bohr's analogies in passages such as the conclusion of his September 1927 address ("the idea of complementarity is suited to characterize the situation, which bears a profound analogy to the general difficulty in the formation of human ideas, inherent in the distinction between subject and object"), as well as passages in a paper of 1929 ("Strictly speaking, the conscious analysis of any concept stands in a relation of exclusion to its immediate application"; "The necessity of taking recourse to a complementary, or reciprocal mode of description is perhaps familiar to us from psychological problems"; "In particular, the apparent contrast between the continuous onward flow of associative thinking and the preservation of the unity of the personality exhibits a suggestive analogy with the relation between the wave description of the motions of material particles, governed by the superposition principle, and their indestructible individuality").[54]

One characteristic trait of Bohr should not be overlooked in this discussion, for without it the necessary predisposition for reaching the complementarity point of view would have been missing. I refer to Bohr's well-known dialectic style of thinking and of working. One of those who worked with him longest, Léon Rosenfeld, attests that Bohr's "turn of mind was essentially dialectical, rather than reflective . . . He needed the stimulus of some form of dialogue to start off his thinking."[55] Rosenfeld also records a well-known dictum of Bohr: "Every sentence I say must be understood not as an affirmation, but as a question." Bohr's habit of work was frequently to develop a paper during dictation, walking up and down the room and arguing both with himself and a fellow physicist whom he had persuaded to be his sounding-board, transcriber, and critic—and whom he was likely to leave in an exhausted state at the end. As Einstein, Heisenberg, Schrödinger, and many others had to experience, it seemed as if Bohr looked for and fastened with greatest energy on a contradiction, heating it to its utmost before he could crystallize the pure metal out of the dispute. Bohr's method of argument shared with the complementarity principle itself the ability to exploit the clash between antithetical positions. We have given earlier only the first line of a couplet from Schiller, reported to have been one of Bohr's favorite sayings: After the line "Only wholeness leads to clarity" there follows "And truth lies in the abyss":

> Nur die Fülle führt zur Klarheit,
> Und im Abgrund wohnt die Wahrheit.

Of Niels Bohr stories there are legions, but none more illuminating than that told by his son Hans concerning the fundamentally dialectic definition of truth. Hans reports that one of the favorite maxims of his father was the distinction between two sorts of truth: trivialities, where opposites are obviously absurd, and profound truths, recognized by the fact that the opposite is also a profound truth.[56] Along the same line, there has been a persistent story that Bohr had been impressed by an example or analogue for the complementarity concept in the mutually exclusive demands of justice and of love. Jerome S. Bruner has kindly given me a first-hand report of a conversation on this point that took place when he happened to meet Niels Bohr in 1943 or early 1944 for the first time. "The talk turned entirely on the complementarity between affect and thought, and between perception and reflection. [Bohr] told me that he had become aware of the psychological depths of the concept of complementarity when one of his children had done something inexcusable for which he found himself incapable of appropriate punishment: 'You cannot know somebody at the same time in the light of love and in the light of justice!' I think that those were almost exactly the words he used. He also . . . talked about the manner in which introspection as an act dispelled the very emotion that one strove to describe."[57]

Complementarity Beyond Physics

We can now ask: what was Bohr's real ambition for the complementarity conception? It certainly went far beyond dealing with the paradoxes in the physics of the 1920's. Not only were some of the roots of the complementarity principle outside physics, but so also was its intended range of application. Let me remind you of Bohr's statement: "The integrity of living organisms, and the characteristics of conscious individuals, and most of human cultures, present features of wholeness, the account of which implies a typically complementary mode of description . . . We are not dealing with more or less vague analogies, but with clear examples of logical relations which, in different contexts, are met with in wider fields."[58] The complementarity principle is a manifestation of a thema in a sense which I have previously developed[59]—one thema in the relatively small pool of themata from which the imagination draws for all fields of endeavor. When we devote attention to a particular thema in physics or some other science, whether it be complementarity, or atomism, or continuity, we must not forget that each special statement of the thema is an aspect of a general conception which, in the work of a physicist or biologist or other scientist, is exemplified merely in a specific form. Thus a general thema, θ, would take on a specific form in physics that might be symbolized by θ_{ϕ}, in psychological investigation by θ_{ψ}, in folklore by θ_{μ}, and so on. The general thema of discontinuity or discreteness thus appears in physics as the θ_{ϕ} of atomism, whereas in psychological studies it appears as the thema θ_{ψ} of individualized identity. One may express a given θ as the sum of its specific exemplifications, as symbolized (without straining for precision) by the expression:

$$\theta = \sum_{n=\alpha}^{n=\omega} \theta_n$$

From this point of view we realize that Bohr's proposal of the complementarity principle was nothing less than an attempt to make it the cornerstone of a new epistemology. When "in general philosophical perspective . . . we are confronted with situations reminding us of the situation in quantum physics,"[60] it is not that those situations are in some way pale reflections or "vague analogies" of a principle that is basic only in quantum physics; rather, the situation in quantum physics is only one reflection of an all-pervasive principle. Whatever the most prominent factors were which contributed to Bohr's formulation of the complementarity point of view in physics—whether his physical research, or thoughts on psychology, or reading in philosophical problems, or controversy between rival schools in biology, or the complementary demands of love and justice in everyday dealings—it was the *universal* significance of the role of complementarity which Bohr came to emphasize.

Moreover, this universality explains how it was possible for Bohr to gain insight for his work in physics from considerations of complementary situations in other fields. For as Léon Rosenfeld accurately remarks, "As his insight into the role of complementarity in physics deepened in the course of these creative years, he was able to point to situations in psychology and biology that also present complementary aspects; and the considerations of such analogies in epistemological respect in its turn threw light on the unfamiliar physical problems."[61] "Bohr devoted a considerable amount of hard work to exploring the possibilities of application of complementarity to other domains of knowledge; he attached no less importance to this task than to his purely physical investigations, and he derived no less satisfaction from its accomplishment."[62]

During the last thirty years of his life, Bohr took many opportunities to consider the application of the complementarity concept in fields outside of physics. Rosenfeld reports that the first important opportunity of this kind offered itself when Bohr was invited to address a biological congress in Copenhagen in 1932.[63] Starting from the idea of complementarity as used for understanding the dual aspects of light, Bohr then proceeded to point to the application of complementarity relations in biology. Rosenfeld's account of the talk is worth citing in detail:

> This had a special appeal to him: He had been deeply impressed by his father's views on the subject, and he was visibly happy at being now able to take them up and give them a more adequate formulation. [His father], in the work of the reaction against mechanistic materialism at the beginning of the century, had put up a vigorous advocacy of the teleological point of view in the study of physiology: without the previous knowledge of the function of an organ, he argued, there is no hope of unravelling its structure for the physiological processes of which it is the seed. At the same time, he stressed, with all the authority of a life devoted to the analysis of the physical and chemical aspects of such processes, the equally imperious necessity of pushing this analysis to the extreme limit which the technical means of investigation would permit us to reach . . .
>
> Such reflections came as near as one would expect at the time to establishing a relation of complementarity between the physico-chemical side of the vital processes, governed by the kind of causality they are accustomed to herald as the truly scientific one, and the properly functional aspect of these processes, dominated by teleological or finalistic causality. In the past, the two points of view, under varying forms, have always been put in sharp opposition to each other, the general opinion being that one of them had to prevail to the exclusion of the other, that there was no room for both in the science of life. Niels Bohr could now point out that this last belief was only the result of a conception of logic which the physicists had recognized as too narrow, and that the wider frame of complementarity seemed particularly well suited to accommodate the two standpoints, and make it possible without any contradiction to take advantage of both of them, quite in the spirit of his father's ideas. Thus an age-long sterile conflict would be eliminated and replaced by a full utilization of all the resources of scientific analysis.[64]

One need not be tempted into imagining Bohr in a Hamletlike striving to establish his father's ideas; but one also need not remain untouched by the closing of the circle. For surely one of the paths leading to complementarity had opened while Niels Bohr was in his father's laboratory and shop club.

In the years following the congress of 1932 Bohr took his point of view before an even wider audience; in addition to his written and spoken contributions before physical scientists, he presented himself at such meetings as the Second International Congress for the Unity of Science in Copenhagen (June 1936) in a discussion on "Causality and Complementarity"; the International Congress for Physics and Biology in October 1937 on "Biology and Atomic Physics"; the International Congress for Anthropology and Ethnology, Copenhagen, 1938, on "Natural Philosophy and Human Cultures"; and on many later occasions of a similar sort.[65]

In each of these lectures Bohr provided a new set of illustrations of the common theme. Thus in his address before the anthropologists in 1938, on the eve of World War II, Bohr stressed complementary features of human societies. He also returned to the problem posed by the student (*licentiate*) in Møller's story. As Rosenfeld writes:

> He could now look back at the duality of aspects of psychical experience with all the mastery he had acquired over the nature of complementarity relations, and point out that this duality corresponded to different ways of drawing a separation between the psychical process which was chosen as the object of observation and the observing subject: drawing such a separation is precisely what we mean when we speak of fixing our attention on a definite aspect of the process; according as we draw the line, we may experience an emotion as part of our subjective feeling, or analyze it as part of the observed process. The realization that these two situations are complementary solves the riddle of the licentiate's egoes observing each other, and is in fact the only salvation from his qualms.[66]

Speaking before the Congress of the Fondation Européenne de la Culture in Copenhagen on October 21, 1960, in an address entitled "The Unity of Human Knowledge," Bohr returned again to the need to search, within the great diversity of cultural developments, "for those features in all civilizations which have their roots in the common human situation." He developed these ideas in sociological and political context, particularly since he was increasingly more preoccupied with helping to "promote mutual understanding between nations with very different cultural backgrounds."[67] Deeply concerned about the dangers of the Cold War, Bohr spent a good part of his later years on political and social questions, including work on plans for peaceful uses of nuclear energy and for arms control. In these and other articles on this topic, one can discern Bohr's dissatisfaction with his own state of understanding; the problems posed by national antagonisms did not seem to be fully understandable in the same terms that had seemed to him successful in physics and psychology. As he confessed at the end of his lecture before the American Academy of Arts and Sciences in 1957, "The fact that human cultures, developed under different conditions of living, exhibit such contrasts with respect to established traditions and social patterns allows one, in a certain sense, to call such cultures complementary. However, we are here in no way dealing with definite, mutually exclusive features, such as those we meet in the objective description of general problems of physics and psychology, but the differences in attitude which can be appreciated or ameliorated by an expanded intercourse between peoples."[68]

Bohr returned to the same theme repeatedly. For example, in the essay quoted earlier, on "The Unity of Human Knowledge," Bohr reexamined the requirement that even the most abstract principles of quantum physics, for example, must be capable of being rendered in commonsense, classical language. "The aim of our argumentation," Bohr wrote, "is to emphasize that all experience, whether in science, philosophy, or art, which may be helpful to mankind, must be capable of being communicated by human means of expression, and it is on this basis that we shall approach the question of unity of knowledge."[69]

The last phrase, used in the title of the essay, suddenly puts into perspective for us that Bohr's manifold and largely successful ambitions place him in the tradition typified by another "philosophizing scientist," one who belonged to the generation before Bohr—a man whom Bohr, like many others, had read early, and whose views Høffding had described in a sympathetic way in his *Moderne Philosophen* and in *Problems of Philosophy*. It is Ernst Mach.

Bohr seems to have mapped out for himself the same grand, interdisciplinary task—in his forceful and innovative influence on physics and on epistemology, in his deep interest in the sciences far beyond physics itself, even in his active and liberal views on social-political questions. And as physicist, physiologist, psychologist, and philosopher, Ernst Mach had also wanted to find a principal point of view from which research in any field could be more meaningfully pursued. This point of view Mach thought to have found by going back to that which is given before all scientific research, namely the world of sensations. On this basis, Mach had established himself as the patriarch of the Unity of Science movement. In his turn, Niels Bohr, starting from the profound reexamination of the problem of sensation and particularly of object-subject interaction, also hoped he had found (in the complementarity point of view) a new platform from which to evaluate and solve the basic problems in a variety of fields, whether in physics, psychology, physiology, or philosophy.

Bohr's achievement, from 1927 on, of attaining such a principal point of view was not an accidental development. On the contrary, it was the fulfillment of an early ambition. A biographer of Bohr records that "as a young student, fired with the ideas Høffding was opening to him, Bohr had dreamed of 'great inter-relationships' between all areas of knowledge. He had even considered writing a book on the theory of knowledge . . . But physics had drawn him irresistibly."[70] In the end, Bohr's attempt to understand the unity of knowledge (a topic on which he wrote nearly two dozen papers) on the basis of complementarity could be seen as precisely the fulfillment of the desire to discover the "great inter-relationships among all areas of knowledge."

Bohr's aim has a grandeur which one must admire. But while his point of view is accepted by the large majority in physics itself, it would not be accurate to say that it is being widely understood and used in other fields; still less has it swept over philosophy the way Mach's views did during the generation of scientists brought up before the theory of relativity and quantum mechanics. Even those who in their professional work in physics have experienced the success of the complementarity point of view at first hand find it hard or uncongenial to transfer to other areas of thought and action, as a fundamental thematic attitude, the habit of accepting basic dualities without straining for their mutual dissolution or reduction. Indeed, we tend to be first of all reductionists, perhaps partly because our early intellectual heroes have been men in the tradition of Mach and Freud, rather than Kierkegaard and James.

Perhaps, also, it is just a matter of time—more time needed to assimilate a new thema widely enough; to sort out the merely seductive and the solid applications; and to learn to perceive the kind of grandeur in the scope of the new notion which Robert Oppenheimer delineated:

> An understanding of the complementary nature of conscious life and its physical interpretation appears to me a lasting element in human understanding and a proper formulation of the historic views called psychophysical parallelism. For within conscious life, and in its relations with the description of the physical world, there are again many examples. There is the relation between the cognitive and the affective sides of our lives, between knowledge or analysis, and emotion or feeling. There is the relation between the esthetic and the heroic, between feeling and that precursor and definer of action, the ethical commitment; there is the classical relation between the analysis of one's self, the determination of one's motives and purposes, and that freedom of choice, that freedom of decision and action, which are complementary to it . . .
>
> To be touched with awe, or humor, to be moved by beauty, to make a commitment or a determination to understand some truth—these are complementary modes of the human spirit. All of them are part of man's spiritual life. None can replace the others, and where one is called for, the others are in abeyance . . .

The wealth and variety of physics itself, the greater wealth and variety of the natural sciences taken as a whole, the more familiar, yet still strange and far wider wealth of the life of the human spirit, enriched by complementary, not-at-once compatible ways, irreducible one to the other, have a greater harmony. They are the elements of man's sorrow and his splendor, his frailty and his power, his death, his passing, and his undying deeds.[71]

REFERENCES

1. After much further work, Bohr published the lecture in 1928 under the title "The Quantum Postulate and the Recent Development of Atomic Theory"; it has been reprinted in several places, for example as one of four essays in the collection by Niels Bohr, *Atomtheorie und Naturbeschreibung* (Berlin: Springer, 1931), also published as *Atomic Theory and the Description of Nature* (Cambridge, Eng.: University Press; New York: Macmillan, 1934).

2. Bohr, "The Quantum Postulate," in *Atomic Theory and the Description of Nature*, p. 55.

3. A. Einstein, "Das Comptonsche Experiment," *Berliner Tageblatt* (April 20, 1924), supplement, p. 1, cited by M. J. Klein, *A Twentieth-Century Challenge to Energy Conservation* (forthcoming).

4. Bohr (in "The Quantum Postulate," *Atomic Theory and the Description of Nature*, pp. 54-55) introduced the need for working out a "complementarity theory" in the following, rather overburdened sentence: "The very nature of the quantum theory thus forces us to regard the space-time coordination and the claim of causality, the union of which characterizes the classical theory, as complementary but exclusive features of the description, symbolizing the idealization of observation and definition respectively." Max Jammer, to whose book *The Conceptual Development of Quantum Mechanics* (New York: McGraw-Hill, 1966, p. 351) we shall frequently refer, adds: "This statement, in which the term 'complementary' appears for the first time and in which spatiotemporal description is referred to as complementary to causal description, contained the essence of what later became known as the 'Copenhagen' interpretation of quantum mechanics."

   Heisenberg's uncertainty principle, formulated early in 1927, had given a first indication of complementary relations between physical concepts, though in a restricted sense. The uncertainty principle tells us that if we attempt to localize a particle in space (or time), we shall, during the measurement process, impart to the particle momentum (or energy) within a range of values that increases as we decrease the size of the space-time region on which we wish to focus attention. Position and momentum are not mutually exclusive notions since both are needed to specify the state of a system and both can be measured in the same experiment. But they are complementary in the restricted sense that they cannot both at the same time be ascertained with arbitrarily high precision; that is, the more precision is obtained in one measurement, the less it is possible to have in the other. In contrast, the wave-particle aspects of matter are complementary *and* mutually exclusive; an atomic entity cannot exhibit both its particle and its wave properties simultaneously. It is for this reason that textbooks often say that Bohr's statement of complementarity at Como transcended the Heisenberg uncertainty principle.

5. Niels Bohr, "Discussion with Einstein on Epistemological Problems in Atomic Physics," in P. Schilpp, ed., *Albert Einstein: Philosopher-Scientist* (Evanston, Ill.: The Library of Living Philosophers, 1949), pp. 209-210; italics in original.

6. Jammer, *Conceptual Development of Quantum Mechanics,* p. 354.

7. Translated from *ibid.,* p. 358.

8. A. Einstein, "Reply to Criticisms," in Schilpp, ed., *Albert Einstein,* p. 674.

9. Bohr, "Discussion with Einstein," in Schilpp, ed., *Albert Einstein,* p. 218.

10. For some aspects of the early history of the theories of light, see the interesting book by Vasco Ronchi, *Optics, the Science of Vision* (New York, 1957), or Johann Müller, *Lehrbuch der Physik* (Braunschweig: F. Vieweg und Sohn, 1926). I have relied on both extensively.

11. For example, see W. H. Bragg, *Philosophical Magazine,* 20 (1910), 358-416.

12. Merle A. Tuve, in Caryl P. Haskins, ed., *The Search for Understanding* (Washington, D.C.: Carnegie Institution, 1967), p. 46.

13. Jammer, *Conceptual Development of Quantum Mechanics,* p. 87.

14. Quoted, with permission, from a letter of Niels Bohr in the American Philosophical Society Library, Philadelphia. I thank Dorothy Goodhue Livingston for having drawn this letter to my attention.

15. N. Bohr, H. A. Kramers, and J. C. Slater, "The Quantum Theory of Radiation," *Philosophical Magazine,* 47 (1924), 785. The German version is in *Zeitschrift für Physik,* 24 (1924), 69.

16. E. Schrödinger, "Quantisierung als Eigenwertproblem," *Annalen der Physik,* 79 (1926), 375.

17. Jammer, *Conceptual Development of Quantum Mechanics,* pp. 271-272; italics in original.

18. *Ibid.,* p. 271.

19. As quoted *ibid.,* p. 272.

20. See "On the Thematic Analysis of Science: The Case of Poincaré and Relativity," *Mélanges Alexandre Koyré* (Paris: Hermann, 1964), II, 257-268; "The Thematic Imagination in Science," in Gerald Holton, ed., *Science and Culture* (Boston: Houghton Mifflin, 1965), pp. 88-108; "Science and New Styles of Thought," *The Graduate Journal,* 7 (Spring 1967), 399-421.

21. Oskar Klein, "Glimpses of Niels Bohr as Scientist and Thinker," in S. Rozental, ed., *Niels Bohr: His Life and Work as Seen by His Friends and Colleagues* (New York: John Wiley, 1967), p. 93.

22. Niels Bohr, *Essays 1958-1962 on Atomic Physics and Human Knowledge* (New York: John Wiley, 1963), p. 7; italics supplied.

23. L. Rosenfeld, "Niels Bohr in the Thirties," in Rozental, ed., *Niels Bohr,* p. 121.

24. Cited in Bohr's essay, "The Unity of Human Knowledge," 1960, in *Essays 1958-1962,* and in L. Rosenfeld, "Niels Bohr's Contribution to Epistemology," *Physics Today,* 16 (1963), 63. In this article and elsewhere, Rosenfeld has insisted on the importance of the story for Bohr; moreover, Rosenfeld believes that the struggle of the student with his many egos was "the only object lesson in dialectical thinking that Bohr ever received and the only link between his highly original reflection and philosophical tradition" (p. 48).

25. Bohr, "The Unity of Human Knowledge," in *Essays 1958-1962,* p. 12.

26. *Ibid.,* p. 14.

27. The permission granted by the estate of Niels Bohr and the American Philosophical Society to reproduce this section of the interview is gratefully acknowledged.

28. Bohr, "The Quantum of Action and the Description of Nature," in *Atomic Theory and the Description of Nature.*

29. Letter to the author, February 28, 1968.

30. *Ibid.* In an interview conducted with Werner Heisenberg by T. S. Kuhn for the History of Quantum Physics project on February 11, 1963, Heisenberg volunteered that James was one of Bohr's favorite philosophers; the chapter on the "stream of thought" seemed to have made a profound impression on Bohr. Heisenberg placed these discussions somewhere between 1926 and 1929, most probably around 1927. When told of doubts about the timing, Heisenberg responded that he could not "guarantee" that these discussions with Bohr had not been after 1932.

31. Rosenfeld, "Niels Bohr's Contribution to Epistemology," p. 49. See also L. Rosenfeld, *Niels Bohr* (Amsterdam: North-Holland Publishing Co., 1945, 1961), pp. 12-13.

32. We follow here the sequence given in Meyer-Abich, *Korrespondenz,* pp. 133ff.

33. Bohr, "The Quantum Postulate," in *Atomic Theory and the Description of Nature,* p. 91.

34. William James, *Principles of Psychology* (New York: Dover, 1950), I, 203; italics in original in all passages quoted from *Principles of Psychology.*

35. *Ibid.,* p. 204.

36. *Ibid.,* pp. 206-207.

37. *Ibid.,* p. 206.

38. In Rozental, ed., *Niels Bohr,* p. 74. One notices here the remarkable similarity of Bohr's experience with that recorded in Einstein's autobiographical notes—the same early religious acceptance in contrast to his parents' beliefs, followed by a loss or rejection of "the religious paradise of youth," as Einstein called it.

39. As quoted in Rozental, ed., *Niels Bohr,* p. 13.

40. Klein, "Glimpses of Niels Bohr," *ibid.,* p. 76.

41. See Jammer, *Conceptual Development of Quantum Mechanics,* p. 349.

42. *Ibid.*, pp. 347, 349.

43. Ruth Moore, in her book, *Niels Bohr* (New York: Knopf, 1966), p. 432, records that on one wall of Bohr's house in Carlsberg, there "were portraits of those nearest to Bohr, grouped reverentially together": Bohr's father and mother, his brother Harald, his grandfather Adler, and "Bohr's teacher Høffding. If any doubts existed of Høffding's influence on Bohr's life, it was settled by the placement of his portrait."

44. H. Høffding, in R. Schmidt, ed., *Die Philosophie der Gegenwart in Selbstdarstellungen* (Leipzig: Felix Melner, 1923), p. 86.

45. For example, *ibid.*, p. 75.

46. Danish edition, 1892; German edition, 1896.

47. In Rozental, ed., *Niels Bohr,* p. 27. See also the account of J. Rud Nielsen, "Memories of Niels Bohr," *Physics Today,* 16 (1963), 27-28. Referring to a visit from Bohr in 1933, Nielsen wrote: "Knowing Bohr's interest in Kierkegaard, I mentioned to him the translations made by Prof. Hollander of the University of Texas, and Bohr began to talk about Kierkegaard: 'He made a powerful impression upon me when I wrote my dissertation in a parsonage in Funen, and I read his works night and day,' he told me. 'His honesty and willingness to think the problems through to their very limit is what is great. And his language is wonderful, often sublime. There is of course much in Kierkegaard that I cannot accept. I ascribe that to the time in which he lived. But I admire his intensity and perseverance, his analysis to the utmost limit, and the fact that through these qualities he turned misfortune and suffering into something good.' "

48. A preliminary treatment of the subject has been made in the section "The Philosophical Background of Non-classical Interpretations," in Jammer, *Conceptual Development of Quantum Mechanics,* pp. 166-180.

49. J. Passmore, *A Hundred Years of Philosophy* (New York: Basic Books, 1966), p. 480.

50. H. Høffding, *A History of Modern Philosophy* (New York: Dover, 1955), II, 286. The work was originally issued in 1893 and intended to cover the ground to 1880. The English translation was published in 1900. Høffding also explored the role of discontinuity in other contexts, for example, in *Moderne Philosophen* (1904), where he contrasts at length the older *Kontinuitätsphilosophie* (as in Taine, Fouillée, Wundt, Ardigò) with the more recent *Diskontinuitätsphilosophie* (for example, Renouvier, "der Nestor der Philosophie der Gegenwart," and Boutroux).

51. *Ibid.*, p. 287.

52. *Ibid.*, p. 288.

53. *Ibid.*, pp. 287-288.

54. Bohr, "The Quantum of Action," in *Atomic Theory and the Description of Nature,* pp. 96, 99.

55. Rosenfeld, "Niels Bohr in the Thirties," in Rozental, ed., *Niels Bohr,* p. 117.

56. Hans Bohr, "My Father," in Rozental, ed., *Niels Bohr,* p. 328.

57. Jerome S. Bruner, private communication to the author, December 25, 1967. Bruner added a comment which will become relevant to us in what follows below: "I knew Bohr for years afterwards and again spent several hours with him when he was at the Institute for Advanced Study at Princeton, and he came to visit. He had an extraordinary sensitivity for psychological problems, and indeed he once repeated Mach's famous remark about basically our only two sciences: one treats sensation as external and is physics, the other treats it as internal and is psychology. He did not cite this old saw of Mach's approvingly, but urged that there was a grain of truth in it."

58. Bohr, "Quantum Physics and Philosophy," in *Essays 1958-1962,* p. 7.

59. See note 20, and G. Holton, "Stil und Verwirklichung in der Physik," *Eranos Jahrbuch,* 33 (Zurich: Rhein-Verlag, 1965), particularly pp. 333ff.

60. Bohr, "Quantum Physics and Philosophy," in *Essays 1958-1962,* p. 7.

61. Rosenfeld, "Niels Bohr in the Thirties," in Rozental, ed., *Niels Bohr,* p. 116.

62. *Ibid.,* p. 120.

63. Niels Bohr, "Light and Life," address at the Second International Congress for Light Therapy, Copenhagen, August 1932, *Nature,* 131 (1933), 421-423, 457-459.

64. Rosenfeld, "Niels Bohr in the Thirties," in Rozental, ed., *Niels Bohr,* pp. 132-133.

65. For a partial bibliography of Bohr's writings, see Meyer-Abich, *Korrespondenz,* pp. 191-199.

66. Rosenfeld, "Niels Bohr in the Thirties," in Rozental, ed., *Niels Bohr,* pp. 135-136. A useful summary of Bohr's views concerning the application of the complementarity conception to physics, biology, psychology, and social anthropology is given in Niels Bohr, "On Atoms and Human Knowledge," *Dædalus* (Spring 1958), pp. 164-175.

67. Bohr, "The Unity of Human Knowledge," in *Essays 1958-1962,* pp. 14-15.

68. Bohr, "On Atoms and Human Knowledge," *ibid.,* pp. 174-175.

69. Bohr, "The Unity of Human Knowledge," *ibid.,* p. 14.

70. Moore, *Niels Bohr,* pp. 406-407. There is a great deal of evidence of the large scale of Bohr's later hopes along these lines. In his 1933 discussion, J. Rud Nielsen ("Memories of Niels Bohr," p. 27) reports: "Bohr talked a good deal about his plans for future publications. 'I believe that I have come to a certain stage of completion in my work,' he said, 'I believe that my conclusions have wide application also outside physics . . . I should like to write a book that could be used as a text. I would show that it is possible to reach all important results with very little mathematics. In fact, in this manner one would in some respects achieve greater clarity.' This book, which Bohr referred to as his testament, was never written."

Similarly, Rosenfeld ("Niels Bohr's Contributions to Epistemology," p. 54) writes: "Bohr had great expectations about the future role of complementarity. He upheld them with unshakable optimism, never discouraged by the scant response he got from our unphilosophical age . . . Bohr declared with intense animation that he saw the day when complementarity would be taught in the schools and become part of general education."

71. J. Robert Oppenheimer, *Science and the Common Understanding* (New York: Simon and Schuster, 1953), pp. 80-82.

Note: An early draft of this essay was presented at the *Tagung* of *Eranos* (August 1968). I have profited from discussions with students in my seminar, particularly Bernard Lo and Kellogg Steele, and with Dr. Arthur Miller.

ALVIN M. WEINBERG

Scientific Teams and Scientific Laboratories[1]

SCIENTIFIC TRUTHS discovered in one age are essential for scientific progress in another: the laws of thermodynamics, discovered in the nineteenth century, will remain relevant and necessary for the scientist of the twenty-second century. Similarly scientific truth discovered in one place is required for scientific progress elsewhere: Lord Rutherford's experiments at Manchester on the scattering of alpha particles led eventually to the prolific investigation of nuclear phenomena throughout the world. To paraphrase Alfred Korzybski, man the scientist is both a time-binder and a space-binder.

In this sense science has always been a cumulative, team activity, more than, say, the arts or literature.[2] To be sure, great individual geniuses, like Newton or Maxwell or Darwin, create the revolutions that punctuate scientific progress. (T. S. Kuhn, in his *The Structure of Scientific Revolutions,* calls these turning points in science "paradigm-breaking."[3] I shall refer to them, along with the more modest "important discoveries," simply as "breakthroughs.") Yet the connections of even such individual geniuses with their predecessors and their contemporaries are surely more direct and demonstrable than is the connection between Beethoven and Mozart, or Picasso and Renoir. As Newton wrote to Robert Hooke, "If I have seen further (than you and Descartes) it is by standing upon the shoulders of giants."[4]

Nineteenth-century science was mainly conducted by geographically isolated, though intellectually interacting, individuals; much of today's science is conducted by large interdisciplinary teams. These teams often center around pieces of expensive equipment and are then said to be part of "big science." Team science is characteristically conducted in the large multipurpose scientific laboratory, an institution that is predominantly a phenomenon of World War II and after. My purpose will be first to trace the origins of big team science and to examine its multipurpose institutions, second to estimate the capacity of this new scientific style to launch and carry off the scientific breakthroughs so necessary for the progress of science, and finally to speculate on the future of team research and its institutions.

I. *The Origins of Big Team Science*

The emergence of the large interdisciplinary scientific team as the landmark of science can be traced to at least three separate developments. First is the extraordinary growth of science and the resulting increase in the amount of scientific information produced; second is the emergence and institutionalization of applied science; third, and possibly most important, is the increasing complexity of scientific machinery.

A. *The Information Crisis and the Rise of Team Science*

The scientific information explosion has caused scientists to become more specialized. Some scientists respond to the information crisis by confining their range of scientific undertakings to those over which they can still retain command of the relevant information sources. Others form interdisciplinary teams in which are represented different though overlapping ranges of expertise or technique. In principle, the problems that can be tackled successfully by such teams ought to be more complex than those tackled by individuals.

This trend in the sociology of science was foreshadowed in an essay, "The Limits of Science," by Eugene Wigner in 1950.[5] Wigner argued that, for the reason I have mentioned, team research in which individual scientists are orchestrated into a productive whole by a scientific leader would become more common. He then asked how this new social structure would change the course of science. Could the theory of relativity or the Schrödinger equation have been discovered by an interdisciplinary team? Or, for that matter, could the mysteries of the "omega-minus" particle and violation of charge-parity invariance (both discoveries of teams of high-energy physicists) have been unearthed by the typical individual scientist of the nineteenth century? I shall return to these questions later.

The information explosion has been the subject of many essays and studies. Here I will mention only how the spawning of the scientific information specialist has affected the organization of scientific research. In previous generations the scientist gathered his information more or less on his own and rather haphazardly. Today scientists of course continue to browse in this manner, but they are now backed by a host of information services, ranging from libraries to abstract services and specialized information centers.

Already one can see the considerable influence of the information specialists on those fields of science such as nuclear physics and high-energy physics where the spectroscopy[6] has become so elaborate as to outrun any single individual's capacity to hold all the relevant data in his mind. As a result, much of the output of the nuclear or high-energy spectroscopist goes to a secondary source—such as K. Way's or A. H. Rosenfeld's centers—where the data are compiled and collated. But in the process the role of the individual scientist who first made the measurements is weakened; the citation now often tends to be to the secondary source rather than to the original experimenter. Could this mean that one of the delicious joys and motivations of science—recognition and approbation by one's peers—will be attenuated? Parts of basic science have already acquired some of the facelessness that characterizes applied science, and this trend, it seems to me, will continue as the information crisis deepens.[7]

B. The Emergence of Applied Science and the Large Industrial Laboratory

A second source of the trend toward team science is the rise of applied science and particularly the growth of the large industrial laboratory. Here the interdisciplinary team has predominated from the first, for reasons that are implicit in the strategy of science—that is, the way that scientists choose what they do. To make this point clearer, I shall digress to consider the strategy of scientific research.

Science is the "art of the soluble" according to Peter Medawar.[8] What a scientist does is largely determined by what he thinks he can do successfully. According to this view, science is a meandering stream that pushes salients out wherever the bank is weak and can be conquered; that such meandering may lengthen and make more tortuous the path to the sea is somewhat irrelevant. The river valley (to push the metaphor) is irrigated more heavily, and becomes greener, as a consequence of the meandering.

Insofar as Medawar is referring to basic science, his view of science as the art of the soluble contains much truth. In basic science, the scientist's criteria for deciding what he ought to try are usually internally generated; that is, they derive from the internal logic of the specific field in which he works and from his assessment of how soluble the problem is. Moreover, in basic science success is achieved if one solves the problem he sets out to solve, if he solves a different problem, or even if he can show that a particular approach is unfruitful. For all these reasons, in basic research it is acceptable to tailor problems to one's capacity for solving them. An expert in nuclear magnetic resonance can confine his researches to that segment of the field of nuclear magnetic resonance over which he can comfortably retain command. Thus basic science (at least before the advent of the big machine) with its internally generated problems, can be pursued adequately within a narrow discipline. If the problem takes the researcher out of his specialty, he is still observing the canons of pure science if he turns to a different problem that is more easily accommodated by his interest and competence. It is for this reason that much of basic science can remain disciplinary and little: the interdisciplinary team is not its social characteristic.

Of course, even as a description of basic little science this is oversimplified; to characterize science as the art of the soluble tells only part of the story. Basic science, at its best, is the art of the soluble *and the important* (as Medawar himself recognizes). Researchers, even poor ones, usually have more ideas than they have resources with which to pursue them; their research strategy always amounts to choosing, from among a variety of soluble problems, the ones they regard as important. What constitutes importance in science? One, though certainly not the only, criterion is the degree to which a given piece of science relates to neighboring sciences. Indeed, the motivation for much basic scientific activity originates outside that activity. Sometimes the motivation lies in a neighboring basic science. For instance, a nuclear physicist may study light-element reactions because these are needed by an astrophysicist who wants to understand the mechanism of stellar evolution. Sometimes the motivation lies in technology: a physicist may investigate the basic properties of plasmas because of their relevance to the controlled release of thermonuclear energy. But the main point is that as soon as a scientist ventures to deal with a question arising in a field outside his discipline he has less control over where to look for a solution. He no longer has the luxury of narrowing the problem to what is soluble with his own expertise. Externally motivated science tends to be interdisciplinary and therefore more of a team activity than internally motivated science.

Applied science is externally motivated par excellence. Its questions are posed from without: from engineering, military, and even social demands. Such questions usually transcend the individual disciplines. The criterion of success in applied science is simply, "Does it work?" not "Does it add to knowledge in a particular discipline?" Thus applied science is characteristically interdisciplinary; it lends itself to—in fact it almost requires—teams of interacting individuals, none of whom by himself commands all the knowledge necessary to make progress, but all of whom, when taken as a whole, hopefully do.

It is therefore no accident that the great institutions of applied science in industry and in government are typically homes for interdisciplinary teams. The jobs of these institutions are set outside the disciplines, even outside science; in consequence their style is interdisciplinary. Though the first of these laboratories, such as the General Electric Research Laboratory, Bell Telephone Laboratories, and the National Bureau of Standards, appeared around the turn of this century, there was an enormous development of them during and after World War II. The best known of the wartime laboratories were the Radiation Laboratory at the Massachusetts Institute of Technology, which developed radar, and the Metallurgical and Los Alamos laboratories of the Manhattan Project, which developed the atomic bomb. My own experience has been almost entirely confined to the atomic energy laboratories, and so I shall draw largely on them to illustrate some characteristics of the big applied scientific institutions.

From its very beginning in late 1941, the Chicago Metallurgical Laboratory, at which the first fission chain reaction was established, was interdisciplinary. Arthur H. Compton, the director of the Metallurgical Project, realized that the technology of the chain reactor would require physicists, mathematicians, chemists, instrument experts, metallurgists, biologists, and the various engineers who could translate these scientists' findings into practice. The chain reactor was much more than a nuclear physicist's experiment. Uranium to fuel the first reactor had to be purified and reduced to metal. Graphite of unprecedented purity was needed to moderate the neutrons. The chemistry of the new element plutonium was largely unknown. The production of plutonium was very hazardous, and the most sophisticated instruments were needed to keep everything under control. The biological effects of the radiation that would be released had to be assessed if not mitigated.

The difference between most interdisciplinary engineering enterprises and the engineering at the Metallurgical Laboratory lay in the incredible speed with which the latest scientific findings at the Laboratory were converted into engineered chain reactors. Eugene P. Wigner, who headed the theoretical physics group, began to engineer the water-cooled Hanford reactors in early 1942, almost ten months *before* the first chain reaction had been established.

There was nothing very complicated or obscure about the function and purpose of the Metallurgical Laboratory. Its output was a specific gadget and a specific process: the nuclear chain reactor and the production and extraction of plutonium. If the reactor succeeded, the Laboratory succeeded; if it failed, the Laboratory failed. Because of this singleness of purpose, which at least for the first two years was evident to all, there was remarkably little difficulty in forging the teams necessary to get on with the job.

Like most institutions of this sort, the Laboratory was organized into divisions. Enrico Fermi and Eugene Wigner were in charge of the physicists; James Frank and then Sam Allison and Farrington Daniels, in charge of the chemists; Charles Cooper, the engineers; and so on. But the over-all project overwhelmed the disciplinary divisions. This was relatively easy because everyone knew the stakes; one could readily submerge his personal aspirations for the sake of achieving the whole objective. This is not to say that there was no tension between the project and the divisions (that is, the disciplines). Even in the dark days of 1943 one could find physicists at the Metallurgical Laboratory working on the spherical harmonic method of solution of the Boltzmann equation (an activity that at the time seemed like an unjustified luxury) instead of estimating more routinely the multiplication constant of the latest reactor design.

This criss-cross organization—with each scientist having a permanent home in a division but being lent out temporarily to an interdisciplinary project—is the usual organization in applied laboratories. The project leaders generally control the funds; the division leaders, the people. The projects maintain pressure on the division managers to keep their outlook and activities relevant as judged by the projects; the disciplinary divisions maintain pressure on the project managers to keep their activities up to the standards of sophistication imposed by the divisions. It is hoped that out of this criss-cross tension between project and division there will come both relevance and sophistication.

The Metallurgical Laboratory was hierarchical. Arthur Compton was boss, but there were many other managers at lower levels, each on top of a pyramid of lesser and usually younger scientists. This pyramidal structure gave very great power to the man on top: he could command information resources; he could order investigations in many directions that would be out of the question had he not had a team at his disposal. In such hierarchical scientific teams, the members lower down must submerge their personalities and to some extent their scientific instincts to those of the boss. One therefore finds genius in such organizations less often than in the universities, where science is conducted more individually. On the other hand, a really good man in a position at the top of a pyramid obviously can get much more done than he can if he works in the usual university setting. Glenn Seaborg at the Metallurgical Laboratory had about thirty chemists working for him, and in only two years his group elucidated much of the chemistry of plutonium in addition to developing a process for extracting plutonium that was used successfully at Hanford!

Though there were many scintillating talents around—Szilard and Fermi and Wigner and Seaborg—the decisions were finally made by Compton. Yet, as in any organization, those with enough energy, confidence, and ability could impose their views in the face of official rejection. At the Metallurgical Laboratory, a showdown of this sort occurred during 1942. The issue was the coolant—and therefore the whole engineering design—of the Hanford plutonium-producing reactors. The prevailing view held that since helium absorbed no neutrons, helium should be used to cool the reactors. With this view Wigner disagreed vigorously; he wanted to cool the reactors with water. To him, the handling of hot and somewhat radioactive helium under pressure seemed much more serious than the loss of nuclear performance caused by the tendency of hydrogen to absorb neutrons. In arguing his case, Wigner commanded all the relevant elements of knowledge—the engineering, the chemistry, the metallurgy, and the physics. And, when the Hanford reactors were actually to be built, the DuPont engineers chose the water-cooled pile rather than the original helium-cooled version.

Writing about these events twenty-seven years after they occurred, I am struck not by their uniqueness but by their generality. The Metallurgical Laboratory, in Anthony Downs's terminology,[9] was a bureaucracy —that is, a large organization that is not governed, or is only indirectly governed, by the feedback from the marketplace. In this sense almost every large laboratory, even if it is part of a big corporation, is a bureaucracy; its connection with the marketplace is usually tenuous. Many of the organizational features and sources of power in the large laboratory are not characteristic specifically of a scientific establishment but rather of any large nonmarket establishment. The hierarchical structure, the possibility of the energetic individual prevailing against the official position, the great logistic power of a big laboratory,[10] above all, the urgent imperative of the various groups to survive and to expand— all these are obvious to students of large organizations, scientific or otherwise.

C. The Influence of the Big Scientific Machine

The third thread in the development of big team science goes back for some of its spirit to the explorations of the fifteenth and sixteenth centuries. To a degree, we would have to regard the great explorers as geographers, and hence scientists of sorts. Their enterprises were on a grand scale by the standards of their time; they required large teams and much money. And at least Columbus among them politicked with John of Portugal and Queen Isabella in much the same way that a promoter of a large accelerator must now politick with the Atomic Energy Commission or the National Science Foundation, or even with the President himself, to sell his project.

Many of today's explorations in basic science involve such elaborate and expensive pieces of hardware that the whole enterprise requires much the same mobilization of resources as was required by the explorers. The most extreme example today of huge mobilization of resources for a purpose that is at least partly scientific is the exploration of space. And even before we began to use rockets, earth-based astronomy had some of the attributes of modern big science: the 200-inch Hale telescope at Mount Palomar, completed in 1948, was one of the largest and most expensive pieces of scientific machinery until the advent of the large research reactors and large accelerators during and after the war.

The new style of big science based on very large pieces of equipment is generally attributed to Ernest O. Lawrence. His 37-inch cyclotron at Berkeley was a monster for its time; this was followed by the 60-inch, the 184-inch, the synchrocyclotron, the proton synchrotron (Bevatron), in ever-increasing size and complexity. To be sure, there had been earlier scientific teams dominated by great leaders: J. J. Thomson and later Rutherford at the Cavendish Laboratory, Fermi and his neutron group in Rome, and, of course, the German institutes. But Lawrence's laboratory was probably the first in which the central piece of equipment was so elaborate, and possibly so temperamental, as to require a more or less full-time engineering staff. The logistics of keeping the place going—whether this means the scientific machinery or the elaborate organization that tends the machinery—becomes an essential ingredient of the activity. There are engineers and instrument technicians and financial people and personnel experts, many of whom identify rather little with the purpose of the entire laboratory, but each of whom is valued for his specialized expertise.

Thus the modern home of big basic science, especially the big accelerator or reactor laboratory, acquires much of the flavor of the industrial laboratory. The time allotted for use of the machine is rigidly scheduled, and this imposes a regularity on the working habits at least of the technicians who tend the machine. There is a division of labor between those who are expert in electronics and computing and electrical engineering; and this requires coordination. The necessity for explicit planning is taken for granted, in much the same way as planning by a project manager is the accepted way of doing business in the applied laboratory.

The typical home of massive basic science, like CERN in Geneva or the Stanford Linear Accelerator, is however more specialized than is the modern home of applied science; this goes back to the aforementioned distinction between basic science, which tends to be internally motivated and disciplinary, and applied science, which tends to be externally motivated and interdisciplinary. The General Electric Research Laboratory covers a wider range of specialties than does the Stanford Linear Accelerator. The Argonne National Laboratory, with its experts ranging from biomedical researchers and ecologists to high-energy physicists, covers a wider range of specialties than does the nearby Fermi

National Accelerator Laboratory. I imagine that this greater specialization will in the long run pose some difficulties if the question of redeploying the large basic laboratories, like Fermi or SLAC, should ever arise.

II. Individual Science and Team Science: Breakthroughs Versus Spectroscopy

A. The Xenon Compounds: A Breakthrough by an Individual

In 1962 Neil Bartlett, a young chemist at the University of British Columbia, stumbled onto the fact that oxygen could be oxidized by platinum hexafluoride. About the same time Bartlett had noticed, while browsing through a table of ionization potentials, that the energy required to strip an electron off xenon (to form Xe^+) was about the same as that required to form the O_2^+ ion. He therefore concluded it was worth trying to oxidize xenon with PtF_6. Almost on his first try he was successful, and, in 1962, sixty-eight years of chemical dogma came to an end: the first stable compound of a noble gas, $Xe(PtF_6)$, was produced. The noble gases were no longer noble.[11]

Immediately after Bartlett's discovery, a group of chemists at the Argonne National Laboratory plunged into the new chemistry of the noble gases. They came to this task well prepared: for many years they had been interested in the chemistry of the fluorine compounds of plutonium and other transuranics. Their laboratories were well equipped for handling treacherous, extremely toxic materials like elemental fluorine and PuF_6. Almost immediately H. H. Claassen, J. G. Malm, and H. Selig discovered that xenon could be oxidized by fluorine alone, and within a year of frenzied activity many compounds of xenon and other noble gases were prepared and characterized. A blank page in inorganic chemistry had been expanded into a good-sized, well-filled book.[12]

This incident serves to illustrate, in almost too perfect outline, the usually held stereotypes as to the strengths and the weaknesses of the traditional individual and the newer team styles of research. The brilliant initial stroke—Bartlett's crazy idea that xenon could be oxidized if only one chose a sufficiently strong oxidant—was very much the doing of an individual. Not that this idea was absolutely new: in 1933 Linus Pauling had suggested that stable xenon compounds exist, and D. M. Yost even tried, unsuccessfully, to prepare them at California. And, even closer to home, at Oak Ridge S. S. Kirslis, F. H. Blankenship, and W. R. Grimes, who were developing a reactor that was fueled with molten uranium fluoride, had noticed that the fission product xenon consistently was missing, whereas the fission product krypton was always present as expected in the gas phase. The question of whether the xenon could be disappearing as a chemical compound did arise and was discussed but of course was rejected, although, as it turned out, XeF_4 was being produced. Scientific dogmas of such strength as the nobility of the rare gases are hard to dethrone.

But, once the brilliant, individual breakthrough had been made, the integrated team and the great logistic power of the National Laboratory moved in, massively and professionally, to fill in the spectroscopic details. A field that in earlier times would have remained fertile and exciting for a decade or more was largely elucidated in little more than a year's time.

In attacking the xenon compounds so massively, the Argonne National Laboratory was working very much in the style of the applied laboratory. There was a group leader with a staff of highly professional people, each of whom was an expert. Bartlett with graduate students probably would have been no match for Argonne with its professionals. And indeed, this extraordinary elaboration of a field of chemistry in just a year has led some to suggest that at least in the field of chemistry the future belongs to the professional team supported with superb equipment and unencumbered by teaching commitments, rather than to the professor whose professionalism in research is diluted by teaching.[13] This view has been sharply criticized by representatives of the university scientific community who insist that only individuals can achieve breakthroughs.

In point of fact, the team can and has achieved breakthroughs, and it is by no means clear that the team will snuff out the fire of scientific revolution. In the table below, as an example, I list the Nobel Prizes in physics during the past twenty years, the time during which team physics has grown so markedly. Of course not every discovery that wins a Nobel Prize breaks a paradigm, but I believe most physicists will agree that these discoveries at the very least represent important breakthroughs. Though the individual winners exceed the team winners, the fact remains that team science has produced several Nobel Prizes in physics. Indeed, examples of teams achieving breakthroughs are not hard to find. I shall describe one that occurred in Oak Ridge in the past few years.

B. Anomalous Losses in Channeling: A Breakthrough by a Team

Charged particles in traversing crystals often become trapped in channels formed by rows of regularly spaced atoms. This phenomenon, called channeling, was predicted theoretically by Mark T. Robinson in 1962, and then was discovered experimentally. In 1964 a team at Oak Ridge, consisting of several nuclear physicists, solid state physicists, and a physical chemist, examined the energy loss of the channeled particles as they emerged from thin crystalline gold foils. They were astonished

Nobel Prizes in physics, 1948–1968.

Year	Winner	Discovery	Team	Individual
1948	P. M. S. Blackett	Development of the Wilson method and discovery by this method of the π- and μ-mesons		x
1949	H. Yukawa	Prediction of mesons		x
1950	C. F. Powell	Development of the photographic method of the study of nuclear processes and discovery concerning mesons		x
1951	Sir J. D. Cockcroft E. T. S. Walton	Cockcroft–Walton accelerator and first disintegration		x
1952	F. Bloch E. M. Purcell	Nuclear magnetic resonance		x*
1953	F. Zernike	Phase-contrast microscope		x
1954	M. Born W. Bothe	Quantum mechanics; coincidence method		x
1955	P. Kusch W. E. Lamb	Lamb shift; anomalous magnetic moment of electron		x
1956	W. Shockley W. H. Brattain J. Bardeen	Transistor	x	
1957	C. N. Yang T. D. Lee	Nonconservation of parity		x
1958	P. A. Cerenkov I. Y. Tamm I. M. Frank	Cerenkov effect		x
1959	E. G. Segrè O. Chamberlain	Antiproton	x	
1960	D. A. Glaser	Bubble chamber		x
1961	R. Hofstadter R. L. Mössbauer	Electron nucleon and nuclear interaction; Mössbauer effect	x	x
1962	L. D. Landau	Liquid helium, etc.		x
1963	E. P. Wigner Maria Goeppert-Mayer J. H. D. Jensen	Shell theory; symmetry in physics		x
1964	C. H. Townes N. Basov A. Prokhorov	Maser and laser		x*
1965	R. P. Feynman Julian S. Schwinger S. Tomonaga	Quantum electrodynamics		x
1966	A. Kastler	Optical pumping		x**
1967	H. Bethe	Nuclear (astro) physics		x
1968	L. W. Alvarez	Giant bubble chamber and resonances obtained with it	x	

* Strongly influenced by the wartime teamwork on radar.

** Small team of students.

to find that the particles lost their energy in discrete jumps: the amount of energy a particle lost depended very sensitively on the angle with the channel axis at which the particle entered the channel. From this quite accidental discovery came a completely new and possibly quite powerful method of probing the details of the interatomic potential in certain crystals.

I tell this story because it illustrates the unique power of an interdisciplinary attack. Here is an instance in which the whole team is much more than the sum of its separate components, where a team as opposed to an individual (as in the case of Bartlett and xenon compounds) achieves a breakthrough. First, the experiments required very thin, perfect gold crystalline foils; these happened to be the specialty of T. S. Noggle, a metallurgist and electron microscopist. Next the 50-MeV (million electron volts) iodine ions had to be accelerated, and their energies after degradation had to be measured with precision; this required experts on Van de Graaff accelerators and particularly on sophisticated time-of-flight techniques. Once the phenomenon was discovered, its full significance required the insight of a young solid state theorist, H. Lutz, as well as a variety of additional experiments that served to corroborate the theoretical predictions. And the team required orchestration: this was supplied by S. Datz, a chemist who had been concerned with the related phenomenon of sputtering.

To be sure, elaborate equipment was needed—a time-of-flight Van de Graaff machine. Nowadays this is not so unusual; there are perhaps two dozen laboratories which possess such instruments. But the number having at the same time an electron microscopist who can make perfect gold crystals a few hundred angstroms thick, an expert on sputtering, and a solid state theorist capable of interpreting the experiments is much smaller. It was very much more the style of research—the willingness of all parties to collaborate fully—that led to the breakthrough. This willingness among professionals to collaborate is actually not to be taken for granted, especially in the academic world. In reading James Watson's *The Double Helix*[14] one is constantly aware of the barriers that were placed between Watson and Francis Crick (who had the major idea about the structure of DNA) and Rosalind Franklin and Maurice Wilkins (who had the means for making the measurements needed). I suppose it is for reasons such as this that I am convinced the success of team science depends on the institution. There must be a tradition of interdisciplinary collaboration between professionals. This is more likely to exist in an institute with hierarchical organization—such as one finds in the applied or project laboratories—than in the typical university.

The example I have given would only marginally qualify as big science: the machines, though large, are not all that large. If one examines the more typical endeavors of big science, particularly those that require unique accelerators or unique reactors, one finds many examples of breakthroughs by teams. One of the most recent is the finding by Val Fitch and his collaborators (many of whom were students) of the nonconservation of charge parity in the decay of the K-meson or, perhaps even more uniquely tied to the capacity of a single machine, the discovery by Segrè of the antiproton, at the Lawrence Radiation Laboratory in Berkeley. In these cases I would argue that it was the machine properly used and good leadership more than the team. The team was needed mainly because the experiment was so complex. In a certain sense, the team tends to be incidental to the machine in very big science; by contrast, the team is central in those cases—such as the work on DNA and channeling—in which the means are more modest. Here what is important is a delicate balancing and interweaving of individual expertise.

So we see that teams can achieve breakthroughs, operating either in the interdisciplinary mode or in the big science mode. Yet the power of the team seems to me to lie primarily in its ability to do spectroscopy; as team science becomes more and more common, so might the emphasis on the spectroscopic style of science. This trend may be accentuated by the weightiness and inertia of modern big science. Where scientific teams have mobilized around very big pieces of machinery there is an understandable incentive to exploit that machine. The path of development, instead of following the logical demands of the discipline, tends to be constrained to directions that are made accessible by the machinery at hand.

Something like this has always happened in science: one exploits whatever tools one has available. But scientists are naturally much less ready to scrap a 400-MeV proton-synchrocyclotron that costs several million dollars, but which no longer can cut at the main edge of high-energy physics, than they are to scrap, say, an optical microscope. The somewhat bureaucratic imperative to exploit expensive machinery circumscribes the direction of scientific growth. The spectroscopic filling in of details tends to crowd out the breakthroughs, simply because the number of breakthroughs possible with a particular machine is very small compared with the practically infinite spectroscopic detail the machine can generate.

It would be foolish to underestimate the importance of spectroscopy in setting the groundwork for important discoveries and conceptual breakthroughs. Quantum mechanics would have been impossible without its underlying detailed optical spectroscopy. Or, more recently, low-energy physics has a strongly spectroscopic flavor. Most experiments seek to measure, in various nuclides, specific properties that already fit into a general theoretical framework. Yet out of this elaborate spectroscopy (conducted, incidentally, by teams) has come a seemingly endless succession of breakthroughs: either in experimental techniques, as in the discovery of the lithium-drifted germanium-detector, or in new insights into nuclear structure, as in the discovery of isobaric analogue states and short-lived isomers.

There is another side to the story which deserves mention. The team, especially around the big machine, is a powerful scientific device. More difficult experiments can be tried with a large team equipped with a unique facility than with a smaller outfit not so equipped. Thus, insofar as breakthroughs flow from difficult experiments, one might expect teams working with powerful and unique apparatus to continue to contribute their share of important discoveries. For example, as soon as the high-flux isotope reactor became available, questions in the phonon distribution in solids that had plagued solid state physicists became answerable.

To make important breakthroughs in science will always require competent, imaginative leadership. But it seems reasonable to expect that the degree of insight required to make such discoveries may be somewhat less than it was in the day of individual science: the team, or the big machine, may offer elements of uniqueness that were formerly supplied by sheer intellectual power. And, since competence is so much more common than genius, the team may be spreading the possibility of significant scientific discovery to many more scientists than in former days. Perhaps this democratization will prove to be one of the main by-products of big team science.

III. The Future of Team Research

A. The Institutional Setting

It seems clear to me that team science in the modern style is done better in the hierarchical, logistically strong institute than it is in the university. This, coupled with the unrest that wracks the university, suggests that we might see a gradual movement of modern science away from the university and toward the national institute—possibly even a growing separation between education and research. To most writers on this subject, especially since the Seaborg Report,[15] the notion that research and education are inseparable and indissoluble, that the one cannot be done without the other, has acquired the ring of holy dogma. But the facts do not really bear this out: certainly insofar as one is elaborating a certain area, such as the chemistry of xenon, professionals are better than students. For many years applied chemistry has been conducted to great advantage in the industrial laboratory without benefit of students. I know that at Oak Ridge some (though not all) of our division directors are convinced that they achieve results more quickly and more reliably with professionals than with students.

The universities have responded to the trend toward team research by setting up institutes, interdisciplinary and logistically strong, where team research can be performed effectively, but largely by students. But on the average these institutes suffer from a mismatch between the social ethos of the university and the social ethos of the institute: the one is individual and democratic, the other collective and hierarchical. When the institute acquires a collective and hierarchical character, which I believe is necessary for its success, its tie with the university department becomes more tenuous.

So we may be going full circle. Science in the seventeenth and eighteenth centuries was practiced predominantly in the academies, not the universities. It moved into the German and English universities in the nineteenth century. And perhaps, with the growth of the large team, it may gradually be moving out again, or at least it may not retain as intimate a connection with the university as it has had in the immediate past.

B. New Fields for Team Research: The Rise of Big Biology

The big interdisciplinary team has generally been confined to the physical sciences and to engineering. The biological sciences have remained the bastion of little, individualistic science, probably because the experimental tools needed to conduct biological experiments have typically been small and relatively inexpensive. Yet there are now important trends toward large interdisciplinary teams in the biomedical sciences and, very recently, teams that include engineers as well as physical scientists.

Part of this trend comes from the rise of molecular biology. In one sense the most important parts of molecular biology are really a branch of crystallography; the double helix model for DNA, for example, is based on a crystal structure deduced from X-ray diffraction data. Watson in his book bemoans his lack of expertise in crystallography, a lack which was made up for him by Crick and by Wilkins' group. It is highly significant that of the two men who made the most important discovery of modern biology, one (Crick) was originally a physicist. Again, the extraordinary elucidation of the working of peripheral nerve, for which A. L. Hodgkin and A. F. Huxley received the Nobel Prize, would have been impossible had it not been for the underlying work on electrical properties of nerves by the physicist K. S. Cole. Many biologists, especially in the most active fields of biochemistry, feel it necessary to rub shoulders with physicists and with physical chemists.

The second trend discernible in biomedical science is the rise of the very large-scale experiment. With our present concern with low-level insults to the biosphere (radiation, pesticides, smog), it becomes necessary to conduct animal experiments on a scale far greater than had hitherto been customary in biology. The husband and wife team of William B. and Liane Russell at Oak Ridge maintains more than 100,000 mice in order to study the mutagenic effect of moderate levels of radiation. Such biological experimentation immediately becomes a team activity: geneticists to manage the entire experiment; statisticians to scan the data for significance; veterinarians to manage the animals; pathologists to look for somatic effects; and, of course, the whole array of animal attendants, janitors, and cage-washers who are needed to keep 100,000 animals alive and thriving.

I would expect this trend toward very large-scale animal experimentation to become increasingly prevalent as we become more sensitive to the widespread influence of seemingly subtle factors in our environment. A call for such experimentation has been made by René Dubos, for example. Should this call be answered by the funding agencies, we may expect to see an increasing fraction of biology being conducted in the style of big science.

Biologists, particularly biochemists, are beginning to learn how to employ engineers and other supporting scientists, notably analytical chemists. This represents a new trend since biological institutes traditionally have not crossed deeply into the physical sciences, even less into engineering. True, the National Institutes of Health is an enormously large complex, but NIH has not had within it a strong tradition in the physical sciences or in process engineering. By contrast, the atomic energy laboratories have from the beginning spanned the biological sciences, the physical sciences, and the engineering sciences. This unusual juxtaposition has now begun to pay off—for example, in the brilliant development of zonal centrifuges under the leadership of Norman Anderson at the Oak Ridge complex. These centrifuges, which were first developed to separate uranium isotopes, have been modified by Anderson, together with a large team of engineers, to handle biological materials. The centrifuges are now being used very widely to separate, on a large scale, various cell moieties; for instance, they have been used to purify flu vaccine of its antigenic protein impurities. Anderson is now exploiting in his Molecular Anatomy Program (MAN) whatever relevant engineering and analytical expertise he can find in the Oak Ridge complex to systematically separate, and then prepare on large scale, the many cellular particles which now can only be seen in the electron microscope. Anderson's success I believe is only the forerunner of future successes that biology will enjoy as it enlists the cooperation of the engineering sciences.

C. Redeployment of the Big Institutions

The modern scientific team arose as an integral part of the great laboratories; whether in basic research or in applied research, the team style is the dominant mode in the modern big laboratory. It seems inevitable then that the future of team research will depend on the fate of the big laboratories. I shall therefore close with a few speculations on these institutions.

Though it is evidently impossible to generalize, one can see limits to the prospects of the big laboratories. For example, in those institutions devoted to high-energy physics, the future is limited by the sheer increase in expense of the necessary gadgets. The Alternating Gradient Synchrotron facility which was completed in 1960 cost $30,650,000; the Stanford Linear Accelerator, completed in 1966, cost $114,000,000; and the National Accelerator, a 200-GeV proton synchrotron, expected to be operating in 1972, is estimated to cost some $240,000,000. Presumably each of these devices will become obsolete, not because there is not always more spectroscopy to be done, but rather because people will eventually get bored with spectroscopy. Unless there are occasional stirring breakthroughs—perhaps a breakdown of quantum electrodynamics—I cannot visualize these institutions forever sustaining themselves, or remaining immortal, by simply amassing spectroscopic details about elementary particles.

The atomic energy laboratories must also face questions of redeployment, though for a different reason. True, the two central problems of nuclear energy—breeding and controlled fusion—have yet to be solved. But even these are questions of finite dimensions; the first because it is not all that difficult, the second because it may prove so difficult that interest in it will wane. There is already evidence that the world's atomic energy laboratories are beginning to adjust to these facts, mainly by expanding their areas of concern beyond nuclear energy.

By contrast, the future of the great biological laboratories seems clear enough: the questions which biomedical science seeks to answer are urgent, massive, timeless until they are solved. It seems likely therefore that the need for redeployment will hardly arise for, say, the instrumentalities of the NIH. My guess is that these institutions will acquire a more interdisciplinary flavor, especially by developing engineering skills, simply because the cross between the physical sciences and biomedical research has been so fruitful.

Thus redeployment of at least some of the big laboratories is in the cards. This realization comes at a time when we hear much about the many social and socio-technological conflicts that plague our modern society—racial unrest, the decay of the city, pollution, overpopulation. John Platt sees modern society on the verge of crises so profound as to warrant launching wartime-like projects to resolve them.[16] The Committee on Government Operations of the United States Senate has been holding hearings during the past couple of years under the chairmanship of Senator Edmund S. Muskie aimed at establishing a Select Committee on Technology and the Human Environment. Everywhere there is a restlessness and concern: the priorities our society has lived with in the postwar world need reassessment; we must abjure our preoccupation with hard science and address ourselves to these subtler, more difficult, and more important human problems.

Whether science can help very much with these social questions is a moot point. Many of us scientists believe that science can help: that almost every one of the conflicts and problems that we face has some technological, as well as social, components, and that therefore science directed specifically at their resolution may be helpful. In this we may be displaying an uninformed naïveté; perhaps racial conflict, urban decay, and overpopulation are beyond help even from science.

Yet this much can be said: if science has something to offer toward resolving these questions, it surely will have to be a broadly interdisciplinary, team type of science. The social components of these problems are more obvious than are the technological ones, but there is always an interaction between the two aspects; it is quite natural to visualize interdisciplinary teams, ranging over social science as well as natural science and engineering, being mobilized to attack some of these desperately troublesome questions. As of now, however, such teams have no natural home: the university is unsuitable because of its prejudice against teams, the national laboratory because of its inexperience in social science. I have therefore suggested the creation of new entities, national socio-technological institutes. Some such institutes might be formed *ab initio;* others by co-opting experts in the social sciences to work in existing hardware-oriented laboratories.

National socio-technological institutes at which one would apply the methods of science to our difficult socio-technological problems might also serve an entirely different purpose: a means of focusing the socially relevant energies of our young people. Many young students seem to be disillusioned with natural science: those who in the previous decades went into physics or chemistry now go into the more "relevant" social sciences, and even the students of the natural sciences are acquiring a taste for socially relevant issues. Yet I can foresee this socially motivated cohort of students being frustrated all over again if, once they are trained, once they are readied to do battle on behalf of society, they find no instrumentalities to which they can attach themselves to carry on their commendable crusade. I should think that just as the institutions of big science for several decades provided a home for the aspiring scientists of the 1950's and 1960's, so the socio-technological institutions might provide a home for the aspiring social engineers of the 1970's.

There have been several suggestions by now for national socio-technological institutes, most recently, in Senate Bill 3410 sponsored by Senators Howard Baker and Edmund Muskie to establish national environmental laboratories. It is premature to really assess such proposals. It could be that the difficulties we face go beyond resolution by the methods of science—hard analysis, empirical observation, engineering design. Yet, before taking so pessimistic a view of man's capacity for self-betterment, I would urge trying the interdisciplinary team attack—an approach that was so notably successful in the generation immediately following World War II and that just may help guide us during the coming generation.

REFERENCES

1. For alternative approaches to this topic see L. Kowarski, "Team Work and Individual Work in Research," in N. Kaplan, *Science and Society* (Chicago: Rand McNally, 1965), pp. 247-255; and Cecil F. Powell, "Promise and Problems of Modern Science," Concluding Address, *Maria Sklodowska-Curie: Centenary Lectures, Proceedings of a Symposium,* Warsaw, October 17-20, 1967 (Vienna: International Atomic Energy Agency, 1968).

2. Dr. Saul Benison pointed out at the Bellagio conference that I may be overdoing this distinction between art and science: Leonardo trained in the atelier of Verrochio, who influenced much of his early style; Melville was much influenced by the Bible and by Shakespeare. Yet the connection between, say, the physicist Hertz and his predecessor Maxwell is, to my mind, far more explicit and continuous than the connection between two artists. Hertz used Maxwell's equations precisely as Maxwell formulated them; his work flows from Maxwell with an inevitability and logic that can never be matched in the work of an artist who follows an illustrious predecessor. As Professor Edward Shils puts it, "There is a coercive element in the tradition of the sciences that is absent in the arts."

3. University of Chicago Press, 1962.

4. "Letter to Robert Hooke, February 5, 1675/6," in John Bartlett, *Familiar Quotations* (Boston: Little, Brown and Company, 1968).

5. *Proceedings of the American Philosophical Society,* 94 (October 1950), pp. 422-427.

6. As will be apparent later in the discussion, I often extend and generalize the

word "spectroscopy" to mean both the activity and the results of filling in the scientific details after a major discovery has broken new ground.

7. The facelessness of big team research has been commented on by others—for example, Gerald Holton, "Scientific Research and Scholarship, Notes Toward the Design of Proper Scales," *Dædalus* (Spring 1962), pp. 362-399.

8. P. B. Medawar, *The Art of the Soluble* (London: Methuen & Company Ltd., 1967).

9. *Inside Bureaucracy* (Boston: Little, Brown and Company, 1967).

10. To anyone who has spent some time in a large laboratory, it must be perfectly clear what I mean by its great logistic power; but to those who are unfamiliar with such institutions, perhaps I can illustrate with the experience of a distinguished demographer who spent a summer at Oak Ridge National Laboratory studying urban decentralization. At the end of his stay I asked him what he thought of ORNL as a possible locale for demographic research. He replied: "Demography would be revolutionized if it were conducted there. It would be converted from a small, rather individualistic enterprise into a big-scale, massive business. There would be huge computers with programmers and mathematicians to help one use them, experts of every sort available at the other end of the hall, as well as editorial assistants, draftsmen, travel agents; above all, they would be ready and willing to help you get on with the job." If the large laboratory possesses so much logistic strength in the eyes of a demographer, one can imagine how much greater is its strength in the fields of science it was originally set up to exploit!

11. Neil Bartlett and N. K. Jha, "The Xenon-Platinum Hexafluoride Reaction and Related Reactions," in H. H. Hyman, ed., *Noble-Gas Compounds* (Chicago: University of Chicago Press, 1963), pp. 23-30.

12. Hyman, ed., *Noble-Gas Compounds;* cf. J. H. Holloway, *Noble-Gas Chemistry* (London: Methuen & Company Ltd., 1968).

13. "Basic Chemical Research in Government Laboratories," Report of the Panel on Basic Chemical Research in Government Laboratories of the Committee for the Survey of Chemistry, Division of Chemistry and Chemical Technology, National Academy of Sciences Report 1292-A (Washington, D.C., 1966).

14. James D. Watson, *The Double Helix* (New York: Atheneum, 1968).

15. "Scientific Progress, the Universities, and the Federal Government," Statement by the President's Science Advisory Committee (Washington, D.C.: U.S. Government Printing Office, November 15, 1960).

16. John Platt, "What We Must Do," *Science,* 166 (November 28, 1969), 1115-1121.

A. HUNTER DUPREE

The *Great Instauration* of 1940: The Organization of Scientific Research for War

THE CONCEPT of the way in which science should be organized underwent a transformation in 1940. The new arrangement between science and society brought to research a flow of support in a volume undreamed of even by Francis Bacon. It also allowed a concentration of talent on a very few practical problems in a way that demonstrated to everyone the power of science, most notably by the explosion of the first atomic bombs. Only this one transforming concept—the organization of science—had its climactic crystallization in phase with the great crisis of civilization that befell all mankind in the year 1940. In contrast, the substantive themes of the disciplines from psychoanalysis through sociology, biology, chemistry, and physics reached their climax of transformation either before or after the historical crisis of 1940.

Scientists, not surprisingly, see their substantive accomplishments as more important and interesting than their organizational arrangements. Linus Pauling was participating in one of the follow-on studies of the great reorganization when he was in New York on February 6, 1945, to work on a section of what became Vannevar Bush's report, *Science—The Endless Frontier*. Yet the occasion was memorable to him because one of his fellow members described some work on sickle-cell anemia.[1] Compared with science itself, the organization of science has always seemed a bore. Yet when the crisis of World War II came, those who fathomed the depths of the problem of science's junctions with society were the ones who shaped the new era. Vannevar Bush and James B. Conant both had distinguished scientific reputations even in 1940, but it was what they were besides being scientists, rather than their linear ranking on a scale with Einstein and Bohr, which made them the instruments of the *Great Instauration* of 1940.[2]

The scientific community in America had formed its essential structure in the period before World War I, drawing support from universities, government, industry, and private foundations.[3] A clear division of labor separated the scientists supported by the various sectors, with the government increasingly limiting itself to applied research and the universities becoming the recognized home of basic research. World War I gave a taste of the potentialities of applying science to military problems as university scientists followed George Ellery Hale and Robert A. Millikan in setting up and manning the projects sponsored by the National Research Council.[4] However, the mobilization experience of the scientists in World War I was so brief and fragmentary that in the 1920's and 1930's they returned to a pattern of activity which made the university department their normal habitat and allowed military research to drift to the periphery of their activity and consciousness. Industrial research did indeed enter prominently into the life of many corporations in the interwar years, but those scientists thought of themselves as creating a more comfortable and efficient peacetime civilization.

The Great Depression was a sufficient shock to call forth some reconsideration of the organization of science in the 1930's, both in the United States and in Great Britain. The generation of scientists who could face the crisis of World War II thus had the benefit both of the experience of 1918 and of the discussions of the 1930's. Yet their institutional models for organizing science for war appeared oddly anachronistic for the magnitude of the crisis they faced. Neither the National Academy of Sciences, founded in 1863, nor the National Research Council appeared to many thoughtful scientists to offer a model for the organization of science.[5]

If the themes of the history of science are few and long-lasting, the themes of the organization of the community of science are just as long-lasting and even fewer in number. Furthermore, the community of science is so weak in its institutional structure that it does not dwell apart as a separate republic, but rather in and among the other institutions of society. Joseph Haberer in his recent *Politics and the Community of Science* has distinguished two models for the community of science which have had clear definition since the seventeenth century.[6] The models stem not from Galileo or Newton but from Francis Bacon and René Descartes.

Bacon's community put science to work. He recognized that science was a socially conditioned enterprise and that it required a new structure of institutions and administration. His house of Salomon was a group enterprise which minimized the importance of the individual member and provided for a division of labor.[7] Bacon's man of science was fallible and could not by himself reach certainty.[8] A lack of towering individual genius had been often charged against American science and might alone have attracted the adjective Baconian to the American scientific community.[9] Add to the lack of Newtons and Darwins in the American record the massive research establishments manned by almost anonymous specialists serving both agriculture and industry, and the temptation to make American science irretrievably Baconian in organization is strong.

The Cartesian model for the scientific community, however, presents elements which were also recognizable on the American scene in the late 1930's. The leading figures of modern science had often exhibited the characteristics which Haberer attributes to those organized by the Cartesian model of community. The scientist of the Cartesian tradition had an inclination to work in solitude and to be secretive, at least until he had completed his work. He had a tendency to be closed-minded in that he minimized the influence of other scientists and lived in a state of rivalry with them for recognition and status. Hence, he did not conceive himself to be in an institutional setting at all. He was reluctant to make commitments to society, and his dislike of mediocrity made him wary of cooperative efforts. He wished the public to support his own research, but without obligation. "When resources are needed to perform some experiments, a public . . . can assist in only one manner: finance him and then stay out of the way."[10] Haberer finds Descartes' model for science comparable to the laissez-faire model in economics.[11]

The key to the political decisions facing the scientific community in the United States at the onset of World War II was not that one could choose either the Baconian or the Cartesian model but rather that both traditions had been essential to the scientific community since the seventeenth century. The coming of the refugee scientists in the 1930's[12] had for the first time provided in America the individual investigators of outstanding reputation who made the Cartesian tradition credible. Many of the newcomers were physicists, whose posture at least on the theoretical side seemed much more Cartesian than Baconian. At the same time the political awareness of the refugees from Nazi Germany and Fascist Italy far surpassed that of their American counterparts. The refugees took a leading role in urging the conversion of science to power which could be used in the conflict with Nazism. They knew that the Nazis had already done much to mobilize scientists.

The conviction that unprecedented steps had to be taken in the organization of science possessed many groups in the late 1930's. Some had formal responsibility for the scientific community, for example, the National Academy of Sciences, which took the unusual step of electing an industrial scientist, Frank B. Jewett, as its president in 1938. Yet an invisible college which had begun meeting under the unlikely guise of the Committee on Scientific Aids to Learning in 1937 felt that no existing institution could effectively reorganize science for the looming emergency. Bush, Conant, and Frank B. Jewett were all members of that committee, and by early 1940 they had reached the point of action which brought science and government into a new partnership. Out of this juncture came a new system of support. In 1939 the system was not there. By 1960 it had become a permanent part of national and international life.

The new system of science support was not the product of revolution. Quite the opposite. The movement has none of the characteristics which would make if fit ready-made figures of speech drawn from the great revolutionary upheavals of history. No Voltaire or Rousseau prepared the way; no Mirabeau or Robespierre or Lafayette played out his role among the American community. No Thermidor followed a Reign of Terror. What happened in 1940 was entirely different. The American political tradition, which dates in its present constitutional form from 1787 and as a part of the Anglo-American legal and parliamentary tradition from much earlier, managed without revolution to adapt itself to the institutions of the scientific community, which as we have seen can trace their own past at least to the seventeenth century of Bacon and Descartes. One complicated pluralistic set of institutions got together with another pluralistic set of institutions. While the result has been drastic in its consequences for the world, one of the oldest continuous political traditions of the Western world has gained immense capacity for deliberate change without revolution.

How to apply science to weaponry in the circumstances of the late 1930's was not clear. At one extreme, the military planners might have reasoned that no research had ever paid off in a war already begun, and since World War II seemed to be a total war, no military aims beyond that present war existed. Hence research should be cut, with the money released spent on procurement of matériel already designed, and the manpower released should be used directly in a regular military way. The 1936 attempted cuts in United States army research argued in this direction, and scientists could remember that the British had lost Henry G. J. Moseley, one of the finest young physicists of Sir Ernest Rutherford's laboratory, as a soldier before Gallipoli. In this solution, the scientific community and the universities as the homes of basic research would simply be disbanded.

At the other extreme, the idea was entirely conceivable, witness the horror evoked by poison gas in World War I, that warfare might become the exclusive affair of scientists creating weapons and counterweapons which would dominate the battlefield. If this premise were correct, the military establishment itself was a major stumbling block to survival in total war. The only hope lay in giving "scientists" a completely free call on the resources of the nation and complete authority to deploy the force they created. The military establishment, and with it the military mind of Colonel Blimp, would disappear. There is at least a faint point of similarity between this view and that of the Marxists, with overtones from Bacon, Thorstein Veblen, Technocracy, and the philosopher-kings of Plato's Republic.

Both of the extreme solutions were revolutionary. Both neglected the resources of tradition, institutions, and training already existing in a great modern industrial state. The United States had the greatest resources of any nation, although it had drawn back as a people from the very thought of converting more than a tiny fraction of those resources into military force. Though both of the extreme solutions had a certain role as ideas, they could hardly take place in the United States under the threat of war, even the threat of total war. The application of science to weaponry had to occur within the framework of existing institutions, both scientific and military. The leadership of Franklin D. Roosevelt was an important factor in preventing a revolutionary solution from becoming a serious choice for the American people. As the architect of the New Deal, he had not dismantled what he considered to be essential institutions of American democracy. As a patrician long interested in military affairs and a President who took his position as commander in chief of the armed forces seriously, he was never even tempted to place responsibility for the security of the nation in the hands of scientists. Thus any application of science to weaponry had to take place within the existing framework of institutions.

The three basic alternatives for deploying the research resources of the country were:

(1) Build up the small government laboratories, especially in the military departments, to include virtually all American scientists.

(2) Use the industrial research of the interwar period as a nucleus to build up laboratories in the war industries with all American scientists so that the widest possible range of items would have the benefit of the application of science.

(3) Organize the scientists of the country into an independent corps. If this road were taken, several alternatives were available.

(a) Arrange the corps by discipline—chemists, physicists, biologists together, as they were accustomed to in their universities—and place it at the disposal either of industry or the military departments.

(b) Arrange the corps according to component problems to be solved—explosives, airplanes, radar, medicine—and place it at the disposal either of industry or of the military departments.

(c) Deploy the corps according to the possibilities presented by the state of science rather than the needs either of the military departments or of industry. Electronics and medicine had already presented themselves, and after January 1939 physicists could advance the release of nuclear energy as a major possibility.

The choices made in the United States in 1940 did not completely close out any of these alternatives, but the leadership which emerged from the scientific community did make a set of conscious decisions which had the effect of quickly mobilizing selected laboratories and scientists for a concentrated attack on a relatively few problems. Plenty of precedents for some of these alternatives existed in American history already, for example the problem approach employed by the Department of Agriculture and the growth of industrial research since World War I. Yet the decisions could not be made solely on the basis of the workings of American democracy.

The emergency of 1939 forced a comparison with the other great contending industrial states and a prediction as to how they were answering the same set of questions. Here the prestige of the dictatorships and the fear that scientific research was the factor which made the panzer divisions invincible in Poland combined to rule out a decision on the part of the United States not to make at least the attempt to mobilize science. The dictatorships, like the monarchies before them, seemed to possess a terrible and possibly decisive advantage over pluralistic democracies because of their ability to deploy the people, hence the intellectual resources of their citizenry, at will. The curious ambivalence of Americans toward the relationship of science and authority, evident since the days of John Quincy Adams, came into play.

In 1937 Vannevar Bush, at the age of forty-seven, was already dean of engineering and vice-president of MIT. He had worked on submarine detection devices for the navy during World War I. He was an electrical engineer, a holder of several patents, one of the founders of Raytheon Corporation. He was a Yankee who spoke with the accent and manner of an earlier New England, saying "patt'n" for pattern and describing a passing upper respiratory illness as the "grippe." His prose was sufficiently precise for a scientific paper but had in addition a cadence and eloquence befitting the son of a clergyman. Even then he was an eminently practical engineer who at the same time had much of the manner, bearing, and outlook of a scientist—a combination which had found proper scope at MIT after 1930 when Karl Compton came as president to make science the pathway to technology.

By 1939 Bush had moved from Cambridge to Washington as president of the Carnegie Institution of Washington and had taken a place on the National Advisory Committee for Aeronautics (NACA), of which he would soon become chairman. The post at the Carnegie Institution was one of the most central in American science. From its headquarters on P Street in Washington, Bush had an overview of such famed institutions of basic research as Mount Wilson Observatory and the genetics experiment station at Cold Spring Harbor. Its board of trustees included Walter S. Gifford of American Telephone and Telegraph; Frederic A. Delano, the President's uncle; Herbert Hoover; Frank B. Jewett; and Elihu Root, Jr., who at crucial times provided Bush with confidential legal advice. Bush may well have had the Carnegie Institution in mind when he later wrote, "This country has produced . . . an extraordinary number of men of wealth who have regarded their accumulations as a public trust and have utilized them intelligently for the public benefit, often furthering good causes that could be furthered in no other way."[13]

Three years younger than Bush, hence forty-six years old in 1939, James B. Conant was a recent Ph.D. in chemistry when he worked for the Chemical Warfare Service during World War I.[14] He was a thoroughly professional and highly regarded organic chemist in 1933 when he became president of Harvard University, a position which automatically carried the prestige of the oldest, most renowned, and most influential of American universities. By the time of the Harvard Tercentenary in 1936, it was already clear that Harvard was moving rapidly toward the head of the list of the world's universities as Germany destroyed its own heritage and the institutions of all western Europe fell under the shadow of the coming war. The problems of academic appointments and of selecting the very best representatives of all fields of learning occupied much of Conant's thought during the 1930's, when large numbers of deserving young men were stranded by the depression in Cambridge. It would have been easy to solve the "young man" problem by applying less demanding standards than Conant did in setting up a system of ad hoc appointment committees in which he personally participated.

Although he normally refrained from joining committees and taking political stands, Conant attacked Franklin D. Roosevelt over the packing of the Supreme Court. In the fall of 1939, when the Harvard *Crimson* was accusing him of "earning an unenviable place on the roadgang that is trying to build for the United States a super-highway straight to Armageddon," he wrote that "I am personally strongly in favor of a modification of the so-called neutrality law so as to permit the sale of implements of war to France and England. . . . I believe that if these countries are defeated by a totalitarian power, the hope of the free institutions as a basis of modern civilization will be jeopardized."[15] On May 29, 1940, he said over the Columbia Broadcasting System: "A total victory of German armies is now well within the range of possibility. Can we as a free nation, considering first and foremost our own best interests, tolerate the overwhelming destruction of the British Fleet?" Urging "rearmament at lightning speed," he said that "the United States should take every action possible to insure the defeat of Hitler. I shall mince no words. The actions we propose might eventuate in war. But the fear of war is no basis for a national policy!"[16]

Having thought his way through to total involvement in behalf of Britain even if the United States should have to go to war, Conant became a major public voice for intervention. Add to this position his pivotal place in the university world as president of Harvard and his unchallengeable credentials as a chemist, good despite his own disclaimers, and the picture of a leader matched to the crisis emerges.

Vannevar Bush wasted no time either with public statements or with becoming enmeshed in the chaos of special interests swarming around Washington during the fall of France. He had reason to feel that science was a separate entity, that the scientific community had something to offer the nation, that he could, through the men with whom he was in touch, speak in the emergency for the scientific community, and that the place for science to join the government was at the level of the President and no lower. By May 25, 1940, Bush had his ideas on paper and in the hands of Frederic A. Delano. Clearly Bush had been thinking about the way in which NACA worked, and John Victory, its secretary, had drafted an executive order establishing what he dubbed the National Defense Research Committee (NDRC).[17] One form of Bush's covering memorandum read as follows:

> The creation and improvement of military devices involves three stages: fundamental research, engineering development, and production. This memorandum deals with the first stage.
>
> The National Advisory Committee for Aeronautics carries on important fundamental research, and correlates military and civil research activities, on aeronautical devices. No similar agency exists for other important fields, notably anti-aircraft devices. The NACA is composed largely of men of aeronautical background, and should not attempt to expand its field.
>
> The National Academy of Sciences assembles groups of distinguished scientists and engineers to advise the government on specific scientific problems. This valuable activity was greatly expanded during the last war, is now increasing, and can again be expanded to similar advantage.
>
> There appears to be a distinct need for a body to correlate governmental and civil fundamental research in fields of military importance outside of aeronautics. It should form a definite link between the military services and the National Academy. It should supplement, and not replace, activities of the military services themselves, and it should exist primarily to aid these services and hence aid in national defense. In its organization it should closely parallel the form which has been successfully employed in the National Advisory Committee for Aeronautics. It could perform a very valuable function indeed in stimulating, extending, and correlating fundamental research which is basic to modern warfare.

It should not be created unless it would be welcomed, and hence supported by the three bodies primarily concerned, the War and Navy Departments, and the National Academy of Sciences. If it has their support, it will also be able to enlist the support of scientific and educational institutions and organizations, and of individual scientists and engineers, throughout the country.[18]

Bush's use of the term "fundamental research" might seem on the surface to be confused and confusing, but in actuality it contained within it much of the genius of the NDRC. The pressure of events and the coupling specifically with the military departments and with modern warfare guaranteed that Bush had in mind applications of science to a limited and consciously selected range of problems. Yet equally there is no doubt that he had in mind for the key personnel the university scientists who he knew would rally only to the banner of fundamental research. Bush's attitude toward cooperation with the government and with other agencies was equally a blend of reassurance that existing organizations would not be displaced by NDRC and uncompromising assertion of freedom of action, subject only to the authority of the President. The National Academy, especially, got both reassurance and a position removed from the chain of decision. Even NACA, which preserved, perhaps partly because of hard bargaining by Victory, an autonomous position, had carefully prescribed boundaries and a specific bar to its expansion into a general scientific agency. Even the pattern of NACA as an analogy had its limitations, for while the administrative arrangement of a committee with both civilian and military members carried over, Victory clearly envisaged that NDRC would "construct and operate research laboratories" as well as "make contracts for research, studies, and reports with educational and scientific institutions, with individuals and with industrial and other organizations for scientific studies and reports."[19]

In two whirlwind weeks, as Hitler knocked France out of the war, Bush touched all the bases of impinging interested organizations and carried his NDRC to Franklin D. Roosevelt in the form of actual drafts of letters and executive orders. He was working closely with Harry L. Hopkins as early as June 6, 1940, and was watching the committee dealing with industrial mobilization in order to, as Jewett reported, "keep things on the proper track and avoid the turmoil and confusion which would result from various Departments of Government setting up hastily conceived groups to organize science and invention."[20] He went with Hopkins to see the President on June 12, and by June 15 the whole structure had taken definite form. Bush had seen both General George C. Marshall and Admiral Harold R. Stark. He had checked his drafts with Karl T. Compton and Jewett. He had secured the services of Conant, who asked only two questions: "Is it real?" and, "Are you to head the committee?"[21] He had already secured the services of Irvin Stewart and of Carroll L. Wilson of the Research Corporation as administrators.

In the letter of June 15, 1940, from Roosevelt to Bush, which set things in motion, and which was probably very close to Bush's own submitted version, occurred the following paragraph:

> The Committee will consist of not less than eight members, and will be attached to the advisory Commission to the Council of National Defense. Through this organization you will be provided with such facilities and funds as may be necessary for the operation of the office of the Committee. It is expected that, in the furtherance of the objective of the Committee, you will arrange, by agreement with research laboratories in education and scientific institutions and in industry, for such studies, experimental investigations and reports, as may be found desirable in order to accelerate the creation or improvement of the instrumentalities of warfare. I feel sure that you will have the hearty support of the scientists of this country in these efforts, and that they will cooperate to the utmost under the guidance of your Committee.[22]

Specific mention of the NDRC's directly operating laboratories had disappeared; the use of both educational and industrial laboratories by contract came to the fore. The connection with the Council of National Defense (and in the flux of the moment, by implication its supporting agency, the Office of Emergency Management) gave Bush a line position in the government and with it the crucial ability to spend money. In contrast to the line arrangement upward and the contract arrangement to laboratories, the scientists both individually and collectively would cooperate from a position of full independence. This conjunction was basic to the whole future course of government-science relationships in the United States. The triple relationship of government, private institutions, and scientist made possible quick formation of research teams which had both adequate support and a great measure of flexibility.

Clearly Bush had led the scientific community into an unprecedented position of independent power. Not since the days of John Quincy Adams had the scientific community had such close touch with the presidency. And the long evolution, beginning with the days of Andrew Jackson, of the government's mechanisms for central scientific organization now found its culmination in a structure which lay largely beyond the reach of both the Congress and the established departments of government. In July 1940, Henry A. Wallace as Secretary of Agriculture wrote to Bush suggesting that because "the Department of Agriculture is one of the greatest of Scientific Research Agencies," a representative should be included on NDRC. Bush attached a memo: "Stewart—this may not be part of the official archives, but it is nevertheless of historic interest, & a matter we may approach later."[23] Bush was too well placed to need Wallace or any other cabinet officer, no matter how powerfully connected politically or how long associated with science in the government.

Bush was not a "science czar," even after the adjustment of the organization in the summer of 1941, which brought into existence the Office of Scientific Research and Development (OSRD), installed him as sole director, brought medicine under his purview, and extended the scope of his activities in the area of development. In the first place he always kept in mind the temporary nature of the OSRD. On a temporary basis he could take on great powers, dispense with cumbersome safeguards, and take swift and drastic action. Always in the background, however, as the ideal to which he aspired, lay the committee of scientists serving as a direct executive over scientific research. The NACA form of committee was not dead but only temporarily superseded, and both the NDRC and the NACA itself continued to act as committees during the entire war, though at least in the case of NDRC not quite as an independent body.

The second reason Bush was not a science czar lay in the limitations on the mission of the OSRD. He consistently fought against the idea that he had plenary responsibility for all research and development in the country. Implicit in his insistence on the right to choose problems, even if the military did not ask for the research, was the right to refuse to follow lines of research which did not appear likely to pay off within the time limits which the concept of the duration of the war imposed. Nor would he take responsibility for using all scientists in the country or get bogged down in supporting either industrial research or the universities as institutions. Hence Bush used his unusual position of power to refuse responsibility as well as to accept it. In this indirect way the OSRD gained the initiative in making priority decisions.

Finally, Bush was not a science czar because of the drastic decentralization of the OSRD. The flow of ideas upward from the scientific community met the flow of requirements from the military services not so much in the office of the director, or even in that of Conant as chairman of NDRC, but rather in the offices and persons of the division chiefs and their staffs. These men, thorough scientists all, comprised the major part of the headquarters staff of the OSRD and the chief link between the scientific community and the government. Bush, who had an appreciation of the role of nonscientist administrators such as Irvin Stewart and Carroll Wilson, was careful to concentrate them largely in his own office and to place scientists on the division and section staffs. The names of the refugees and many of the heroic names of the wartime research effort are largely absent from these posts, and one may guess that they were chosen for scientific and administrative judgment rather than for research ability in the abstract. There may even have been in the key staffing decisions by Bush, Conant, and Karl Compton an implicit set of criteria which was quite different from those applied to the selection of heads of research teams. The OSRD thus became the spawning ground of a type of administrator-scientist who could serve in the intermediary role between science and government precisely because he could not be distinguished from the investigators.

Purpose: Criteria for Scientific Choice

The basic purpose of the OSRD was to keep the exercise of scientific choice in the hands of scientists, who alone were in a position to judge the merits of a given line of research. Lay leaders, both civilian and military, had no means of making such decisions. Alvin Weinberg, looking at the 1960's,[24] sees scientific choice coming into play when money becomes scarce, and suggests that criteria measure technological merit, scientific merit, and social merit. Because of the arrangements made in the summer of 1940, money was not the limiting factor at any time during World War II, but the severe foreshortening of the time scale available meant that limitations of both knowledge and trained people forced scientific choice upon the leaders of the OSRD.

The social merit of research the OSRD saw in stark simplicity. Military power was the one social criterion which must reign for the duration. In practice much of the energy of the leaders of the OSRD had to go into the infinitely subtle task of judging the potentialities of existing scientific knowledge, the quality of individual investigators, and the facilities available in existing institutions. The strategy of the allocation of scientific resources to weaponry, then, became essentially one of judging the technological merit of the various lines of scientific research as applied to weapons within the confines of the duration, a concentration of the judgment of the scientific community on technological issues. The scientific community actually arrogated to itself the crucial judgments which determined the major technological innovations in weaponry during World War II. To a very large degree the OSRD made these essential judgments before Pearl Harbor. Indeed, the entry of the United States into the war wrought none of the changes in the scientific program that it did in either economic mobilization or in the military services.

The first scientific choice made by the leaders of the NDRC-OSRD was to create an electronic environment for warfare. The state of electronics in all the major countries of the world, and the progress which they all had made on such devices as radar, meant that the achievements of both friends and enemies dictated an all-out effort. Radar, a measuring instrument which cut through both night and fog and which by operating on a line-of-sight principle extended the horizon for "seeing" aircraft and ships far beyond visual range, provided the major means of bringing about an electronic environment. But radar alone would never have accomplished what the total system of radar and radio connected to weapons

—guns and aircraft and ships carrying guns—actually accomplished. To help develop an electronic environment the OSRD established the Radiation Laboratory at MIT, the radar countermeasures laboratory at Harvard, and the programs for developing sonar. Characteristic of the trend was the work on the development of the proximity fuse. None of the other belligerents had such a device, which was in effect a miniature radar in the nose of a shell to signal the proximity of any nearby object—for instance, aircraft, or the ground—and to detonate the shell at the distance calculated to do the most damage. Merle Tuve of the Carnegie Institution of Washington was the chairman of the section, working through Johns Hopkins University and the National Bureau of Standards, to bring via the proximity fuse an electronic solution to one of the most intractable problems of the World War II era.

In contrast to the positive choice on the electronic environment, the United States government in effect made two major negative choices in not emphasizing jet engines for aircraft and missiles. Several sections of the OSRD did indeed work on rockets for tactical missions and for the assistance of take-off of aircraft, though nothing resembling the German team at Peenemunde ever emerged, and it could be argued that the difference in the geographic position of the allies relative to Germany dictated this low strategic priority.[25]

As to jet engines, which were known to the British in practice and to the Americans at least in theory, the real decision made by the OSRD was not to enter into a consideration of the problem at all. The independence of the NACA meant that the OSRD never even took up the problem directly. In addition the NACA, which was very small and until 1939 had concentrated on aerodynamics to the relative exclusion of research on engines,[26] made the decision in 1940 that the small supply of competent engineers available must not be diverted from the important task of improving the airplanes already scheduled for production in order to undertake jet development.[27] The Army Air Corps elected to concentrate on a mass fleet of piston aircraft whose designs already existed, although it supported small-scale research and development on the British-designed jet engine beginning in 1941.

In between the positive choice of the electronic environment as a field of research and the negative choices of rockets and jet engines lay the puzzle of nuclear fission. Since the story has been often told, most adequately in Richard Hewlett and Oscar E. Anderson, *The New World: 1939/1946*,[28] only a few remarks about the OSRD's role are necessary. Before the creation of the NDRC in 1940 the Briggs Committee on Uranium could not assure a flow of money into research on nuclear fission. The assignment of the uranium problem to the NDRC in 1940 had started the flow of money into the program under Bush's guidance, but what was needed in the summer of 1941 was a decision on whether atomic energy in the form of bombs was within reach scientifically and technologically before the end of the war. Indeed, some of the scientists who had participated in the very early stages of the research in the United States were actually leaving the field of nuclear physics to go into the radar program, as in the case of Luis Alvarez and E. M. McMillan from Lawrence's laboratory in Berkeley, and Merle Tuve took over leadership of the proximity fuse program after his initial research in nuclear physics at the Carnegie Institution.

Bush and Conant probed hard for a decision in 1941, depending on a National Academy committee under the chairmanship of Arthur Compton, and later adding engineers to it to get a better view of the technological choices. Both Bush and Conant seemed balky to the scientists working on the committees because of their awareness of the immense advantages which would flow from a negative choice. If a bomb was not possible within the duration, the Germans could not get it either, and the decision not to go ahead, to relegate nuclear fission to the basic research end of the spectrum, would free large numbers of the scarcest resources—physicists—for other tasks.

Reversal of the OSRD-NDRC top command's negative tendency came not from the military or political leaders above them but from the scientists below. The British, whose urgency about the war was still differentially much greater than the Americans' in the summer of 1941, were confident that a U-235 bomb could be achieved within the duration. Furthermore, the increase in chance of success promised by the appearance of an alternative route to the bomb—plutonium—tipped the balance of scientific and technological choice. Lawrence's Berkeley laboratory and Lawrence's own energies both influenced the decision and give promise of an important resource. The enthusiasm of Lawrence in turn strengthened the faith of Arthur Compton in the possibility of rapid achievement of a fission bomb. Thus the leaders of American science made their most critical decision well before Pearl Harbor. The threat of a bomb at the disposal of Hitler was an especially sobering thought to American scientists, with their historic veneration for German science and research technology. A negative choice not only seemed crucially dangerous; it gave the proponents of going ahead a unique esprit de corps. With the opportunity of exploiting the energy of the nucleus open before them, combined with the threat of Hitler, Lawrence and Compton pushed the decision to go ahead. They had with them the refugees and a swarm of young, bright, able men. The decision not to go ahead had no alternative focus around which all this talent could gather.

The alliance of Bush and Conant with the scientific community which they served had had a severe test in the turning-point decision to go on for a fission bomb. As administrators they might have been tempted to give it only token support. Once the decision was made, they not only accepted its consequences to follow all promising routes to success within the duration, but also started the process of adjusting other choices in the light of the major commitment to the atomic bomb. Further, they recognized that for their research choice to be effective within the duration, they would have to extend their effort not only to the technology of producing fissionable material and a workable bomb, but they would have to create a whole industry and possibly even a military delivery system. Thus they decided even before Pearl Harbor to spin off the whole program from the OSRD-NDRC complex and place it directly under the military, where it took the form of the Manhattan District of the Corps of Army Engineers. Coordination with the OSRD would ultimately be accomplished almost entirely by Bush and Conant's maintaining positions with both organizations. Thus while the OSRD was responsible for the basic decision to go ahead with the fission bomb, it was also responsible for the separation of nuclear science and its dependent technologies into a universe of policy sealed off except at the top from the rest of American science.

A look at the scientific choices made by all the major warring nations shows a distinct similarity between the British and American priorities. The electronic environment and nuclear fission were judged worthy of the bulk of research as the areas most likely to be effective within the duration. Germany, on the other hand, insofar as Hitler had a conscious research policy, put more emphasis on rockets and jet engines than did the Allies. Her radar program was more concerned with countermeasures to Allied radars than with outperforming them. Thus the British and Americans had a tendency to consider one another as rivals. The German research pattern, on the other hand, was complementary to that of the Allies. In the postwar world this pattern would give German-American relations in missilery an attractiveness that British-American relations in atomic energy did not possess.

Mechanism: The Contract

The OSRD, as a nonrevolutionary and limited-objective organization, did not take upon itself the job of making over American science. It did not even take upon itself the job of long-term institution building within the government. Rather, having made some major choices, it attempted to find existing institutions in the universities, in industry, and even in government to do the research, and to bring together the laboratories and the personnel of American science to work on specific problems. The glue which held the whole system together was not the headquarters staff of the agency nor its organization chart, but rather the contracts which it made. One of the great inventions of the NDRC-OSRD was the research contract, and the inventors were not scientists but lawyers.

Oscar Cox, first in the Treasury Department and then in the Office of Emergency Management, had the ingenuity to devise legal forms for the realities of crisis in his design for lend-lease.[29] He was also responsible in a general way for providing legal advice to the NDRC-OSRD and for giving guidance to the general counsel he provided the agency, John T. Connor, and later to Connor's successor, Oscar M. Ruebhausen. The standard procurement contracts in use by the armed forces were designed to purchase goods produced by industry and often put out for bids. The architects of the research contract had an entirely different conception of its function.

If the NDRC-OSRD had stayed in any way with the concept of procurement, it would have been saddled with the requirement that the product to be procured was the physical paper and ink that constituted a report. The requirement of performance would have been the delivery of the paper and ink in the form of a report as the end item of research. Such a concept would have specified a result from the research in the terms of the contract. Instead, the research contract assumed that the end item was research and development itself. Hence the performance of research, not the achievement of a specific result, was the basic requirement. The key phrase in the preamble of the contract was, "Whereas, the Government desires that the Contractor conduct studies and experimental investigations as hereinafter specified requiring the services of qualified personnel; and whereas, the Contractor is willing to conduct such studies and experimental investigations on an 'actual cost' basis . . ."[30]

The sum total of the descriptions in general terms of the projects contracted for was the real description of the scope of the OSRD's operation. Basic to the relationship was the assumption of responsibility for research by an existing institution, whether a business corporation or a university. The scientific personnel would then be in the employ of the contractor and not of the government. The system provided for the mobilization of both institutions and personnel in a flexible way. Both concentration and distribution was possible. Large numbers of scientists could be moved into the MIT Radiation Laboratory from colleges and universities without individual government contracts. Yet where time would be saved by leaving an investigator in place, the research contract went to the investigator's institution.

The whole range of OSRD policy was spelled out in the contract. Classification and security became the responsibility of the contractor by its provision. The patent policy, giving discretion to the OSRD contracting officer as to whether rights would go to the contractor or to the government, was spelled out in the contract. The same contract could be used for corporations and universities, but the no-loss-no-gain principle had a different effect on the two types of institutions because the corporations had accounting departments which could determine the cost of overhead on the contracts, while the universities did not. Hence the determination of indirect costs on research performed by universities required the creation of standards which had not before existed.

One of the great pressures on the OSRD from the military was use of its technical talents for production once a prototype had been achieved. In some cases, where no trained personnel existed in industry to produce radars urgently needed for immediate combat use, for instance, the OSRD reluctantly let itself be used for production, and a procurement contract was available for such cases. Yet Bush resisted such pressure, thereby preserving research and development as a separate area of activity in which the government could act more flexibly than in ordinary procurement. A major consequence of the separation of research and development on the one hand and production on the other was the relatively slight cost of research under the OSRD. In fiscal years 1941 through 1946 its expenditures somewhat exceeded $500,000,000,[31] as compared to the several billion dollars which went into the atomic bomb and radar procurement alone. Indeed, one of the major reasons for the unquestioning flow of money from the Congress was the relatively small amount which research and development cost when separated from production. At the same time the policy of separation was in effect a device for protecting scientific research personnel against possible alternative uses in production which would be reminiscent of the scientist's position in American industry in the late nineteenth century.

In summary, the research contract was the device by which the government tied the other sectors of science support to research on weaponry and medicine, in line with the strategic choices made early in the emergency. Equally important, the contract was the device by which the universities and industrial research laboratories were preserved as institutions even while their social role was temporarily but radically changed. Any solution which brought direct government operation of the laboratories where the OSRD's work was done would have had much more revolutionary effects on American scientific institutions, even if there had been prompt return of facilities at the end of hostilities.

Trends as the War Continued

"In an early meeting of NDRC, [Karl T.] Compton observed that NDRC could either do its job or get credit for doing it, but not both."[32] The director and the executive secretary handled personally the burden of public relations and publicity for the NDRC and OSRD. Congressional relations were hardly more burdensome. Since most of the research was classified, its results were not published in the general scientific literature during the war. As long as the single-minded goal of weapons for this war remained predominant and the duration stretched out ahead, these informal arrangements, unique in both the history of American science and American democracy, sufficed.

Only in the area of scientific manpower did the otherwise advantageous policy have drawbacks, because many draft boards had no appreciation of research as a priority use of trained manpower outside the military services. The talk of a science reserve corps went on throughout the wartime period, but nothing ever came of it in practice, and only in the later stages did manpower loom as the major unsolved problem of the OSRD.

It required a historical project after 1945 to see clearly the major accomplishments of OSRD.[33] In addition to its period of decision-making on the uranium program and its famous participation in the development of radar and the proximity fuse, the OSRD presided over a vast array of individual projects. Just a list of the divisions indicates something of the scope. In NDRC the titles of the divisions were: ballistic research, effect of impact and explosion, rocket ordnance, ordnance accessories, new missiles, subsurface warfare, fire control, chemical engineering, transportation development, electrical communication, radar, radio coordination, optics, physics, war metallurgy, and miscellaneous weapons. Applied mathematics and applied psychology accomplished much as panels assisting all the divisions and the services. The titles of the divisions of the Committee on Medical Research were: medicine, surgery, aviation medicine, physiology, chemistry, and malaria. Even allowing for a certain number of projects which proved unproductive, the record of the OSRD was, given its mission, impressive compared to both earlier American efforts and those of other nations.

In general the organization, personnel, and policies of the OSRD were remarkably stable throughout the war, and the leaders were able to accomplish their swift results in part because they were relieved from the endless justifications required by public scrutiny. All of these forces had been reconciled with the American democratic tradition by Bush's determined stand that the OSRD was a temporary agency which would go out of existence at the end of hostilities. As long as the goal of the end of the war in Europe seemed far off, the OSRD was a smoothly working organization which dominated American science by its élan rather than by discipline or comprehensiveness.

Indeed, the speed which the OSRD gained by refusing to do all possible jobs open to scientists, by making scientific choices, by refusing to take comprehensive responsibility for long-range institution building or fundamental reforms in society, by tapping only those scientists who would be of immediate use on specific projects, by making unsparing judgments of the quality of individuals—this speed was purchased with the creation of an almost automatic opposition to OSRD. The opposition consisted of three strains.

The first strain was made up of those scientists who were left out of the war effort and could not understand why. Among these were geologists and biologists whose skills were in less demand, and few of whom were known personally to the major leaders. Prominent among this group were the straight engineers, who might have been used more extensively as the swing from research to development and production gathered force. But the scientists, who had become familiar with the military problems, simply shifted toward the applied end of the spectrum with their research.

The second strain of opposition came from those concerned about the quality of the American production effort and the seeming indifference of Bush and his colleagues to the problem of research in war industry. The NDRC had hardly come into existence when Secretary of the Interior Harold L. Ickes, acting on a plan volunteered to him by Lyman Chalkley, had tried to bring about a comprehensive mobilization of scientists in behalf of industry within the permanent framework of government.[34] From another point of view, the Lea bill, introduced into Congress in 1941, had tried to create a system of general government support for science in order "to promote industry and commerce through research in the physical sciences."[35]

The War Production Board (WPB), in contrast to the OSRD, came into existence only in response to Pearl Harbor and the great surge of national determination to overcome the shortages of matériel which seemed so huge in early 1942. Maury Maverick of the WPB was one who tried to stimulate a strong link between scientific resources and industry within the framework of the WPB. Bush evaded these attempts to organize science on a basis other than the OSRD in order to save the resources he wanted for the projects he chose to undertake.

The third strain of opposition to the OSRD came from those who wished to use the war as an occasion for reforming the relation of science to society. Harry Grundfest, secretary of the American Association of Scientific Workers, was an effective spokesman for this point of view, as was Waldemar Kaempffert, science editor of the *New York Times*. Kaempffert's interest in increased planning for scientific research led him to disagree with Bush's determination that the OSRD go out of existence at the end of the duration. Its permanence would make a fundamental change in the whole way of doing business in American science.

As long as the war continued, Bush easily contained all the strains of opposition, and the OSRD became a way of life for so many American scientists that they could not conceive of ever going back to the style of life and the scale of research of the 1930's again. Yet as the months and years of intense effort began to pay off, the tension on the leaders of the OSRD actually began to increase, because their design of war research became less well adapted to the problems of reconversion. Not only did Bush show every intention of carrying out his desire to dismantle the OSRD at the end of the war; the scientists who made up both the staff and the research teams seemed to agree with him. They wanted to go back to the universities if not to their old penury. But who was to decide when demobilization should begin? Should there be a slackening at the end of the European war? Or should the research effort carry on until the Japanese war was over as well?

The whole problem of the shape of postwar science was to be determined not only by the accomplishments of the OSRD but by the manner of its withdrawal as well. By 1944, as the number of new research projects leveled off, the leaders of the OSRD had to think increasingly of the world beyond "the duration," a world for which their brilliant improvisations gave them few guidelines.

As the United States emerged from the war and the general lines of the reorganization of science became dimly perceptible to the world outside the OSRD and the Manhattan Project, the first impression was that after three hundred years Bacon's House of Salomon had become a reality. An organization which put science into an effective partnership of power and knowledge effected the relief of mankind's estate. Both in securing support from the general society and bequeathing the fruits of research to it, the system of science support which emerged after the war met all the Baconian specifications, especially in the direction of organizing large-scale research carried out by near-anonymous workers.

At the same time, however, the scientists who emerged from Oak Ridge and the Radiation Laboratory at MIT and other research sites were fired anew with the desire to be independent individualists choosing their own problems and following the internal imperatives of their research no matter whether it led to even more esoteric reaches of abstraction. Thus the new system of science support bequeathed by the OSRD gave resources and vindication to the heirs of Descartes as much as to those of Bacon. If team research has come of age as a concept, so has that of the lone individual. If the application of science to the arts of war and peace has become institutionalized in both government and society, so has basic research.

The fate of the OSRD in postwar America was determined not only by its formally going out of existence but also by its survival. The mechanism of the contract and the OSRD way of doing business lived on in weapons research, in medicine, and, in addition, as support for basic research in science itself. Basic research had been no part of the wartime operation, but by 1944 Bush was already at work through committees on the report which became *Science—The Endless Frontier*. When the National Research Foundation that the report proposed failed to materialize until 1950, other agencies took up the burden—the National Institutes of Health for medicine and the Office of Naval Research for basic unclassified research in the universities. Indeed, the insistence that the government freely support pure research in the universities became an article of faith with those scientists who benefited from the contracts and built within their own institutions the unprecedentedly strong research programs which made the 1950's and 1960's seem to many brilliant decades in the history of science.

At the same time the OSRD system was operating with contracts between the government—the Department of Defense and the Atomic Energy Commission—and industrial firms which specialized in performing research in aerospace and electronic fields. Thus the new system was interrelated in the sense that it tied together all the sectors of science support—the government, the universities, industry, and the private foundations. The new system was also plural in the sense that several different agencies in several departments of the government participated in supporting research in many institutions in all the sectors. After the National Science Foundation had come into existence and the army and air force were supporting research in universities through offices modeled on the Office of Naval Research, it was easy to forget that the basic research system was the sibling of the military research program of contracts to industry.

At the very end of his administration President Dwight D. Eisenhower warned the nation against the military-industrial complex. The warning may have originated from anxiety about the segment of industry and education not dominated by the interrelated system. Since it could hardly at the time have indicated the conversion of President Eisenhower to a New Left position, the only real puzzle in the farewell address was: Where did it leave the scientific community, who had stayed with the interrelated system for twenty years without a break?

George Kistiakowsky, as Eisenhower's science adviser, lost no time in addressing himself to the problem, presumably calling Eisenhower's attention to the fact that, however inadvertently, the President had hit the scientists hard, and to expect them to remain in harness with such rewards was unrealistic.

In the farewell address on January 17, 1961, President Eisenhower had said:

> Today, the solitary inventor, tinkering in his shop, has been overshadowed by task forces of scientists in laboratories and testing fields. In the same fashion,

the free university, historically the fountainhead of free ideas and scientific discovery, has experienced a revolution in the conduct of research. Partly because of the huge costs involved, a government contract becomes virtually a substitute for intellectual curiosity. For every old blackboard there are now hundreds of electric computers.

The prospect of domination of the nation's scholars by federal employment, project allocations, and the power of money is ever present—and is gravely to be regarded.

Yet, in holding scientific research and discovery in respect, as we should, we must also be alert to the equal and opposite danger that public policy could itself become the captive of the scientific-technological elite.[36]

A news story, which attributed the authorship of the farewell address to Malcolm Moos, called it Eisenhower's "most memorable utterance as President."[37] No wonder Kistiakowsky wanted to know what the farewell address meant. *Science* magazine on February 10, 1961, carried his statement.

I would like to comment briefly about President Eisenhower's reference to science and technology in his farewell address. Several questions have been directed to me about it, and since Mr. Eisenhower talked to me at some length later in the week, others may be interested to know more about his views than could be developed in a short talk.

The major point, I believe, which he wanted to convey was his conviction that the part of science which is engaged in research for armaments purposes must never be allowed to dominate all of science or curtail basic research. He was concerned to see so many pages of advertisements identifying "science" with armaments, asserting to the people that research means just bigger and better missiles, etc., while very little is said about the true nature of basic research as a cultural endeavor and a source of advancing welfare to the people. And he was particularly anxious that educational institutions, whose task he sees as the support of free intellectual inquiry and the acquisition of new scientific knowledge, should not concentrate on large-scale military research and development contracts at the expense of their true scientific endeavors.[38]

For the 1960's the subtle balance as sketched out by Kistiakowsky involved recognition of the scientific community's commitment to the national security goal of the government. The threat to entrap the universities into short-run military missions, however, was to be kept in check not by an established ceiling on the military component, but by using the nation's response to the missile age as a lever for increased basic research.

What had begun as a perpetuation of the ancient traditional models of scientific organization associated with the names of Bacon and Descartes had now become a single tightly-bound system which comprised both large-scale military research and the most abstract of university projects. The loyalties which made the OSRD generation feel that the combination was a natural one wore thin in the late 1960's. In addition, the leaders who had emerged in 1940 and had shaped the scientific community's response to the government's call during the quarter-century after World War II were retiring. The younger scientists were buffeted by emotions which came from issues seemingly very different from those of 1940, and the response even of those within the policy circles was very different. The retirement of Lee du Bridge as President Nixon's science adviser in the summer of 1970 marked the end of an era in American science policy—the era of the OSRD. Du Bridge had been director of the Radiation Laboratory at MIT and was the last of those who had borne high responsibility under Bush to hold a significant post in the making of science policy.

The passing of the era of the OSRD does not mean that its legacy is no longer being felt. Rather, it has now spread so far that the tradition of the *Great Instauration* of 1940 is now no longer able to guide either the government or the scientific community in adjusting to changing problems.

REFERENCES

1. Linus Pauling, "Fifty Years of Progress in Structural Chemistry and Molecular Biology," *Dædalus* (Fall 1970), p. 1011. See p. 281, above.

2. The autobiographical literature has been enriched by the appearance of James B. Conant, *My Several Lives: Memoirs of a Social Inventor* (New York: Harper and Row, 1970) and Vannevar Bush, *Pieces of the Action* (New York: Morrow, 1970).

3. A. Hunter Dupree, *Science in the Federal Government: A History of Policies and Activities to 1940* (Cambridge, Mass.: Harvard University Press, 1957), chaps. 14 and 15.

4. *Ibid.*, chap. 16.

5. Irvin Stewart, *Organizing Scientific Research for War: The Administrative History of the Office of Scientific Research and Development* (Boston: Little, Brown, 1948), pp. 5-6.

6. Joseph Haberer, *Politics and the Community of Science* (New York: Van Nostrand Reinhold, 1969), pp. 15-78.

7. *Ibid.*, pp. 31-36.

8. Moody E. Prior, "Bacon's Man of Science," in Philip P. Wiener and Aaron Noland, eds., *Roots of Scientific Thought: A Cultural Perspective* (New York: Basic Books, 1957), pp. 382-389.

9. George H. Daniels, *Science in the Age of Jackson* (New York: Columbia University Press, 1968) develops the idea of the Baconian philosophy as characteristic of American science in detail.

10. Haberer, *Politics and the Community of Science*, pp. 60–73.

11. *Ibid.*, p. 72. The same analogy is strongly present in Michael Polanyi, "The Republic of Science: Its Political and Economic Theory," *Minerva*, 1 (1962), 55-73.

12. See D. F. Fleming and Bernard Bailyn, eds., *The Intellectual Migration: Europe and America, 1930-1960* (Cambridge, Mass.: Harvard University Press, 1969). For my analysis of that work see A. H. Dupree, "The Coming of the Refugees," *Science,* 166 (December 19, 1969), 1495-1497.

13. Vannevar Bush, *Modern Arms and Free Men: A Discussion of the Role of Science in Preserving Democracy* (New York: Simon and Schuster, 1949), p. 92.

14. An account of part of Conant's life, shorter and more tentative than a biography but much more extended than *Who's Who in America* or similar sources, appears in Paul F. Douglass, *Six Upon the World: Toward an American Culture for an Industrial Age* (Boston: Little, Brown, 1954), pp. 325–409.

15. *Ibid.,* pp. 367, 371-372.

16. *Ibid.,* p. 372.

17. James Phinney Baxter, *Scientists against Time* (Boston: Little, Brown, 1946), p. 14.

18. Draft Memorandum, n.d., OSRD papers, National Archives Record Group 227, Central Classified File, Organization, Washington, D. C.

19. *Ibid.*

20. F. B. Jewett to V. Bush, June 6, 1940, OSRD papers, NARG 227, Office file of F. B. Jewett, folder 49.01. See also Robert E. Sherwood, *Roosevelt and Hopkins: An Intimate History* (New York: Harper, 1948), pp. 153-154.

21. Baxter, *Scientists against Time,* p. 16.

22. F. D. Roosevelt to V. Bush, June 15, 1940, File OF4010, Franklin D. Roosevelt papers, Hyde Park, N. Y.

23. H. A. Wallace to V. Bush, OSRD papers, NARG 227, Central Classified File, Organization (July 1940), filed July 11, 1940.

24. Alvin Weinberg, "Criteria for Scientific Choice," *Minerva,* 1 (1963), 159-171.

25. See, for example, H. H. Arnold, *Global Mission* (New York: Harper, 1949), pp. 260-261.

26. Jerome C. Hunsaker, "Forty Years of Aeronautical Research," *Annual Report of the Board of Regents of the Smithsonian Institution, Showing the Operations, Expenditures, and Condition of the Institution for 1955* (Washington, D. C., 1956), p. 263.

27. Jerome C. Hunsaker, *Hearings* before the U.S. Senate Special Committee Investigating the National Defense Program, 79th Congress, 2d Session, Pt. 33 (February 27, 1946), p. 16821.

28. University Park, Pa.: Pennsylvania State University Press, 1962.

29. Edward R. Stettinius, *Lend-Lease: Weapon for Victory* (New York: Penguin Books, 1944), pp. 74-77.

30. Irvin Stewart, *Organizing Scientific Research for War: The Administrative History of the Office of Scientific Research and Development* (Boston: Little, Brown, 1948), p. 339.

31. *Ibid.*, p. 322.

32. *Ibid.*, p. 285.

33. Baxter, *Scientists against Time.*

34. Memorandum for the President, re: A proposed Office of Scientific Liaison, from Harold L. Ickes, August 19, 1940, File OF2240, Franklin D. Roosevelt papers, Hyde Park, N. Y.

35. H.R. 3366, 77th Congress, 1st Session, introduced February 17, 1941.

36. Quoted in [Graham DuShane], "Footnote to History," *Science,* 133 (February 10, 1961), p. 355.

37. Robert J. Donovan, copyrighted story of Los Angeles Times News Service, *Providence Journal,* March 31, 1969.

38. [DuShane], "Footnote to History," p. 355.

R. R. WILSON

My Fight Against Team Research

TEAM RESEARCH! Most of us have a preconditioned response to these words. We have a suspicion that team research is superficial, uncreative, and dull; that it is overorganized and overfinanced. But research done by individuals receives a different response. Individual research evokes a happy image of dedicated and inspired scientists in white coats (not grey flannels). These men are doing creative, poetic, and enduring work —true intellectuals they, not bureaucrats enslaved by a computer. Team research, the cliché tells us, is bad; individual research is good.

I have come to think differently. As a young man, I accepted the cliché, and I worked hard to attain that exalted image of scientific purity—the lone scientist in pursuit of truth. But my search for truth led me deep into the nucleus of the atom, and it is almost as hard to reach the nucleus by oneself as it is to get to the moon by oneself. To reach the moon, one must join a large team, and to reach the nucleus, one must also use the help of others. I have resisted joining a team, but in the end, I have succumbed. It would seem as hard for a nuclear physicist to attain the nirvana of independent research as for a rich man to pass through the eye of a camel. Now, in my later years, I find myself the director of a large government laboratory, the compleat (team) organization man—and even enjoying it.

"I am a part of all that I have met," sings Tennyson's Ulysses. In my own time I have seen physics in the United States change from a parochial establishment to something of world-wide significance. I have seen the attitude of the government toward science change from indifference to a concern which manifests itself in billions of dollars. It is my fortune to have been a witness—and even something of a participant—in the growth industry, physics. Perhaps mine is a typical history of a mid-twentieth-century adventurer in the field, confessing ambivalence, prejudice, the pressures of expediency—vacillating between the me and the us.

As a youth, I read *Arrowsmith* by Sinclair Lewis. That romantic idealization of a man dedicated to research made a deep impression upon me. As I remember it, Lewis portrays in his novel a young and devoted medical scientist who, working long hours in solitude, finally experiences the ultimate exaltation that comes to any creative person. I was a lonely boy in Wyoming, and for some reason it was natural for me to relate to Lewis' hero. At that time of my life my job, when I wasn't in school, was riding the range for cattle. It, too, was an isolated life, but one of delightful independence. It was necessary to live by myself in a little cabin, do my cooking, shoe my horse, repair broken equipment. I loved that isolated life, and felt in some way that I had something in common with and could empathize with Arrowsmith on his lonely research frontier. Perhaps the common element, if indeed there was one at all, might have been a passionate love of nature. I loved my section of Wyoming and studied, unsystematically, everything I could about its mountains and deserts, its flora and fauna, its legends. The next step to more basic knowledge would have to be scientific in character.

It was experimental physics, rather than medicine, that attracted me—perhaps because contriving instruments in a laboratory is not so different from contriving farm equipment in the blacksmith shop of a ranch. Also, in elementary school, I had heard about atoms, about the atomic theory, and I was tremendously thrilled by such a simple explanation of matter. My real introduction to the mysteries of atoms, or rather I should say of physics, philosophy, and mathematics, occurred at the University of California, where I somehow managed to enroll. Physics is not easy, and soon I found myself completely immersed in study. Because of the intensity of that study, my life at the university became almost as isolated as was my life in Wyoming.

Some time during my junior year, I began a life of research by becoming apprenticed to Ernest Lawrence. Lawrence had developed, with Jesse Beams, a method of using an electro-optical shutter to measure the very short interval of time between the application of a voltage to a spark gap and the occurrence of a spark. Of course, this was just the time that Lawrence had first demonstrated the feasibility of his cyclotron invention, and he was completely involved with its further development. He had turned his spark research over to Harry White, and he assigned me to assist White. Harry White turned out to be a good friend, and it was he who initiated me into many of the fine points of experimental techniques. At the end of the year he left the University to go into industry. This was a lucky break for me. I was now in the best of all possible positions for a student—I had a good problem, had fallen heir to a wealth of experimental equipment, and was entirely on my own. Because Lawrence was so completely occupied with his new cyclotron, it was necessary for me to formulate my own research plans. I saw my professor only a few times during that year. My first taste of success came at the end of my senior year, when I developed a new method for making the measurement. It was my fortune to get some rather unexpected results, and then I was able to concoct a theory which explained these results and at the same time provided a little insight into the mechanisms of the spark discharge process. I had worked pretty hard to do this, and the experience in my then romantic view seemed to parallel that of my hero, Arrowsmith. I continued to have in my laboratory much the same kind of independence that I had valued so much in Wyoming.

At the end of the year, the work was published, and I had to choose whether to continue with research on gaseous discharges or to go on to something else. I had studied enough by then to know that as far as I was concerned the fundamental problems of gaseous discharge were essentially solved. I wanted to move on to what I regarded as being the basic problem of physics, the nucleus of the atom, which was then virtually unexplored. At the University of California, in order to study nuclear physics, one had to be accepted by Lawrence to work in his Radiation Laboratory. I applied and was accepted.

Lawrence had already assembled a team of outstanding physicists. It has been suggested that team research in physics originated in that laboratory at that time. Perhaps that is true for research in this country, but ten years earlier, Ernest Rutherford, and ten years before that J. J. Thomson, had gathered comparable, if not larger, groups of scientists at the Cavendish Laboratory in England. I hesitate, however, to equate Cambridge University with team research since the Cavendish is noted for the individuality and independence of its researchers. One might better search for the genesis of team research in the cooperative efforts of the astronomers to build large telescopes, or in the large commercial laboratories of General Electric, Westinghouse, or Bell Telephone.

Not long ago, Eduardo Amaldi told me about some very early team research. A group was formed in Florence just after Galileo, but before Newton, and worked in the Pitti Palace under the auspices of Prince Leopold Medici. Borelli and Toricelli were members of the team. What is extraordinary is that they published as a group, without any of their names appearing, the only name being that of the Academy of Florence. Thus they established a kind of anonymity that has only recently been achieved by modern scientists where the individual's name is lost in the long list of authors appearing on their articles. Having worked in Italy, I find it hard to believe that this was really the beginning of organized research as we know it now. My confidence in Italians was restored on further learning that the whole business came to a climactic end by the individuals writing up and then signing their own work. Of course, now we know of Borelli's and Toricelli's work as their own and not as a group effort.

To return to my story, I did become a full-fledged member of a laboratory notable for the kind of effort which has since become known as team research. My first response to working with a group of such outstanding physicists was enthusiastic, but it was not long before my enthusiasm was somewhat dampened because my own methods came into conflict with the procedures already traditional in the laboratory. As the youngest member of the laboratory, it was I who had to yield and to watch the team go its way rather than my way. I soon learned that the real Arrowsmith in the Radiation Laboratory was Ernest Orlando Lawrence. It was he who was independent and creative. In some degree, the members of the team just carried out his ideas. However, by simply shifting my working hours from day to night I discovered that I could be in the laboratory when most of the team (as well as the leader) was away and hence I could have the best of both worlds. In the late hours of night, I found that I could go my own way pretty much as I pleased, and yet with nearly all of the facilities of a great laboratory at my disposal.

Team research at the Radiation Laboratory in those days was really not as lock-step as I have implied. In fact, it meant that the members of the laboratory cooperated in the construction of the cyclotron according to plans laid out by Lawrence. However, research utilizing the cyclotron was a different thing. Ernest had a passion and a drive to make cyclotrons that was marvelous, but his interest and participation in the research carried out with the cyclotron, though sympathetic and encouraging, was not as overpowering. For example, as a thesis problem, I decided that I wanted to study the scattering of protons by protons. When I mentioned this to Lawrence, he said, "Fine, fine, most interesting," but as far as I could recall, that was the last of his participation as my research professor in the experiment, except occasionally to express encouragement. Even at that time, 1938-1940, it was necessary for me to construct all of the equipment for the experiment, mechanical as well as electronic. (We got electronic parts in those days by taking old radios apart.)

On the other hand, I did have responsibilities to the laboratory. These consisted of having to operate the cyclotron when we were making radioactive iron for George Whipple's experiments at Rochester, and in helping to make improvements upon the cyclotron. I regarded that as the way I made a living, although actually my stipend was given to me for being a teaching assistant. (Eventually I got a scholarship.) Most of the other research men in the laboratory had a similar situation. Although research collaborations of two or three people were formed spontaneously for carrying out the actual research, nevertheless, we all had a very real sense of identification with the Radiation Laboratory.

We knew that the work was important. The sense of history being made permeated the place. There were also the intense, if informal, discussions with other members of the laboratory by which the younger students learned from the older students as well as from the faculty. Important, too, was a rather formal Journal Club which met every Monday night. On these evenings, theorists and experimentalists would discuss the research going on in the lab as well as the research which was being done elsewhere in the world.

The Journal Club played an important part in the process of research and education. Oppenheimer and Lawrence as well as other members of the faculty of the University of California were very much "on" during these seminars, partly, I suppose, for the benefit of graduate students, but in a deeper sense because it was at these meetings that a feeling of coherence and meaning was given to the research. I have observed similar seminars at other universities, and I am sure it represents a universal pattern. In one way, at these research seminars, the whole department acts as a team. I doubt that this corresponds to any of the usual connotations of team research, for in this basic sense, every scientist who has ever done research is a member of the "big team."

When my thesis was finished in 1940, I had to choose between staying at the Radiation Laboratory as a research fellow or going to Princeton University as an instructor. In urging me to stay on, Lawrence outlined to me his dream of the future development of the Radiation Laboratory. He expressed confidence that the laboratory would become much larger and that, in his words, it would become "like an astronomical observatory"—a place where millions of dollars would be spent on research every year instead of the thousands then being spent, and a place where there would be adequate technical help for the researchers as well. There is no doubt that Lawrence had team research in mind, and exactly the kind of organized research that eventually evolved at the Radiation Laboratory.

As opposed to all this was research at Princeton. Harry Smyth, who represented a completely different tradition, offered me an instructorship there. He had worked in Rutherford's Cavendish Laboratory, and although a cyclotron had already been built at Princeton by Milton White, the research procedures were in more traditional and individualistic patterns. Given the choice, I found it very easy to decide to go to Princeton. In retrospect, I just hope that it was not the magnificent salary of $2,000 a year that was offered to me at Princeton, compared to the less munificent $1,800 at Berkeley, which affected my decision.

Princeton was everything Harry Smyth had painted it to be. I was able to continue to study the scattering of protons by protons almost as though I were a lone worker in the laboratory. It was also true that I had to continue to do most of the work with my own two hands. This would not have been true at Berkeley had I stayed there. Again, the group of a few nuclear physicists at Princeton acted as a team in keeping the cyclotron together and running, but tended to do their research quite independently of each other.

All of this changed with the war. I was soon caught up in a collaborative effort between Columbia and Princeton that was directed toward the realization of a self-sustaining nuclear reaction. This was intrinsically team research from the beginning. A group of nuclear physicists, dominated by the force of Enrico Fermi's genius, did essentially what he wanted them to do. I do not remember Fermi ever issuing orders about exactly what should be done or specifying who should do it. In a formal sense, he was not even in charge of the work. However, because of his brilliant theoretical analyses of our problems it simply seemed to be self-evident what was to be done next and who was to do it. Apart from Fermi himself, a motivating and unifying force was provided by the urgency and necessity of winning World War II. Although we were not yet in the war, the Battle of Britain was on, and it became clear to me that individualism, if it were to survive at all, had to be subjected to the common good.

Although, even at the beginning, the project to develop nuclear energy had to take on the form of organized research, it was in fact pretty far from the present organization of research teams. We at Princeton would come together about once a week for a consultation with Fermi, who was at Columbia; then we would go our own happy ways for the rest of the week and do what might occur to us in the light of what had happened at that meeting. I remember getting the idea, after some very rough measurements, that the fission process would show a resonancelike behavior, that is, that it would occur more strongly when induced by neutrons having certain energies than at other energies. This was contrary to the prevalent theory, according to which fission should vary smoothly with the neutron energy. Although few of my colleagues were at all convinced by my analysis, it seemed strange neither to them nor me that I should go off on a tangent to try to prove the hypothesis. In this I was not successful (although later my supposition turned out to be true), but in the floundering around I invented a new technique that was designed to decide the question of resonance fission. The idea was that, instead of bombarding uranium atoms with neutrons of variable energy (for at that time we could not vary the energy of the neutrons), I proposed to turn the method around and to produce a beam of uranium ions whose energy could be changed because of their electric charge, and then to bombard very slow neutrons, that is, neutrons at rest, with these ions—an exactly equivalent process but just reversed. In retrospect, it was a lousy idea, but just at that time, Franz E. Simon visited Princeton and reported on some new fission measurements at Cambridge which showed that just a few grams of U^{235} would make a bomb. Well, my beam of uranium atoms was to be intense enough to deposit a few grams of U^{235} if only I could figure out how to separate the U^{235} from the U^{238}. An idea, based on the differing time of flight of U^{235} and U^{238}, came to me as a revelation a few nights later. For several weeks following this, I became obsessed with the thought that I personally was going to make a nuclear bomb which would win the war—this from a man who less than a year earlier had been a dedicated pacifist. I did not go to Chicago with the team, but remained at Princeton to try to realize the new invention.

It soon became apparent to me that I could not develop the invention, the isotron, by myself. It would be necessary to gather a group —a team if you will—and then to direct them in this grim task. Without a moment's hesitation, for whatever motives, I abandoned all thought of being a peaceful lone researcher, and gave myself over to forming a new team. Harry Smyth, who had brought me to Princeton so that I could remain an individual physicist, was as anxious as I to abandon the Princeton traditions of individualism. He was my mentor, colleague, boss, and co-conspirator in all this. It was a big change for me; I soon learned that being a director of some twenty or so physicists involved much more than physics. Raising money, getting people to come, finding places for them to live, gas rationing, draft exemptions, spending a good part of a million dollars through the archaic business office of Princeton, building a new building, marital difficulties of the staff, and on and on. In retrospect, I realize that I accepted all of the problems as matter of factly as I would have accepted any difficulty in an experiment. An experimentalist learns that "A physicist does what has to be done," and then he approaches any problem with the sublime arrogance that he can do anything. The hardest part for me now is to confess that I was soon enjoying this new kind of life. I also found to my surprise that I was developing an intense loyalty to my happy group of colleagues and that there were elements of this that had nothing to do with physics.

It turned out that we at Princeton were in competition with a group of physicists under Lawrence at the Radiation Laboratory at Berkeley who were trying to separate U^{235} from U^{238} by the use of a magnetic method known as the Calutron. The competition grew in intensity, but before long Lawrence wanted to close us down. We resisted with considerable warmth. Ernest wanted to cannibalize our group, our equipment, and our money to make the Calutron process go faster. (I now think he was right.) We resorted to every device of politics and rhetoric to forestall the take-over. We fought like tigers (Princeton Tigers, of course), but eventually it became evident that we were outgunned on all fronts. The last bitter blow was that the Cambridge University measurements turned out to be wrong—our method couldn't produce nearly enough U^{235} to make a bomb.

We became then what I suppose is the worst of all possible things, a research team without a problem, a group with lots of spirit and technique, but nothing to do. Like a bunch of professional soldiers, we signed up, en masse, to go to Los Alamos which was just then being formed. I justified my enthusiasm for all that team effort at the time because of the necessities of war, but friendship and loyalty were important factors. Power and excitement were also ingredients making my loss of individualism more palatable.

At the time I was about twenty-eight years old, which does not now seem very old to be a research director. I remember some of my managerial techniques. I had, of course, closely observed Lawrence and Fermi, both of whom I had studied under and both of whom I had idolized. In part, they could lead because they were established personalities. Not being an established personality, I could not and did not ever even try to give a direct order to a colleague. Instead, I tried to emulate Fermi and Lawrence who, by the intensity and clarity of their thinking, automatically provided leadership. Of course, I wouldn't compare myself to either of them, but I found that a group of capable physicists could supplement my own abilities, and that if I listened carefully to them, my own ideas as well as their ideas and objectives could be clarified and improved. Thus, in some measure, I followed my team, but made sure I knew whither I was following them. I also found out that a research director has a big advantage. Because he is in a position to know everything about a project, and because most of the decisions must be made by instinct, the director tends to give an impression of infallibility. This impression is fortified by the fact that in most matters it is the director who decides what is fallible. Most important of all (except for plain hard work) is a quality which allows one to make an arbitrary decision rapidly. I suspect that a fair degree of arrogrance is involved in this. One soon finds that a bad decision is better than no decision, for even a bad decision is a basis of action and eventually it can be corrected. In his relationship to his team, the director ought genuinely to respect them, and thus to make decisions in a manner that adds to their mutual respect. Above all, there is nothing like success to weld a team together.

So much has been written about Los Alamos that I doubt that I can add very much to it. I don't remember any particular innovations in team research that were made there. The organization into teams followed the organization worked out at the radar Radiation Laboratory at Massachusetts Institute of Technology which, along with its name, assumed the style that had earlier been developed at Berkeley. The difference was that at Berkeley there was but one team centered about Lawrence while at the Radiation Laboratory at MIT, and later at Los Alamos, there were dozens of research teams centering about various personalities. I believe that Rabi, working within the framework of a steering committee, was tremendously influential at MIT. At Los Alamos, Oppenheimer used the committee as a weapon. On the other hand, I doubt that Lawrence was even particularly aware of committees in his laboratory.

There is little doubt that by the end of the war American physicists had learned how to do team research very effectively. But most of us wanted to return to university life, and we were divided as to how to proceed with the research there. We were agreed that basic research, rather than the applied kind of physics we had been doing during the war, was the only kind appropriate for a university. In basic research most of the battle lies in delineating a question or specifying a goal for the research. This process normally involves a single individual, or at most a few, working on an intuitive level. The physicists who believed in the individualistic approach generally returned to the universities where there had not been large war projects. They looked forward to a quiet academic life with a nice balance between research and teaching. Others foresaw a different future at the universities. They had liked their wartime experience with team research, and they saw it as a wave of the future. They saw that a strong team of research individuals chosen to complement one another would attract money from the government, could provide mutual inspiration, and might do for basic research what team research had done for practical problems of radar and the atomic bomb. The Enrico Fermi Laboratory in Chicago grew naturally out of the old Metallurgical Laboratory and represented an exceedingly strong group of physicists, chemists, and metallurgists. Similar teams were set up at MIT and at Columbia. Of course the Radiation Laboratory at Berkeley, greatly augmented during the war, epitomized team research at a university. A word was even coined—Berkeleitis—to describe the syndrome that existed there. A third group of physicists, in some cases for reasons of simple patriotism, in other cases because of a social point of view, chose to stay in the already-established laboratories such as Los Alamos or the Lincoln Laboratory where the team approach was to be used for further weapons development or to develop peaceful applications of techniques that grew out of weapons research.

As the war was coming to a close, I decided to pull myself together, to go straight, to eschew the blandishments of team research. Harvard was the sanctuary to which I would fly, there to join with Ken Bainbridge in bringing back the Harvard cyclotron that had been transported to Los Alamos (by my Princeton group, incidentally) during the war. We hoped in this way that we could get on immediately with fascinating, and fundamental, problems of the nucleus. In particular, I wanted to return to my study of the scattering of protons by protons that had been interrupted by the war. We were stymied from the start. The physicists at Los Alamos had a continuing need for the Harvard cyclotron, and it would have been a severe blow to them had it been returned to Harvard, even though such a bargain had been made when it was taken from Harvard. Furthermore, General Leslie Groves was more than generous in offering recompense for allowing it to stay at Los Alamos. Then, too, Lawrence, with the same enthusiasm he brought to anything pertaining to cyclotrons, promised to give us the full use of his laboratory in building a cyclotron at Harvard "ten times" larger. Ken Bainbridge and I had spent five long years as professional patriots. We would be patriotic a few more years. We resignedly left the cyclotron at Los Alamos and accepted the riches "beyond all belief" from the government. For a happy year, I had the best of two worlds—being a Harvard professor, but living in Berkeley while the new cyclotron was being designed there. What was to have been private and very personal research blossomed into a sizable project of team research. Then, very shortly after arriving in Cambridge, I was offered the directorship of the newly formed Laboratory of Nuclear Studies at Cornell University and, deciding that I might as well be hanged for a sheep as a lamb, I accepted the job.

There I tasted research at its best as far as I was concerned. We were a small, compatible group of theorists and experimentalists who had worked together during World War II, and we were to spend some twenty years in a close and productive collaboration. The fact is, we worked out an interesting mixture of team effort which also allowed for a large amount of individual research. In general, we collaborated as a group in the construction and operation of the large accelerator, but each faculty member worked on his own individual research surrounded of course by a few graduate students. I remember with great pleasure my almost nonexistent duties as the administrator of the lab. In the end, we had worked so long together and so intimately, and we all so complemented one another, that everyone did without being asked what was necessary to be done. An informal word or two at casual meetings of a few of the staff in the hallway was sufficient for most of the administrative decisions. Even so, occasionally I felt victimized by team research, and continued to fight the system. At one time, I remember feeling so fed up with cooperating with other people, however angelic, that I withdrew to a corner of the laboratory for two years while I tried to fathom the mysteries of potential scattering—a phenomenon in which the electric field of the nucleus causes a gamma ray to be deflected. It is closely related to the scattering of one photon by another photon and it is a problem that is likely to fascinate any physicist. In fact, it should interest anyone who has observed a starry winter sky in all its clarity and beauty. When one considers that the light from those stars has traveled so far in reaching the eye and that, in doing this, it has had to pass through a tremendous amount of light from other stars on its way, he should be amazed that the stars still appear so sharp—that the light has not bumped into and so been deflected by the myriads of other photons through which it must pass and which would cause the image of the stars to be blurred. There is some ambiguity as to whether or not my measurements were the first to detect potential scattering, but I do know that I was thoroughly gratified both by my theoretical contemplations and my experimental investigations. That the subject was also close to the heart of quantum electrodynamics, just then being developed, did not detract from its interest. Incidentally, I could do all of the work entirely by myself and that was not the least attractive aspect of it.

I was not all that pure of heart, though, for right in the midst of the work, an idea occurred to me as to how the original 300 MeV electron synchrotron at Cornell might be converted without much cost or effort so that it would become a synchrotron that could give over a billion electron volts. Without further thoughts about doing a really thorough job on the potential scattering, I threw myself completely and with abandon again into building an accelerator—an activity that epitomizes team research at its worst.

Pure physicists tend to look down their noses at accelerator builders. From the lofty point of higher mathematics, the man who builds an accelerator is some kind of glorified technician who is building an instrument that will be used later on by someone else more qualified in its application to research. I don't accept this position for the accelerator builders. The contribution that anyone, unless he happens to be an Einstein or Newton, makes by his research is bound to be a tiny fragment of the whole body of understanding. In this sense, almost anyone who contributes to science can be called a mere technician. In the pre-World War II days at the Radiation Laboratory, one did not distinguish research physics as separate from accelerator building. Everyone involved was a research physicist and everyone was engaged in both activities at the same time. Building the accelerator is the first half of the experiments. It is true that accelerator building is a team effort, and many who build accelerators do tend to specialize on that end of the experiments. Many physicists, and I count myself among them, alternate their activities—sometimes building, sometimes experimenting. Thus there are long periods during which I have devoted myself entirely to machine construction. But with the completion of the construction phase, a new period starts. Thus it was that when the 1000 MeV Cornell synchrotron was finished, a thoroughly team effort, I interested myself in scattering of electrons by protons. Under the excuse of a shortage of personnel at Cornell, I started that experiment as an independent effort of my own. After so many years of cooperating closely with others, I indulged myself almost orgiastically in assembling and building the experiment and in taking all the data by myself—even though it meant rather long periods of continuous effort. In that particular research, and in the excitement of the initial results, and in being able to interpret them in terms of a picture of the inside of the proton, I felt that I had realized my rendezvous with Arrowsmith.

In my continuing struggle against team research at Cornell, one thing that was effective was to adapt what I had learned in Berkeley and to shift my working hours to night time. Unless I had an early class, I would work from about noon till six or seven in the evening. Then I would enjoy a regular family life until the kids went to bed. Then I'd return to the lab and work until the early morning hours. In that way, if someone wanted to see me, he could do so during the afternoon when my university administrative life was lived. The late hours were mine alone.

The electron scattering experiment was a failure insofar as it contributed to my fight against team research. Ironically, this came about because of the success of the early phases of the experiment which meant that, from a laboratory point of view, I could not indulge myself by keeping the experiment all to myself and be a responsible director. Soon a couple of graduate students were involved, and finally some younger members of the faculty, who were considerably more competent than I, took over the whole endeavor. The pendulum then swung back and I reverted to being an accelerator builder, with the idea, now almost becoming tiresome, of increasing the energy of our synchrotron to 10 BeV. This idea consumed me, as well as several of my colleagues, for the next few years. At the end of the period, I found that the pendulum was stuck. The rhythm was broken because this time, instead of using the 10 BeV machine for some experiments I had been anticipating, I found myself indentured to the construction of the 200 BeV proton synchrotron at Batavia, Illinois, where I, now the compleat bureaucrat, lead what is perhaps the *ne plus ultra* of team research in high energy physics. *Sic transit gloria.*

W. O. BAKER

Computers As Information-Processing Machines in Modern Science

How Science Is Helping Itself

SCIENCE, THE vast producer and consumer of knowledge, has until lately held mostly to handwork for helping headwork. Of course, there have been mechanical aids to printing and publishing the record of scientific advance and more recently a host of ways of sensing and recording the measurements of scientific experiments by instruments. But in between the original observation and its insertion into the great corpus of scientific record there has persisted a wide, in some ways a primitive, personalized domain. Of course, this gap in mechanization does happen to contain a most awesome function of mankind—the ability to think. Now, suddenly, digital machines have come into this realm of the scientist, supplementing mysterious talents to perceive and to recognize patterns, to organize data, and, above all, to envision in the mind's eye new concepts and new understanding. How science and research are being changed by those machines and methods will be described in some of the examples which follow.

Indeed, we are now seeing the centuries-old pattern of practical progress in physical mechanization of science widening toward what we hope can be a technology in support of *how* we know.

Encoding of Information Scientifically

Thus, we now know that any information can be encoded adequately to convey its content by the use of binary digit signals; that is, zero or one, electrically plus or minus, on or off, magnetically north or south. This is made practical for doing things with information by the electrical digital computer. A first step in making this possible was taken by George Stibitz, at Bell Telephone Laboratories, in creating an electrical Complex Number Calculator, in 1937. Independently, Howard Aiken, at Harvard, made a general purpose Automatic Sequence Controlled Calculator. Soon, John Eckert and John Mauchly, at the University of Pennsylvania, extended these ideas, and computers were made embodying them and working at electronic speeds. The concept was thrust ahead by the stored program and other principles of John von Neumann. Hence, with the analog computer systems of Vannevar Bush and others, there evolved new capabilities of calculating, analyzing data, and indeed of organizing and manipulating knowledge. As usual in science, these machines reflected many historic concepts: binary numbers perhaps coming from the Chinese around 2000 B.C., Blaise Pascal's adding machine, Charles Babbage's efforts on an Analytical Engine (1812-1871), George Boole's symbolic logic algebra (1850's).

Then, the discovery of the transistor by John Bardeen, W. H. Brattain, and William Shockley (and soon the related solid state components and circuits) and the magnetic stores of Jay W. Forrester enabled the construction of such devices with the capacity and flexibility necessary for changing the course of much of modern science and technology. We need hardly mention how high-speed computers have been essential to the design of nuclear reactors and weapons, the exploration of outer space, and the realization of ballistic rockets and missiles.

But computers have become part of science because they permit altogether new ways of obtaining and using information. Information is "knowledge communicated or received," the very essence of human action.

So indeed it is necessary to accept that the symbols by which we express information through language, words, letters, as well as patterns, figures, graphs, and, of course, numbers can be encoded in binary digits. With a sufficiently extensive code, all of any language or pattern can be represented by combinations of such digits, but a further question is how many are necessary just satisfactorily to represent the information involved? Is this number manageable in even the large digital machines which technology has now provided? Well, there are many ways of ascertaining this, and a discussion of them is far beyond the scope of our present discourse. But the experience of electrical communications gives some direct ways of seeing how this binary processing of information is applied. Thus, we know that human speech is satisfactorily transmitted over a telephone channel by oscillating electrical currents, in what is called an analog fashion. But information theory,[1] through its sampling theorem, states that such a continuous signal as the voice coming electrically over the telephone is completely contained and capable of perfect reconstruction, by taking samples of the wave no less than twice as often as the highest frequency present in the variation of the voltage itself. Thus in a familiar telephone channel of 4,000 cycles per second there must be acquired 8,000 samples per second of amplitude. The size of each of these samples, however, can then be encoded digitally, as is shown in Figure 1.

In the case of specific separate signals, rather than a continuous range of values of the waves noted above (meaning discrete values such as from the ten number digits or twenty-six letters in our alphabet), the representation is still easier. In English texts there are about 4½ letters per word, and there are spacings between the words, so in a simple form we might average about 5.5 letter characters per word. This representation is easily done by using 27½ binary digits (each *bit* is just 0 or 1) per word, if we decided to encode character by character. But if we wanted to encode word by word for a vocabulary, say, of 16,384 English words, it turns out there are just that number of 14-digit combinations of binary numbers, that is, combinations of zeros and ones, each combination being 14 digits in size. So there we would get by with about 14 binary digits per word. Now actually the way in which we finally decide to do this in order to achieve the most efficient system, of course, has to depend on a much more detailed application of information theory. Nevertheless, this has been done, and many experiments, and nowadays actual operating systems, have shown that this holds not only for all forms of language but also for pictures, music, and so on. Much of the method in deciding what has to be done depends on the *entropy* of the information or, speaking generally, the probability of occurrence of certain sequences of letters and words in language, of shapes and figures in drawings, and so forth.

For instance, for equal probabilities, which we generally do not find in language, the entropy for some number n of symbols is equal to minus the logarithm of $1/n$. This applies to information in the sense that information is really the resolution of uncertainty, in communication theoretic terms.

Now that we can represent codes for information in forms which electrical circuits can easily recognize (on or off, pulse or no-pulse, magnetized one way or the other) we can see the general scheme by which the machine is organized to process the information. This is done by shifting around to various parts of the machine the electrically represented ones or zeros so that they can be combined under rigid control. Thus, from the *memory*, in which each item is separately stored in a place which has its own distinct *address*, movement is controlled in a register. From this, arithmetic operations can be carried out most simply by following rules of adding or subtracting the binary digits, in which the circuits sequentially react as a 0 or 1 is encountered and come up with the result of the whole sequence.

These results can be combined in very many ingenious ways, according to the *instructions*, which are also put in storage to be used as needed.

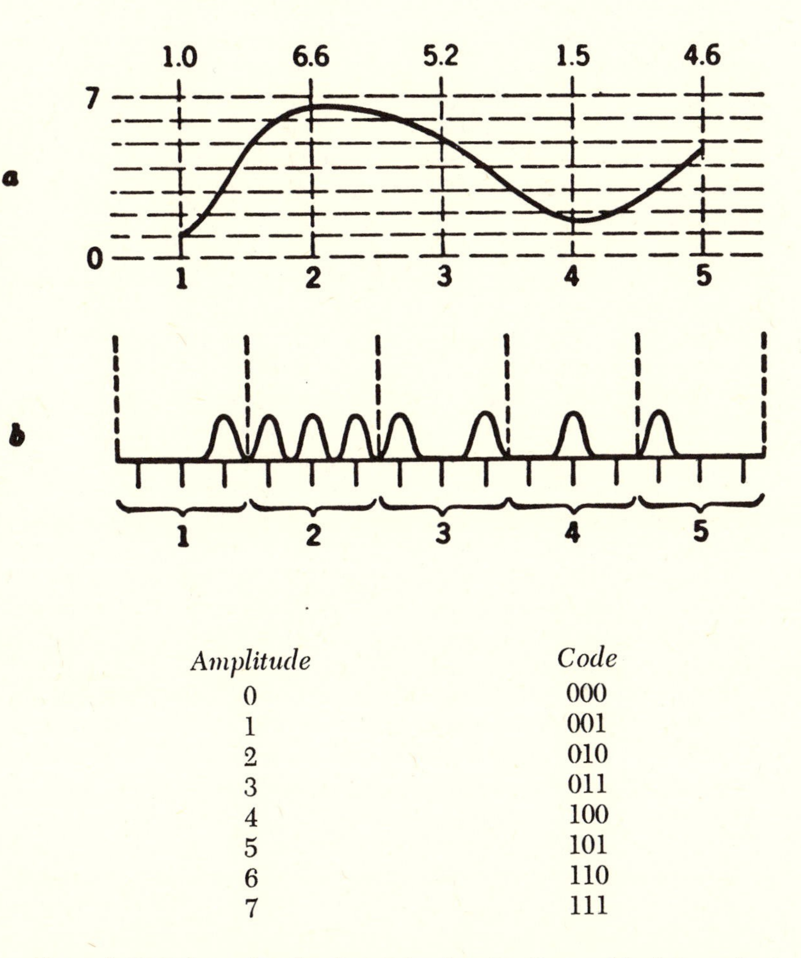

Amplitude	*Code*
0	000
1	001
2	010
3	011
4	100
5	101
6	110
7	111

Figure 1. Digital encoding by electrical pulses (in diagram *b*) of the analog wave of a voice signal varying over time and amplitude, with time range 1 through 5, amplitude 0 to 7. (From W. O. Baker, "Some Social Meanings of Communications Science and Technology," in Morton Leeds, ed., *Washington Colloquium on Science and Society*, 2d ser. [Baltimore: Mono Book Corp., 1967].)

Each of these actions is extremely simple by itself, but can be done with very great speed. Thus in the actual arithmetical additions, subtractions, and multiplications in the accumulator, speeds of 357,000 additions or 178,000 multiplications per second are common in the older systems, and rates up to billions per second are now at hand. With microelectronics, *access* to the memory matches this magnitude, and indeed had achieved about two-millionths of a second speed in the older generation referred to above, which would contain in its magnetic cores about 33,000 words. The combinations of binary signals which produce the arithmetic are arranged by *logic gates*, three in kind, which along with memory units, such as *flip-flop* circuits, carry out the rules for any arithmetic that the machine is instructed to do. Thus, in summary, combinations of binary symbols are subject to logical operations, just as we do calculation in our brains, but of course with a different instrumentality.

All these results come out in some manageable form, just as the problem went in, with punched cards or paper tape or magnetic tape or other signal. Because of the ultimate simplicity of each operation, speed is the keynote of every other function, in addition to the dominant arithmetical processing and access to memory. Thus we shall describe later how printing can be done by the direct connection of the computer to an electron beam working in a cathode ray tube. Recently an electrostatic printer has been announced that can print up to 1,200 pages per minute, although about 60 pages per minute is presently in good balance with conventional computer outputs.

Range and Qualities of Information in Modern Science

So we have at least touched on the way all information can be handled by digital machines, and what some of the basic considerations in encoding it are. We can now consider what this means for new science, in which progress still depends above all on the integrated thoughts of the single human mind. We say the single human mind because that is where the completed and critical concepts must arise. For in science, although a hundred minds may seize upon the same or a similar problem, each does not contribute one-hundredth of a crucial insight or concept. Yet we must somehow deal in science with increasingly complex systems and interactions. The days when, as Victor Weisskopf so poignantly put it, "If you understand hydrogen, you understand everything that is understood" were fine for the early period of quantum mechanics, even quantum statistics, of descriptive biology, of terrestrial mechanics, and of synthetic chemistry. Now, however, understanding is often gained by dealing with very large assemblies of things, such as the states of energy in a perfect crystal, the molecular configurations in a helix of desoxyribonucleic acid, the neural responses from hearing due to activation of the spiral innervation of the cochlea. Even in the interactions of elementary particles, three is already a large number. In some ways even more challenging, modern science would like to deal with complex social situations, wherein very difficult statistical ensembles, new correlations, and interdependences are to be discerned. It may be that the future well-being of man on earth will depend, in fact, on such explorations in the behavioral sciences, involving masses of observations and complexities of interactions almost undreamed of in the simple(!) problems of dealing with 10^{23} virtually identical atoms in a mole of a gas or of a nearly perfect crystal.

And we may no longer be able to afford the luxury of limiting science to situations that fit our earlier (and successful) ideas of simplicity. Niels Bohr spoke of this in his essay on "The Unity of Human Knowledge," published just shortly after his death.[2] On the one hand he said, "The aim of our argument is to emphasize that all experience, whether in science, philosophy, or art, which may be helpful to mankind, must be capable of being communicated by human means of expression, and it is on this basis that we shall approach the question of unity of knowledge." So, accordingly, we must also accent in the language of Bohr that information processing with machines does not imply in any form restriction to numerical tabulations or the expression of formalized operations alone. As Bohr indeed asserted in this same essay, "Mathematics is therefore not to be regarded as a special branch of knowledge based on the accumulation of experience, but rather as a refinement of general language, supplementing it with appropriate tools to represent relations for which ordinary verbal expression is imprecise or too cumbersome." We must be mindful of the special interest of the natural philosophers and scientists of this century in the issue which Bohr so greatly advanced through his discussion of complementarity. This is summarized in his comment that "Far from giving rise to confusing complications, the recognition of the extent to which the account of physical experience depends on the standpoint of the observer proved most fertile in placing fundamental laws valid for all observers."

This framework of what amounts really to a central phase of scientific method in our century clearly challenges our ancient methods of accumulating and dealing with information by written accounts and graphs and even high-speed, machined tabulations of numeric compilations. Beyond even this, however, there is the present-day challenge to the kind of advances and kind of over-all insights which twentieth century science in the latter half of the century is able to give us. The voices of discontent are arising around the earth, but nowhere more strongly than in our own technologic western civilization. They express dismay at what we once have held as the prime tactic of modern physical science. This is, that we do what seems possible and where curiosity and logic invite us to go, rather than to pursue a course apparently more directly attached to human needs or even to the aspirations of people and societies.

Now we shall try to show how digital logic machines in their most generalized forms can augment our advance along the paths to new findings in science, into deeper complexions which Bohr's visions have so illuminated for us, as well as to new applications of science in society. But first we should assert emphatically that we shall not be considering *thinking machines* or those which operate as imitators of human sense or desires. Our information-processing machines are artificial servants of man, and we are only now learning how broad that service can be.

Humanists are not yet very comfortable about this. We shall try to answer some of their concerns in the examples now to be given, concerns which have been thoughtfully expressed by many, such as by the poet Robert Graves in the twelfth Arthur Dehon Little Memorial Lecture at the Massachusetts Institute of Technology a few years ago. In this lecture he said, in speaking of the movements of scientific and humanistic thought, "And some answers given by the latest computers to the latest questions make merely cosmic sense—that is to say, they can no longer be imaginatively grasped by the human brain in any predictable stage of development, and refer to processes, not to phenomena." Graves thought of this as a new era of nonsense and went on to say, "Well, now that electronic computers have passed the limits of the brain's imaginative grasp, physicists should consult an anthropologist who still speaks their own scientific language and let themselves be reassured that vast tracts of human thought remain to be explored, which the computer knows nothing of and which call for no complex apparatus."

But it is precisely the invasion of these challenging regions of complex thought which we shall show can be explored by the computer, these "vast tracts of human thought." And this can be to the great benefit of the most searching curiosity and inquiry of man. In fact, because of the relief of drudgery by use of computers in the routines of mental information processing, we shall see greater and greater adventures into now distant scientific realms, such as the behavior of elaborate natural systems of the weather, of geophysics, of the oceans, of biology, of ecology, and, of course, of the sociology of man himself.

Further, however, these machines do also heighten the perception and meaning of what has been regarded as basically mathematical, computational, information processing as well. Thus, to realize the transition between the commonly appreciated role of digital logic machines (as accessories in mathematical functions) and their new and ever-deepening part in the facilitation of non-numeric and conceptual effort, we should examine first a few typical cases where the new applications have brought out exciting aspects of essentially mathematical operations, which were long ago thought already to have been highly refined. Then, we shall consider progressively less defined matters of scientific work, where progress is more and more dependent on automata. Indeed, much of the strategy of a large research center is coming to be guided by what computers make possible. As always, I shall have to speak of cases most intimately known to me, which will thus focus heavily on work in my own laboratories; actually, the total world effort in these fields is still very modest indeed. Nevertheless, the influence on scientists and on other scholars is already changing their ways and places of work.

Computers in Mathematics Beyond Numbers

Let us think about symbolic algebra, for example, one of the earliest successes of abstract thought. W. S. Brown has now discovered a language and a system, the ALPAK system and the ALTRAN language, for efficient operations in large algebraic problems. In this system, one man-hour by sophisticated algebraic experts doing not only the usual operations of addition, subtraction, multiplication, division, and integral exponents, but also substitution, differentiation, derivation of greatest common divisors, and various simplification moves, equals one second of time for the same actions on the older generation of digital computers, such as the IBM 7094. Indeed, the available storage capacity of such a machine, arranged according to Brown's system, holds up to 8,000 polynomial terms. Further, the ALPAK system handles variables with values of rational numbers, real-valued variables, those whose values are polynomials and those whose values are irrational functions. Brown has created methods for instructing the machine how to manage each of these types explicitly, and has dealt with such important details as characterization of real or complex constants, so that round-off errors and similar slips can be avoided. Thus algebra, one of man's oldest and most exacting intellectual tasks, is being revolutionized by computers. In fact, machines will do so many "pure" operations in novel ways that we need to think of new ways of communicating with them.

Computing to See Newton Plain

Consider next traditional problems of analytic mechanics wherein the motions of nonlinear dynamical systems can be formulated, as began with Galileo, Newton, and Maxwell, but have only recently been widely and usefully evaluated. This advance is possible not only because the differential equations of motion themselves can often be approximated, even though not solved analytically, and evaluated to close approximation by the main force of the computer. (Even this, of course, has been an immensely useful function, starting with the early analog computer applications two or three decades ago.) But what we are discussing now is a different approach, in which the information about these equations of motion for a particular dynamical system is rigorously processed in the machine, so that there is numerical integration of the equations for any particular choice of conditions. The conditions can be varied continuously, and many states of the system can be specified by the values produced by the machine. However, E. E. Zajac worked these programs out for a particular case of the method for stabilization of earth satellites (used for communication) in a particular earth-pointing position by the use of gyroscopes. Under traditional methods, he would have derived plots from the numerical integration of three angles necessary to describe the orientation of the satellite at any particular time in its orbits about the earth. Limitations in the usefulness of this form of information were many, however, so Zajac produced instructions for the computer which caused it to regulate an electron beam so that it traced figures on the face of a cathode ray tube in the form of continuously time-varying diagrams of the satellite, drawn schematically (Figure 2). This movement of the satellite, as derived by the machine, revealed precisely the results of numerical integration of the equations of motion at each particular instant. The result was easilv photographed as a continuous moving picture. Thus, photographs of the figures on the tube face show by wholly computer-produced pictures how the earth-orbiting vehicle oscillates, tumbles, and eventually stabilizes under influence of the gyroscopes. The scientific impact of this kind of revelation on understanding of orbital motion, and on ensuring the necessary properities for desired performance of the radio system, is extensive. Effects extend from moon landings to global telephony and meteorology. The student of classical mechanics and the engineer attempting to improve designs both acquire a new dimension of insight and stimulus to invention from such continuous visualization of exceedingly obscure and intricate differential equations.

Charged Particles—Around the Earth and the Nucleus

There have been various other applications of information-processing machines to problems of historic import in outer space. One is the interpretation of signals from satellite exploration of the Van Allen belts of radiation, which are characteristically present a few earth radii distances from this planet. The discovery and development of solid state semiconductor radiation counters and their effective adaptation to the

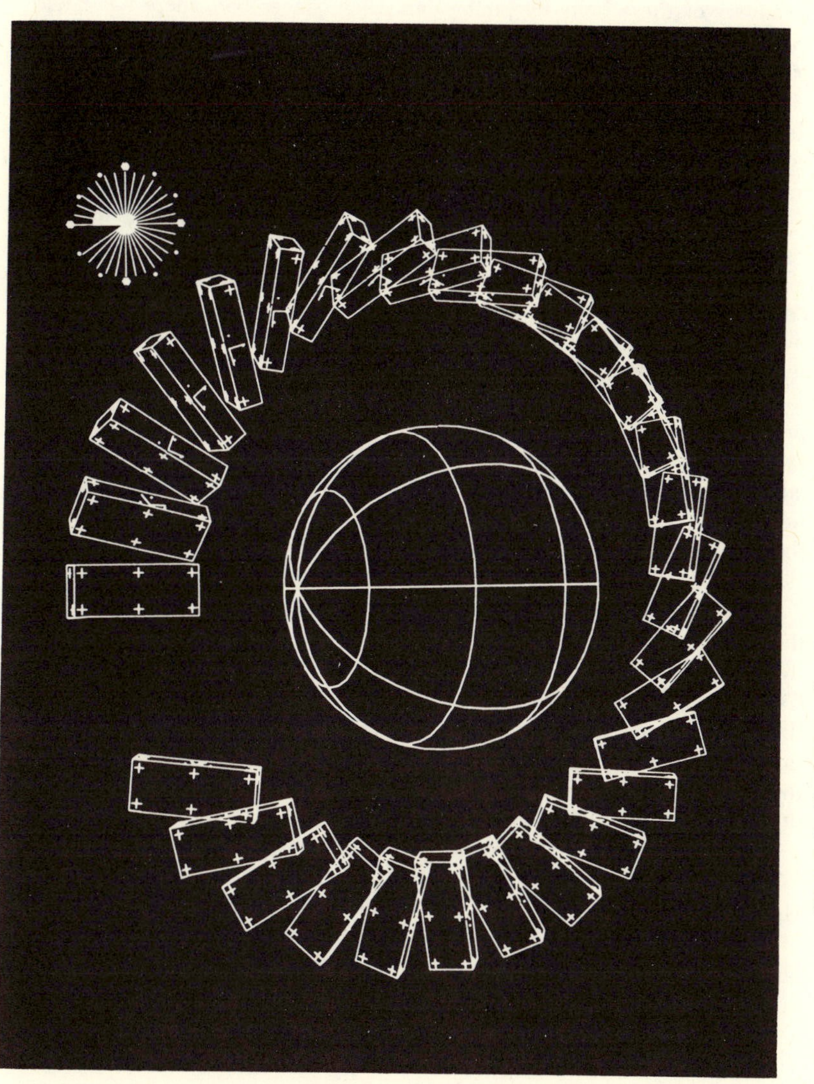

Figure 2. Composite sketch of a series of scenes from the moving picture, produced by computer, of the stabilizing orbit of an earth satellite. The orientations of the satellite are analyzed mechanically as each position is computed, and the whole phenomenon of stabilization is then displayed in the movie, as produced by E. E. Zajac. The tumbling of the satellite is eventually damped out by computer-designed qualities.

study of pa:·:cles in outer space, early by use on the TELSTAR® satellites, and later and continuously on other scientific satellites of NASA, have led to a new insight of the environs of our earth. Following the discovery of these belts by James Van Allen, since 1958, there has been intensive effort to understand the behavior and also the changes in population of electrons and protons in the various shells. For relativistic electrons above about 0.5 Mev in energy, special information was derived from a narrow shell created by the Argus tests and the Russian nuclear test of November 1, 1962. It is necessary, in comprehending the interactions of these electron belts with the magnetic field (and especially regarding the pitch angle diffusion which might result from fluctuating electromagnetic fields around the earth), to study enormous quantities of data. Programs have been evolved by W. L. Brown, C. S. Roberts, and J. D. Gabbe so that such findings are completely analyzed and plotted directly by the computer, as shown in Figure 3. This has given remarkable impetus to knowledge of these radiation belts and to their role in space physics and space technology. Also the machines acquire, process, and plot out directly the telemetered information on such things as electron density as a function of time, yielding important information on the duration of magnetic influences and the general causes for population decay. Obviously, the sheer volume of information received from these space physics experiments makes nonmechanized information processing exhaustingly tedious.

Also, these methods are of comparable significance in the study of the microcosmos of the nucleus. An early use of nuclear data acquisition directly in a computer was achieved by Edwin Norbeck and Clarence Carlson at the State University of Iowa. There, energy for single coincident events, position information, and certain related results were obtained directly by an on-line machine processing. They also used a direct display on a cathode ray tube, so that, for example, in the reaction of lithium-7 with beryllium-9, each resultant particle caught by an array of solid state detectors is automatically placed as a printed band of events.

An especially interesting application of information-processing machines to experimentation in nuclear physics is provided by the studies of J. F. Mollenauer, J. V. Kane, and associates at Bell and the Brookhaven Laboratories of two particle coincidence experiments. Here, the nuclear reaction, such as deuteron plus proton, yields two protons plus a neutron, and two of the resultant three particles are detected and their possible energy values are, of course, constrained by conservation of energy and momentum.

In order to analyze the experiments, Mollenauer and associates have superimposed simulated data conditions in a large grid pattern, as produced by computer, onto the observed situation. They are able con-

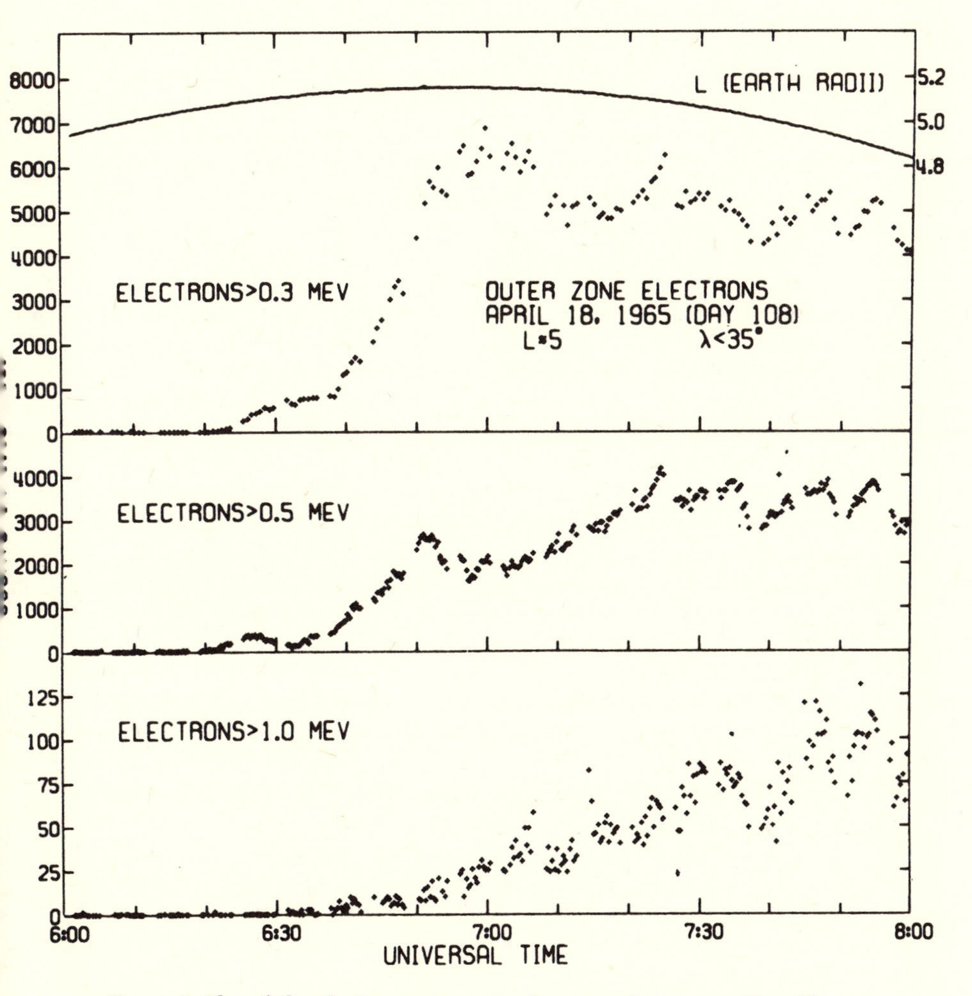

Figure 3. Plot of the electron content as a function of time in a Van Allen belt about 5 earth radii out. The raw data telemetered from the detecting satellite are processed and plotted by the computer directly.

stantly to monitor the reasonableness or general consistency of their experimental conditions in this way and to close in on meaningful interpretation.

Up the Scale of Complexity—to the Molecule

Still another area of modern science, molecular structure, has challenged conventional study. Yet the knowledge of the shape and atomic arrangements of molecules is basic to understanding, ranging from the units of living matter to the behavior of the earth's atmosphere. Now, both the examination of hypotheses and the analysis of observations are being aided by computers, in various revelations of molecular structure. On the experimental side, many basic physical effects, such as the radio frequency or microwave resonance of magnetic and electric dipoles in the presence of external fields, have led to whole new realms of spectroscopy. These reflect, often with great delicacy, the positions and interactions of groups as well as single atoms in molecules, and even also during chemical change. Coupling the computer directly with the experiments, as well as with continuous refinements of the underlying theories, is under way.

An interesting example is in the use of nuclear magnetic resonance spectroscopy by S. Meiboom and his associates. A small general purpose digital computer is connected to the high resolution NMR spectrometer consisting of an electromagnet, a field control probe, and stabilizing circuits. Included as well is a sample probe with its transmitter and receiver for detection of the nuclear magnetic resonance in the sample. The experimental array is shown in Figure 4. The output of the spectrometer itself is converted electronically into digital form, and at the same time the computer reacts back by setting a frequency synthesizer, a system which yields a radio frequency output precisely controlled by digital input. Thus, rather than the usual pen recorder trace of the amplitude of the nuclear signal, the spectrum is recorded in digital form and transferred to the memory of the computer. In addition to the great convenience and speed of examining the spectra themselves that are provided, the computer actually processes information from the experiment in a fashion that had earlier been restricted to a separate special purpose averaging computer. (This last was developed so that a particular frequency scan of the sample could be done many times over and the amplitudes of each successive scan could be added in to those previously acquired by the computer memory. Of course, the signal itself increases linearly with the number of scans, but the random noise, previously a limiting factor in the experiments, increases only as the square root of the number of scans.) Hence we are immediately

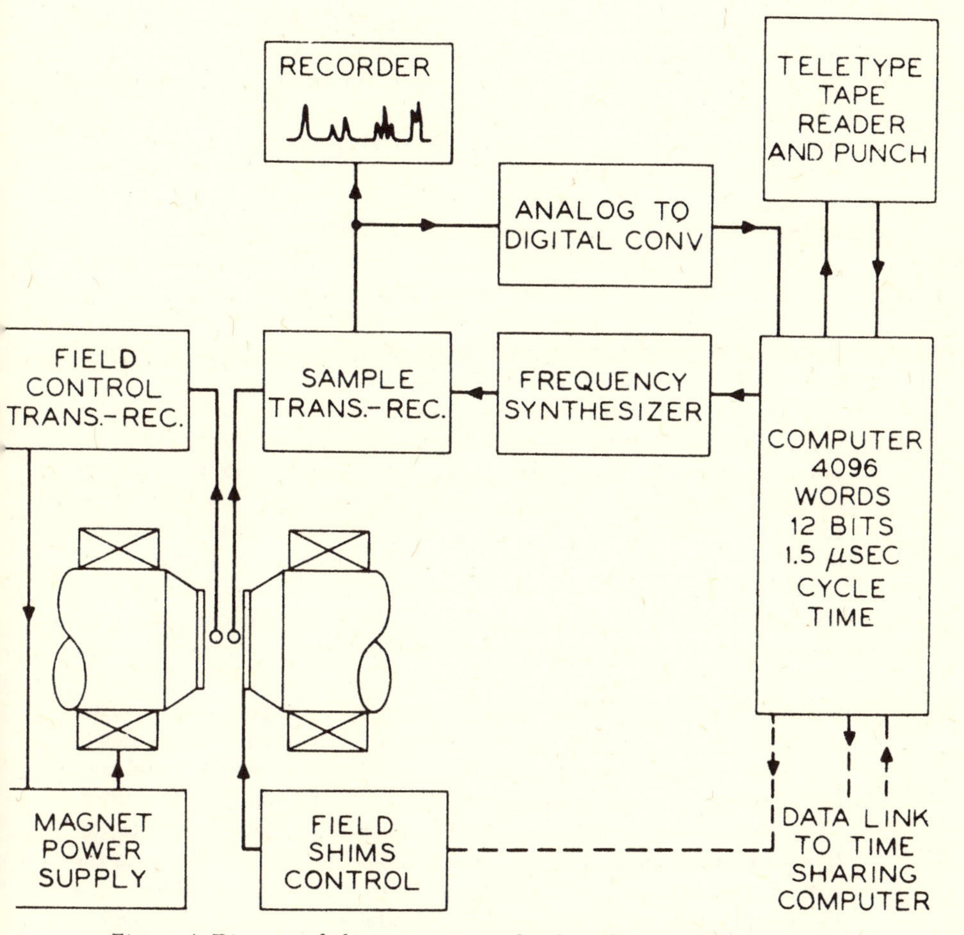

Figure 4. Diagram of the computer-regulated nuclear magnetic resonance experiments of S. Meiboom shows how the spectra are produced and analyzed and then displayed on the recorder, with immediate feedback to adjust the apparatus for subsequent and optimized experimental conditions.

enabled to process extremely weak or confused signals, learning, as in the case of the nuclear physics experiments cited above, just what the experimental conditions must be as the experiment proceeds.

And in some cases it is also desirable to obtain the spectrum in which the amplitude is not noted as a function of frequency but rather as a function of time, or to observe the behavior of the specimen after it gets a pulse of input radiation. Again, this can be done repeatedly and conveniently with the on-line computer. Results are immediately converted by a Fourier transform to conventional frequency domain observations, which it turns out appeal more intuitively to the interpreter. This sequence is outlined in Figure 5. Meiboom is also extending his system so that the computer will eventually control uniformity of magnetic field in the experimental setup, and perhaps eventually by the use of a very large time-sharing system it will be possible simultaneously to produce and compare theoretical spectra derived from detailed models of molecules with those from actual molecules being observed. An early example is shown in Figure 6, in which the upper spectrum is experimental, the lower, theoretical. For instance, L. C. Snyder has carried out an extensive theoretical synthesis of the nuclear magnetic resonance spectrum to be expected from the simple but puzzling molecule cyclopropane. Jointly, Meiboom using cyclopropane molecules with and without the carbon 13 isotope, and by "double irradiation" techniques, has obtained experimental spectra of such quality that it is now possible to derive very closely the stereo structure of cyclopropane, as represented in the upper part of Figure 7. In a similar fashion, with the techniques of solutions of molecules to be studied in nematic or liquid crystal solvents, as initiated at the University of Freiburg, our workers have detected by this refined information processing by computer, of magnetic resonance data, as little as 1/10 of a degree of distortion in the bond angles of the symmetrical molecules neopentane and tetramethyl silane. In the Brookhaven National Laboratory's chemistry group, computer-generated color display has recently been adapted to 3-dimensional imagery on a television tube, so that highly complex crystal configurations have been represented, and experimental confirmations aided.

So again we are seeing the new course in the growth of science, for though I cannot here elaborate on the details which fully evaluate the role of structural information processing by machine, it is doubtful if the insights now being acquired could otherwise have been achieved. But naturally these matters become even more meaningful as science attempts to invade more and more complex and mysterious realms. Thus I shall round out my theme by pursuing the train of complexity of scientific study further into the bio-sciences, and ultimately into the behavioral and social science domains.

DIGITAL OPERATIONS

TIME AVERAGING

INTEGRATION
(LINE INTENSITIES)

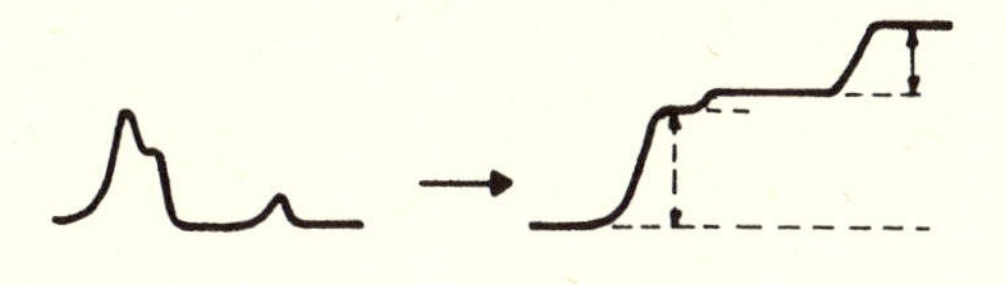

FOURIER TRANSFORMS
(TIME DOMAIN
SPECTROSCOPY)

THEORY

EXPERIMENT

THEORY

SPECTROMETER
CONTROL
(FIELD UNIFORMITY)

DATA STORAGE
(SPECTRA LIBRARY)

Figure 5. Examples of the digital processing of nuclear magnetic resonance experiments on molecular structure, in which spectra from theoretical models are calculated concurrently with experiments to guide interpretation during the obtaining of data.

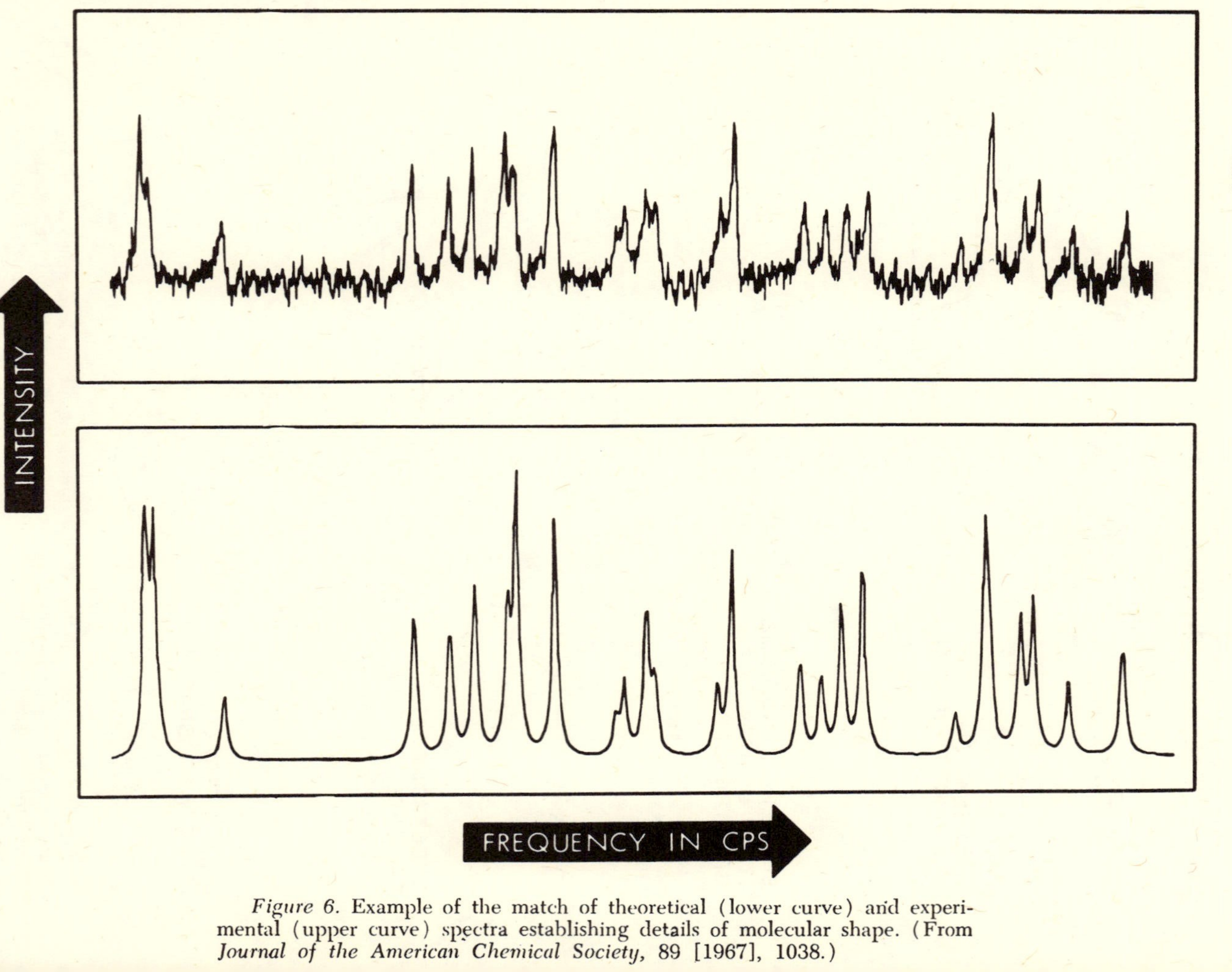

Figure 6. Example of the match of theoretical (lower curve) and experimental (upper curve) spectra establishing details of molecular shape. (From *Journal of the American Chemical Society*, 89 [1967], 1038.)

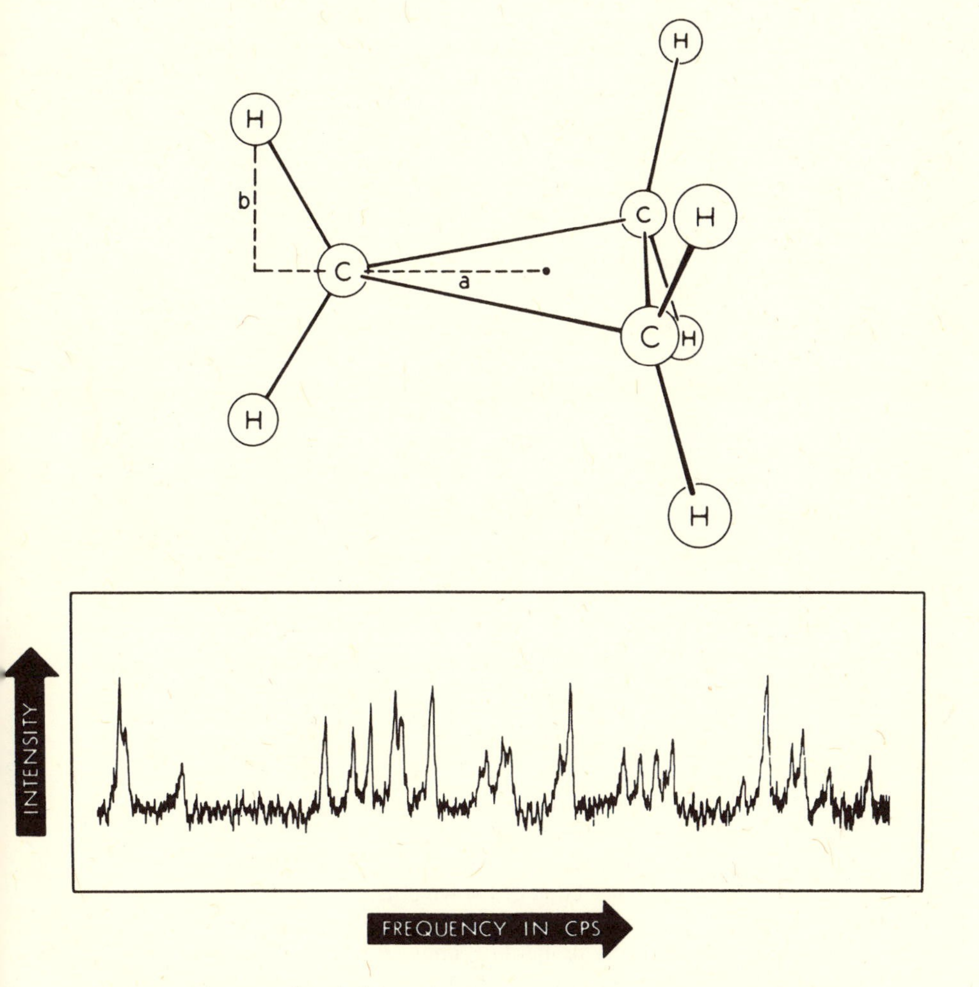

Figure 7. Example of the application of computer-generated information in determination of the detailed configuration of the organic molecule cyclopropane, of special significance in hydrocarbon fuels and its use as an anesthetic.

Logic Machines in the Life Sciences

In the bio-sciences, building on the basis of the understanding of matter itself, as already sampled, the support of information-processing machines promises a new scale of exploitation of the most powerful modes of discovery. These are based on *simulation* and the testing of models derived from the vast cumulative resource of earlier science and earlier understanding. One beginning in those sciences has been the work of W. E. Blumberg on the assembling of simplified models of transition metal ion complexes in such structures as the cupric enzymes, whose role is central to our understanding of vital processes. Again, the extensive use of high capacity digital machines affects the whole mode of thought which is being applied. For Blumberg has pointed out that atomic qualities of transition metal ion complexes have now been probed by physical science with great detail. Thus, it should be possible to transfer some of this elegant learning of solid state physics and chemistry into the intriguing questions of biology. However the matter of trying either a direct rigorous application of existing knowledge to new complex structures or of attempting the necessary approximations and correlations by what might be called a "manual" mental approach has so far proved impractical. Thus, magnetic susceptibility, spectroscopic splitting factors and hyperfine coupling constants of electron paramagnetic spectra, the optical spectral transitions themselves, optical rotatory dispersion, measurements of optical dichroism in single crystals, and so forth are all important leads. They represent a large array of information which up to now could not be focused on the scientific phenomena of the greatest import, that is the functioning of the transition metal ion, or indeed its position in the bio-structure. Blumberg actually designed a strategy in which he took twelve physical measurements as principal parameters for the structure simulations to be attempted in some enzyme molecules. These include (and do not permit satisfactory translation for the unspecialized reader) three spectroscopic splitting factors of the EPR spectrum, the positions of the 3 d-d optical transitions highest in energy in the complex, and the oscillator strength and the optical rotatory dispersions of these three transitions. These twelve parameters permit the selection of twelve others to define field conditions, particularly the electrostatic potential on the copper ion. Blumberg then proceeded to explore the probable geometric adjustments in the enzyme which would maximize the consistency of the diverse physical measurements. Then following these adjustments and using crystal field theory, it was necessary to compute values for all the twelve physical measurements which characterize any particular configuration of the complex. Blumberg then succeeded in synthesizing by the computer a set of models of the fields as a function of position of various copper cupric ion sites. He went further, and constructed these as they were expected to appear in copper sulphate pentahydrate where the ion locations are known through X-ray crystal structure studies. In this crystal, the copper sites have only a slight distortion from the elementary concept of a perfect square planar complex, and the model so conceived does indeed agree closely with that synthesized by computer. A schematic diagram is shown in Figure 8.

The power of these syntheses is further reflected in the fact that the model fitting procedure yields a spin-orbit coupling constant of 683 wave numbers. If this is used in Hans Bethe's early theory, in which it represents 82 per cent of the free ion value, we could conclude that the complex is about 36 per cent covalent. This is a remarkable implication to be able to draw so directly.

Finally, all these experiences were brought together in the suggestions for ceruloplasmin. Despite the special nomenclature required, some idea of the meaning of the study may be conveyed as follows: The diagrammatic representation (rather than a physical model) consolidates the information which must be applied to any detailed understanding of this elaborate structure. Likewise, new lines of thought are introduced. The copper ion seems to be a little above the average position of the four strongest ligands; two ligands on one side are stronger than the others. Peter Hemmerich had indicated that a rigid square planar cupric complex in plasmin does not allow the cupric ion to be reduced to *cuprous* without being liberated from its binding in the protein.

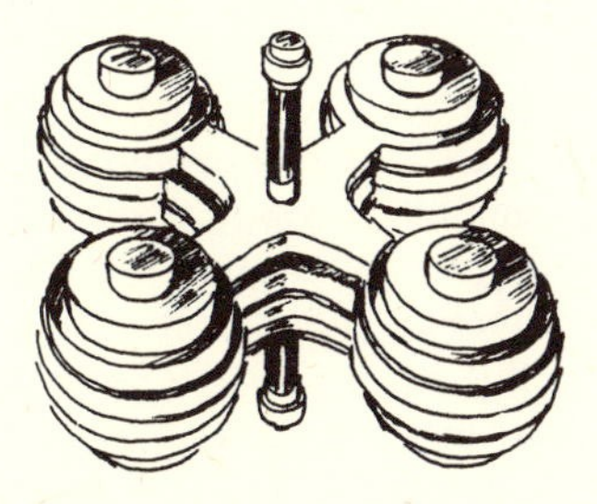

Figure 8. Computer-generated model of the coordination of copper ions in enzymes, in which the bonding and placement of the copper is shown by the central rod, as conceived by W. E. Blumberg. (From J. Peisach, P. Aisen, and W. E. Blumberg, eds., *The Biochemistry of Copper* [New York: Academic Press, 1966].)

However, from the model synthesized it looks as though the four strongest ligands do have sufficient tendency toward tetrahedral array so that with the disengagement of the very weakly bound, apical, ligand the *cuprous* complex would become stabilized tetrahedrally. There are certain other speculations, too, which remain to be explored. But over-all, new elements of flexibility and new ranges of possibilities have been introduced into this whole set of problems.

Computing for Psychology—Perception

Perhaps now the intricate mosaic of learning is visible—of how mechanized information processing is pervading basic patterns of scientific inquiry, inquiry whose successes have long ago been established but whose future growth has been until recently less assured. But what about the far vaster realms of nature and man, in which the application of scientific thought has been very limited, difficult, and in many cases unsatisfactory? Is our use of new machine aid to information handling giving much promise of evolution or revolution there? It seems to be clear that it is, in the biological, psychological, social sciences. As before, I shall be able here merely to sample the activities I see coming along. Let us think for instance of the whole question of human perception, an issue absolutely central to the study of communications among men and to the understanding of our most vital social functions. Simulation of the mechanism of hearing, in both mechanical and neural form, has recently been accomplished and depicted in moving diagrams synthesized by the computer itself. Thus, as in our earlier case of classical mechanics elucidated by the production of visual displays by the machine, we have now an example of physiological and psychological function similarly explicated especially through the work of A. M. Noll. Much the same sort of new insight is being provided in studies of speech, in which already some years ago, by methods based on information theory and with a sampling rate of about 50,000 times per second, the phoneme reservoir stored in a large computer was tapped. By suitable new programs, intelligible, even if colorless, words were spoken, for the first time completely separated from the vocal exercise or analogy to the sound system of a human being. Currently, C. H. Coker and Mrs. N. Umeda have carried this further, to the production of quite natural speech *from written text,* without human participation.

Nor is vision now being neglected. B. Julesz is actively recasting much of the understanding of stereopsis and our beautiful capabilities of form and depth perception. This is happening through the crucial circumstance that he was able to synthesize, by computer, abstract images completely devoid of cues from experience. Two matrices of random marks are produced, as indicated in Figure 9, and these then are super-

Figure 9. Random dot patterns containing no subjective clews, created by B. Julesz for new understanding of stereoptic vision. (From B. Julesz, *Foundations of Cyclopean Perception* [Chicago: University of Chicago Press, forthcoming].)

imposed, by regulation of the machine. Indeed, these images of random dots are precisely shifted by the computer program so that fascinating and utterly unique stereo images are formed, as in Figure 10. Thus it has been established that in absence of all familiar cues and monocular patterns, stereopsis is achievable. Indeed, the personal experience of perceiving depth under these conditions is compelling, and one believes the conception, first forwarded by Heinrich Dove in 1841, that stereopsis is a function of central nervous system processes. The studies have already indicated a whole range of neurophysiological experiments which should be undertaken, such as a search for an isotropic orientation of cortical units fired when binocularly stimulated with regularly growing inequality. In the remarkable physiological studies of David Hubel and Torsten Wiesel, an important range of binocular units in the cortex about the lateral geniculate nucleus has already been established. Indeed, again a new and formerly unimplied range of thinking is under way. For example, the physiology work in animals now indicates so general anisotropy in receptor field, with no concentric receptor field organization, so that if similar structures were true for man there is an interesting puzzle about how we could perceive a single dot. This

Figure 10. Superimposed patterns which yield special depth perception, used by Dr. Julesz in studies of vision.

may occur, of course, from some sort of intersecting edge detections, but this sounds like a crude process. The differentiation between ordinary vision and stereopsis is apparently going to force acute examination of many casual assumptions about the essence of visual perception.

Sorting in Complex Systems—the Rth Dimension

Now, what are crucial issues in future scientific progress in psychological, behavioral, and social science, and perhaps the borders of economics and political science? These are, of course, usually some form of trying to isolate variables or perceive relationships among an exceedingly large number of cooperative variables. This scholarly search for structure is extensive, and often apparently uncorrelated, so that deriving information is naturally very difficult. In the last few years R. N. Shepard, J. B. Kruskal, and J. D. Carroll have discovered new methods of revealing structure in highly involved assemblies of data, by what they call reducing dimensions. One way of saying that there are N situations with R measurements for each is that the data can be put into N row vectors and then regarded as N points in R dimensional space. Evidently, for $R = 3$ a direct representation of this situation is cumbersome, and for $R = 4$ any visualization is unlikely. Nevertheless by appropriate computer application, M. D. McIlroy and Kruskal have carried through analysis for $N =$ about 10,000 row vectors with $R = 657$ measurements each, or 657 coordinates. This was for an important engineering application, and thus was not simply a feasibility game.

This idea that a spatial structure is the way to manage information derived from exceedingly profuse observations seems to be a powerful application of large-scale computing. Shepard has proposed that the rank ordering of interpoint distances in computing space of R dimensions can be manipulated by iteration, such that there arises a monotonic relation between similarity or cross-connection of various parameters and the interpoint distances. Structure of data from many human behavioral studies has been nicely organized in this way, using the enormous power of the high-speed computer. Highly specific relationships have been obtained from much apparently nonmetric data. Indeed, these methods can apply to nonmetric systems, such as preference studies. Shepard has illustrated this in the case of ordering properties of 30 random cylinders varying in altitude a and base area b, as shown in Figure 11. The twelve physical variables listed in the table are defined in terms of these two primary variables a and b. Thus the cylinders actually have only these two degrees of freedom, a and b, so the situation ought to be represented as a two-dimensional problem. However, the two-dimensional quality is quickly submerged in looking at the

VARIABLE	FORMULA
1. ALTITUDE	a
2. BASE AREA	b
3. CIRCUMFERENCE	$(2\sqrt{\pi})\,b^{1/2}$
4. SIDE AREA	$(2\sqrt{\pi})\,ab^{1/2}$
5. VOLUME	ab
6. MOMENT OF INERTIA	$(1/2\pi)\,ab^2$
7. SLENDERNESS RATIO	$(1/\sqrt{2\pi})\,ab^{-1/2}$
8. DIAGONAL-BASE ANGLE	$\tan^{-1}[(\sqrt{\pi}/2)\,ab^{-1/2}]$
9. DIAGONAL-SIDE ANGLE	$\cot^{-1}[(\sqrt{\pi}/2)\,ab^{-1/2}]$
10. ELECTRICAL RESISTANCE	ab^{-1}
11. CONDUCTANCE	$a^{-1}b$
12. TORSIONAL DEFORMABILITY	$(2\pi)\,ab^{-2}$

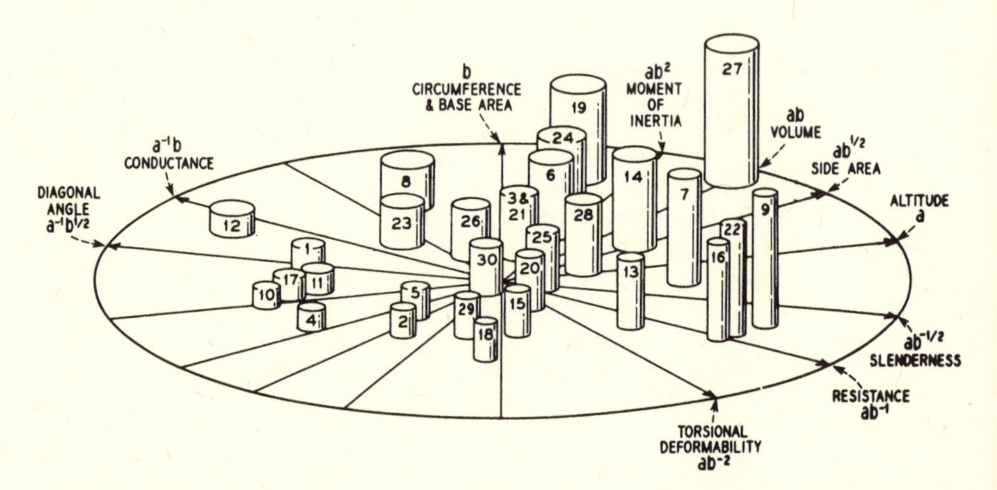

Figure 11. Example of the ordering of cylinders of matter embodying twelve properties dependent on dimensions. The multidimensional scaling techniques of R. N. Shepard enable computer processing in multidimensional space so that ultimately the ranking of properties displayed in the lower sketch shows the proper relationships among all the properties.

12×30 matrix, giving a value for each of the thirty cyclinders on each of the twelve variables. In that, of course, there are many (more than two) different ranks ordering of cylinders and many of the variables reflect nonlinear combining properties, such as products, powers, and trigonometric forms. As a result, the twelve variables are not expressible as linear combinations of any two, as ordinary factor analysis would require. However, with amazing discrimination the 12×30 matrix was put in the computer with an appropriate algorithm; the solution illustrated shows the thirty cyclinders placed on the two-dimensional plane in a well-ordered way. The machine has done this finding of two-dimensional structure by its tracing of the order in which the various points are met in space in each of the many dimensions, twelve in this case, of the space defining the problem. It is hard to overestimate the potential of such methods for discerning most subtle meanings in systems of involved and highly interacting variables.

A Natural Science for Natural Language?

Information processing in science involves primarily the organization of natural language. The meaning of this for the science of linguistics is probably the most important part, but it is too early to say much about this.

Various recent surveys have established that both recognition-of-language machines, and also "mechanical translation," are presently not practical. One of the principal limits seems to be our lack of organizing principles for sentence structure. L. E. McMahon has been grappling with this problem, and has shown some simple rules which do permit at least crude machine manipulation of English sentences, so that rough reconstruction by computer preserves much of the meaning. One example of such procedures, called FASE (*F*undamentally *A*nalyzable Simplified *E*nglish), yields machine qualified features, as illustrated in Figure 12. With these features, the machine can recognize and reconstruct the essence of the sentence, and of course can recognize the key features of several sentences as in a paragraph, so that intelligible although very crude abstracts might be produced.

However, the chief result of these studies up to now has been emphasis on how difficult is the rationalization for computer processing of the usage of English words. Indeed, the manipulation of ideas, as has been represented in the various examples of scientific studies through mathematical and graphical modes discussed above, seems in the verbal mode to be so far beyond what computer-processed natural language provides that there would seem to be little immediate prospect of drastic changes in the verbalization of scientific or other creative thought.

This sentence is written in FASE.

SENTENCE	ANALYSIS ØUTPUT		EDITØR ØUTPUT
THIS	ADJ	SB	MØDIFIES SUBJECT
SENTENCE	NNN	MSB	SUBJECT ØF SENTENCE
IS	PBE	VP	BEGINS PREDICATE ØF SENTENCE
WRITTEN	PTI	VP	HEAD ØF PASSIVE VERB PHRASE
IN	PRE	VPM	INTRØDUCES PREPØ- SITIØNAL PHRASE; MØDIFIES PRECEDING VERB
FASE	NNN	ØP	ØBJECT ØF PRECEDING PREPØSITIØN
	PRD	EØS	END ØF SENTENCE

Figure 12. Example of the coding for computer of a particularly adapted form of natural English language, conceived by L. E. McMahon. With this, the computer is able to parse and process simple sentences and paragraphs of English, as an experimental step toward automatic abstracting and synthesis of text. (From Baker, "Some Social Meanings," in Leeds, ed., *Washington Colloquium.*)

PLIFIER) STABILITY OF ACTIVE TRANSMISSION	LINES WITH ARBITRARY IMPERFECTIONS. (OPTICAL AM	63-1347--
PPRESS SUBSIDIARY RESONANCE IN FERRITES. SPIN WAVE	LINEWIDTH AND OPTIMUM MODULATING FREQUENCY TO SU	63-4214-5
MEASUREMENT OF FERROMAGNETIC RESONANCE	LINEWIDTH AT X-BAND. PART-1 - THEORY.	63-4214-10
TEST PROCEDURE. MEASUREMENT OF FERROMAGNETIC RESONANCE	LINEWIDTH AT X-BAND. PART-2. INSTRUMENTATION AND	63-4214-11
ION TEMPERATURE. ANTIFERROMAGNETIC RESONANCE	LINEWIDTH IN MANGANESE FLUORIDE NEAR THE TRANSIT	63-1115-13
THE	LINEWIDTH OF THE OPTICAL MASER OSCILLATOR.	63-2814-10
DIELECTRIC	LINING PROCESSES FOR WAVEGUIDES.	63-1361-9
TRAFFIC STUDIES - BLOCKING IN POSITION	LINK FRAMES FOR TRAFFIC SERVICE POSITIONS.	63-3121-2
ELF-CONTAINED ELEC/ A CIRCUIT TO PERFORM SWITCHING NETWORK	LINK SUPERVISION OVER THE TRANSMISSION PATH. (S	63-5333-11
2108 ANC 383A EQUALIZERS FOR TYPE A WIRE LINE ENTRANCE	LINKS.	63-2182-13
THE EKS MULTILINK INTERCOM ALLOTTER AND TALKING	LINKS. (KEY TELEPHONE SYSTEMS)	63-5335-5
REDGE-MOLINA FORMULA. PART-1. DEPENDENCE AMONG FIRST-STAGE	LINKS. (NETWORKS, BLOCKING, BALKING) /THE KITT	63-1391-1
NG AND FAILURE PERIODS IN TRANSMISSION SYSTEMS COMPOSED OF	LINKS - THEORY. STATISTICS OF WORKI	63-3241-6
BLOCKING IN A SMALL CROSSBAR NETWORK WITH SWITCHABLE	LINKS. (TRAFFIC ENGINEERING)	63-3313-3
DEAF.	LIP-PHONE - A COMPROMISE VIDEOTELEPHONE FOR THE	63-2447-1
PLASMAS) FERMI-	LIQUID EFFECTS IN CYCLOTRON RESONANCE. (METALS,	63-1115-12
OR PRECISE MEASUREMENT OF FIELD CHANGE IN SMALL VOLUMES AT	LIQUID HELIUM TEMPERATURES. /ANCE MAGNETOMETER F	63-1353-19
TERM " ZONE LENGTH " AND THE RATIO OF DENSITY OF SOLID TO	LIQUID IN THE MATHEMATICS OF ZONE MELTING. / THE	63-1166-3
TRIC AMPLIF/ ACCURATE MEASUREMENT OF THE INSERTION LOSS OF	LIQUID NITROGEN TEMPERATURE CIRCULATOR. (PARAME	63-2813-4
ICONDUCTORS, EPITAXY) VAPOR-	LIQUID- SOLID MECHANISM OF CRYSTAL GROWTH. (SEM	63-1166-5
ROACH TO EXPLANATION OF THE RIGID DISK PHASE TRANSITION. (	LIQUIDS) A SYSTEMATIC APP	63-1124-40
THE DEPOLARIZATION OF THE RAYLEIGH SCATTERING FROM SIMPLE	LIQUIDS. USE OF A GAS LASER IN STUDIES OF	63-1123-30
/DIFFRACTION IN MEASURING THE PARAMETER OF NONLINEARITY OF	LIQUIDS AND THE PHOTOELASTIC CONSTANTS OF SOLID/	63-1217-8
AN INSTRUMENT FOR MEASURING CONTACT ANGLES OF	LIQUIDS ON SOLID SURFACES.	63-2835-24
/BLISHEC-NON-PUBLISHED " - A PROPOSED NEW TELEPHONE NUMBER	LISTING SERVICE TO ALLEVIATE THE " NON-PUB " PR/	63-2365-3
/ OPERATORS. PART-4. COMBINATIONS OF DETAILS IN RETRIEVING	LISTINGS IN A SEMI-AUTOMATIC INFORMATION BUREAU/	63-3224-8
NUCLEAR MAGNETIC RESONANCE - REVIEW OF 1962	LITERATURE.	63-1132-11
OF METEORIC DAMAGE TO TELSTAR-II - A PRELIMINARY STUDY AND	LITERATURE SURVEY. THE PROBABILITIES	63-2122-9
NEUTRON DI/ THE ANTIFERROMAGNETIC AND CRYSTAL STRUCTURE OF	LITHIUM CUPRIC CHLORIDE DIHYDRATE. (MAGNETISM,	63-1114-18
ORS) HALL EFFECT INVESTIGATION OF	LITHIUM DIFFUSED GALLIUM ARSENIDE. (SEMICONDUCT	63-1132-24
ACCEPTORS IN DONOR-DOPEC GALLIUM ARSENICE RESULTING FROM	LITHIUM DIFFUSION.	63-1132-4
AN X-BAND COINCICENCE FERRIMAGNETIC LIMITER USING	LITHIUM FERRITE.	63-2615-17
MAGNETOACOUSTIC RESONANCE IN	LITHIUM FERRITE. (MAGNETOELASTIC MATERIALS)	63-2615-19
THE BETA DECAY OF	LITHIUM-8. (NUCLEAR PHYSICS)	63-1131-17
GRAMS. ELECTRON GUN ANALYSIS BY MEANS OF THE	LITTON NETWORK BOARD AND ASSOCIATED COMPUTER PRO	63-2814-19
SEE BOTH	LLL AND LOW-LEVEL LOGIC	
AN ANALYSIS OF MARGINAL TESTING IN TRL AND	LLL CIRCUITS.	63-6325-1
LTIPLEX - CHARACTERIZATION OF BACK-TO-BACK TRANSMISSION OF	LMX1 MASTERGROUPS. L MU	63-3222-5
ENDING OF A CORRUGATEC CIAPHRAGM SUBJECT TO A CONCENTRATED	LOAD AT THE CENTER. B	63-6326-4
(SUBMARINE CABLE/ ACVANTAGE OF RE-EVALUATING THE REQUIRED	LOAD CAPACITY OF A MULTICHANNEL CARRIER SYSTEM.	63-3213-2
(PROBABILITY STUDIES, TRAFFIC RESEARCH)	LOAD CARRIED ON EACH TRUNK OF AN ORDERED GROUP.	63-3122-2
/C FREQUENCY (17.5KCPS) AND CONVENTIONAL (60CPS) AXIAL	LOAD FATIGUE TESTS ON 2024-T4 ALUMINUM ALLOY RO/	63-1163-5
GS. (TRAFFIC STUDIES) ERLANG-B	LOAD VS. LOSS CURVES TO 500 TRUNKS AND 500-ERLAN	63-3343-7
BAND-LIMITING OF HIGH-INDEX NOISE	LOADED FM SYSTEMS.	63-2122-10
SION LI/ CONCERNING OPTIMIZATION FOR ATTENUATION IN A LINE	LOADED INDEPENDENTLY WITH INDUCTANCE. (TRANSMIS	63-1381-1
MPLEX VARIABLE METHODS APPLIED TO THE PROBLEM OF LATERALLY	LOADED PERFORATED PLATES. CO	63-4253-15
STRIPLINE NETWORKS. NARROW BAND, DIELECTRIC	LOADED STRIPLINE- WAVEGUIDE FILTER FOR USE WITH	63-2813-26
DESIGN OF BUILD-OUT NETWORKS FOR VOICE FREQUENCY	LOADED TRUNKS.	63-3152-3
DEFLECTIONS OF AN INCLINED BEAM UNDER GRAVITY	LOADING.	63-4253-35
IDEO TRANSMISSION SYSTEM - EFFECTS OF 1860 CHANNEL MESSAGE	LOADING. MODIFIED A2A V	63-3211-3
TED ON COPPER CONDUCTOR/ EXPLORATORY STUDIES OF CONTINUOUS	LOADING BY MEANS OF THIN LAYERS OF PERMALLOY PLA	63-5411-2
OLSTICS) BLAST WAVE	LOADING OF SPHERES AND HEMISPHERICAL DOMES. (AC	63-4253-11
A NEW LOCK AT NEGATIVE RESISTANCE	LOADING OF TELEPHONE LINES.	63-5163-6
ILITY. (ALUMINUM ALLOYS) INFLUENCE OF THE MECHANICAL	LOADING SYSTEM ON LOW-TEMPERATURE PLASTIC INSTAB	63-1165-19
RESPONSE OF A THIN ELASTIC CYLINDRICAL SHELL TO DYNAMIC	LOADINGS. (BLAST EFFECTS)	63-4253-26
URFACE TEMPERATURES OF TANTALUM RESISTORS UNDER ELECTRICAL	LOADS. S	63-2624-9
SIGNALING COMPATIBILITY OF T1 CARRIER IN THE	LOCAL EXCHANGE AREA. (PCM)	63-2362-4
SIGNALING COMPATIBILITY OF T1 CARRIER IN THE	LOCAL EXCHANGE AREA. (PCM SYSTEMS)	63-2362-1
COMMON SYSTEMS - V4 REPEATER - COMPATIBILITY WITH	LOCAL EXCHANGE AREA TRUNKS USING DX SIGNALING.	63-2362-2
OF 4-WIRE WIDEBAND SWITCHING CONNECTIONS THROUGH A 2-WIRE	LOCAL OFFICE. / TOLL SWITCHING SYSTEM - ORDERING	63-2322-8
THEORY OF A	LOCAL SUPERCONDUCTOR IN A MAGNETIC FIELD.	63-1111-20
ED THROUGH THE DIRECT DISTANCE DIALING NETWORK. (TOLL AND	LOCAL SWITCHING) /ATICALLY CONNECTED CALLS PLAC	63-5311-10
ELLITE. CALCULATION OF TRANSIENT	LOCAL TEMPERATURE ON THE SKIN OF A SPHERICAL SAT	63-2812-3
) TO IMPROVE ARTIFICIAL CABLE MATCH/ SUBMARINE CABLE FAULT	LOCALIZATION TEST SETS - ENGINEERING TEST METHO	63-2134-6
ITANCE BRIDGE USED FOR NETWORK ADJUSTMENTS IN J68864 FAULT	LOCALIZATION TEST SETS. SLOW PULSE EXCITED CAPAC	63-2134-2
OF FOAMED POLYMERS. PART-1. NUCLEATION OF DISSOLVED GAS BY	LOCALIZED HOT SPOTS. (PROPYLEX CABLE) /UCTION	63-1122-20
THEORY OF VIBRATIONAL STRUCTURE IN OPTICAL SPECTRA OF	LOCALIZED IMPURITIES IN SOLIDS.	63-1155-22
THE ELECTRONIC SPECIFIC HEAT ASSOCIATED WITH	LOCALIZED STATES IN ACTINIDE METALS.	63-1112-15
INTERACTION BETWEEN	LOCALIZED STATES IN METALS. (FERROMAGNETISM)	63-1111-18
SD SUBMARINE CABLE SYSTEMS - SD FAULT	LOCATING TEST SET B.	63-2132-2
RANDOM ROOT	LOCATION.	63-1215-3
STAL OSCILLATOR FOR THE SF SUBMARINE CABLE SYSTEM. (FAULT	LOCATION) A HIGH STABILITY CRY	63-2131-1
TROUBLE	LOCATION PROCEDURE FOR THE UNICOM MODEMS.	63-5311-3
TEST SET FOR 500-KC AND 5.9-MC CRYSTAL FOR SF CABLE FAULT	LOCATION SYSTEM. OVEN	63-2632-16
APPARATUS - HANDSETS - G-8-TYPE HANDSETS FOR USE IN NOISY	LOCATIONS. STATION	63-5133-3
ITCHING REGULATORS. LOW-VOLTAGE	LOCK-OUT IN SATELLITE POWER SYSTEMS EMPLOYING SW	63-5151-7
TRUNK IDENTIFICATION, SELECTION AND	LOCKOUT FOR A PBX ANI SYSTEM. (SCANNERS)	63-5332-13
OPTIMUM DESIGN OF MAGNETS FOR SATELLITE	LODESTONES.	63-2813-7
. CONVERSION) A NONLINEAR ENCODER AND DECODER WITH	LOGARITHMIC COMPANDING. (PCM, ANALOG-TO- DIGITA	63-2111-26
OF TWO INDEPENDENT RANDOM VARIABLES WHICH ARE DEFINED ON A	LOGARITHMIC DOMAIN. (CONVOLUTION, PROBABILITY /	63-1359-4
HIGH-SPEED SWITCHING THROUGH CURRENT ROUTING. (TRANSISTOR	LOGIC)	63-2111-2
SEE BOTH TRL AND TRANSISTOR RESISTOR	LOGIC	
SEE BOTH LLL AND LOW-LEVEL	LOGIC	
SIMPLIFIED VERSION OF A FULL BINARY ADDER USING THRESHOLD	LOGIC. A	63-2111-14
LITE - AN EIGHT-MEGACYCLE RING COUNTER USING CURRENT MODE	LOGIC. TSX-1 SATE	63-4252-9
ATOR. (PNPN TRANSISTOR MAGNETIC LOGIC (PTML), MAGNETIC	LOGIC) A SOLID-STATE PRETRANS	63-5332-22
ODE SERIES TRANSLATOR. (PCM, CODE TRANSLATION, HIGH-SPEED	LOGIC) A 110-MEGABIT GRAY-TO-BINARY C	63-2111-25
MARGINAL TESTING ANALYSIS OF TRANSISTOR RESISTOR	LOGIC AND LOW-LEVEL LOGIC CIRCUITS.	63-6222-1
DESIGN OF HIGH-SPEED DCTL	LOGIC CIRCUITS.	63-4252-18
COST FOR A 36-BIT HIGH-SPEED ADDER. (DIGITAL COMPUTERS,	LOGIC CIRCUITS) SPEED AN	63-4254-23
STING ANALYSIS OF TRANSISTOR RESISTOR LOGIC AND LOW-LEVEL	LOGIC CIRCUITS. MARGINAL T	63-6222-1
COST AND DELAY ANALYSIS OF LOW-LEVEL	LOGIC COMPLETE DECODING NETS.	63-2443-3
NO. 1 ESS - NOTES ON SIGNAL PROCESSOR	LOGIC DESIGN.	63-2421-2
NIKE-ZEUS -	LOGIC DESIGN OF A DIGITAL COMPARATOR.	63-6411-1
AN IMPROVED SILICON DIRECT-COUPLED TRANSISTOR	LOGIC DESIGN PROGRAM.	63-4252-13
DATA PROCESSING UNIT. TITAN FORWARD-LOOK -	LOGIC DESIGN SUITED TO THIN FILM PACKAGING OF TH	63-4211-1

Figure 13. Example of the permutation indexing of a list of technical memoranda titles so that key words can be rapidly scanned by the reader in searching for items of interest. This system is already widely used in surveying the content of scientific and technical titles. (From Baker, "Some Social Meanings," in Leeds, ed., *Washington Colloquium.*)

Nevertheless, automated improvements in the sheer mechanics of handling single words, as contrasted to complex assemblies, are great. These assistances to language manipulation certainly must not be undervalued, for as in the case of mathematics and data analysis the relief of the mind through logical support of the machine can vastly empower the simpler actions of thought. Thus, in matters of associative memory and search for key words, the convenience of a title listing in which the words are permuted by machine is now widely being recognized. Thus as in Figure 13, in a sample listing of technical memorandum titles, the scanner can see readily words which may be otherwise buried.

Nor is the special value of word and character manipulation by computers restricted to such position shifts. In various laboratories the computer has been programmed to take carefully structured sequences which form letters patterned after any desired kind of English type. These letters are formed on the screen of a cathode ray tube by an electron beam, as noted before with respect to graphs and pictures. As shown in Figure 14, the printing thus composed can easily be picked up on a photographic or similar plate, and used for production. Figure 15 shows some of the early letters and print thus produced by M. V. Mathews and Miss J. E. Miller, and the methods have been consistently refined. Now all kinds of sub- and superscripts can be readily manipulated, as well, with speeds of 100,000 characters per second, conveniently, but with special performance of as much as 300,000 per second. Obviously a great step since Gutenberg is at hand. The influence on the recording and distribution of scientific and technical studies, as well as that of literature in general, may be profound.

Further, the initial writing can itself be greatly aided through editing programs which take care of justification and arrangement. Especially, corrections and insertions into earlier text can be done, as shown by Mathews and co-workers in Figure 16. More generally, the influence of this capability, along with the prior printing facility, can provide texts which are easily revised and updated, so that the somewhat sacred restrictions of the published book can be relaxed. This is useful in the teaching of science and mathematics, to say nothing of its role in pedagogy in general. Present-day distress at the inflexibility of instructional material required in higher education alone, with more than 7,000,000 college students in the nation, may well be modulated by these systems.

"The Purpose of Computing Is Insight, Not Numbers"— R. W. Hamming

Thus we have sought to show how, within less than three decades, the digital logic machine has altered the ways and means of scientific

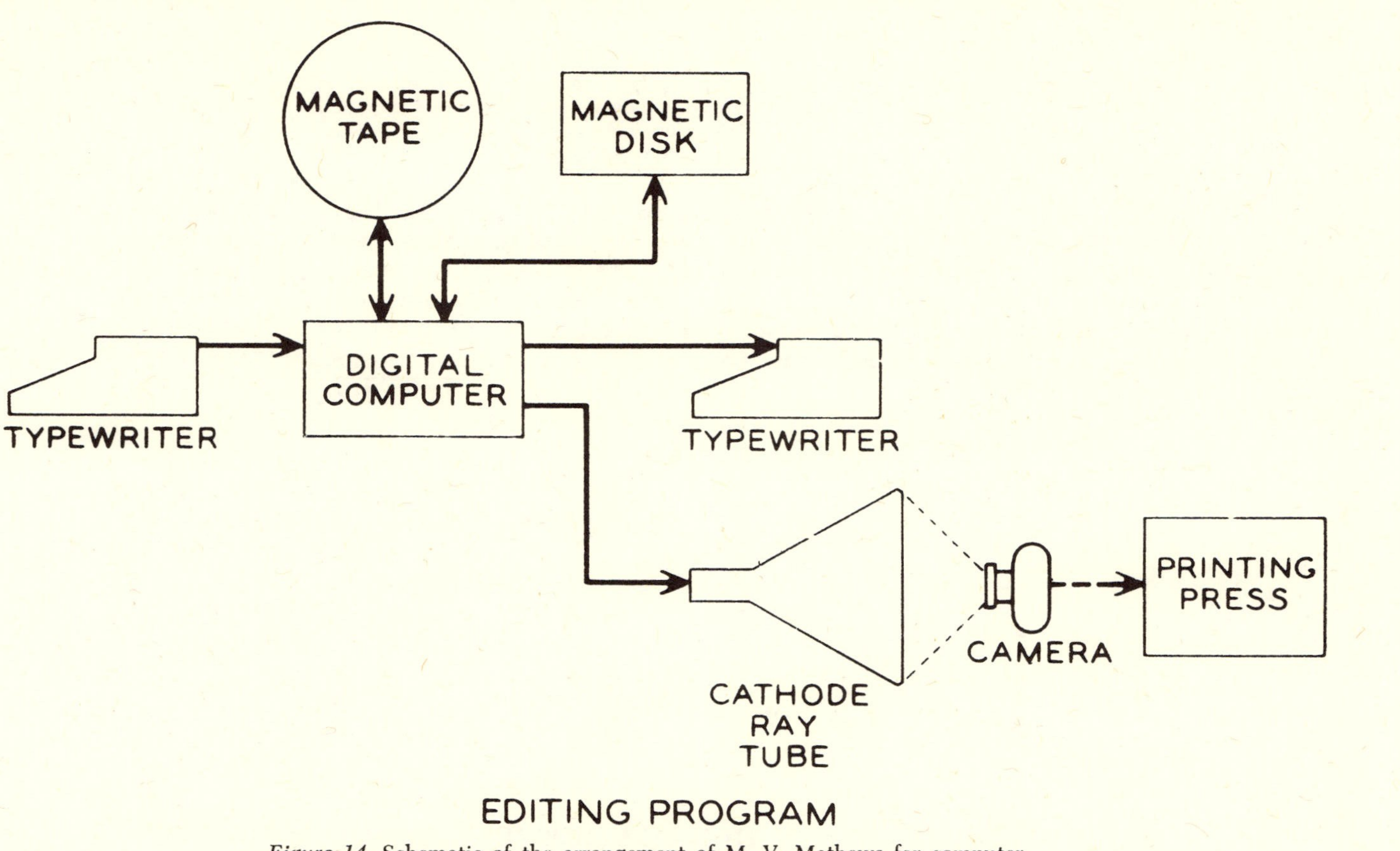

Figure 14. Schematic of the arrangement of M. V. Mathews for computer use in editing and creating print by movement of an electron beam by a digital computer. A hundred thousand characters per second can be produced in this way. The typewriter, of course, is used to signal the letters that are desired. Its speed is significant only for input. (From Baker, "Some Social Meanings," in Leeds, ed., *Washington Colloquium.*)

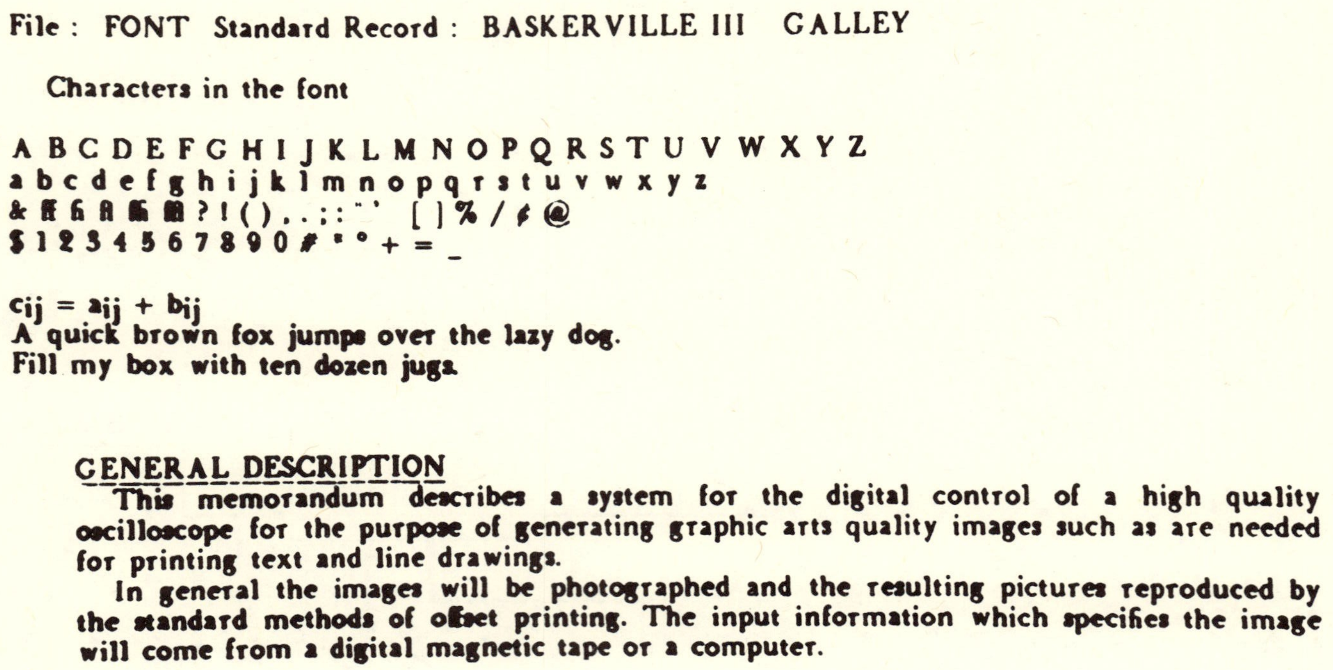

File : FONT Standard Record : BASKERVILLE III GALLEY

Characters in the font

A B C D E F G H I J K L M N O P Q R S T U V W X Y Z
a b c d e f g h i j k l m n o p q r s t u v w x y z
& ff fi fl ffi ffl ? ! ( ) , . ; : " ' [ ] % / ¢ @
$ 1 2 3 4 5 6 7 8 9 0 # * ° + = _

$c_{ij} = a_{ij} + b_{ij}$
A quick brown fox jumps over the lazy dog.
Fill my box with ten dozen jugs.

GENERAL DESCRIPTION

This memorandum describes a system for the digital control of a high quality oscilloscope for the purpose of generating graphic arts quality images such as are needed for printing text and line drawings.

In general the images will be photographed and the resulting pictures reproduced by the standard methods of offset printing. The input information which specifies the image will come from a digital magnetic tape or a computer.

Figure 15. The type designs can duplicate those of any conventional system, but there is no geometrical source of type shape except the influence of computer software on the beam in the oscilloscope. (From Baker, "Some Social Meanings," in Leeds, ed., *Washington Colloquium.*)

Vol. 10, No. 15 April 14, 1965

MATHEMATICS DEPARTMENT BULLETIN

*11 *F

Computer Seminar

Tuesday
April 27

Speakers: (M.V. Mathews,) J. Kohut, Joan E. Miller,
Carol C. Lochbaum

Location: MH 3A 112-116

Time: 2:00 PM

Topic: Editing and Typography by Computer

*12 *J *P

A program for editing (typewriter) input, storing writtfn e

text on disc for furrier editing and processing, and producing

high quality typographical output

is descried.
is described. *E#,-1/ *S2,3/on (disk for further)editing/
*C#, 2/
*U1,1/
*C1,2/

Before

Computer Seminar

Tuesday
April 27

Speakers: M. V. Mathews, J. Kohut, Joan E. Miller,
Carol C. Lochbaum

Location: MH 3A 112-116

Time: 2:00 PM

Topic: Editing and Typography by Computer

A program for editing typewriter input, storing written text on disk for further editing and processing, and producing high quality typographical output is described.

Figure 16. Example of a technical seminar announcement composed and edited by Dr. Mathews' computer program which can serve directly as a rapid and precise method for producing clean and corrected "manuscripts." (From Baker, "Some Social Meanings," in Leeds, ed., *Washington Colloquium.*)

discovery. Only now are generations of new students and teachers arising who will use these means fully and extend the reach of new thought with their help. Dr. George Stibitz, creator of the first electrical digital computer (based on the logic of telephone relay switches), said in describing his work beginning in 1937, "I had observed the similarity between the circuit paths through relays and the binary notation for numbers and had an idea I wanted to work out." Thus began a new path in science. In his endeavors, and those of Aiken, Eckert and Mauchly, von Neumann, Shannon, Wiener, and the others, began a fateful chapter of empowering the reach of the mind. Already the scientific concepts of our age—quantum theory and the wave mechanics, the structure patterns of molecules and crystals, and indeed the whole vast sweep of inorganic, organic, and social science which is surveyed in this volume—are peculiarly matched to a need for massive computation and logical manipulation.

Likewise, the applications of science for the benefit of man through industry and government increasingly depend on the conception and design of large interacting systems, in which many of the elegant uses of computers that we have sampled must be combined in the most ingenious ways. Far from seeing a plateau or decline in the rate of progress in a golden age of science and technology, I see boundless opportunities through the use of logic machine systems to resolve the puzzles of cosmology, of life, and of the society of man. And, above all, I see that automation, once viewed as an ominous, Frankensteinlike threat to personalism and humaneness, turns out to augment especially the bold individualism of new thought.

REFERENCES

1. Information theory was discovered by Claude Shannon, and has roots in the electrical communications ideas of Harry Nyquist and the cybernetics of Norbert Wiener and others.

2. Bohr, "The Unity of Human Knowledge," in *Essays 1958-1962 on Atomic Physics and Human Knowledge* (New York: Interscience Publishers, 1963).

Notes on Contributors

W. O. BAKER, born in 1915, is vice-president, research, at Bell Telephone Laboratories, Murray Hill, New Jersey. He is a member of the President's Foreign Intelligence Advisory Board and consultant to the Department of Defense and to the Special Assistant for Science and Technology. Mr. Baker holds thirteen patents and is a frequent contributor to scientific publications.

SAUL BENISON, born in 1920, is professor of history, graduate faculties, at the University of Cincinnati. He is the author of *Tom Rivers: Reflections on a Life in Medicine and Science* (Cambridge, Mass., 1967). Mr. Benison pioneered with Allan Nevins in oral history at Columbia University. He received the William H. Welch Medal from the American Association for the History of Medicine in 1968 for his contributions to the history of medicine.

A. HUNTER DUPREE, born in 1921, is George L. Littlefield Professor of American History at Brown University. He is the author of *Science in the Federal Government: A History of Policies and Activities* (Cambridge, Mass., 1957) and *Asa Gray: 1810-1888* (Cambridge, Mass., 1959). For the past decade Professor Dupree has been working on a sequel to *Science in the Federal Government* for the period since 1940.

ERIK H. ERIKSON, born in 1902, is professor emeritus of human development and lecturer in psychiatry at Harvard University. Mr. Erikson's publications include *Young Man Luther* (New York, 1958), *Childhood and Society* (2d ed., New York, 1963), *Insight and Responsibility* (New York, 1964), *Identity: Youth and Crisis* (New York, 1968), and *Gandhi's Truth: On the Origins of Militant Nonviolence* (New York, 1969).

GERALD HOLTON, born in 1922, is professor of physics at Harvard University and a member of the history of science department. From 1957 to 1963 he was editor of the American Academy of Arts and Sciences and, until 1961, editor of *Dædalus*. His book publications include *Introduction to Concepts and Theories in Physical Science* (Reading, Mass., 1952) and (as editor) *Science and the Modern Mind* (Boston, 1958).

ROBERT OLBY, born in 1933, is lecturer in history and philosophy of science at the University of Leeds. He is the author of *Origins of Mendelism* (London and New York, 1966) and *Charles Darwin* (Oxford, 1967). Mr. Olby is presently working on a history of nucleic acid researches and molecular biology.

TALCOTT PARSONS, born in 1902, is professor of sociology at Harvard University. His publications include *Toward a General Theory of Action* (Cambridge, Mass., 1951), *Family, Socialization, and Interaction Process* (Glencoe, Ill., 1955), *Structure and Process in Modern Societies* (Glencoe, Ill., 1959), *Societies: Evolutionary and Comparative Perspectives* (Englewood Cliffs, N. J., 1965), and *Politics and Social Structure* (New York, 1969). Mr. Parsons was coeditor, with Kenneth B. Clark, of the *Dædalus* Library volume *The Negro American* (Boston, 1966) and has served as president of the American Academy of Arts and Sciences since 1967.

LINUS PAULING, born in 1901, is professor of chemistry at Stanford University. He is the author of *The Nature of the Chemical Bond* (3d ed., Ithaca, 1960), *General Chemistry* (3d ed., San Francisco, 1970), and *No More War!* (rev. ed., New York, 1962). Mr. Pauling was awarded the U. S. Presidential Medal of Merit, 1948, the Nobel Prize in Chemistry, 1954, the Nobel Peace Prize, 1962, and the Lenin International Peace Prize, 1970.

PAUL A. SAMUELSON, born in 1915, is Institute Professor at Massachusetts Institute of Technology. He is the author of *Foundations of Economic Analysis* (Cambridge, Mass., 1947), *Economics* (New York, 1948), *Readings in Economics* (New York, 1955), *Linear Programming and Economic Analysis* (New York, 1958), and *The Collected Scientific Papers of Paul A. Samuelson* (Cambridge, Mass., 1966——). Mr. Samuelson was awarded the Alfred Nobel Memorial Prize in Economic Science in 1970.

EDWARD SHILS, born in 1911, is professor of sociology and social thought at the University of Chicago and a Fellow of Peterhouse, Cambridge University. His books include *The Present State of American Sociology* (Glencoe, Ill., 1948) and *The Intellectual Between Tradition and Modernity* (The Hague, 1961). He is also editor of *Minerva,* a quarterly review of the relations of science, learning, and policy.

GUNTHER S. STENT, born in 1924, is professor of molecular biology at the University of California at Berkeley. He is the author of *Molecular Biology of Bacterial Viruses* (San Francisco, 1963), *The Coming of the Golden Age* (New York, 1969), and *Molecular Genetics* (San Francisco, 1970). Mr. Stent was visiting professor of neurobiology at Harvard Medical School, 1969-1970, in order to prepare for a new research career in the study of the nervous system.

CURT STERN, born in 1902, is professor of zoology and genetics at the University of California at Berkeley. He is the author of *Principles of Human Genetics* (2d ed., San Francisco, 1960) and *Genetic Mosaics*

(Cambridge, Mass., 1968). Mr. Stern is the recipient of the Kimber Genetics Medal of the National Academy of Science, 1963.

ALVIN M. WEINBERG, born in 1915, is director of the Oak Ridge National Laboratory, Oak Ridge, Tennessee. He is the author of *The Physical Theory of Neutron Chain Reactors* (Chicago, 1958) and *Reflections on Big Science* (Cambridge, Mass., 1967). Mr. Weinberg was a member of the President's Science Advisory Committee, 1960-1963, and is the recipient of the Atoms for Peace Award, 1960, and the Atomic Energy Commission's E. O. Lawrence Award, 1960.

R. R. WILSON, born in 1914, is director of the National Accelerator Laboratory, Batavia, Illinois, and professor of physics at the University of Chicago. Since 1947, Mr. Wilson has been involved in the construction of a series of particle accelerators with which to explore the structure of the proton. He has had formal training as a sculptor in the United States and at the Academia Belli Arte in Rome. A large sculpture of his in bronze, commissioned by the Institute for Advanced Studies at Princeton, was installed in 1965.

INDEX

DATE DUE